Studies in Systems, Decision and Control

Volume 579

The series "Studies in Systems, Decision and Control" (SSDC) covers both new developments and advances, as well as the state of the art, in the various areas of broadly perceived systems, decision making and control–quickly, up to date and with a high quality. The intent is to cover the theory, applications, and perspectives on the state of the art and future developments relevant to systems, decision making, control, complex processes and related areas, as embedded in the fields of engineering, computer science, physics, economics, social and life sciences, as well as the paradigms and methodologies behind them. The series contains monographs, textbooks, lecture notes and edited volumes in systems, decision making and control spanning the areas of Cyber-Physical Systems, Autonomous Systems, Sensor Networks, Control Systems, Energy Systems, Automotive Systems, Biological Systems, Vehicular Networking and Connected Vehicles, Aerospace Systems, Automation, Manufacturing, Smart Grids, Nonlinear Systems, Power Systems, Robotics, Social Systems, Economic Systems and other. Of particular value to both the contributors and the readership are the short publication timeframe and the worldwide distribution and exposure which enable both a wide and rapid dissemination of research output.

Indexed by SCOPUS, DBLP, WTI Frankfurt eG, zbMATH, SCImago.

All books published in the series are submitted for consideration in Web of Science.

Alexander S. Filatyev · Vyacheslav G. Petukhov

Through Optimization of Aerospace Vehicle Trajectories by the Pontryagin Maximum Principle

Springer

Alexander S. Filatyev
Faculty of Space Research
Lomonosov Moscow State University
Moscow, Russia

Applied Mathematics Department
Moscow Aviation Institute (National
Research University)
Moscow, Russia

Vyacheslav G. Petukhov
Research Institute of Applied Mechanics
and Electrodynamics of MAI (RIAME
MAI)
Moscow, Russia

ISSN 2198-4182 ISSN 2198-4190 (electronic)
Studies in Systems, Decision and Control
ISBN 978-3-031-80755-8 ISBN 978-3-031-80756-5 (eBook)
https://doi.org/10.1007/978-3-031-80756-5

This Springer imprint is published by the registered company Springer Nature Switzerland AG
The registered company address is: Gewerbestrasse 11, 6330 Cham, Switzerland

АНГАРА А5
Optimal pitch angle
90
60
30
0
L/D max
0
1
2
3
4
5
6
traditional
novel
Time

Es gibt nichts Praktischeres als eine gute Theorie

Immanuel Kant

Foreword

This book is devoted to the applications of Pontryagin maximum principle to the optimization of aerospace vehicle trajectories. The authors consider various problems of launch to orbit, transfer between orbits, and deep space missions. Complicated optimal control problems are analyzed including cases with many extrema, bifurcations, state and control constraints. The approaches developed by the authors make it possible to simplify significantly the procedure of solving specific boundary value problems.

I consider this book very useful for students and wide range of researchers and engineers in the fields of optimal control, aerospace flight mechanics, and astrodynamics.

September 2024

Prof. Felix Chernousko
Academician of the Russian Academy of Sciences
Moscow, Russia

Preface I

Launching spacecraft (SC) into orbits is one of the important problems of the current rocket dynamics. Its practical solution so far is mainly associated with the use of space transportation systems (STS) of a ballistic type with a vertical take-off. For such STS, the atmosphere is considered as a medium with a drag, which increases expenses for the launch. In this regard, the control program is traditionally chosen so that the flight in dense atmosphere layers occurs with an almost zero angle of attack (the gravity turn program) and STS quickly overcomes the atmospheric section at a relatively low level of aerodynamic loads.

The possibilities of further increasing the efficiency of this type of STS by reducing aerodynamic losses are practically exhausted. A qualitative leap in the development of STS is possible with a fundamental change in the attitude towards the role of aerodynamic forces in the launch process: from reducing the influence to their optimal use. In general, the optimization of the trajectory control is one of the important reserves for a practical increase in the STS efficiency, since due to the significant excess (by one to two orders of magnitude) of the mass of the consumed fuel over the payload, even a small improving control variation without changing the STS design can lead to a noticeable increase in the payload mass.

Inserting SC into an orbit is carried out, as a rule, by composite (staged) launchers. This raises the problem of dropping spent parts (SP) in the admissible drop fields (DF). For example, in Russia, by the end of the 1990s, the total DF area amounted to hundreds of thousands of square kilometers. In countries where spaceports are located in coastal areas and launches are carried out towards a sea, there is also the problem of alienation of water areas, both from the point of view of ecology and restrictions on shipping and fishing. This could not but affect the policy of companies engaged in the provision of space launch services. NPO Molniya, developer of the SC Buran, whose general designer, G. E. Lozino-Lozinsky, in the 60s of the last century, designed and conducted flight tests of the SC Spiral for the aerospace transportation system (ASTS) project with a hypersonic booster aircraft; in the 80s he developed a new project for the MAKS multi-purpose aerospace system with the An-225 Mriya carrier aircraft. At the turn of the century Khrunichev Space Center began the development of the reusable winged booster (RWB) “Baikal”, returning to the launch site, as the 1st stage of STS Angara. In 2019, a Special Design Bureau was created at TsNIIMash to design a similar RWB. In 1999, Polet Airlines Company participated actively in the development of the international commercial ASTS project AirLaunch with the An-124 carrier aircraft.

In the United States, which was the first to implement the idea of an air launch in the Pegasus project, in 2019, there were began flight tests of the twin-body Stratolaunch aircraft designed to the SC take-off. SpaceX, which since its inception, has taken a course towards the development of reusable STS, in 2017 for the first time implemented a re-launch of the reusable 1st stage booster, although the use of an offshore platform for its landing seems excellent but eclectic due to unnatural synthesis in one system of vehicles moving from speeds that differ by several orders of magnitude.

One of the most important requirements for SC launching is to ensure flight safety. From the point of view of trajectory control of composite STS, taking into account safety requirements implies the fulfillment of restrictions at various flight phases, for example, in terms of loads in the active launch phase, the safe separation of SPs and their return to specified DFs or landing fields. The fulfillment of the restrictions depends both on the ascent trajectory, which actually sets the initial conditions for the SP return sections, and on the control on the SP movement section itself. Thus, the conditions of STS ascent and SP return are linked into an interdependent complex, and we need to consider the STS injection as a branching process with essential constraints on all branches.

The most complete and objective analysis of STS capabilities during insertion into an orbit can be carried out on the basis of through consideration of all flight phases using optimal control methods. One of such effective methods is the Pontryagin maximum principle (PMP).

For the STS-related optimization problems considered here, the use of methods other than indirect ones, such as PMP, and gradient methods is very limited due to the possibility of ensuring the required accuracy. There are objective reasons for this. First, a high degree of responsibility is required developing actual products. Second, the payload mass delivered into an orbit is orders of magnitude less than the start mass, and the ascent trajectory itself is unstable. Third, when conducting a qualitative analysis of optimal solutions, even on simplified models, a strict error control is required, without which an objective study of optimal solution dependences on parameters and disturbances is impossible.

The PMP copes successfully with this complex of problems. As will be shown in the monograph, in particular, in the first chapter, there are objective reasons for this, which lie in the physical sense of the conjugate variables introduced in the PMP as functions of sensitivity of the optimal functional value to small variations of the phase vector. As a result, the PMP provides a whole complex of versatile objective information with controlled accuracy about capabilities of the object both under nominal conditions and under the action of a wide range of disturbances based on the obtained nominal solution without any noticeable additional computational costs.

In accordance with the dialectics laws, advantages, as a rule, are inseparable from the difficulties of their implementation. So, two reproaches can most often be heard about the implementation of PMP. One is that PMP, based on the necessary optimality conditions, can only give local extremals. Another is that PMP reduces the optimal control problem to a two-point boundary value problem (TPBVP), which is very sensitive to the choice of the initial approximation for varied parameters, has a poor convergence, and is allegedly difficult to solve.

Accusing PMP of being able to find only local extremals seems far-fetched, since most other optimization methods, with a few exceptions (dynamic programming and convex problems), in particular the widely used direct methods, often cannot guarantee even this. If we take into account that local extremals can become global under certain conditions, such a "shortcoming" of PMP only expands our knowledge of possible types of optimal solutions. The second argument is more important. The regularization of a numerical procedure for solving TPBVP requires special attention, since the formal application of even objectively efficient algorithms, such as the continuation method, can turn a simple problem into an incorrect one.

The monograph contains a description of many pitfalls in solving a wide range of problems of optimizing the trajectories of aerospace vehicles and recipes for effectively overcoming them. Practice is the criterion of truth, or, in any case, its serious argument. Therefore, the monograph provides not only a theoretical explanation of the approaches used, but also an extensive gallery of the results of practical numerical studies. The number of extremals presented here amounts to tens of thousands, and all of them are obtained with very high accuracy in automatic mode without the intervention of a researcher in the process of solving the corresponding TPBVP. For these purposes, a unique *ASTER* software package with an open architecture has been developed, which makes it possible to continuously increase its capabilities to solve new problems, both methodological and applied. In particular, below (mostly in Chap. 3) are the results of a qualitative analysis of various types of local extremals and their bifurcation transformations into global ones. Examples of practical optimization of branching trajectories of complex dynamics systems with finite and continual sets of branches, in deterministic and stochastic statements, with control constraints and phase constraints of various orders are demonstrated.

Chapters 1–6 were written by Alexander S. Filatyev, Professor of Lomonosov Moscow State University, and Chaps. 7–12 were written by Vyacheslav G. Petukhov, Professor of Moscow Aviation Institute.

Moscow, Russia Alexander S. Filatyev

Preface II

The relevance of using electric propulsion system (EPS) to support space transportation operations is only increasing with time. In fact, the idea of using electric propulsion (EP) and its main advantage—high specific impulse—appeared at the beginning of the 20th century. Research on the use of EPS for interorbital and interplanetary flights has been actively carried out since the early 1950s to the present. EPS themselves have been used on board spacecraft (SC) since the 1960s in experimental mode, and since the 1980s—as regular propulsion systems for stationkeeping of geostationary SC. Since the 1990s, EPS have been used for space transportation.

Both from the point of view of applications and from the point of view of mathematical models used, it is possible to distinguish 3 significantly different types of space transportation operations of SC with EPS: interorbital flights, space flights in an environment with significant perturbing accelerations (including, first of all, flights to the Moon, to the libration points and other problems of low-thrust flight in the problem of three or more bodies), and interplanetary flights. All these types of problems are considered to one degree or another in this monograph.

Typical inter-orbital flights of SC with EP are characterized by an extremely low ratio of thrust acceleration to gravitational acceleration of the primary—usually in the range of 10^{-5}–10^{-4}. Under such conditions, the trajectories contain a large number of orbits, and the solutions to the interorbital transfer problem are close to asymptotic (averaged) solutions. In this regard, most of the problems of interorbital transfer of a SC with an EPS can be completely characterized as problems of low-thrust flight.

Problems of flight to the Moon and libration points (as well as other highly perturbed problems) are characterized by the fact that the perturbing accelerations are comparable to, and often significantly exceed, the control thrust acceleration. The presence of strong perturbations leads to specific features of the trajectories of a SC with an EPS, which, together with the absence of an analytical solution (unlike the case of motion in a central Newtonian field) for the coasting motion of a SC, leads to the need to use specific methods for solving such problems.

On interplanetary trajectories, the ratio of the level of thrust acceleration of the EPS to the solar gravity acceleration in typical problems is of the order of 0.01–0.2, and therefore we should talk about the problem of flight not with low, but with finite thrust. Over a large part of the trajectory, the perturbations are weak, therefore, the method of spheres of influence, including the method of zero spheres of influence, can be effectively used to calculate interplanetary trajectories of SC with EPS. In many cases, it is necessary to consider the problems of flights along complex routes—either to fulfill the target tasks of a SC (the study of several celestial bodies by one SC from the flyby trajectory or from the orbit around the celestial body, round-trip missions) or to ensure the possibility of space transportation in conditions of limited SC resources (gravity-assist maneuvers). It should be noted that the possibility of using gravity-assist maneuvers is considered in the analysis of most modern interplanetary missions using EPS.

The applied significance of the problems of calculating the low-thrust trajectories is the support of space mission analysis and trajectory design, preparation of initial data for the design of SC systems (on-board control complex, telecommand, power supply system, EPS, thermal control system, etc.), development of on-board and ground software for the mission control, preparation of the mission control data, support of the mission control and analysis of telemetry and scientific data received from the SC.

The purpose of space mission analysis and trajectory design is to select the main design and trajectory parameters of the SC. To achieve this goal, it is necessary to develop appropriate methodology and software based on the following:

- A mathematical model of SC motion that allows calculation of SC trajectories with the required accuracy for given initial conditions, control program and values of the key design parameters of the SC.
- A SC design model that enables the calculation of the main design parameters from a limited set of initial data, including characteristics of launch vehicles, thrust, specific impulse and lifetime of the propulsion system, etc.
- Methods for optimizing SC trajectories.

At the stage of space mission design, the trajectory optimization problems play a special role, which is associated not only with the natural requirement to achieve the best characteristics of the SC, but also with the need to reliably assess the impact of variations in various design parameters of the SC on the main characteristics of the mission. In fact, a correct assessment of the influence of variations in any design parameter on the space mission indicators (payload mass, flight duration, etc.) can be obtained only if the SC trajectory options with the compared design parameters are optimal. Otherwise, the difference in these indicators may be caused not by changes in the design parameters, but by a different degree of non-optimality of the trajectories in the compared variants.

In many cases, the tasks of optimizing the trajectory and the main design parameters cannot be separated, which leads to the necessity of solving the problem of joint optimization of the trajectory and the main design parameters during the space mission design. The SC flight trajectories can be divided into sections based on the motion conditions, the SC composition, and the mathematical models used to describe the SC motion. In this case, end-to-end trajectory optimization is required. Usually there are many trajectories that satisfy the necessary optimality conditions, so it is necessary to develop methodology and software for local and global optimization of trajectories.

Traditionally, methods for solving trajectory optimization problems are divided into direct and indirect methods. When using direct methods, some control parameterization is performed, or the trajectory is divided into small sections with a given type of control program in each section. As a result, the trajectory optimization problem is reduced to the problem of minimizing the cost function in a high-dimensional space in the presence of a significant number of constraints, i.e., a high-dimensional nonlinear programming problem. This problem, in turn, can be solved by various methods, the most common of which are currently the methods of sequential quadratic programming (SQP) and differential dynamic programming (DDP), as well as their modifications: static-dynamic programming (SDC), hybrid dynamic programming (HDDP) and others. This class of methods has been successfully applied to solve many problems of optimization of the trajectories of SC with EP. Differential dynamic programming methods and their modifications use some ideas of Bellman's optimality principle; therefore, in some cases, they can be classified as hybrid methods. Another large family of direct methods uses various modifications of the gradient method. Direct methods are usually insensitive to the choice of initial guess and, in many variants, allow to easily introduce constraints on the control and state coordinates into the mathematical model of motion, as well as change the composition of the motion model itself by including additional perturbing accelerations in it. Typical disadvantages of direct methods are extremely high requirements on computing power, low convergence rate and, in many cases, low accuracy associated with simplified modeling of the dynamics of SC motion in individual sections of the trajectory (in many methods the Euler integration scheme is actually used; in some others—EP thrust modeled by a sequence of instantaneous velocity increments over a small section of the trajectory, etc.).

Indirect methods use necessary and/or sufficient optimality conditions in the classical variational form, the form of the Pontryagin maximum principle, or of the Bellman or Krotov optimality conditions. When using indirect methods, the trajectory optimization problem is typically reduced to solving a boundary value problem (BVP) for a system of ordinary differential equations. This BVP is solved either by a specific method (e.g., the shooting method or quasi-linearization), or by general methods for solving systems of nonlinear equations (e.g., modifications of the Newton-Raphson method or the continuation method), or by methods for minimizing the weighted sum of residuals of the BVP (e.g., variants of gradient methods). Indirect methods are fast, have a good (typically quadratic) convergence rate, and provide high accuracy of computations. However, typical indirect methods have a limited domain of convergence and therefore their successful operation usually requires the choice of a good initial guess for the decision vector of the BVP.

Finally, many hybrid methods have been developed that combine features of direct and indirect methods in which the necessary optimality conditions are used to parameterize the control through conjugate variables, but then direct minimization of the cost function is performed using nonlinear programming methods. The main incentive for the development of hybrid methods is the desire to combine the advantages of direct and indirect methods and get rid of their disadvantages.

It should also be noted that there are a number of specific methods based on the linearization of the equations of motion of a SC with an EPS, including a well-known transport trajectory method, its development for problems with an arbitrary angular range, a well-known solution for the equations of motion of a SC with a perfectly controlled engine linearized in the vicinity of a circular orbit, and a method for computing the control based on the solution of linear programming problems. Another interesting direction is the use of inverse dynamics problems to compute initial approximations of trajectories and the use of various heuristic methods (genetic programming, swarm optimization, annealing method, etc.). Using such approaches, it is possible to construct effective heuristic global optimization algorithms in a number of problems.

This study investigates the optimization problems of low- and finite-thrust SC trajectories using an indirect method based on the Pontryagin maximum principle. It should be noted that throughout the remainder of this monograph, trajectories will be referred to as optimal if they satisfy the necessary conditions for the optimality of the maximum principle, that is, the Pontryagin extremal. The maximum principle allows us to reduce the infinite-dimensional problem of computing an optimal control program for the thrust vector of the SC EPS to a finite-dimensional BVP of small dimension for a system of ordinary differential equations. Thus, the maximum principle provides a methodological basis for the construction of fast (compared to direct methods) numerical methods for solving optimal control problems.

However, numerical methods based on the application of the maximum principle also have a number of significant disadvantages, such as

- Limited area of convergence of the solution, which leads to the necessity of searching a good initial guess value for the decision vector before solving the optimization problem.
- High sensitivity of the trajectories to variations of the decision vector of the BVP of the maximum principle.
- Lack of information (in the vast majority of cases) about the existence and number of extreme solutions before starting to solve the problem.
- The existence of many solutions satisfying optimality conditions, among which it is often necessary to find the best.
- Jumps in the matrix of partial derivatives of the residuals of the BVP on the decision vector of the BVP when changing the structure of the control program on the trajectory (sequences of coasting and thrusting in trajectory optimization problems with thrust switching), which can lead to failures of numerical methods for solving the problem.
- the complexity of the mathematical model of the optimal motion of the SC (a joint system of differential equations for phase and conjugate variables) for perturbed optimal control problems, since it is necessary to calculate the partial derivatives of the perturbing accelerations with respect to the state variables, while the perturbing

accelerations themselves often have to be calculated not using finite formulas, but using some iterative algorithms.

The listed disadvantages are still the main reasons why many authors prefer to use direct methods for the optimization of low-thrust SC trajectories. In general, when using direct methods, it is quite easy to introduce arbitrary perturbing factors into consideration and it is easier to ensure the convergence to the solution. However, these advantages are achieved at the expense of a significant increase in the computational cost (and, consequently, the computation time—it is usually orders of magnitude longer than when solving by indirect methods) and a decrease in the accuracy of the problem solution. When performing mission analysis and trajectory design, it is often necessary to calculate hundreds and thousands of optimal trajectories, so the computational performance of the method used is one of its most important characteristics.

Recent advances in applied computational mathematics allow a new approach to indirect methods of solving optimal control problems based on the maximum principle and, to a large extent, correct the above-mentioned shortcomings of numerical methods built on their basis, while maintaining high computational performance and accuracy of calculations.

For example, to extend the convergence domain of the numerical method for solving the BVP of the maximum principle, it is proposed to use the method of continuous continuation (or the Newtonian homotopy method), which makes it possible to formally reduce the BVP to the Cauchy problem. Various implementations of this method are the basis of this work. Based on these methods, techniques have been developed that allow optimizing certain types of problems without choosing an initial guess values (more precisely, when using trivial initial guess values) for unknown parameters of the BVP.

As noted above, when solving the problem of optimizing the trajectory of a low-thrust SC, in many cases there is no a priori information about the existence of an optimal solution. However, the solution to the problem of the flight of a SC with EPS with limited power (LP), which exists for arbitrary nonsingular boundary conditions, in many cases provides sufficient information about the existence of a solution to a similar problem for a SC with limited-thrust (LT) EPS. Therefore, the method proposed in this work for continuing a LP solution into a LT solution makes it possible in many cases to resolve the question of the existence of a LT solution after the computation of the LP solution before starting the next continuation stage. In this work, the LP problem is the problem of the flight of a SC with a perfectly controlled thruster, when only the jet power of the EPS is limited, and the LT problem is the problem of the flight of a SC with an EPS, the control characteristics of which depend on the power consumption (a special case of the LT problem is the problem of the flight of a SC with an EPS with fixed thrust and flow rate). If the maximum thrust acceleration on the optimal LP trajectory does not exceed the thrust acceleration in the LT problem, and the characteristic velocity on the LP trajectory does not exceed the maximum allowable increment of the characteristic velocity in the LT problem, then there is obviously a solution to the LT problem. However, a solution to the

LT problem can exist even with some violation of these constraints. Therefore, to diagnose the existence of an optimal LT trajectory, it is proposed to first solve the thrust minimization problem ($T_{\min}$-problem). The solution to this problem makes it possible to calculate the region of existence of LT trajectories with minimum fuel consumption for a given flight duration or angular range: such trajectories exist if the minimum thrust found does not exceed the thrust value specified in the LT problem.

To reduce the sensitivity of trajectories to variations in the unknown parameters of the maximum principle BVP, a number of problems use the division of the trajectory into sections (for example, between gravity maneuvers) and propose a special scheme for selecting the composition of the unknown parameters of the BVP.

The trajectory optimality conditions can be satisfied on a certain set of decision vectors of the BVP, and each decision vector from this set corresponds to a locally optimal trajectory, characterized by a certain quality indicator (flight duration, final mass of the SC, etc.). For example, for multi-orbit flights of a SC with constant thrust, there are many local extrema corresponding to different angular flight ranges. To compute trajectories with an optimal ratio of duration and angular range, either the angular range continuation method or the problem statement with a fixed angular range and free flight time is used.

Jumps in the elements of the sensitivity matrix (partial derivatives of the boundary conditions with respect to the parameters of the BVP) due to the appearance or disappearance of active or passive sections are one of the main reasons for the failure of traditional numerical methods in solving the problem of trajectory optimization with thrust switching. To overcome this difficulty, a method of smoothing the relay thrust function based on the properties of the switching function is used, which ensures the smoothing of jumps in the elements of the sensitivity matrix and thus allows the reliable determination of the optimal trajectory structure.

To solve the problem of the complexity of deriving differential equations of optimal motion in perturbed problems it was proposed to use the complex step method, which is used for high-precision calculation of partial derivatives of the Hamiltonian with respect to state variables. Thus, when using the proposed method to develop a mathematical model of optimal control in a perturbed problem, all that is additionally required is the development of algorithms for calculating perturbing accelerations, as when using direct methods. Also, a continuation method was proposed for solving optimization problems of LT trajectories using the complex step method for highly accurate calculation of the matrix of partial derivatives of the residuals of the maximum principle BVP with respect to its decision variables. When using the maximum principle and the continuation method for the optimization of perturbed trajectories, an accurate calculation of the second mixed derivatives of the perturbing accelerations with respect to states and unknown parameters of the BVP is required. Various approaches are possible to solve this problem, such as the use of multicomplex, hyperdual or complex dual numbers.

The combination of these new developments has made it possible to significantly expand the scope of use of indirect optimization methods and overcome their typical disadvantages.

In this monograph, the main emphasis is placed on the consistent application of methods of numerical continuation with respect to a parameter to solve various problems in the flight mechanics of a SC with an EPS. The problems considered include the optimization of direct interplanetary flights of SC with thrusters having limited power or limited thrust, of interplanetary flights of such SC using gravity maneuvers, of multi-orbit interorbital transfers of SC with limited-thrust EPS. As a result, it was possible to develop a set of numerical continuation methods for solving a wide range of optimization problems for SC trajectories with EPS with improved characteristics in terms of convergence and performance. Some of such methods use a trivial initial guess values for the unknown parameters of the maximum principle boundary-value problem; i.e. a practical automation of the process of determining the solution has been achieved.

Moscow, Russia Vyacheslav G. Petukhov

Acknowledgments and In Memoriam

The author is fully aware that the completion of the complex work presented, bringing it to practical results, would not have been possible without the active and selfless support of fellow scientists, Dr. Olga V. Yanova and Mr. Alexander A. Golikov, for which I am sincerely grateful to them.

At the final stage of work on the manuscript, invaluable assistance was provided by Alexander E. Sagalakov from Lomonosov Moscow State University. Many graphic illustrations were brought to a high artistic level through the efforts of a great publishing professional Nataliya Mazaleva at TsAGI. Great thanks to them.

The author is grateful to Prof. V. A. Ilyin, Prof. G. L. Grodzovsky and Academician of the Russian Academy of Sciences (RAS) Prof. G. S. Byushgens from TsAGI, to Prof. A. K. Platonov from Keldysh Institute of Applied Mathematics and Corresponding Member of RAS Prof. N. M. Ivanov from Roscosmos Mission Control Center for support in the work and participation in the discussion of some results; to Head of the Ballistic Department of RSC Energia Prof. R. F. Appazov, Academician of RAS Prof. T. M. Eneev from Keldysh Institute of Applied Mathematics, RSC Energia expert Mr. V. A. Vysokanov, First Deputy General Designer of Khrunichev Space Center Prof. S. A. Petrokovsky, Professor Emeritus N. X. Vinh from University of Michigan, Professor Emeritus J. Shinar from the Technion and Prof. Paolo Santini from the Sapienza University of Rome for fruitful discussing the statements and results of studying some problems. Unfortunately, most of them passed away, leaving an important mark on my scientific research and good memories of meetings and conversations with these outstanding personalities.

Boundless gratitude and fond memory to my loved ones who gave me life and their love: father Sergei Ivanovich Filatyev, mother Raisa Konstantinovna Filatyeva and wife Tatyana Alexandrovna Bugaeva.

Alexander S. Filatyev
Moscow, Russia

Contents

Acronyms

ASTS	AeroSpace Transportation System
ASV	AeroSpace Vehicle
BVP	Boundary Value Problem
CA	Carrier Aircraft
DF	Drop Field
DP	Drop Point
EO	Earth Orbit
EPS	Electric Propulsion System
EPT	Electric Propulsion Thruster
FA	Feasibility Area
GAM	Gravity Assisted Maneuver
GEO	Geostationary Orbit
GTO	GeoTransfer Orbit
HEO	High Earth Orbit
HF	Half-Fairings
IO	Intermediate Orbit
LEO	Low Earth Orbit
LLO	Low Lunar Orbit
LO	Lunar Orbit
LP	Limited Power
LT	Limited Thrust
LV	Launch Vehicle for ascent to an orbit
NEPS	Nuclear Electric Propulsion System
NFA	Nominal Feasibility Area
NPP	Nuclear Power Plant
OCP	Optimal Control Problem
ODE	Ordinary Differential Equations
OEL	Okhotsimsky-Eneev-Lawden (see Sect. 2.2.1)
PLM	Payload Module
PM	Propellant Module
PMP	Pontryagin Maximum Principle

PPS	Power and Propulsion System
PS	Propulsion System
PSFS	Propellant Storage and Feeding System
RLTV	Reusable Lunar Transfer Vehicle
RV	Reentry Vehicle
RWB	Reusable Winged Booster
SC	Spacecraft
SEPS	Solar Electric Propulsion Unit
SM	Service Module
SOFC	Suboptimal Feedback Control
SRP	Solar Radiation Pressure
STS	Space Transportation System
TPBVP	Two-Point Boundary Value Problem
US	Upper Stage

Symbols

α	Angle-of-attack
$\alpha_{K_{max}}$	$= \arg\max_{\alpha}(C_L/C_D)$
β	Rotation angle of the asymptotic velocity at GAM
γ	Bank angle
Γ^{-} and Γ^{+}	Upper and lower boundaries of the «ascent corridor» (fail-safety phase domain), respectively
γ_{PPS}	Specific mass of PPS
δ	Throttling function
$\eta = T/T_{max}$	Thrust throttling factor
η_{EPS}	EPS efficiency
θ	Flight path angle
$\theta_{i\ opt}$	Optimal LV initial path angle
ϑ	Pitch angle
ρ	Atmosphere density
τ	Continuation parameter; duration of pre-release ASTS maneuver
ω	Angular velocity of the daily Earth rotation; argument of perigee
λ	Geographic longitude
μ	Fuel mass consumption rate; Lagrange function; gravity parameter of primary
ν	Lagrange multiplier; true anomaly
φ	Geographic latitude
φ_E	Engine installation angle
ψ	Yaw angle

Ω	Right ascension of ascending node; force function of gravity field
a	Thrust acceleration; semi-major axis
$\mathbf{a}$	Vector of thrust acceleration
a_τ, a_r, a_n	Circumferential, radial, and binormal components of thrust acceleration
$\mathbf{a}_p$	Vector of perturbing acceleration
a_{PSFS}	PSFS specific mass
a_r	$= \frac{\rho_0 R_0 F_{\text{ref}}}{2m_i}$
a_s	Sound speed
$\mathbf{b}$	Vector of BVP initial residuals
c	Exhaust velocity
C_D	Aerodynamic drag coefficient
C_L	Aerodynamic lift coefficient
C_L^α	Derivative of C_L with respect to α
D	Feasible area
$\mathbf{E}$	Unit matrix
$\mathbf{e}$	Unit vector
$\mathbf{e}_\tau$	Unit vector of vehicle longitudinal axis
$\mathbf{e}_T$	Unit thrust vector
$\mathbf{e}_v$	Unit velocity vector
e	Eccentricity
e_x, e_y	Components of eccentricity vector
E	Eccentric anomaly
$\mathbf{f}$	Vector of right-hand sides of the phase system of equations; BVP residuals vector
F_{ref}	Reference area
g_0	Standard gravity acceleration (9.80665 m/s^2)
h	Flight altitude
h_O	Apogee altitude in osculating RV return trajectory with negligible aerodynamic loads
H	Hamiltonian
h, e_x, e_y, i_x, i_y, L	Equinoctial orbital elements
i	Inclination
i_{orb}	Orbit inclination
i_x, i_y	Components of inclination vector
I_{sp}	Specific impulse
J	Functional
$\tilde{J}$	Extended functional
K	$= C_L/C_D$, aerodynamic quality
K_{max}	$= \max\limits_{\alpha}(C_L/C_D)$, maximum aerodynamic quality
k	Reflectivity coefficient
L	Range from a launch site; true longitude
ΔL	Angular distance measured in true longitude

M	Mach number
M_h	Actuator hinge torque
m	Mass
m_{orb}	Mass on a target orbit
m_p	Propellant mass
m_{pl}	Payload mass
n	Mean motion
n_A	$= A/G$, aerodynamic load factor
n_r	Number of revolutions around the primary
n_T	$= T/G$, thrust-to-weight ratio
n_x	Longitude load factor
n_z	Normal load factor
$\boldsymbol{p}$	Conjugate (costate) vector
$\mathbf{p_r}, \mathbf{p_v}, p_m$	Conjugate variables to **r**, **v** and *m*, respectively
p	Solar radiation pressure; semi-latus rectum
P_e	Electrical power
P_b	Beam power
q	Dynamic pressure
$\mathbf{Q}$	Earth rotation matrix
$\mathbf{r}$	Radius vector from origin to vehicle mass center
r	Distance of spacecraft from primary
$\mathbf{r}_{pl}$	Distance of distant celestial body from primary
r_p	Radius of pericenter
R	Radius from Earth's center to vehicle mass center
$\mathbf{R}$	$= \mathbf{r} + \mathbf{R}_o$
$\mathbf{R}_o$	Radius vector from Earth's center to origin
s	Switching function
S_{SRP}	Reflectivity area
S_m	Reference area
$\mathbf{s}$	Unit vector along SC-to-sun direction
T	Thrust; temperature in the stagnation point
t	Time
Δt	Transfer duration
Δt_b	Burning time
$\mathbf{u}$	Control vector
U	Perturbing force function due to high-order harmonics of geopotential
v	Speed
v_0	Apogee speed in osculating RV return trajectory with negligible aerodynamic loads
$\mathbf{v}$	Velocity vector
$\mathbf{V}_\infty$	Vector of asymptotic velocity
V_{ch}	Characteristic velocity
$\mathbf{x}$	Phase (state) vector

x	Longitudinal range
$\mathbf{z}$	Vector of BVP unknown parameters
z	Lateral range

Superscripts

*	Reference value
i	Initial
f	Final

Subscripts

∞	Asymptotic (at influence sphere bound)
τ	Solution continuation parameter
+	Right limit
−	Left limit
i	Initial
α	At apocenter
π	At pericenter
adm	Admissible
bif	Bifurcation point
dim	Dimensional
f	Final
GEO	Geostationary satellite orbit
lamb	Related to the Lambert problem
max	Maximum
min	Minimum
orb	Orbital
p	Perturbing
pl	Payload; planet
SC	Related with SC
LO	Related with LO
IO	Related with IO
PO	Related with PO
EPS	Related with EPS

Chapter 1
The Pontryagin Maximum Principle for Trajectory Optimization

1.1 Introduction

Today, there is a wide arsenal of means for optimizing dynamic systems, and there is no single method for effectively solving all optimization problems. The choice of an optimization method depends on characteristics of the object under investigation and the study target. The monograph considers predominantly powered spacecraft trajectories in a gravitational field that are characterized by instability. The presence of aerodynamic forces in dense atmosphere layers enhances this property, introducing additional significant non-linear effects due to the exponential dependence of the forces on the flight altitude and uncertainty due to random atmosphere disturbances. All this makes the solutions of optimization problems very sensitive to the initial conditions, parameters of a problem, and computational errors.

Therefore, before solving the class of problems considered here, the pros and cons of the main known optimization methods were analyzed. Below, we briefly share a number of reasons based on the results of such a comparative analysis, which led the authors to an objective choice of one of them: the Pontryagin maximum principle [1].

The classification of optimization methods is an independent difficult task, which in different periods of time and with different goals was solved by many authors, for example, in [2–5]. At the same time, one must be aware that any strict classification is conventional, and in practice it is often necessary to use some kind of symbiosis of methods for a more effective (as authors always want to believe) achievement of goals of a particular study.

Let us highlight some features of optimization methods that are important for our purposes. Consider, for example, the following two main classes: indirect and direct methods.

Indirect methods search for the optimum using necessary conditions, as, for example, in the calculus of variations and PMP [1, 6], and sufficient conditions, as, for example, in the dynamic programming [7–9].

A. S. Filatyev and V. G. Petukhov, *Through Optimization of Aerospace Vehicle Trajectories by the Pontryagin Maximum Principle*, Studies in Systems, Decision and Control 579, https://doi.org/10.1007/978-3-031-80756-5_1

Direct methods build an improving sequence by analyzing the change in the objective function itself, and differ, first of all, in the order of the derivatives of the objective function with respect to the varied parameters [10, 11]. Second-order methods for determining the direction and step of an improving variation use the Hessian matrix of second partial derivatives, as, for example, in Newton's method. First-order methods rely on calculating the objective function gradient. The most extensive class is formed by direct zero-order methods (search methods), such as bisection (dichotomy), Fibonacci, Golden ratio, random search [12], or genetic [13] methods.

Qualitatively, the main difference between direct and indirect methods can be illustrated (Fig. 1.1) by a simple example of finding the maximum of the objective function $J(x)$ using the search method. The direct method builds a maximizing sequence $\ldots J(x_{j+1}) > J(x_j) > \ldots$ based on a comparison of the values of the objective function at neighboring iterations and ends the procedure when condition $\left|\frac{J(x_{j+1})}{J(x_j)} - 1\right| < \varepsilon_J$ or similar is met. In a vicinity of the maximum, direct methods quite accurately define the value of objective function J, but cannot remain as "accurate" in finding the optimal value of varied parameter x (here, $x_{\text{opt}} = 0$), especially taking into account a calculation error. Indeed, the increment of a sufficiently smooth objective function in a vicinity of the extremum is $\delta J \sim (\delta x)^{2k}$, where k is a positive integer, $\delta x = x - x_{\text{opt}}$. Let us denote the relative error in calculation of the objective function ε_J. Then relative error ε_x of determining the optimal value of x by a direct method allows the estimate $\varepsilon_x \sim \varepsilon_J^{1/(2k)}$, that is, the number of significant digits in the x_{opt} estimation is at least half that in J_{opt}.

If, however, an indirect optimization method is used that identifies the position of the desired maximum point according to the necessary condition $J'(x_{\text{opt}}) = 0$, then finding the optimal value of the parameter with any given accuracy, obviously, does not present any particular difficulties (we assume $J''(x_{\text{opt}}) < 0$), and the relative errors ε_J and ε_x are of the same order. Thus, a fundamentally important advantage of

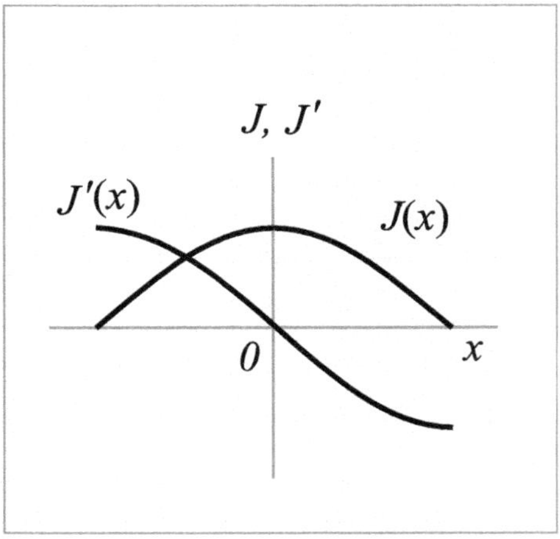

Fig. 1.1 Illustration of the search for the maximum of the objective function $J(x)$ by direct and indirect methods

indirect methods is a higher accuracy in determining the optimal values of the problem parameters or, similary, the optimal control program in optimization problems for dynamic systems.

Why is high accuracy in determining the optimal values of the problem parameters, and not just the objective function, so important? The fact is that when designing, it is important not so much the value of the objective function as its sensitivity to variations in design parameters. In practice, sensitivity functions are estimated on the basis of finite-difference derivatives, in the calculation of which close values are subtracted. But in this case, the error of such an estimate is inversely proportional to the increment of the objective function and therefore increases by orders of magnitude compared to the error of calculating J_{opt}. The "shift" of the reference value relative to x_{opt} caused by the error in determining the optimal point increases this error due to the non-fulfillment of the necessary extremum conditions. We will return to this important fact later when we analyze PMP advantages.

The use of PMP [1, 14, 15] gave a new impetus to the work in the field of optimization of ASV trajectories. On the basis of PMP, the authors of [16–20] studied the structure of optimal control of the rocket thrust vector in the absence of aerodynamic forces. In particular, the conditions for the existence of arcs of singular thrust control in a homogeneous plane-parallel gravitational field in the absence of constraints on the range were studied [18–20]. An analysis of the singular control of the orientation of the thrust vector and chattering arcs when optimizing rockets motion in a vacuum is contained in [21–24]. Attention is paid to singular optimal controls in monographs [6, 25–29]. In [30], the problem of climbing to the maximum altitude was studied, taking into account the dependence of the mass of the vehicle structure on the maximum thrust.

There are a large number of works containing the results of the numerical solution of optimization problems and, in particular, problems of trajectory optimization (see, for example, [31–40]). In many cases, this implies the existence, to some extent, of independent developments of the corresponding computational programs. However, there are relatively few descriptions in the literature of autonomous software systems that allow not only obtaining individual "unique" (optimal) trajectories, but also freely conducting broad parametric studies of extremals, including, taking into account practical limitations. Works [41–48] (see also the detailed review [2]) are among them. It should be noted that in countries with more powerful computing technology, applied packages of optimization programs based on direct methods have received particular development (the literature mentions the possibility of solving problems with the number of optimized parameters up to tens of thousands [2]). Therefore, the vast majority of these software packages are based on the use of non-linear programming methods [10, 49–57].

We can also note an original approach to the study of optimal solutions based on the convexification method [58–61], which allows, in case of success in reducing to the convex optimization problem, it is guaranteed to find the global extremum. However, this will be a solution only for the modified dynamic model, and the main difficulty for the researcher shifts towards justifying its adequacy to the original statement.

The use of indirect optimal control methods, such as PMP, is mentioned comparatively less frequently in numerical studies of the same level. In addition to the works of the authors who developed the *ASTER* software package for through optimization of Space Transportation System (STS) branching trajectories [62–65], in this area the works of H.J. Pesch. He previously published a series of theoretical papers on the application of PMP conditions for optimizing the STS motion, taking into account the discontinuity of the phase vector and the right parts of the motion equations, constraints on the phase and control variables (see, for example, [66, 67]), as well as on the application of the method neighboring extremals for the formation of trajectory control algorithms for such systems [68]. In [69], the necessary optimality conditions of PMP for the case of composite vehicles are written in the general form. The software package [70, 71] is also designed to optimize STS trajectories, but using direct methods.

For optimization problems, considered here and associated with STS, the use of other methods, except for indirect and gradient methods, is very limited. There are at least two main reasons for this. First, when developing an actual space vehicle, the movement of which is described by complex models, a high degree of responsibility for possible errors is required. Second, when conducting a qualitative analysis of optimal solutions on simplified models, strict error control is required to objectively study the dependences on parameters.

The problem of insertion into orbit of a horizontally launched single-stage spacecraft with a combined propulsion system is considered in [33] with an original idea to identify generalized variables, using which, instead of conjugate variables, it is possible to reduce the order of the integrated phase and conjugate system and, accordingly, the TPBVP. However, in this case, the class of problems is significantly limited: only single-stage vehicles are allowed to move in the vertical plane without constraints on trajectory in internal points and control; and the boundary conditions should not contain the range. Due to these limits, a number of important advantages of PMP as controlling the accuracy of the solution and the possibility of assessing the impact of the parameters on the functional are also lost. Special approximate methods for solving problems of optimal ascent into orbit for certain ASV classes, based on the use of the results of [72, 73] and various methods, taking into account small corrections for the influence of aerodynamic forces, are considered, for example, in [74]. In [6, 67, 68, 75], PMP and method of neighboring extremals are used to form algorithms for optimal trajectory control of ASV inserting into orbit. Algorithms for terminal guidance in the exoatmospheric ascent segment, based on solving TPBVP on board using approximate optimal solutions [72, 73], are described in [76, 77].

The monograph describes the indirect optimization by PMP, taking into account phase constraints [1, 6, 16, 78–87] and discontinuous dynamics systems [6, 80, 87–89].

The choice of the optimization method was determined by the following objective advantages of PMP over direct methods, including gradient and heuristic ones, and approximate engineering techniques based on simplified control programs with varied parameters:

- No a priori setting of the control structure is required; it is obtained from the solution of the problem;
- The optimal control in the functional space is determined with high accuracy from the solution of BVP, the number of varied parameters in which does not exceed the dimension of the phase vector;
- There are objective criteria for the operational control of the correctness and accuracy of the software implementation of the method and, if necessary, the localization of errors in blocks responsible for the formation of the phase and conjugate equations, optimal control and boundary conditions;
- Each solution contains additional information about the effect on the functional of small variations in parameters of aerodynamic layout, propulsion system, boundary conditions, constraints, etc.;
- Each solution contains additional information about the effect on the functional of small systematic and random external perturbations (wind and variations in the thermodynamic parameters of the atmosphere) and deviations from the nominal characteristics of the STS (propulsion and control system), which are distributed along the trajectory.

Thus, PMP provides obtaining a entire complex of diverse objective information, with controlled accuracy, about the maximum capabilities of the vehicle both under nominal conditions and in a wide range of disturbances based on the nominal solution, i.e. without solving the corresponding BVPs and other noticeable additional computational costs.

The reasons hindering the widespread use of PMP are often explained with myths (which have a basis of half a century ago) about the difficulty of solving the BVP and strict satisfaction of PMP formal conditions. The practical overcoming of these difficulties in application to the problems of SC injection into an orbit was one of the main and primary goals of the study carried out by the author. The *ASTER* computer program developed as a result of such study operates with a wide range of motion models: from simplified training models to those detailed in terms of the physical models of the shape, gravitational field, atmosphere of the Earth and STS design features, which are accepted in design firms and specialized research institutes. The software package is equipped with a user-friendly interface for Windows and an extensive database with optimal solutions for a wide range of tasks and types of launch vehicles (LV) which made it possible to automate the solution process to a large extent and, thereby, made it possible to conduct an in-depth qualitative analysis of the behavior of extremals depending on parameters, and to improve the efficiency of existing and future LV.

When solving BVP, to which optimization problems are reduced by PMP, the modified Newton method [18, 20, 90–92] is used, the features of the numerical calculation of the Jacobi matrix for residual functions of varied parameters are taken into account [93, 94]. To significantly reduce the influence of the choice of the initial guess on the convergence of the iterative procedure for solving TPBVP, the continuation method was used [91, 95–98]. The monograph includes examples of successful automated solution of complex optimization problems for branching processes, which are reduced to TPBVP with several dozens of varied parameters that determine constraints at the internal and boundary points of branches, including branching points themselves.

It is worth mentioning another common opinion that PMP does not allow one to obtain the global extremal, since it is based on the necessary optimality conditions only. First, the vast majority of other methods (except, perhaps, dynamic programming and convex problems with known limitations) do not at all guarantee obtaining at least any extremal. The main real motive for using, for example, heuristic or, in general, simplified direct algorithms, is the guarantee to complete the solving process by the appointed time. In this case, the last functionally improved variant is assigned as the extremal or even the global extremal. Secondly, even local extremals, if they obtained with a good accuracy, provide important information about the existing structures of optimal solutions, since they can become global when changing the problem parameters. Such examples are widely demonstrated in the monograph. Thirdly, sometimes with the help of PMP it is possible to prove the global extremity of a solution, if you find the entire set of local extremals (see examples in [99, 100]). Then one or more of them are global.

When using the results of the theory of non-equilibrium reentry into the atmosphere with sub-circular velocities, which generalizes problems associated with the return of LV spent parts (SPs) and crew rescue in emergency situations during STS ascent, the author relied on fundamental works on the problems of reentry into the atmosphere with near- and supercircular velocities [101–104], optimal spatial maneuvering in the atmosphere at quasi-stationary regimes [105, 106].

The number of publications devoted to the study of optimal spacecraft trajectories is enormous. More detailed references are given in corresponding chapters. Despite the extensiveness of the list of references in the monograph, it is obvious that it can be significantly expanded.

The first six chapters are devoted to the study of the problem of the optimal motion of ASVs during ascent into satellite orbits, emphasizing the qualitative influence of aerodynamic forces even then they are much less than the thrust and gravitation forces. Since constraints can be set not only on the active launch phase, but also on the return phases of STS SPs, it is necessary to consider all flight segments in a single complex, i.e. taking into account the branching nature of these trajectories and SPs motion.

The main attention is paid to the following areas [62]:

- development of methods, algorithms and a set of programs for through optimization of composite vehicle motion in the atmosphere using aero and rocket forces, taking into account the branching nature of the process and mutual influence of branches;
- the study of qualitatively new phenomena, including the bifurcation type, due to the presence of even a relatively small aerodynamic lift, and reflecting the fundamental laws of optimal motion in atmosphere;
- development of a method for complex optimization of STS trajectory control and parameters.

When optimizing LV trajectories, as a rule, the mass inserted into orbit is the functional. This is quite traditional for ballistics problems and makes it possible to compare with the results of other authors. Other criteria, for example, the payload mass or range are also considered. Obviously, the class of criteria is significantly expanded taking into account mutual problems.

The application of the results of the Chapter 1 will be demonstrated with numerous examples in subsequent chapters.

1.2 The Pontryagin Maximum Principle

Here is the formulation of PMP, adapted to the purposes of this study.

Let the motion of an object be given by a normal system of ordinary differential equations (ODE):

$$\frac{d\mathbf{x}}{dt} = \mathbf{f}(\mathbf{x}, \mathbf{u}), \tag{1.1}$$

where $\mathbf{x} = (x_1, ..., x_n)^{\mathrm{T}}$ is the phase vector; t is the independent variable ("time"), $t \in [t_i, t_f]$; $\mathbf{u} = (u_1, ..., u_m)^{\mathrm{T}}$ is the control vector, $\mathbf{f} = (f_1, ..., f_n)^{\mathrm{T}}$ is the vector function of the right part.

Remark 1.1 The phase system (1.1) is autonomous (the right part does not depend on the independent variable). This does not limit the generality of the statement, since if it is necessary to take into account the dependence of the right part on the time, the actual time τ can be introduced as an additional phase variable $x_{n+1} = \tau$ using equation $\frac{dx_{n+1}}{dt} = 1$. With this in mind, the independent variable will be conditionally called the time.

To begin with, we assume that $\mathbf{f}(\mathbf{x}, \mathbf{u})$ is continuous and continuously differentiable in all variables.

At the initial moment of time t_i, which we will consider fixed, some constraints of the equality type are imposed on the admissible phase vector $\mathbf{x}^i = \mathbf{x}(t_i)$:

$$t = t_i = \text{fix}, g_1^i(\mathbf{x}^i) = 0, ..., g_{k_i}^i(\mathbf{x}^i) = 0 \text{ or } \mathbf{g}^i(\mathbf{x}^i) = \left(g_1^i(\mathbf{x}^i), ..., g_{k_i}^i(\mathbf{x}^i)\right)^{\mathrm{T}} = 0 \tag{1.2}$$

At the final moment t_f, which we consider free,

$$t = t_f = \text{free}, g_1^f(\mathbf{x}^f) = 0, ..., g_{k_f}^f(\mathbf{x}^f) = 0 \text{ or } \mathbf{g}^f(\mathbf{x}^f) = \left(g_1^f(\mathbf{x}^f), ..., g_{k_f}^f(\mathbf{x}^f)\right)^{\mathrm{T}} = 0 \tag{1.3}$$

The number of these constraints should not exceed the dimension of the phase vector: $k_i + k_f \leq n$, but in a more general form, for the solvability of equalities (1.2) and (1.3), we require that the gradients of these functions at the initial and final points be linear-independent. In other words, the matrices whose rows are formed by the gradients of these functions must have the maximum ranks:

$$rang\begin{pmatrix}\nabla g_1^i(\mathbf{x})\\ \dots \\ \nabla g_{k_i}^i(\mathbf{x})\end{pmatrix}_{t=t_i}=k_i,\quad rang\begin{pmatrix}\nabla g_1^f(\mathbf{x})\\ \dots \\ \nabla g_{k_f}^f(\mathbf{x})\end{pmatrix}_{t=t_f}=k_f. \tag{1.4}$$

Remark 1.2 The accepted conditions (1.2), (1.3) at the moments $t_i,\ t_f$ do not limit the generality of the statement. Time is relative, so one of the moments (here it is the initial moment of time) must be specified. If the final time must also be fixed, then, for example, we can take the actual time τ as an additional phase variable $x_{n+1} = \tau$ using the equation $\frac{dx_{n+1}}{dt} = 1$ and add one more constraint on the extended phase vector to (1.3): $g_{k_f+1}^f\left(\mathbf{x}^f\right) = x_{n+1} - \tau_f = 0$, where τ_f is the given final time.

So, we formulate the problem as follows. It is required to find such a vector control function $\mathbf{u}(t),\ t \in \left[t_i, t_f\right]$, so that the functional, in the general case, depending on the initial and final values of the phase variables, reaches the maximum value:

$$J\left(\mathbf{x}^i, \mathbf{x}^f\right) \Rightarrow \max \tag{1.5}$$

Remark 1.3 The functional in the form (1.5) also allows us to consider problems with a time-dependent functional, in particular, time-optimal problems. To do this, one can, for example, introduce time as a new phase variable (see Remarks 1.1 and 1.2).

Remark 1.4 In the classical presentations of PMP [1] and the calculus of variations [6], the problem is formulated, in contrast to (1.5), as a problem for the minimum of the functional. This leads to a change of the sign in transversality conditions while maintaining the form of the control optimality condition. Here it is done so that the conjugate variables and Lagrange multipliers coincide with the functions of the influence of the corresponding variables on the functional not only in absolute value, but also in sign.

To formulate PMP conditions, along with the phase system (1.1), we consider the conjugate system:

$$\frac{d\mathbf{p}}{dt} = -\left(\frac{\partial H}{\partial \mathbf{x}}\right)^{\mathrm{T}} = \begin{pmatrix} -p_1\frac{\partial f_1}{\partial x_1} - & \dots & -p_n\frac{\partial f_n}{\partial x_1} \\ \dots & \dots & \dots \\ -p_1\frac{\partial f_1}{\partial x_n} - & \dots & -p_n\frac{\partial f_n}{\partial x_n}\end{pmatrix}, \tag{1.6}$$

where $H(\mathbf{p}, \mathbf{x}, \mathbf{u}) = \mathbf{p}^{\mathrm{T}}\mathbf{f}(\mathbf{x}, \mathbf{u})$ is the Hamiltonian, $\mathbf{p} = (p_1, \dots, p_n)^{\mathrm{T}}$ is the conjugate vector.

PMP THEOREM. Let the control function $\mathbf{u}_{\mathrm{opt}}(t),\ t \in \left[t_i, t_f\right]$ be a solution to problem (1.5), which, by virtue of (1.1)–(1.3), corresponds to the phase trajectory

$\mathbf{x}(t)$, satisfying conditions (1.2), (1.3) at the initial and final points. Then, there exists a nontrivial (not identically zero) conjugate vector function $\mathbf{p}(t), t \in [t_i, t_f]$, such that the following two conditions are satisfied.

(1) At each moment of time $t \in [t_i, t_f]$, the Hamiltonian $H(\mathbf{p}(t), \mathbf{x}(t), \mathbf{u}_{\text{opt}}(t))$ reaches its maximum in $\mathbf{u} \in \mathbf{U}$ with other variables fixed (the control optimality condition):

$$H(\mathbf{p}(t), \mathbf{x}(t), \mathbf{u}_{\text{opt}}(t)) = \max_{\mathbf{u}\in\mathbf{U}} H(\mathbf{p}(t), \mathbf{x}(t), \mathbf{u}(t)). \tag{1.7}$$

(2) At the initial and final moments of time, the transversality conditions are satisfied:

$$\mathbf{p}^i = \mathbf{p}(t_i) = -\left(\frac{\partial \mathbf{g}^i(\mathbf{x}^i)}{\partial \mathbf{x}^i}\right)^{\mathrm{T}} \boldsymbol{\lambda}^i - \left(\frac{\partial J}{\partial \mathbf{x}^i}\right)^{\mathrm{T}}$$

$$= -\begin{pmatrix} \frac{\partial g_1^i(\mathbf{x}^i)}{\partial x_1^i}\lambda_1^i + \dots + \frac{\partial g_{k_i}^i(\mathbf{x}^i)}{\partial x_1^i}\lambda_{k_i}^i \\ \dots \quad \dots \quad \dots \\ \frac{\partial g_1^i(\mathbf{x}^i)}{\partial x_{k_i}^i}\lambda_1^i + \dots + \frac{\partial g_{k_i}^i(\mathbf{x}^i)}{\partial x_{k_i}^i}\lambda_{k_i}^i \end{pmatrix} - \begin{pmatrix} \frac{\partial J}{\partial x_1^i} \\ \dots \\ \frac{\partial J}{\partial x_{k_i}^i} \end{pmatrix}, \tag{1.8}$$

$$\mathbf{p}^f = \mathbf{p}(t_f) = \left(\frac{\partial g^f(\mathbf{x}^f)}{\partial \mathbf{x}^f}\right)^{\mathrm{T}} \boldsymbol{\lambda}^f + \left(\frac{\partial J}{\partial \mathbf{x}^f}\right)^{\mathrm{T}}$$

$$= \begin{pmatrix} \frac{\partial g_1^f(\mathbf{x}^f)}{\partial x_1^f}\lambda_1^f + \dots + \frac{\partial g_{k_f}^f(\mathbf{x}^f)}{\partial x_1^f}\lambda_{k_f}^f \\ \dots \quad \dots \quad \dots \\ \frac{\partial g_1^f(\mathbf{x}^f)}{\partial x_{k_f}^f}\lambda_1^f + \dots + \frac{\partial g_{k_f}^f(\mathbf{x}^f)}{\partial x_{k_f}^f}\lambda_{k_f}^f \end{pmatrix} + \begin{pmatrix} \frac{\partial J}{\partial x_1^f} \\ \dots \\ \frac{\partial J}{\partial x_{k_f}^f} \end{pmatrix} \tag{1.9}$$

$$H(t_f) = 0. \tag{1.10}$$

The constant vectors $\boldsymbol{\lambda}^i \in \mathbf{R}^{k_i}$, $\boldsymbol{\lambda}^f \in \mathbf{R}^{k_f}$ in (1.8), (1.9) are indefinite Lagrange multipliers. In the special case when the left end of the trajectory is free, it follows from (1.8) that conjugate vector $\mathbf{p}(t_i)$ is completely defined. And vice versa, if the phase vector is fixed at the initial point ($k_i = n$), then, according to (1.8), the initial conjugate vector is undefined. Similar remarks are also valid for the conditions at the right end of the trajectory.

PMP conditions reduce the original problem of finding the optimal control vector function $\mathbf{u}_{\text{opt}}(t), t \in [t_i, t_f]$, in an infinite-dimensional (functional) space, to solving TPBVP for a system of $2n$ ODEs (1.1), (1.6) with no more than n varied parameters and n residuals.

1.3 The Functional Variation Analysis

PMP differs from the calculus of variations in the class of admissible control functions, that does not affect the formation of the conjugate system and boundary conditions. Following [6, 14], conjugate system (1.6) and transversality conditions (1.7)–(1.10) can be obtained based on the analysis of the functional variations. For simplicity, in this section we assume that there are no constraints on the control. In subsequent chapters, the procedure described below will be used to obtain the conjugate system and transversality conditions in the presence of constraints and discontinuous right-hand sides of the phase system and in problems with branching trajectories.

Let us consider extended functional

$$\tilde{J} = J\left(\mathbf{x}^i, \mathbf{x}^f\right) + \boldsymbol{\lambda}^{i\,\mathrm{T}}\mathbf{g}^i\left(\mathbf{x}^i\right) + \boldsymbol{\lambda}^{f\,\mathrm{T}}\mathbf{g}^f\left(\mathbf{x}^f\right) + \int\limits_{t_i}^{t_f} \mathbf{p}^{\mathrm{T}}[\mathbf{f}(\mathbf{x}, \mathbf{u}) - \dot{\mathbf{x}}]dt, \tag{1.11}$$

which, on admissible trajectories (1.1)–(1.3), obviously coincides with the functional (1.5). The increment of the extended functional caused by the variation of the control, phase vector at the left and right ends of the trajectory and the final moment of time has the following form:

$$\begin{aligned} d\tilde{J} = {} & \frac{\partial J\left(\mathbf{x}^i, \mathbf{x}^f\right)}{\partial \mathbf{x}^i} d\mathbf{x}^i + \frac{\partial J\left(\mathbf{x}^i, \mathbf{x}^f\right)}{\partial \mathbf{x}^f} d\mathbf{x}^f + \boldsymbol{\lambda}^{i\,\mathrm{T}}\frac{\partial \mathbf{g}^i\left(\mathbf{x}^i\right)}{\partial \mathbf{x}^i} d\mathbf{x}^i + \boldsymbol{\lambda}^{f\,\mathrm{T}}\frac{\partial \mathbf{g}^f\left(\mathbf{x}^f\right)}{\partial \mathbf{x}^f} d\mathbf{x}^f \\ & - \int\limits_{t_i}^{t_f} \mathbf{p}^{\mathrm{T}}\delta\dot{\mathbf{x}}dt + \int\limits_{t_i}^{t_f} \mathbf{p}^{\mathrm{T}}\frac{\partial \mathbf{f}(\mathbf{x}, \mathbf{u})}{\partial \mathbf{x}}\delta\mathbf{x}dt + \int\limits_{t_i}^{t_f} \mathbf{p}^{\mathrm{T}}\delta_{\mathbf{u}}\mathbf{f}(\mathbf{x}, \mathbf{u})dt, \end{aligned} \tag{1.12}$$

where $\delta_{\mathbf{u}}\mathbf{f}$ is the variation of $\mathbf{f}$ due to the control variation.

The first integral in (1.12) can be taken by parts. Then, taking into account the equation for the conjugate vector (1.6), we obtain

$$\begin{aligned} \int\limits_{t_i}^{t_f} \mathbf{p}^{\mathrm{T}}\delta\dot{\mathbf{x}}dt &= \int\limits_{t_i}^{t_f} \mathbf{p}^{\mathrm{T}}d(\delta\mathbf{x}) = \left(\mathbf{p}^{\mathrm{T}}\delta\mathbf{x}\right)_{t_i}^{t_f} - \int\limits_{t_i}^{t_f} \delta\mathbf{x}^{\mathrm{T}}d\mathbf{p} = \left(\mathbf{p}^{\mathrm{T}}\delta\mathbf{x}\right)_{t_i}^{t_f} - \int\limits_{t_i}^{t_f} \delta\mathbf{x}^{\mathrm{T}}\dot{\mathbf{p}}dt \\ &= \mathbf{p}^{f\,\mathrm{T}}\delta\mathbf{x}^f - \mathbf{p}^{i\mathrm{T}}\delta\mathbf{x}^i - \int\limits_{t_i}^{t_f} \dot{\mathbf{p}}^{\mathrm{T}}\delta\mathbf{x}dt \end{aligned} \tag{1.13}$$

Substituting (1.13) into (1.12), we get

$$d\tilde{J} = \frac{\partial J(\mathbf{x}^i, \mathbf{x}^f)}{\partial \mathbf{x}^i} d\mathbf{x}^i + \frac{\partial J(\mathbf{x}^i, \mathbf{x}^f)}{\partial \mathbf{x}^f} d\mathbf{x}^f + \boldsymbol{\lambda}^{i\,\mathrm{T}} \frac{\partial \mathbf{g}^i(\mathbf{x}^i)}{\partial \mathbf{x}^i} d\mathbf{x}^i + \boldsymbol{\lambda}^{f\,\mathrm{T}} \frac{\partial \mathbf{g}^f(\mathbf{x}^f)}{\partial \mathbf{x}^f} d\mathbf{x}^f$$
$$- \mathbf{p}^{f\,\mathrm{T}} \delta \mathbf{x}^f + \mathbf{p}^{i\,\mathrm{T}} \delta \mathbf{x}^i + \int_{t_i}^{t_f} \left[\dot{\mathbf{p}}^{\mathrm{T}} + \mathbf{p}^{\mathrm{T}} \frac{\partial \mathbf{f}(\mathbf{x}, \mathbf{u})}{\partial \mathbf{x}} \right] \delta \mathbf{x} dt + \int_{t_i}^{t_f} \mathbf{p}^{\mathrm{T}} \delta_{\mathbf{u}} \mathbf{f}(\mathbf{x}, \mathbf{u}) dt. \quad (1.14)$$

Since

$$\mathbf{p}^{\mathrm{T}} \frac{\partial \mathbf{f}(\mathbf{x}, \mathbf{u})}{\partial \mathbf{x}} = \frac{\partial [\mathbf{p}^{\mathrm{T}} \mathbf{f}(\mathbf{x}, \mathbf{u})]}{\partial \mathbf{x}} = \frac{\partial H}{\partial \mathbf{x}}, \quad (1.15)$$

then the expression in square brackets in (1.14) vanishes if the conjugate system is given in the form (1.6).

In the linear approximation, the total increment of the phase vector $d\mathbf{x}(t^*)$ is the sum of variation $\delta \mathbf{x}(t^*)$ at a fixed time moment t^* and the increment $\dot{\mathbf{x}}(t^*)dt^*$ caused by the increment of time (if it is not fixed):

$$\begin{aligned}
d\mathbf{x}(t^*) &= \delta \mathbf{x}(t^*) + \dot{\mathbf{x}}(t^*)dt^* = \delta \mathbf{x}(t^*) + \mathbf{f}(\mathbf{x}(t^*), \mathbf{u}(t^*))dt^*, \\
\delta \mathbf{x}(t^*) &= d\mathbf{x}(t^*) - \mathbf{f}(\mathbf{x}(t^*), \mathbf{u}(t^*))dt^*, \\
&\mathbf{p}^{i\,\mathrm{T}} \delta \mathbf{x}(t^i) = \mathbf{p}^{i\,\mathrm{T}} d\mathbf{x}^i, \mathbf{p}^{f\,\mathrm{T}} \delta \mathbf{x}(t^f) \\
&= \mathbf{p}^{f\,\mathrm{T}} d\mathbf{x}^f - \mathbf{p}^{f\,\mathrm{T}} \mathbf{f}(\mathbf{x}(t^f), \mathbf{u}(t^f))dt^f = \mathbf{p}^{f\,\mathrm{T}} d\mathbf{x}^f - H(t^f)dt^f. \quad (1.16)
\end{aligned}$$

Taking into account (1.15) and (1.16), we rewrite increment (1.14):

$$\begin{aligned}
d\tilde{J} = &\left[\frac{\partial J(\mathbf{x}^i, \mathbf{x}^f)}{\partial \mathbf{x}^i} + \boldsymbol{\lambda}^{i\,\mathrm{T}} \frac{\partial \mathbf{g}^i(\mathbf{x}^i)}{\partial \mathbf{x}^i} + \mathbf{p}^{i\,\mathrm{T}} \right] d\mathbf{x}^i \\
&+ \left[\frac{\partial J(\mathbf{x}^i, \mathbf{x}^f)}{\partial \mathbf{x}^f} + \boldsymbol{\lambda}^{f\,\mathrm{T}} \frac{\partial \mathbf{g}^f(\mathbf{x}^f)}{\partial \mathbf{x}^f} - \mathbf{p}^{f\,\mathrm{T}} \right] d\mathbf{x}^f \\
&+ H(t^f)dt^f + \int_{t_i}^{t_f} \delta_{\mathbf{u}} H(\mathbf{p}, \mathbf{x}, \mathbf{u}) dt. \quad (1.17)
\end{aligned}$$

It is easy to see that from the stationarity condition: $d\tilde{J} = 0$, for independent increments of the control, final time and trajectory boundary points, we obtain the transversality conditions (1.8)–(1.10).

1.4 Constraints on the Control and Phase Vector

Let the control and phase vector be constrained:

$$G(\mathbf{x}, \mathbf{u}) \le 0. \quad (1.18)$$

Following [6, 14], we introduce an extended functional

$$\tilde{J} = J\left(\mathbf{x}^i, \mathbf{x}^f\right) + \boldsymbol{\lambda}^{i\,\mathrm{T}}\mathbf{g}^i\left(\mathbf{x}^i\right) + \boldsymbol{\lambda}^{f\,\mathrm{T}}\mathbf{g}^f\left(\mathbf{x}^f\right) + \int\limits_{t_i}^{t_f}\left\{\mathbf{p}^{\mathrm{T}}[\mathbf{f}(\mathbf{x},\mathbf{u}) - \dot{\mathbf{x}}] + \mu G(\mathbf{x},\mathbf{u})\right\}dt, \tag{1.19}$$

where μ is the Lagrange function:

$$\mu = 0, \text{ if } G(\mathbf{x},\mathbf{u}) < 0. \tag{1.20}$$

The extended functional increment

$$\begin{aligned} d\tilde{J} = {} & \frac{\partial J\left(\mathbf{x}^i,\mathbf{x}^f\right)}{\partial \mathbf{x}^i}d\mathbf{x}^i + \frac{\partial J\left(\mathbf{x}^i,\mathbf{x}^f\right)}{\partial \mathbf{x}^f}d\mathbf{x}^f + \boldsymbol{\lambda}^{i\,\mathrm{T}}\frac{\partial \mathbf{g}^i\left(\mathbf{x}^i\right)}{\partial \mathbf{x}^i}d\mathbf{x}^i + \boldsymbol{\lambda}^{f\,\mathrm{T}}\frac{\partial \mathbf{g}^f\left(\mathbf{x}^f\right)}{\partial \mathbf{x}^f}d\mathbf{x}^f \\ & - \int\limits_{t_i}^{t_f}\mathbf{p}^{\mathrm{T}}\delta\dot{\mathbf{x}}dt + \int\limits_{t_i}^{t_f}\left[\mathbf{p}^{\mathrm{T}}\frac{\partial \mathbf{f}(\mathbf{x},\mathbf{u})}{\partial \mathbf{x}} + \mu\frac{\partial G(\mathbf{x},\mathbf{u})}{\partial \mathbf{x}}\right]\delta\mathbf{x}dt \\ & + \int\limits_{t_i}^{t_f}\left[\mathbf{p}^{\mathrm{T}}\delta_{\mathbf{u}}\mathbf{f}(\mathbf{x},\mathbf{u}) + \mu\delta_{\mathbf{u}}G(\mathbf{x},\mathbf{u})\right]\delta\mathbf{u}dt \\ = {} & \left[\frac{\partial J\left(\mathbf{x}^i,\mathbf{x}^f\right)}{\partial \mathbf{x}^i} + \boldsymbol{\lambda}^{i\,\mathrm{T}}\frac{\partial \mathbf{g}^i\left(\mathbf{x}^i\right)}{\partial \mathbf{x}^i} + \mathbf{p}^{i\,\mathrm{T}}\right]d\mathbf{x}^i \\ & + \left[\frac{\partial J\left(\mathbf{x}^i,\mathbf{x}^f\right)}{\partial \mathbf{x}^f} + \boldsymbol{\lambda}^{f\,\mathrm{T}}\frac{\partial \mathbf{g}^f\left(\mathbf{x}^f\right)}{\partial \mathbf{x}^f} - \mathbf{p}^{f\,\mathrm{T}}\right]d\mathbf{x}^f \\ & + \tilde{H}\left(t^f\right)dt^f + \int\limits_{t_i}^{t_f}\left[\dot{\mathbf{p}}^T + \frac{\partial \tilde{H}(\mathbf{p},\mathbf{x},\mathbf{u})}{\partial \mathbf{x}}\right]\delta\mathbf{x}dt \\ & + \int\limits_{t_i}^{t_f}\delta_{\mathbf{u}}\tilde{H}(\mathbf{p},\mathbf{x},\mathbf{u})dt, \end{aligned} \tag{1.21}$$

where $\tilde{H} = H + \mu G(\mathbf{x},\mathbf{u})$. Whence we obtain the known relations for the conjugate system and the optimality condition for the control in the presence of constraint (1.18):

$$\frac{d\mathbf{p}}{dt} = -\left(\frac{\partial \tilde{H}}{\partial \mathbf{x}}\right)^{\mathrm{T}}, \tag{1.22}$$

$$\mathbf{u}_{\mathrm{opt}} : \tilde{H}\left(\mathbf{p}(t),\mathbf{x}(t),\mathbf{u}_{\mathrm{opt}}(t)\right) = \max_{\mathbf{u}\in\mathbf{U}}\tilde{H}(\mathbf{p}(t),\mathbf{x}(t),\mathbf{u}(t)). \tag{1.23}$$

The Lagrange function μ can be obtained from joint consideration of condition (1.23) and condition of motion along boundary (1.18): $G(\mathbf{x}, \mathbf{u}) = 0$.

The conjugate vector at the moment of reaching boundary (1.18) remains continuous.

1.5 Phase Constraints

Consider the features of PMP in the presence of phase constraint

$$G(\mathbf{x}) \leq 0, \tag{1.24}$$

where $G(\mathbf{x})$ is a smooth function.

We will assume that (1.24) is a first-order constraint, that is, the control enters explicitly into the first time derivative of $G(\mathbf{x})$:

$$\dot{G}(\mathbf{x}) = G^1(\mathbf{x}, \mathbf{u}). \tag{1.25}$$

Such constraints include, for example, constraints on the dynamic pressure, heat flow, Mach number, velocity, etc.

Following [6, 14], we introduce the extended functional

$$\tilde{J} = \ldots + \lambda G(\mathbf{x}^*) + \int_{t_i}^{t_*} \{\mathbf{p}^{\mathrm{T}}[\mathbf{f}(\mathbf{x}, \mathbf{u}) - \dot{\mathbf{x}}]\} dt + \int_{t_*}^{t_f} \{\mathbf{p}^{\mathrm{T}}[\mathbf{f}(\mathbf{x}, \mathbf{u}) - \dot{\mathbf{x}}] + \mu G^1(\mathbf{x}, \mathbf{u})\} dt \tag{1.26}$$

where $\mathbf{x}^* = \mathbf{x}(t_*)$; t_* is the implicitly set time moment of reaching boundary

$$\Gamma = \{\mathbf{x} : \quad G(\mathbf{x}) \equiv 0; \mathbf{u} = \overline{\mathbf{u}} : G^1(\mathbf{x}, \overline{\mathbf{u}}) \equiv 0\}. \tag{1.27}$$

In (1.26), only the part of the extended functional was left, which is related to constraint (1.24).

The functional increment is

$$\begin{aligned} d\tilde{J} = \ldots + [\lambda \nabla G(\mathbf{x}^*) - \mathbf{p}^{-\,\mathrm{T}} + \mathbf{p}^{+\,\mathrm{T}}] d\mathbf{x}^* + [H^-(t^*) - H^+(t^*)] dt_* + \\ + \int_{t_*}^{t_f} \left[\dot{\mathbf{p}}^{\mathrm{T}} + \frac{\partial \tilde{H}(\mathbf{p}, \mathbf{x}, \mathbf{u})}{\partial \mathbf{x}}\right] \delta \mathbf{x} dt + \int_{t_*}^{t_f} \delta_{\mathbf{u}} \tilde{H}(\mathbf{p}, \mathbf{x}, \mathbf{u}) dt = 0, \end{aligned} \tag{1.28}$$

where $\tilde{H} = H + \mu G^1(\mathbf{x}, \mathbf{u})$. Whence we obtain the conjugate system and control optimality condition in form (1.22), (1.23) and the following conditions for the jump of the conjugate vector at moment t_*:

$$\mathbf{p}^+ = \mathbf{p}^- - \lambda \nabla G^{\mathrm{T}}(\mathbf{x}^*), \tag{1.29}$$

$$H^+ - H^- = (\mathbf{p}^+, \mathbf{f}^+) - (\mathbf{p}^-, \mathbf{f}^-) = 0. \tag{1.30}$$

where the superscripts $-$ and $+$ correspond to the limits to the left $(t_* - 0)$ and to the right $(t_* + 0)$ of t_*.

Note that the factor λ cannot be found directly from conditions (1.29) and (1.30). Indeed, substituting expression (1.29) for the jump of the conjugate vector (1.30), we obtain:

$$(\mathbf{p}^-, \mathbf{f}^+) - (\mathbf{p}^-, \mathbf{f}^-) - \lambda \nabla G \cdot \mathbf{f}^+ = 0$$

But

$$\nabla G \cdot \mathbf{f}^+ = \nabla G \cdot \left. \frac{d\mathbf{x}}{dt} \right|_{t_*+0} = \left. \frac{dG}{dt} \right|_{t_*+0} = \dot{G}^+ = 0,$$

and (1.30) implies the transversality condition, which explicitly does not contain λ:

$$\mathbf{p}^{-\mathrm{T}}(\mathbf{f}^+ - \mathbf{f}^-) = 0. \tag{1.31}$$

Let us prove that on the optimal trajectory, boundary (1.27) is reached smoothly, that is, at the moment, when the function G reaches its local maximum.

Indeed, since the phase vector at point t_* is continuous: $\mathbf{x}^- = \mathbf{x}^+ = \mathbf{x}_*$, then

$$\mathbf{f}^+ = \mathbf{f}\left(\mathbf{x}^*, \overline{\mathbf{u}}_{\mathrm{opt}}^+\right), \quad \mathbf{f}^- = \mathbf{f}\left(\mathbf{x}^*, \mathbf{u}_{\mathrm{opt}}^-\right), \tag{1.32}$$

where $\overline{\mathbf{u}}_{\mathrm{opt}}$ is the optimal control when moving along boundary (1.27):

$$\dot{G} = G^1\left(\mathbf{x}, \overline{\mathbf{u}}_{\mathrm{opt}}\right) \equiv 0. \tag{1.33}$$

In view of (1.32), condition (1.31) means that

$$H^-\left(\overline{\mathbf{u}}_{\mathrm{opt}}^+\right) = H^-\left(\mathbf{u}_{\mathrm{opt}}^-\right) = \max_{\mathbf{u} \in \mathbf{U}} H^-(\mathbf{u}). \tag{1.34}$$

If the problem is well-posed, and $\mathbf{u}_{\mathrm{opt}}$ is uniquely determined by optimality condition (1.23), then according to (1.34)

$$\mathbf{u}_{\mathrm{opt}}^- = \overline{\mathbf{u}}_{\mathrm{opt}}^+$$

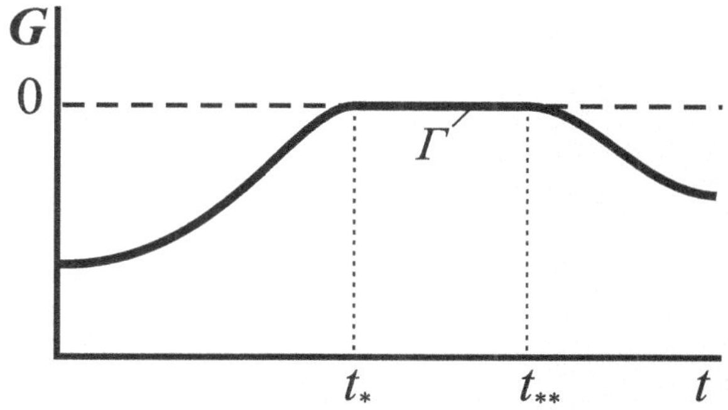

Fig. 1.2 Illustration of the first-order phase constraint $G(\mathbf{x}) \le 0$ on the optimal trajectory

and due to (1.33)

$$\dot{G}\left(\mathbf{x}_*, \mathbf{u}_{\mathrm{opt}}^-\right) = \dot{G}^- = 0. \tag{1.35}$$

This is what was required to be proved: if constraint (1.24) is essential, then function $G(t)$ smoothly exits to boundary $\boldsymbol{\Gamma}$ at moment $(t_* - 0)$ under the optimal control (Fig. 1.2).

When moving along boundary (1.27), the conjugate system is determined by (1.22), (1.23), where

$$\tilde{H} = H + \mu \dot{G}(\mathbf{x}, \mathbf{u}) = \mathbf{p}^{\mathrm{T}}\mathbf{f} + \mu \nabla G \cdot \mathbf{f}(\mathbf{x}, \mathbf{u}) = \left(\mathbf{p}^{\mathrm{T}} + \mu \nabla G\right) \cdot \mathbf{f}(\mathbf{x}, \mathbf{u}). \tag{1.36}$$

It follows from (1.36) that when moving along boundary (1.27), the control optimality condition is written in the same way as outside this boundary, if only vector $\tilde{\mathbf{p}} = \mathbf{p} + \mu \nabla G^{\mathrm{T}}$ substitutes conjugate vector $\mathbf{p}$.

Moment t_{**} of departure from boundary $\boldsymbol{\Gamma}$ is determined by the following condition:

$$t_{**} : \dot{G}\left(\mathbf{x}, \mathbf{u}_{\mathrm{opt}}\right)\Big|_{t_{**}} = 0 \left(\dot{G}\left(\mathbf{x}, \mathbf{u}_{\mathrm{opt}}\right)\Big|_{t_{**}+0} < 0 \right), \tag{1.37}$$

where $\mathbf{u}_{\mathrm{opt}}$ is the optimal control without the constraint. According to (1.37), the departure from boundary (1.27), as well as the exit to it, proceeds smoothly.

Let us note the features of the TPBVP statement in case of phase constraints. In the general case, constraint (1.24) leads to an increase in the TPBVP dimensionality (see Sect. 1.10) by one for each exit point to the phase constraint. Time moment t_* cannot be determined by a seemingly natural condition $G(\mathbf{x}(t_*)) = 0$, because, even if this constraint is essential (active) on the optimal trajectory, it may not be reached at all on trial trajectories during the iterative shooting procedure for solving TPBVP, which will lead to a loss of convergence. Therefore, to determine t_*, it is convenient to accept the necessary condition for the local maximum of $G(\mathbf{x})$:

$$\dot{G}(\mathbf{x}(t_*)) = 0. \tag{1.38}$$

Value $\max_t G(\mathbf{x}) = G(\mathbf{x}(t_*)) = G_{\max}$ reached at this moment serves as an additional residual in the TPBVP. In order for the TPBVP to remain closed, it is advisable to take Lagrange multiplier λ, which determines the jump of conjugate variables (1.29), as the additional varied parameter.

Note the apparent paradox that arises here. In this pair of additional parameters, the varied parameter λ cannot have a direct effect on residual $G(\mathbf{x}(t_*))$, it appears when determining the jump of conjugate variables immediately after reaching the phase constraint, that is, *after* the formation of this residual. The apparent paradox is resolved by the fact that parameter λ affects the residuals calculated in subsequent sections of the trajectory, including the final point, and the additional residual is affected by varied parameters specified in the previous section of the trajectory, in particular, at the initial point. Thus, the introduced additional parameters turn out to be connected indirectly within the framework of one closed BVP.

The above features of accounting phase constraints are also reflected in the procedure for forming the initial estimate of the BVP solution. To ensure the correctness of the problem, it is necessary to verify the existence of its solution. It means confirmation, firstly, that the constraint under consideration is essential, and, secondly, that it is realizable (the system is controllable). Taking into account these conditions, the procedure for solving the problem with a phase constraint consists of two stages. At the first stage, the optimization problem is solved without the phase constraint. If $\max_t G(\mathbf{x}) = G^0_{\max} > 0$ on the optimal trajectory, then constraint (1.24) is essential, and we proceed to the second stage: solving the problem with the constraint. In this case, we use the obtained solution as an initial estimate and continue it with respect to parameter $\tau = 1 - G_{\max}/G^0_{\max}$ from $\tau = 0$ to $\tau = 1$. The optimal solution continuation up to the desired parameter value is possible only if the constraint is realizable within the given problem statement.

1.6 Discontinuity of Right-Hand Sides of Motion Equations

Let the vector of the right parts of the phase system $\mathbf{f}(\mathbf{x}, \mathbf{u})$ undergoes a discontinuity:

$$\mathbf{f}(\mathbf{x}, \mathbf{u}) = \begin{cases} \mathbf{f}^1(\mathbf{x}, \mathbf{u}), & t < t_*, \\ \mathbf{f}^2(\mathbf{x}, \mathbf{u}), & t > t_*, \end{cases} \tag{1.39}$$

at moment t_* given implicitly by equality

$$G(\mathbf{x}(t_*)) = 0. \tag{1.40}$$

Following [6], we consider the extended functional

$$\tilde{J} = \ldots + \lambda G(\mathbf{x}^*) + \int_{t_i}^{t_*} \{\mathbf{p}^{\mathrm{T}}[\mathbf{f}^1(\mathbf{x}, \mathbf{u}) - \dot{\mathbf{x}}]\}dt + \int_{t_*}^{t_f} \{\mathbf{p}^{\mathrm{T}}[\mathbf{f}^2(\mathbf{x}, \mathbf{u}) - \dot{\mathbf{x}}]\}dt, \quad (1.41)$$

where λ is the Lagrange multiplier, $\mathbf{x}^* = \mathbf{x}(t_*)$. Equation (1.41) shows only the part of the extended functional that is related to the discontinuity conditions of the right-hand sides.

Extended functional increment

$$d\tilde{J} = \ldots + [\lambda \Delta G(\mathbf{x}^*) - \mathbf{p}^{-\mathrm{T}} + \mathbf{p}^{+\mathrm{T}}]d\mathbf{x}^* + [H^-(t^*) - H^+(t^*)]dt_* = 0. \quad (1.42)$$

From where we obtain the conditions for the jump of the conjugate vector at the discontinuity moment:

$$\mathbf{p}^+ = \mathbf{p}^- - \lambda \nabla G^{\mathrm{T}}, \quad (1.43)$$

$$H^+\left(\mathbf{p}^+, \mathbf{x}^*, \mathbf{u}_{\mathrm{opt}}^+\right) = H^-\left(\mathbf{p}^-, \mathbf{x}^*, \mathbf{u}_{\mathrm{opt}}^-\right). \quad (1.44)$$

Multiplier λ can be obtained from system (1.43) and (1.44). It should be kept in mind that this system, as a rule, is non-linear with respect to λ, since it is contained implicitly in (1.44) through the control optimality condition $\mathbf{u}_{\mathrm{opt}}^+ = \max_{\mathbf{u}}[(\mathbf{p}^{-\mathrm{T}} - \lambda \nabla G)\mathbf{f}^2(\mathbf{x}^*, \mathbf{u})]$.

1.7 Discontinuity of the Phase Vector and Trajectory Branching

Let us consider the features of solving PMP problems in the presence of a discontinuity of the phase vector at some implicitly specified moment t_*:

$$\mathbf{g}^*(\mathbf{x}^-, \mathbf{x}^+) = \left(g_1^*(\mathbf{x}^-, \mathbf{x}^+), \ldots, g_{k_*}^*(\mathbf{x}^-, \mathbf{x}^+)\right)^{\mathrm{T}} = 0, \mathbf{x}^- = \mathbf{x}(t_* - 0), \mathbf{x}^+ = \mathbf{x}(t_* + 0). \quad (1.45)$$

Following [6, 14], we introduce extended functional

$$\tilde{J} = \ldots + \boldsymbol{\lambda}^{*T}\mathbf{g}^*(\mathbf{x}^-, \mathbf{x}^+) + \int_{t_i}^{t_*-0} \{\mathbf{p}^T[\mathbf{f}(\mathbf{x}, \mathbf{u}) - \dot{\mathbf{x}}]\}dt + \int_{t_*+0}^{t_f} \{\mathbf{p}^T[\mathbf{f}(\mathbf{x}, \mathbf{u}) - \dot{\mathbf{x}}]\}dt \quad (1.46)$$

in which only the part associated with constraint (1.45) was left.

The increment of functional (1.46) associated with constraint (1.45) is

$$d\tilde{J} = \ldots + \left[\boldsymbol{\lambda}^{*\mathrm{T}} \frac{\partial \mathbf{g}^*(\mathbf{x}^-, \mathbf{x}^+)}{\partial \mathbf{x}^-} - \mathbf{p}^{-\,\mathrm{T}}\right] d\mathbf{x}^-$$

$$+ \left[\boldsymbol{\lambda}^{*\mathrm{T}} \frac{\partial \mathbf{g}^*(\mathbf{x}^-, \mathbf{x}^+)}{\partial \mathbf{x}^+} + \mathbf{p}^{+\mathrm{T}}\right] d\mathbf{x}^+ + \left[H^-(t^*) - H^+(t^*)\right] dt_* = 0 \qquad (1.47)$$

From where we obtain the transversality conditions at t^*:

$$\mathbf{p}^- = \left[\frac{\partial \mathbf{g}^*(\mathbf{x}^-, \mathbf{x}^+)}{\partial \mathbf{x}^-}\right]^{\mathrm{T}} \boldsymbol{\lambda}^*, \quad \mathbf{p}^+ = -\left[\frac{\partial \mathbf{g}^*(\mathbf{x}^-, \mathbf{x}^+)}{\partial \mathbf{x}^+}\right]^{\mathrm{T}} \boldsymbol{\lambda}^*, \qquad (1.48)$$

$$H^+\left(\mathbf{p}^+, \mathbf{x}^*, \mathbf{u}^+_{\mathrm{opt}}\right) = H^-\left(\mathbf{p}^-, \mathbf{x}^*, \mathbf{u}^-_{\mathrm{opt}}\right). \qquad (1.49)$$

Note that the conditions given here for conjugate variables at the discontinuity point of the phase vector allow for a discontinuity of its dimension:

$$\mathbf{x}^- \in R^{n_-}, \mathbf{x}^+ \in R^{n_+}. \qquad (1.50)$$

Generalization (1.50) implies the possibility of changing the physical meaning of the phase and, accordingly, conjugate variables at moment t_*:

$$\mathbf{p}^- \in R^{n_-}, \mathbf{p}^+ \in R^{n_+}. \qquad (1.51)$$

As a consequence, the possibility of discontinuity of the right-hand sides of the phase (and, hence, conjugate) system of equations is simultaneously allowed. With this generalization taken into account, conditions (1.48), (1.49) correspond to the transversality conditions at the branching point of the trajectory.

1.8 Invariants

Invariants play an important role in the practical solution of optimization problems. On the one hand, each first integral of the unified system (1.1) and (1.6) makes it possible to express one of the variables in terms of the others and, thereby, reduce the dimensions of this system and the BVP.

On the other hand, and this is the main thing, when system (1.1) and (1.6) is numerically integrated, invariant is observed with an error, according to which it is possible to verify the computational program in terms of the correctness of the ODE system and PMP conditions, as well as to control the integration accuracy and adjust objectively the integration step.

Note that in practice, the important second role of invariants is used more often. As for the first role, although it contributes to some reduction in the integration time, however, it can also lead to a significant deterioration in the convergence of the procedure for solving the BVP, since reducing the dimension of the space of varied parameters means projecting onto a subspace, in which, for example, isolated solution curves can intersect to form singular points. This will be discussed in more detail with a demonstration of examples in the sections below.

Thus, let us consider some useful invariants in solving optimization problems (1.1)–(1.3).

As follows from (1.1) and (1.6),

$$\frac{dH(\mathbf{p},\mathbf{x},\mathbf{u})}{dt}=\mathbf{p}^{\mathrm{T}}\frac{\partial\mathbf{f}(\mathbf{x},\mathbf{u})}{\partial\mathbf{x}}=\frac{\partial\left[\mathbf{p}^{\mathrm{T}}\mathbf{f}(\mathbf{x},\mathbf{u})\right]}{\partial\mathbf{x}}=\frac{\partial H}{\partial\mathbf{p}}\dot{\mathbf{p}}+\frac{\partial H}{\partial\mathbf{x}}\dot{\mathbf{x}}+\frac{\partial H}{\partial\mathbf{u}}\dot{\mathbf{u}}$$
$$=\left(\dot{\mathbf{p}}^{\mathrm{T}}+\frac{\partial H}{\partial\mathbf{x}}\right)\mathbf{f}(\mathbf{x},\mathbf{u})+\frac{\partial H}{\partial\mathbf{u}}\dot{\mathbf{u}}=\frac{\partial H}{\partial\mathbf{u}}\dot{\mathbf{u}}. \tag{1.52}$$

It follows from (1.52) that under optimal control ($\frac{\partial H}{\partial\mathbf{u}}=0$) or under constant control ($\dot{\mathbf{u}}=0$), the first integral takes place:

$$H\equiv\text{const}. \tag{1.53}$$

On the optimal trajectory, taking into account the condition at free right end (1.10), we obtain

$$H\equiv 0. \tag{1.54}$$

The first integral (1.53) under constant control is of interest for verification of the computational program, to which we will return later in special sections.

To obtain another important integral, consider the equation for the variation $\delta\mathbf{x}$ of the phase vector. Based on (1.1) we have:

$$\frac{d\delta\mathbf{x}}{dt}=\frac{\partial\mathbf{f}}{\partial\mathbf{x}}\delta\mathbf{x}+\delta_{\varepsilon}\mathbf{f}. \tag{1.55}$$

Here, the variation $\delta_{\varepsilon}\mathbf{f}$ is not related to the variation of the phase vector. It can be due to a variation of the control or parameters of system (1.1), (1.6). Let $\delta\mathbf{x}$ be some solution of this equation. Consider the time-derivative of scalar product $\mathbf{p}^{\mathrm{T}}\delta\mathbf{x}$:

$$\frac{d\left(\mathbf{p}^{\mathrm{T}}\delta\mathbf{x}\right)}{dt}=\frac{d\mathbf{p}^{\mathrm{T}}}{dt}\delta\mathbf{x}+\mathbf{p}^{\mathrm{T}}\frac{d\delta\mathbf{x}}{dt}=\left[-\frac{\partial H}{\partial\mathbf{x}}+\mathbf{p}^{\mathrm{T}}\frac{\partial\mathbf{f}}{\partial\mathbf{x}}\right]\delta\mathbf{x}+\mathbf{p}^{\mathrm{T}}\delta_{\varepsilon}\mathbf{f}=\mathbf{p}^{\mathrm{T}}\delta_{\varepsilon}\mathbf{f}. \tag{1.56}$$

Integrating Eq. (1.56) on interval $[t_i,\ t]$, we get

$$\left(\mathbf{p}^{\mathrm{T}}\delta\mathbf{x}\right)_t=\mathbf{p}^{i\mathrm{T}}\delta\mathbf{x}^i+\int_{t_i}^{t}\mathbf{p}^{\mathrm{T}}\delta_{\varepsilon}\mathbf{f}dt. \tag{1.57}$$

Equation (1.57) is called the Bliss formula [107, 108]. For $\delta_\varepsilon \mathbf{f} = 0$, which takes place, for example, under optimal or constant control, from (1.57) we obtain the first integral

$$\mathbf{p}^{\mathrm{T}} \delta \mathbf{x} \equiv \text{const.} \tag{1.58}$$

For a fixed final time, it follows from (1.58) that if we specify the conjugate vector at the right end in the form

$$\mathbf{p}^f = \left(\frac{\partial J}{\partial \mathbf{x}^f} \right)^{\mathrm{T}}, \tag{1.59}$$

then

$$\mathbf{p}^{\mathrm{T}} \delta \mathbf{x} \equiv \mathbf{p}^f \delta \mathbf{x}^f = \left(\frac{\partial J}{\partial \mathbf{x}^f} \right)^{\mathrm{T}} \delta \mathbf{x}^f = dJ$$

and at any time moment

$$\mathbf{p}(t) \equiv \left(\frac{\partial J}{\partial \mathbf{x}(t)} \right)^{\mathrm{T}}, \tag{1.60}$$

that is, the conjugate variable is equal to the influence function of the corresponding phase variable at the same moment of time on the maximum value of the functional (under optimal control).

If the final time is free and implicitly specified by the condition

$$g\left(\mathbf{x}(t_f)\right) = 0, \tag{1.61}$$

then

$$dg\left(\mathbf{x}^f\right) = \frac{\partial g\left(\mathbf{x}^f\right)}{\partial \mathbf{x}^f} \left[\delta \mathbf{x}^f + \frac{d\mathbf{x}(t_f)}{dt} dt_f \right] = \frac{\partial g\left(\mathbf{x}^f\right)}{\partial \mathbf{x}^f} \delta \mathbf{x}^f + \frac{\partial g\left(\mathbf{x}^f\right)}{\partial \mathbf{x}^f} \mathbf{f}(\mathbf{x}(t_f), \mathbf{u}(t_f)) dt_f = 0,$$

whence, if $\frac{\partial g(\mathbf{x}^f)}{\partial \mathbf{x}^f} \mathbf{f}\left(\mathbf{x}(t_f), \mathbf{u}(t_f)\right) \neq 0$, we find the increment of the final time

$$dt_f = -\frac{\partial g\left(\mathbf{x}^f\right)}{\partial \mathbf{x}^f} \delta \mathbf{x}^f \Big/ \frac{\partial g\left(\mathbf{x}^f\right)}{\partial \mathbf{x}^f} \mathbf{f}\left(\mathbf{x}(t_f), \mathbf{u}(t_f)\right). \tag{1.62}$$

Then the total functional increment with the variation of the phase vector at moment t is

$$dJ = \frac{\partial J}{\partial \mathbf{x}^f}\left(\delta\mathbf{x}^f + \mathbf{f}\left(\mathbf{x}\left(t_f\right), \mathbf{u}\left(t_f\right)\right)dt_f\right) = \left[\frac{\partial J}{\partial \mathbf{x}^f} - \frac{\frac{\partial J}{\partial \mathbf{x}^f}\mathbf{f}\left(\mathbf{x}\left(t_f\right), \mathbf{u}\left(t_f\right)\right)}{\frac{\partial g\left(\mathbf{x}^f\right)}{\partial \mathbf{x}^f}\mathbf{f}\left(\mathbf{x}\left(t_f\right), \mathbf{u}\left(t_f\right)\right)}\frac{\partial g\left(\mathbf{x}^f\right)}{\partial \mathbf{x}^f}\right]\delta\mathbf{x}^f = 0$$

Thus, if we define the conjugate vector at the right end as

$$\mathbf{p}^{f\mathrm{T}} = \frac{\partial J}{\partial \mathbf{x}^f} - \frac{\frac{\partial J}{\partial \mathbf{x}^f}\mathbf{f}\left(\mathbf{x}\left(t_f\right), \mathbf{u}\left(t_f\right)\right)}{\frac{\partial g\left(\mathbf{x}^f\right)}{\partial \mathbf{x}^f}\mathbf{f}\left(\mathbf{x}\left(t_f\right), \mathbf{u}\left(t_f\right)\right)}\frac{\partial g\left(\mathbf{x}^f\right)}{\partial \mathbf{x}^f}, \tag{1.63}$$

then by virtue of integral (1.58) we obtain

$$\mathbf{p}^{\mathrm{T}}\delta\mathbf{x} \equiv \mathbf{p}^{f\mathrm{T}}\delta\mathbf{x}^f = dJ,$$

and at any time:

$$\mathbf{p}(t) \equiv \left(\frac{\partial J}{\partial \mathbf{x}(t)}\right)^{\mathrm{T}}, \tag{1.64}$$

that is, the conjugate variable, as before (see 1.60), at any time is equal to the function of influence of the corresponding phase variable at the same time on the maximum functional value.

1.9 The Functional Sensitivity to Perturbations of the Optimal Trajectory

It is not uncommon to hear skeptical about the use of rigorous optimization methods. Allegedly, the costs are large, and the output is a solution that is not much superior in the functional to that obtained using simpler engineering methods. Before to object, we note that such the statement has a strict mathematical basis. Indeed, small variations of the control $\delta\mathbf{u}(t)$ (needle variations can also be included here) with respect to the optimal program obviously lead to a change in the functional of a higher order of smallness:

$$dJ = \overline{o}(\|\delta\mathbf{u}\|). \tag{1.65}$$

However, the main advantage of rigorous solutions to optimization problems, no matter how paradoxical it may seem, is not only and not so much in determining the optimal control and maximum capabilities of the object under study, but in the ability to obtain objective information about the influence of the problem parameters on them, including parameters of the vehicle and environment. Using non-rigorous engineering techniques, the influence of parameters on the functional can only be obtained by numerical differentiation, in which close values of the functional have to be subtracted and, thus, the error of the influence functions can exceed the errors of calculating the functional by orders of magnitude. In the case of applying PMP, the conjugate variables, and hence the influence functions, have an error of the same order as the functional.

The influence of small perturbations of right-hand side $\delta_\varepsilon \mathbf{f}$ and constraints on the functional is determined using the Bliss formula, which, accounting for the transversality conditions, takes the following form:

$$dJ = \int_{t_i}^{t_f} \delta \tilde{H}\, dt = \int_{t_i}^{t_f} \left(\mathbf{p}^{\mathrm{T}} \delta_\varepsilon \mathbf{f} + \lambda \delta_\varepsilon g\right) dt \tag{1.66}$$

where the phase and conjugate variables correspond to the nominal (without perturbations) solution of the problem. Perturbation $\delta_\varepsilon \mathbf{f}$ of the right-hand side can be due to variations in the parameters of the vehicle, control program, or external disturbances. Perturbations can be deterministic or random. In accordance with (1.57), the derivative of the maximized functional with respect to parameter ε on the optimal trajectory, not related to the boundary conditions, can be written as [6, 108]:

$$\frac{\partial J_{\max}}{\partial \varepsilon} = \int_{t_i}^{t_f} \frac{\partial \tilde{H}\left(\mathbf{x}_{\mathrm{opt}}, \mathbf{u}_{\mathrm{opt}}, \varepsilon\right)}{\partial \varepsilon} dt, \tag{1.67}$$

where $\mathbf{x}_{\mathrm{opt}}$ and $\mathbf{u}_{\mathrm{opt}}$ are respectively the phase and control vectors on the nominal (at $\varepsilon = 0$) optimal trajectory. Note that parameter ε, in the general case, characterizes disturbances distributed over the entire trajectory.

Consider extended functional (1.19) on the optimal trajectory with constraint (1.18). The increment of the functional with variation of this constraint is

$$d\tilde{J} = dJ = \int_{t_i}^{t_f} \mu \delta G dt. \tag{1.68}$$

If $\delta G = \mathrm{const}$, then

$$\frac{\partial J_{\max}}{\partial G} = \int_{t_i}^{t_f} \mu dt \tag{1.69}$$

Thus, the effect of constraint (1.18) on the functional is determined by the time integral of the Lagrange function on the trajectory section where this constraint is essential. From (1.69) the physical meaning of the Lagrange function follows: at each moment of time it is equal to the increment of the influence function of the constraint on the functional per unit time, and the area of the figure, limited by the graph of the function $\mu(t)$ and the abscissa axis, is equal to the function of the influence of the constraint on the functional.

As can be seen, when using PMP, the partial derivatives (1.36), (1.39) and (1.41) are calculated with the same specified accuracy as the phase and conjugate variables on the nominal trajectory. In this case, it is not necessary to use numerical differentiation procedures and solve additional BVP for perturbed trajectories. Thus, obtaining the sensitivity functions of the functional to a wide range of small perturbations is practically not associated with any noticeable increase in calculations.

For comparison, we note that when using approximate methods for calculating optimal trajectories, each of the components of the vector of partial derivatives (1.36), (1.39) and (1.41) is determined by numerical differentiation, i.e. solving BVP for two close values of the varied parameter. It is obvious that obtaining statistical estimates of the influence of small random perturbations would require, when using approximate control laws, a sufficiently large number of tests with BVP solution for each of them. In addition to a simple increase in calculations, approximate and direct methods of optimal control in comparison with PMP in this case also have a fundamental drawback due to a significant and poorly controlled decrease in the accuracy of the obtained influence functions. Indeed, in numerical differentiation, due to the subtraction of close numbers, there is a sharp loss of accuracy. When using approximate control laws, this loss of accuracy cannot be compensated, for example, by reducing the trajectory integration step, since in this case the result error is determined by control errors. As will be shown in subsequent chapters, these errors (differences from optimal programs) can distort the influence functions not only in orders of magnitude, but also in sign, which can lead to erroneous layout decisions during the design process.

Thus, the use of PMP can significantly reduce the amount of required calculations and qualitatively improve their accuracy compared to direct and approximate optimization methods.

1.10 The Objective Control of Correctness of PMP Application and Software Operation

One of the important advantages of PMP is the possibility of obtaining objective information about the correctness of writing PMP conditions and the operation of various blocks of the software and the calculation accuracy. Namely, the unique possibility of reliable verification of a computational program may turn out to be a decisive factor in the use of PMP for on-board guidance of future STS [109]. The methods for diagnosing the correct operation of all the main blocks of the program responsible for the formation of the phase and conjugate systems of equations, optimal control and transversality conditions in the general optimization problem with constraints on control and phase variables are described below. The use of these procedures in the following order:

1. Analysis of the correctness of the conjugate system,
2. Analysis of the correctness of the control optimality conditions,
3. Analysis of the correctness of the transversality conditions,

allows not only to reliably diagnose the presence of methodological or programming errors, but also to localize them in the program.

With the help of the influence functions (conjugate variables and Lagrange functions), it is possible to establish an objective criterion for the formation of the residual norm and for setting the allowable errors of integration and solution of the BVP, subordinating conditions that are heterogeneous in physical meaning and orders of magnitude to a single requirement: ensuring the specified accuracy of calculating the functional.

The described procedures were used in full to test and verify the *ASTER* software package for the through optimization of STS branching trajectories, which was used in all calculations given below in Chaps. 1–6.

1.10.1 Analysis of the Integration Accuracy

As follows from Bliss formula (1.38), the Hamiltonian variation characterizes the dependence of the optimal functional value on the time:

$$\delta H = \frac{\partial J_{\max}}{\partial t}. \tag{1.70}$$

Therefore, to choose a rational step of integrating the phase and conjugate systems of equations with respect to time, an objective criterion can be the error in the fulfillment of the first integral (1.53), (1.54), i.e. conditions for the Hamiltonian constancy. The change of the Hamiltonian on the optimal trajectory must be limited according to the following estimate from (1.70):

$$|\Delta H| \le \varepsilon_J / \Delta t, \tag{1.71}$$

where ε_J is the allowable error in calculating the functional, $\Delta t = t_f - t$ is the time remaining until the end of the flight. Let us consider, as an example, the active leg of the STS optimal trajectory calculating by the *ASTER* software package. The motion model was specified in a dimensionless form so that the functional J (final mass m_f) is referred to the starting mass $m_i \sim 10^6$ kg, and the value $\sqrt{\frac{R_0}{g_0}} \approx 800\ s$ is taken as the reference time, where R_0 is the average radius of the Earth, g_0 is the gravitational acceleration at zero altitude. Dimensionless duration of the active leg is $\Delta t \sim 1$. The permissible error in the calculation of the final mass on the optimal trajectory, as a rule, was from 0.1 kg to 0.01 kg, which corresponds to the dimensionless value $\varepsilon_J \sim 10^{-7} - 10^{-8}$. In this case, formula (1.71) gives an estimate of the allowable deviation of the Hamiltonian (also dimensionless) over the entire optimal trajectory within $|\Delta H| < 10^{-8}$.

Following (1.71), when choosing the current integration step h_t, it is expedient to control the fulfillment of the condition

$$|H(t + h_t) - H(t)| \leq \frac{h_t \cdot \varepsilon_J}{\Delta t^2}. \quad (1.72)$$

In doing so, one must, of course, keep in mind the limitation of the minimal achievable error. Therefore, the process of reducing the integration step cannot be allowed to "run by its own accord", providing for the forced termination of the procedure of determination of h_t from the implicit relation (1.72).

As experience shows, doubling the integration step in the 4th order Runge–Kutta method increases the error in fulfilling invariant (1.53), (1.54) by approximately one to one and a half orders in the absence of other sources of errors. A typical dependence of the maximum deviations of the Hamiltonian on the integration step in the 4th order Runge–Kutta method in the *ASTER* software is shown in Table 1.1 for the active leg of the optimal STS trajectory.

The change in the Hamiltonian H on a typical optimal STS trajectory with a vertical take-off is shown on Fig. 1.3. Calculations were performed for three integration grids (trajectory integration step is variable): nominal ($h = h_{\text{nom}}(t)$), twice sparse ($h = 2h_{\text{nom}}$), and twice condensed ($h = 0.5h_{\text{nom}}$). It can be seen that the deviations of the Hamiltonian from zero do not exceed $10^{-8} - 10^{-12}$, and a halving of the integration step leads to a decrease in the integration errors accumulated in the Hamiltonian by $1 - 2$ orders of magnitude.

A similar numerical estimate of the error of this integration method is given by the Runge formula [110, 111]:

$$\left|x\left(t_f, h\right) - x(t)\right| \approx \frac{\left|x\left(t_f, h\right) - x\left(t_f, 2h\right)\right|}{15}, \quad (1.73)$$

where $x\left(t_f, h\right)$ is the approximate value of variable x at the right end of the integration interval with step h.

The fundamental difference between estimate (1.72) and (1.73) is that (1.72) makes it possible to control the error in terms of the key parameter, i.e., variations of the functional on the **optimal** trajectory **taking into account given constraints**. In this case, error control is carried out at the integration process, and no additional costs are required for calculating the trajectory with a double step as in (1.73).

Table 1.1 Maximum deviations of the Hamiltonian from zero on the optimal trajectory at different integration steps

The constant integration step	1s	2s	3s
$\max\limits_{t \in [t_i, t_f]} \lvert\Delta H\rvert$	$1 \cdot 10^{-10}$	$2 \cdot 10^{-9}$	$1 \cdot 10^{-8}$

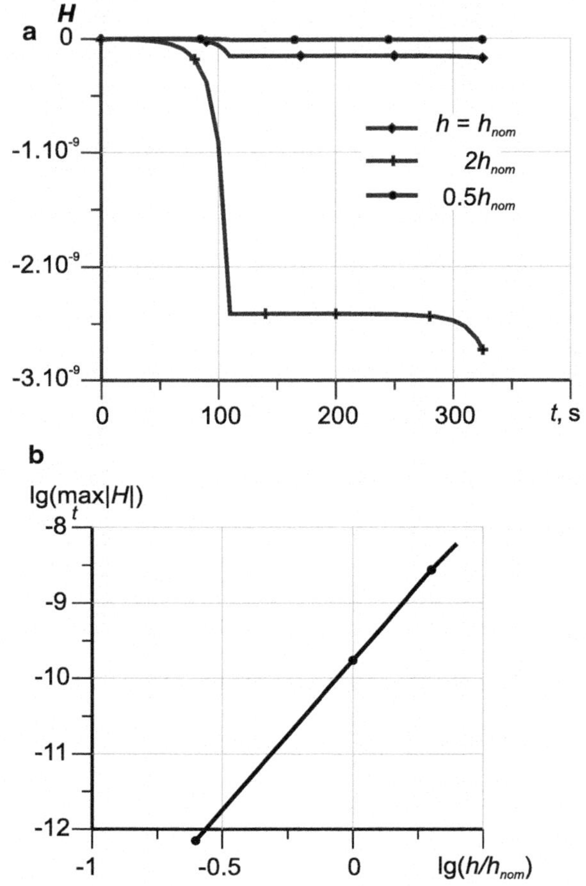

Fig. 1.3 The typical time-dependence of Hamiltonian H on the optimal trajectory of the vertically launched STS for three integration grids corresponding to the nominal steps $h_{\mathrm{nom}}(t)$, twice sparse and twice condensed

If the dependence of the Hamiltonian increment on the integration step is weaker than it pointed above, this indicates the presence of a systematic error, the source of which, most likely, is an error in the notation of the conjugate system or the control optimality condition.

1.10.2 Analysis of the Correctness of the Optimal Control Formation

Due to the autonomy of phase system (1.1) the Hamiltonian must be constant along the trajectory under optimal and constant-in-time controls. There is only one requirement for choosing a constant control: it should not significantly change the range of flight modes that enhance nonlinear effects on the right-hand sides of the equations

Table 1.2 The diagnostics of a possible error in the program for calculating the optimal trajectory based on PMP

	The control		The diagnosis of the correctness of writing		
	Optimal	Constant	Conjugate system	Control optimality conditions	Constant control
The Hamiltonian increment on the trajectory decreases sharply with a decrease in the integration step (Yes/No)	Yes	Yes	True	True	True
	No	No	False	Possible error	Possible error
	No	Yes	True	False	True
	Yes	No	True	True	Possible error

of motion (for example, to a significant increase in dynamic pressure and overloads) compared to the optimal one.

If, for both versions of the control programs, the Hamiltonian increment on the trajectory decreases sharply with decreasing integration step (see the previous section), then this is a good sign of the correctness of the conjugate system and control optimality conditions. Otherwise, if for both variants of control programs the Hamiltonian increment on the trajectory weakly depends on the integration step, then an error may be in the conjugate system. All possible options are presented in Table 1.2.

1.10.3 Analysis of Correctness of the Conjugate System and Transversality Conditions

To control the correctness of writing the conjugate system, it is convenient to use the first integral (1.53), (1.54) for autonomous system (1.1). If the error in fulfilling this condition on the continuity intervals of the phase and conjugate systems is greater than the allowable one and does not decrease noticeably with a decrease in the integration step, then this is a sure sign of an error in the formation of the conjugate system.

To further localize the error and determine specific equations of the conjugate system containing incorrect entries or operators, one can use the verification of the equations of the conjugate system (1.6), (1.22) by comparing values of the right-hand sides of the conjugate system with the Hamiltonian gradients obtained by numerical differentiation at some reference points on the trajectory. When performing numerical differentiation, it is necessary to free as much as possible from the errors inherent in this operation. First of all, it is necessary to take into account the structure of the dependence of the error on the step h of the grid nodes on which the numerical differentiation operation is performed.

Let the Hamiltonian be determined with an error ε, and let the partial derivative $\frac{\partial H}{\partial x_j}$ satisfy the Lipschitz condition with constant M. Then the approximate formula for the partial derivative

$$\frac{\partial H}{\partial x_j} \approx \frac{\Delta H}{\Delta x_j} = \frac{H(x_j + h) - H(x_j - h)}{2h}, \tag{1.74}$$

which has a methodical error of the order of h [93, 112], allows the following estimate:

$$\left| \frac{\partial H}{\partial x_j} - \frac{\Delta H}{\Delta x_j} \right| \le \frac{M}{2} h + \varepsilon \frac{1}{h}. \tag{1.75}$$

Thus, the dependence of the error of numerical differentiation on the step is nonlinear and has a minimum equal to $\sqrt{2M\varepsilon}$, with optimal step $h_{\text{opt}} = \sqrt{\frac{2\varepsilon}{M}}$.

The practical determination of the error of numerical differentiation and the optimal grid step, as follows from the above estimates, requires an analysis of the dependence of the finite-difference partial derivative of the Hamiltonian on the step h. On the typical dependence, three characteristic sections can be selected:

- in interval $(0, h_1)$ the estimate of the derivative is unstable due to rounding errors;
- in interval (h_1, h_2) the numerical derivative stabilizes (the zone of the most reliable estimates);
- at $h > h_2$, the error of numerical differentiation increases due to the methodical error.

To minimize the numerical differentiation error, it is necessary to choose a step in interval (h_1, h_2).

The next stage of the program verification is connected with the examination of the correspondence of the conjugate variables to their physical meaning as the sensitivity functions of the functional J to variations of the current phase vector (1.60), (1.64). In this case, the numerical differentiation error should be taken into account, as described above. In Table 1.3, for example, the coordinates of the conjugate vector $\mathbf{p} = (\mathbf{p_r}, \mathbf{p_v})^T$, where $\mathbf{p_r}$ and $\mathbf{p_v}$ are vectors corresponded to radius vector $\mathbf{r}$ and velocity vector $\mathbf{V}$, are compared with the coordinates of gradient $\frac{\partial J_{\max}}{\partial \mathbf{x}} = \left\{ \frac{\partial J_{\max}}{\partial \mathbf{r}}, \frac{\partial J_{\max}}{\partial \mathbf{V}} \right\}$ on the optimal STS branching trajectory just after the 1st stage booster separation (at the beginning of the booster return flight). The gradient is calculated approximately by numerical differentiation according to the equation for central differences (1.74).

Table 1.3 Comparison of values of the sensitivity functions, obtained by numerical differentiation, and the corresponding conjugate variables

$\mathbf{p_r}$	(0.00542, 0.453, −0.000400)
$\frac{\partial m_{f\,\max}}{\partial \mathbf{r}}$	(0.00543, 0.454, −0.000416)
$\mathbf{p_v}$	(0.141, 0.0938, −0.000435)
$\frac{\partial m_{f\,\max}}{\partial \mathbf{V}}$	(0.141, 0.0940, −0.000436)

In *ASTER* software, for testing the block for calculating the right parts of the conjugate system, another numerical differentiation formula is used with a smaller methodical error than in (1.74):

$$\frac{\Delta H}{\Delta x_j} = \frac{1}{12 \cdot h} \cdot \left[(H_{-2} - H_{+2}) - 8 \cdot (H_{-1} - H_{+1}) \right], \tag{1.76}$$

where $H_k = H\,(x_j + k{\cdot}h)$, $k = \pm 1, \pm 2$. Equation (1.76) was obtained by differentiating the polynomial of the best mean square approximation and makes it possible to smooth out random errors in experimental data [113].

Note that the diagnostics of the conjugate system described above simultaneously checks the adequacy of the used phase system. Moreover, the described procedure of numerical differentiation with respect to the phase variable x_j makes it possible to localize possible errors in those phase equations that contain this variable.

To verify the transversality conditions at the initial and final points, one can also use the operations of numerical differentiation under conditions of type (1.8), (1.9) with the error control of these operations described above. To check the transversality conditions at internal points of the trajectory (e.g., branching points), it is convenient to use the Hamiltonian continuity property. If the Hamiltonian increment upon passing through the boundary point is greater than the permissible error and does not decrease noticeably with an increase in the accuracy of reaching the given boundary point, then this is most likely due to errors in the formation of the transversality conditions.

To further localize the error in specific transversality conditions at a given boundary point that contain incorrect entries or operators, one can use checking the fulfillment of equalities (1.60), (1.64) to the left and right of the boundary point. In this case, the partial derivative on the right-hand side of (1.60), (1.64) is determined by numerical differentiation of $J(\mathbf{x})$ under optimal control, taking into account the features of the dependence of the error of such an operation on grid step h.

1.11 Two-Point Boundary Value Problem

The problem of integrating an ODE system with completely given initial conditions (given initial values for all variables) is called the Cauchy problem. Such a problem, for example, for continuous functions of the right parts of the ODE system, always has a unique solution. In the case of PMP, the initial values for the combined (phase plus conjugate) system, having order $2n$, are usually not fully specified. The missing number of conditions are set at the other end of the trajectory. Such a problem is called the two-point boundary value problem (TPBVP).

The principal scheme for solving the TPBVP (known as the shooting method) is as follows. Let us define the missing initial conditions in some way (we will discuss the methods of such extension below in a special section, but for now we will conditionally call it "arbitrary") in order to obtain the Cauchy problem. The

Cauchy problem is uniquely integrated up to the final time. Since the final value of the independent variable is not defined in the adopted formulation, value t_f is usually extended on the same basis as the initial conditions or the integration is carried out until one of conditions (1.3), (1.9) is satisfied. This condition is considered as an implicit form of specifying the final time. It is obvious that the choice of such a condition requires a detailed analysis (one-to-one correspondence, monotonicity) so that, given "arbitrarily" free initial variables, this condition is necessarily satisfied on a limited time interval. As a result of integration, we obtain finite values of phase and conjugate variables, which, however, do not satisfy conditions (1.3), (1.9) in the general case. Next, a special iterative procedure for adjusting "arbitrarily" predefined initial conditions is launched. Ensuring the convergence of such an iterative procedure is one of the main tasks for automating the PMP application, the solution of which is devoted to a special section below.

Let's calculate the number of unknowns and the number of conditions in the TPBVP when using PMP.

At the initial time, k_i conditions (1.2) are specified for phase vector $\mathbf{x}^i$ of dimension n plus n conditions (1.8) for conjugate vector $\mathbf{p}^i$ of the same dimension, which, however, include k_i unknown Lagrange multipliers that form the vector $\boldsymbol{\lambda}^i$.

So,

$(k_i + n)$ conditions (1.2), (1.8) are specified at the left end of the trajectory for the $(2n + k_i)$ unknowns $\left(\mathbf{x}^i, \mathbf{p}^i, \boldsymbol{\lambda}^i\right)$ and, therefore, $(2n + k_i) - (k_i + n) = n$ unknowns must be further defined. The additionally defined variables form a vector of varied parameters $\boldsymbol{\nu} = (\nu_1, \ldots, \nu_n)^T$

(1.77)

At the right end of the trajectory, k_f conditions (1.3) are specified for phase vector $\mathbf{x}^f$ plus $n + 1$ conditions (1.9), (1.10) for conjugate vector $\mathbf{p}^f$ of the same dimension, which, however, include k_f unknown Lagrange multipliers forming vector $\boldsymbol{\lambda}^f$.

So,

$(k_f + n + 1)$ conditions (1.3), (1.9), (1.10) are specified at the right end, containing k_f indefinite Lagrange multipliers plus one condition for determining final time t_f, excluding which we obtain $(k_f + n + 1) - k_f - 1 = n$ linearly independent conditions at the right end, which we will call residuals.	(1.78)

Thus, as follows from (1.77) and (1.78), when solving TPBVP, at the beginning it is necessary to set n "arbitrarily" varied parameters and then, by means of a special procedure for their sequential adjustment, achieve zeroing at the right end of n residuals. The number of varied parameters is equal to the number of residuals, therefore, TPBVP is closed. The number of varied parameters (residuals) is called the TPBVP order.

The order of TPBVP can often be reduced by using the first integrals of the combined system of ODEs (1.1), (1.6). Therefore, in the general case, it is bounded from above by the phase vector dimension:

$$n_{\text{TPBVP}} \leq n. \tag{1.79}$$

Boundary conditions can be specified both at the initial and final points, and at intermediate points, for example, at branching points (for example, at the STS staging). In case of intermediate conditions one speaks of a multipoint BVP. However, the intermediate conditions can always be transferred to the right end of the trajectory and, thus, the multipoint BVP can be reduced to a TPBVP. As will be shown in Chaps. 3–6, sometimes (for example, under phase constraints) it is convenient from a computational point of view to single out also the internal BVP along with the external one, which in this case has a smaller dimension and better convergence.

1.11.1 Modified Newton's Method for Solving TPBVP

Let us supplement the initial vector $\{\mathbf{x}, \mathbf{p}\}_{t_i}$ with the missing varied parameters $\boldsymbol{\nu} = \{\nu_1, ..., \nu_n\}$ for solving the Cauchy problem from t_i to t_f so that all n residuals are set to zero:

$$Q\boldsymbol{\nu} \equiv \{Q_j(\mathbf{x}_f(\boldsymbol{\nu}), \mathbf{p}(\boldsymbol{\nu})), j = 1, ..., n\} = 0. \tag{1.80}$$

The solution of nonlinear operator Eq. (1.80) is found by the modified Newton's method (MNM) [18, 90] using iterations

$$\boldsymbol{\nu}^{k+1} = \boldsymbol{\nu}^k - \lambda_k \mathbf{J}(\boldsymbol{\nu}^k, \delta\boldsymbol{\nu}^k)^{-1} Q\boldsymbol{\nu}^k, \ (k = 0, 1, ...), \tag{1.81}$$

where $\boldsymbol{\nu}^k$ is the k-th approximation of $\boldsymbol{\nu}$; λ_k is a scalar; $\mathbf{J}$ is the finite-difference Jacobi matrix of transformation Q:

$$\begin{cases} \mathbf{J}(\boldsymbol{\nu}^k, \delta\boldsymbol{\nu}^k) = \frac{\delta Q\boldsymbol{\nu}^k}{\delta\boldsymbol{\nu}^k}, \\ \delta Q\boldsymbol{\nu}^k = \{\delta_j Q\boldsymbol{\nu}^k = Q(\boldsymbol{\nu}^k + \delta_j\boldsymbol{\nu}^k) - Q\boldsymbol{\nu}^k, j = 1, ..., n\}, \\ \delta_j\boldsymbol{\nu}^k = \left\{ \underbrace{0, ..., 0}_{j-1}, \delta\boldsymbol{\nu}_j^k, 0, ..., 0 \right\}. \end{cases} \tag{1.82}$$

Variations $\delta_j\boldsymbol{\nu}^k$ in the general case are proportional to the residual norm $\|Q\boldsymbol{\nu}^k\|$, which, without taking into account error $Q_\varepsilon\boldsymbol{\nu}$ of the calculation $Q\boldsymbol{\nu}$, ensures the quadratic convergence of the iterative process (1.82) in the vicinity of the solution [90].

Since the difference formulas (1.82) for calculating partial derivatives $\frac{\delta Q v_j}{\delta v_j}$ in the presence of the error $Q_\varepsilon \boldsymbol{v}$ have an error of the order of $(\delta v_j)^{-1}$ when $\delta v_j \to 0$ [93], then for a relatively small residual norm $\|Q\boldsymbol{v}^k\|$, the value $|\delta \boldsymbol{v}_j^k|$ must be limited from below by some small value ε_{v_j} proportional to the calculation error $Q_\varepsilon \boldsymbol{v}$:

$$\delta \boldsymbol{v}_j^k = \begin{cases} æ_j^k \boldsymbol{v}_j^k \|Q\boldsymbol{v}^k\|, & |\boldsymbol{v}_j^k| > \varepsilon_{vj}, \\ \varepsilon_{vj}, & |\boldsymbol{v}_j^k| \le \varepsilon_{vj}. \end{cases}$$

Parameter λ_k is selected from the condition of decreasing the residual norm by $k + 1$ iterations (1.82):

$$\lambda_k : \|Q(\boldsymbol{v}^k - \lambda_k \mathbf{J}^{-1} Q\boldsymbol{v}^k)\| < \|Q\boldsymbol{v}^k\|. \tag{1.83}$$

Value λ_k in (1.83) is selected starting from $\lambda_k^o = 1$. If λ_k^{j-1} does not satisfy (1.83), then $\lambda_k^j = \lambda_k^{j-1}/2$. The number of iterations is limited to prevent the procedure from looping.

1.11.2 *The Solution Continuation Method*

The main problem for solving the nonlinear Eq. (1.80) is the choice of initial estimate v°, sufficient for the convergence of process (1.81) in a finite number of $k \le K$ steps:

$$\|Q\boldsymbol{v}^K\| < \varepsilon_Q << 1.$$

An efficient procedure for constructing such an approximation is provided by the continuation (homotopy) method [90, 95–98].

Let operator Q depend on the parameter τ: $Q = Q_\tau$. It is required to find a solution $v_{\tau f}$ to problem (1.80) ($Q_{\tau f}\overline{\boldsymbol{v}}_{\tau f} = 0$) using a known solution $\overline{\boldsymbol{v}}_{\tau i}$, for $\tau = \tau_i : Q_{\tau i}\overline{\boldsymbol{v}}_{\tau i} = 0$. Let us divide the segment $[\tau_i, \tau_f]$ into N intervals $\tau_o = \tau_i < \tau_I < \ldots < \tau_j < \ldots < \tau_{iN} = \tau_f$ and use procedure (1.81) to find solutions $\overline{\boldsymbol{v}}_{\tau j} : Q_{\tau j}\overline{\boldsymbol{v}}_{\tau j} = 0$, successively for $j = 1, 2, \ldots, N$. Initial guess $\boldsymbol{v}_j^o$ for (1.82) can be obtained based on linear extrapolation of the solutions obtained at the previous steps $\boldsymbol{v}_j^o$ and $\boldsymbol{v}_{j-1}^o$:

$$\boldsymbol{v}_{\tau_j}^o = \overline{\boldsymbol{v}}_{\tau_{j-1}} + \frac{\overline{\boldsymbol{v}}_{\tau_{j-1}} - \overline{\boldsymbol{v}}_{\tau_{j-2}}}{\tau_{j-1} - \tau_{j-2}} (\tau_j - \tau_{j-1}), \; \boldsymbol{v}_{\tau_1}^o = \overline{\boldsymbol{v}}_{\tau_o}. \tag{1.84}$$

If the initial guess $\boldsymbol{v}_{\tau_j}^o$ turns out to be so rough that iterative process (1.81) either does not converge in a given number of iterations, or converges to a deliberately "false" local extremal, then corresponding interval $[\tau_{j-1}, \tau_j]$ is divided in half and for each new interval the above operations are repeated. The number of such splits is limited to avoid looping the procedure.

If Q does not depend on parameter τ, it can be introduced artificially. For example, instead of Q, consider operator $\tilde{Q}_\tau$:

$$\tilde{Q}_\tau = \tau Q + (1 - \tau)\overline{Q} \tag{1.85}$$

where $\overline{Q}$ is a transformation for which solution $\overline{\boldsymbol{\nu}}$ of equation $\overline{Q}\overline{\boldsymbol{\nu}} = 0$, $\tau \in [0, 1]$, is known.

When switching to the optimization of a new class of trajectories, one can use the well-known analytical or numerical solution of a similar problem with a simplified motion model as initial guess $\mathbf{v}^\circ$.

Consider, for example, the following mathematical models:

1 is a model of motion in a uniform gravitational field without atmosphere and Earth's rotation;

2 is the same as *1*, but with the isothermal model of the atmosphere;

3 and *4* are models of motion in the central Newtonian gravitational field with the isothermal atmosphere without and with the Earth's rotation, respectively.

For model *1*, the integrals of the optimal motion are known [72, 73], and problem (1.80) is reduced to solving a system of no more than three nonlinear equations, for example, by the method of differentiation with respect to a parameter [95].

The transition from model *1* to model *2* is carried out using the procedure (1.84), where the parameter τ is artificially introduced as a factor for aerodynamic forces, and also for the specific impulse of main engines that depends on atmosphere pressure. Then for $\tau = 0$ we have a model without atmosphere, and for $\tau = 1$ we have the given atmosphere model.

If the vector of gravitational acceleration in the motion equations is set in the form $\mathbf{g} = -g_0 \frac{|\mathbf{R}_0|^2}{|\mathbf{R}|^3}\mathbf{R}$, where $\mathbf{R}$ is the vector from the Earth's center to the LV center of mass, g_0 is the gravity acceleration value at initial point $\mathbf{R}_0$, then we can artificially introduces parameter τ: $\mathbf{R} = \tau\mathbf{r} + \mathbf{R}_0$, where $\mathbf{r}$ is the radius vector of the center of mass from point $\mathbf{R}_0$, then we will have a model of the homogeneous gravitational field (model *2*) at $\tau = 0$ and the central gravitational field (model *3*) at $\tau = 1$. To take into account the Earth's rotation (model *4*), it is natural to use the angular velocity ω of the Earth's rotation as parameter τ, since at $\omega \to 0$, model *4* passes into model *3*.

Processes of convergence of iterative procedures (1.84) for each of the models used in optimizing LV injection into the Low Earth Orbit (LEO) are reflected by the data of Tables 1.4, 1.5, where values of varied parameters $\boldsymbol{\nu} = \mathbf{p}(t_i) = (\mathbf{p}_r(t_i), \mathbf{p}_v(t_i))^{\mathrm{T}}$ (conjugate variables at the initial point) are given at the beginning ($\boldsymbol{\nu}^0$) and end ($\boldsymbol{\nu}^K$) of the procedure of parameter continuation, as well as changing the norm of residuals over iterations.

As can be seen, the largest change in the vector of varied parameters for optimal trajectories compared to the model *1* is due to the influence of aerodynamic forces and centrality of the gravitational field.

Table 1.4 The initial $\boldsymbol{\nu}^o$ and optimal $\boldsymbol{\nu}^K$ values of conjugate variables at the left end of the trajectory for four motion models

$\boldsymbol{\nu}^o/\boldsymbol{\nu}^K$	Model number			
	1 (homogeneous gravitational field without atmosphere and Earth's rotation)	2 (homogeneous gravitational field with atmosphere and without Earth's rotation)	3 (central gravitational field with atmosphere and without Earth's rotation)	4 (central gravitational field with atmosphere and rotating Earth)
p_1^o/p_1^K	0/0	0/0	0/5.6·10^{-3}	5.6·10^{-3}/ 6.0·10^{-3}
p_2^o/p_2^K	2/0.28455	0.28455/0.55719	0.55719/5.69300	5.69300/ 6.04041
p_3^o/p_3^K	0/0	0/0	0/0	0/−1.2·10^{-3}
p_4^o/p_4^K	1/0.12565	0.12565/−0.40932	−0.40932/−0.42303	−0.42303/ −0.43974
p_5^o/p_5^K	1/0.11635	0.11635/0.81143	0.81143/0.83144	0.83144/ 0.87073
p_6^o/p_6^K	0/0	0/0	0/0	0/0

Table 1.5 The change in the norm of residuals $\|Q\boldsymbol{\nu}\|$ during iterations (1.84)

Model number	Iteration number					
	0	1	2	3	4	5
1	20.1	10.1	$7 \cdot 10^{-2}$	$8 \cdot 10^{-3}$	$4 \cdot 10^{-5}$	–
2	60.1	3.2	0.8	0.3	$5 \cdot 10^{-2}$	$2 \cdot 10^{-3}$
3	0.27	0.11	$3 \cdot 10^{-3}$	$2 \cdot 10^{-5}$	–	–
4	$5 \cdot 10^{2}$	$3 \cdot 10^{-4}$	$2 \cdot 10^{-6}$	–	–	–

Note that the procedure (1.84) is justified under the assumption that the dependence of solution $\overline{\boldsymbol{\nu}}$ on parameter τ is continuous. However, this condition may not always be satisfied for optimal systems (see, for example, [114, 115]). To bypass the singular points, you can try to change the method of introducing the parameter τ by considering another operator $\tilde{Q}_\tau$ that satisfies the following conditions: $\tilde{Q}_{\tau=0} = \overline{Q}$; $\tilde{Q}_{\tau=1} = Q$. For example, you can swap the varied parameter and the residual. Often, it turns out to be effective to expand the dimension of the space of varied parameters and bypass the singular points entering the additional dimension. Examples illustrating these possibilities are provided in the following chapters.

1.11.3 Selection of Extremals

The use of PMP based on the necessary optimality conditions can lead to several local extremals. The structures of some types of local extremals revealed in the process of studying the trajectories of various LV classes are shown in Figs. 1.4, 1.5, 1.6 and 1.7 with an indication of the relative mass m_{orb} inserted into LEO. (Local extremals with an analysis of the reasons for their occurrence are described in more detail later in the corresponding sections of the monograph). In most cases, extremals have a certain set of formalized characteristic features that allow them to be eliminated in the process of numerical solution without the intervention of a researcher.

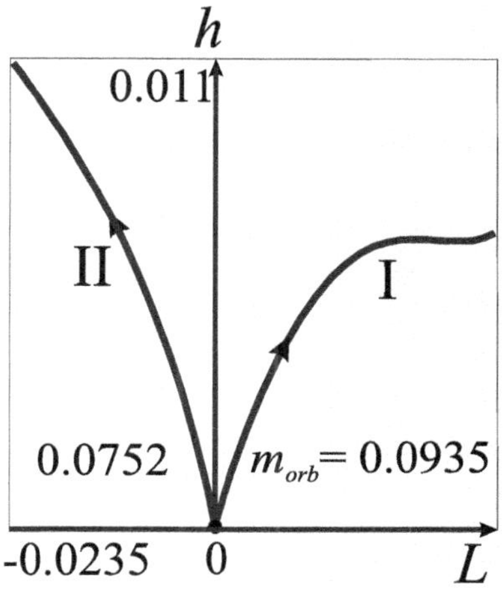

Fig. 1.4 Two qualitatively different types of local extremals with opposite direction of motion along the target orbit

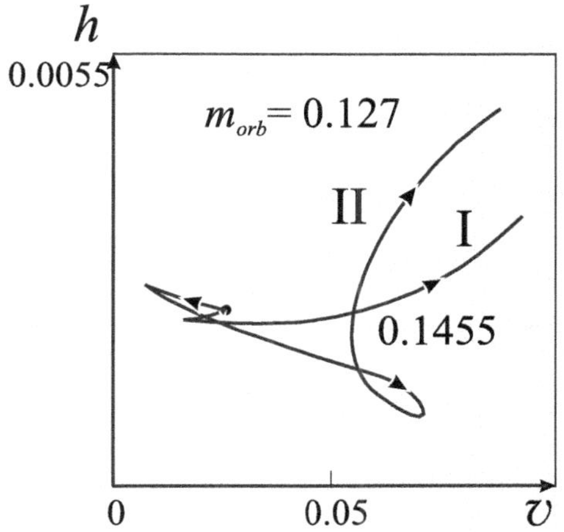

Fig. 1.5 Two qualitatively different types of local extremals, generated by influence of atmosphere

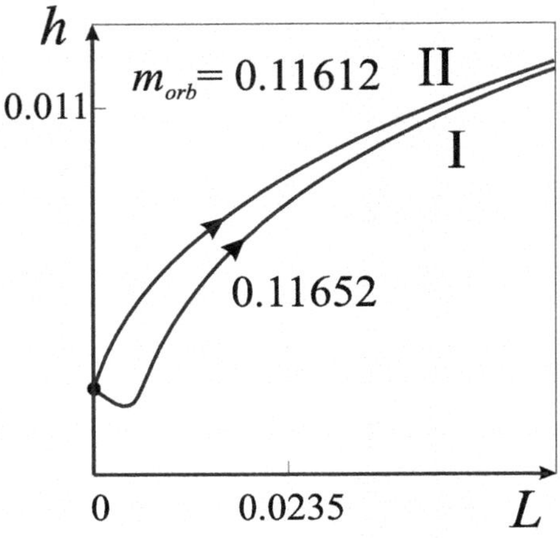

Fig. 1.6 Two qualitatively different types of local extremals, generated by influence of atmosphere at injection of STS with lift-to-drag ratio at subsonic/ supersonic speeds of $K_{\text{sub}} = 1.6/K_{\text{sup}} = 1.0$

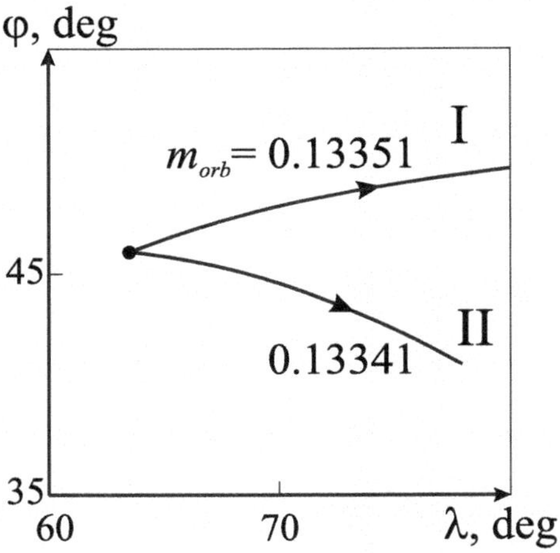

Fig. 1.7 Two qualitatively different types of local extremals corresponding to the SC insertion in the ascending and descending tracks of the target orbit

So, for example, in Fig. 1.4, the worst (in terms of the functional) extremal *II* corresponds to insertion into orbit with the direction of the orbital motion opposite to the optimal one (extremal *I*). On Fig. 1.5, the worst extremal *II* implements the injection with "swinging" in dense atmosphere layers with a loss of the characteristic velocity to overcome aerodynamic drag. On Fig. 1.6 extremal *II*, which is characterized by lower energy losses to overcome aerodynamic drag, but noticeably larger gravitational losses compared to extremal *I*, is generally inferior in terms of the functional to optimal solution *I*, in which the LV aerodynamic lifting properties are actively used. On Fig. 1.7, local extremals *I* and *II*, shown in the projection onto the plane of geographical coordinates, realize the spatial trajectories of launching, respectively, into the ascending and descending LEO turns.

In the above examples, calculated using *ASTER* software, the worse extremals were eliminated "automatically" in the process (1.84) of continuing the known solution with respect to parameter τ. The selection of extremals is based on a comparison of injected mass $m_f = m_{\text{orb}}$ for neighboring values τ_{j-1} and τ_j of parameter τ. The found solution $\overline{\boldsymbol{v}}_{\tau_j}$ is rejected if m_f decreases sharply at $\tau = \tau_j$, i.e., inequality

$$m_f\left(\overline{\boldsymbol{v}}_{\tau_j}\right) - m_f\left(\overline{\boldsymbol{v}}_{\tau_{j-1}}\right) < -\varepsilon_m, \tag{1.86}$$

where $0 < \varepsilon_m << m_f\left(\overline{\boldsymbol{v}}_{\tau_{j-1}}\right)$, is violated. Note that condition (1.86) admits a "jump" to another extremal when passing from τ_{j-1} to τ_j, if the functional m_f increases in this case.

Information about the "rejected" local extremals is stored to identify possible classes of extremals with a qualitatively different structure. Some types of local extremals, rejected during some calculations, can deliver the global maximum with other LV layouts or flight conditions. Therefore, a purposeful search for qualitatively new solutions can be based on the continuation with respect to the parameter of the identified local extremals belonging to different classes. For example, extremal *II* in Fig. 1.5, which implements the LV injection with its preliminary swinging in dense atmosphere layers, is a particular case of oscillatory trajectories, which can deliver a global extremum with sufficiently high LV aerodynamic lifting properties. As studies show, which will be discussed in detail below, type *II* of extremals in Fig. 1.6 have an advantage in the functional over local extremals of type *I* with a decrease in LV lifting properties. Type *II* of extremals in Fig. 1.7 deliver a greater value to the functional compared to type *I*, if the inclination of the target orbit is close to the geographical latitude of the launch site or less than it.

1.11.4 The Convergence Area

One of the main criteria for the effectiveness of the developed numerical procedures is their regularity in the investigated ranges of parameter values and performance. Here, regularity is understood as the ability of the program to find BVP solution with required accuracy without researcher interfering in the process and in absence of any serious requirements for the closeness of the given initial estimate to the optimal solution. The presence of such a property, strictly speaking, requires the proof of the corresponding existence and uniqueness theorems. However, in practice, to check the performance of complex software packages, a non-strict, but rather descriptive, numerical testing procedure is often used. No less important is the overall experience of operating of the corresponding *ASTER* software package (see Sect. 1.12). In most publications, the main difficulties in solving a system of nonlinear equations (to which TPBVP is reduced) are associated precisely with the selection of a sufficiently good initial guess, which ensures the convergence of the proposed iterative procedures. The selection of a sufficiently good initial guess, according to many authors, seems to be such a difficult task that, as a rule, it is taken out of the scope of algorithms for TPBVP solving and belongs to the area of an individual art of the researcher. However, a systematic study of fundamental and purely applied optimization problems cannot be based on "chance" and "luck". Therefore, formalization of the initial guess search procedure implemented in *ASTER* software described below, which ensures its regularity for the wide class of problems under consideration, should be attributed to the important characteristics of the package.

To estimate the regularity and speed of the numerical procedure for solving a system of nonlinear equations based on MNM implemented in *ASTER*, its efficiency was compared with widely known techniques [10] using the example of solving one of the classical tests of finding the minimum of the Rosenbrock function

$$f = 100\left(x_2 - x_1^2\right)^2 + (1 - x_1)^2 \to \min_{\{x_1, x_2\}}. \tag{1.87}$$

Problem (1.87) has an obvious exact solution:

$$x_1 = x_2 = 1, \min f(x_1, x_2) = 0. \tag{1.88}$$

A characteristic feature of the Rosenbrock function is that the minimum point lies at the bottom of a narrow canyon with almost sheer walls [10]. Such a structure of the objective function complicates the work of numerical algorithms for finding the extremum.

The choice of problem (1.87) for comparing algorithms is connected with the availability of sufficiently detailed information about its solution by the most popular methods and programs [10]. It should be noted that the numerical procedure of *ASTER* is not directly intended for a direct search for the extremum of a function. In order to use *ASTER*, problem (1.87) must be formulated as a problem for solving a non-linear operator equation of type (1.80). This is achieved by using the necessary conditions for function minimum (1.87):

$$\begin{cases} \frac{\partial f}{\partial x_1} \equiv y_1 \equiv -400\left(x_2 - x_1^2\right)x_1 - 2(1 - x_1) = 0, \\ \frac{\partial f}{\partial x_2} \equiv y_2 \equiv 200\left(x_2 - x_1^2\right) = 0. \end{cases} \tag{1.89}$$

Two groups of calculations were carried out. The first is to evaluate the *ASTER* performance. The second is to check its regularity. In the first group, the calculations were carried out without using a block that implements the continuation method for the following initial conditions mentioned in the publications:

$$\mathbf{x}^{(0)} = \{-1.2; 1.0\} \tag{1.90}$$

and

$$\mathbf{x}^{(0)} = \{-2.547; 1.489\}. \tag{1.91}$$

All the algorithms with which *ASTER* numerical procedure is compared, when choosing the direction and size of the descent step in the space of varied parameters $\{x_1, x_2\}$, use the calculation of objective function f. Therefore for the correctness of the comparison in *ASTER*, the assessment of the success of the descent in a given direction in first group of calculations was also carried out on the basis of the calculation of function f, rather than residual norm (1.83), as in the standard MNM algorithm of *ASTER*. The convergence rates of various algorithms according to [10] and the *ASTER* numerical procedure can be estimated by the number of calls to the block for calculating the objective function (residuals in *ASTER*), since in practical problems the calculation of the objective function ($Q\boldsymbol{v}$ vector in *ASTER*) requires the greatest computational costs. Of particular interest for evaluating the effectiveness of algorithms is such an indicator as the number of calls to the block for calculating the direction of descent (the Jacobi matrix in *ASTER*). The values of these indicators in solving problem (1.87) with initial guesses (1.90) and (1.91) are given in Table 1.6. It can be seen that the numerical procedure implemented in *ASTER* is almost as fast as any of the known algorithms [10].

To check the regularity of the numerical procedure of the software package, solutions of problem (1.87) were obtained with initial guesses, which were specified parametrically in the following form:

$$\mathbf{x}^{(0)} = \{1 + r \cdot \cos\phi,\ 1 + r \cdot \sin\phi\}, \tag{1.92}$$

where r changed on the interval [2, 6] with the step of 2, φ changed from 0 to 2π with the step of $\pi/4$.

Table 1.6 Comparison of the convergence efficiency of iterative procedures for finding the Rosenbrock function minimum for two variants of initial values

The method	Number of calls to the block for choosing the direction of descent		Number of objective function evaluations		The found minimum value of the function (exact $\min f = 0$)	
	Variant of $\mathbf{x}^{(0)}$		Variant of $\mathbf{x}^{(0)}$		Variant of $\mathbf{x}^{(0)}$	
	(1.90)	(1.91)	(1.90)	(1.91)	(1.90)	(1.91)
MNM (*ASTER*)	**20**	**26**	**67**	**90**	$\mathbf{9 \cdot 10^{-15}}$	$\mathbf{3 \cdot 10^{-12}}$
Davidon-Fletcher-Powell method	23	16	120	87	$1 \cdot 10^{-12}$	$6 \cdot 10^{-16}$
Newton method (Wortman)	17	20	98	130	$3 \cdot 10^{-13}$	$6 \cdot 10^{-16}$
Hook and Jeeves method	87	142	353	588	$2 \cdot 10^{-7}$	$2 \cdot 10^{-9}$
Powell method (Wortman)	15	17	192	216	$4 \cdot 10^{-14}$	$9 \cdot 10^{-14}$
Powell method (Powell)	13		151		$7 \cdot 10^{-10}$	
Davis-Sven-Kempy method (Fletcher)	21		187		$1 \cdot 10^{-12}$	

To implement the continuation method, the objective function f was replaced by the function Φ, depending on the parameter λ, along which the solution was continued:

$$\Phi(x, \lambda) = \lambda \cdot f + (1 - \lambda) \cdot f^{(0)}, \tag{1.93}$$

where $f^{(0)} = \left(x_1 - x_1^{(0)}\right)^2 + \left(x_2 - x_2^{(0)}\right)^2$, $\mathbf{x} = (x_1, x_2)$.

It is easy to see that $\Phi(x, \lambda)$ satisfies the following conditions. At the beginning of the process, at $\lambda = 0$, the specified initial estimate $\mathbf{x}^{(0)}$ is certainly the minimum point of the function $f^{(0)}$. At the end of the continuation process, when $\lambda = 1$, the solution of the problem $\Phi(x) \to \min\limits_{\{x_1, x_2\}}$ coincides with the original problem (1.87).

When solving the problem (1.87) with the initial estimate $\mathbf{x}^{(0)}$ taking values in the range (1.92), the initial value of the objective function reached $f(\mathbf{x}^{(0)}) \approx 10^5$. However, as the testing showed, the numerical procedure of *ASTER* provided convergence of the process for all $\mathbf{x}^{(0)}$ in this range, including when the initial estimate was located far enough from the optimal solution.

Studies of the regularity of the developed procedures for solving the problem of optimal LV injection into LEO were also carried out using the LV example of the Proton-type with a variable initial thrust-to-weight ratio. Using a random number generator with a uniform distribution, the initial thrust-to-weight ratio n_i and altitude of the target orbit h_{orb} were generated in the following ranges: $n_i \in [1.0, 2.0]$, $h_{\text{orb}} \in [100 \text{ km}, 500 \text{ km}]$. The obtained solution of the problem for $n_i = 1.284$, $h_{\text{orb}} =$ 200 km was used as a reference in all variants. The number of iterations N_{Jacobi} was recorded, including the calculation of Jacobian (1.82) and the formation of the direction of descent in the space of varied parameters, which was required to solve TPBVP using a set of developed procedures and, for comparison, using only the MNM. The required accuracy of fulfilling the specified conditions at the end of the injection active leg and calculating the maximum injected mass was 0.01 m in altitude, 0.001 m/s in velocity, 10^{-7} rad in trajectory angle, and 0.1 kg in injected mass. To avoid looping, the admissible number of iterations was limited: $N_{\text{Jacobi}} <$ 64.

In Figs. 1.8 and 1.9 in plane (n_i, h_{orb}), the circles mark the values of the parameters for which the solution was found, and the crosses mark unsuccessful attempts. It can be seen that, despite significant perturbations (especially critical in the direction of decreasing the initial thrust-to-weight ratio), *ASTER* did not give a single failure. At the same time, MNM, left "alone", repeatedly experienced problems with large deviations associated with a decrease in the initial thrust-to-weight ratio (Fig. 1.8). The number of iterations required for the solution when specifying random points in plane (n_i, h_{orb}) is shown in Fig. 1.10. It can be seen that, basically, the increase in the number of iterations is associated with a change in the thrust-to-weight ratio, especially when it decreases. Variations in the altitude of the target orbit in the range of 100–500 km are rather easily parried already by the continuation method without a significant increase in the number of iterations.

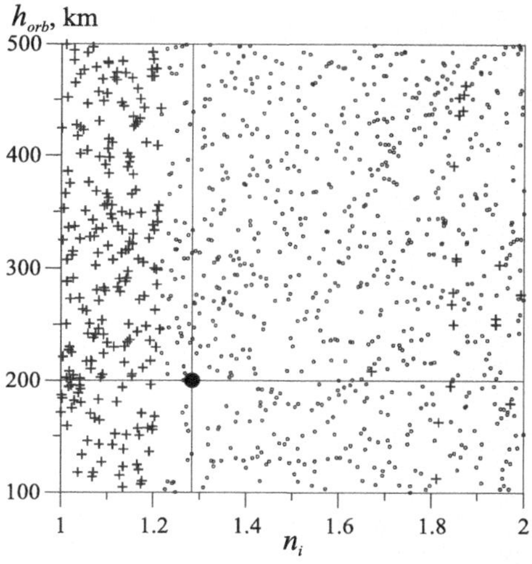

Fig. 1.8 Test results on the TPBVP convergence using the only MNM. In automatic mode the solution was found (○) or not found (+). The initial thrust-to-weight ratio n_i and the altitude of target orbit h_{orb} generated by a sensor of random numbers

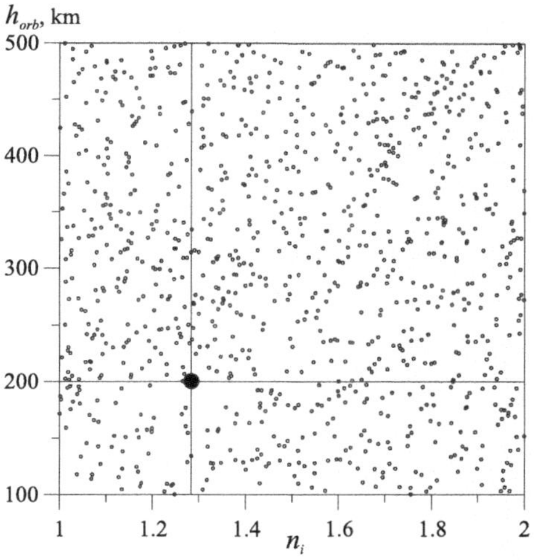

Fig. 1.9 Test results on TPBVP convergence using entire *ASTER* package. In automatic mode the solution was found (○) or not found (+). Initial thrust-to-weight ratio n_i and the altitude of target orbit h_{orb} generated by the sensor of random numbers

Another demonstration of the successful operation of *ASTER* is given in Fig. 1.11 [63], which shows the process of reducing the residual norm from the initial $Q\mathbf{v}^0$ to the final $Q\mathbf{v}^k$ one at each step of the solution continuation with respect to parameter τ when solving the TPBVP for the case of the through optimization of injected trajectories of a vertically launched STS of the Energiya-Buran type ($\tau = 1$). For methodological purposes, the initial ($\tau = 0$) here was the optimal solution for the ASTS with a horizontal take-off, which, obviously, differs significantly from the studied STS not only in values of the design parameters, but also in the mathematical model of motion (see Chap. 6).

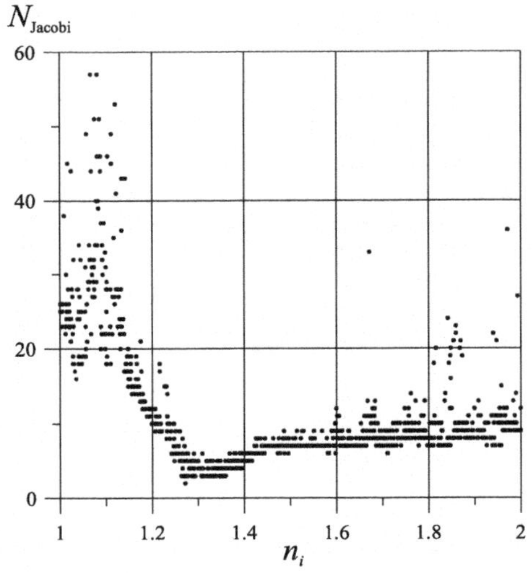

Fig. 1.10 Number N_{Jacobi} of iterations for solving TPBVP for random points $\{n_i, h_{\text{orb}}\}$ shown in Fig. 1.9

Another indirect criterion of the regularity and accuracy of solving optimization problems is the "smoothness" of the dependences of the characteristics of the optimal solution (functional, time of movement along the constraint, maximum loads, etc.) on problem parameters. To control the accuracy of the solution according to this criterion, parametric studies were carried out, as a rule, with a search of a much larger number of options than is necessary for practical applications or ensuring the convergence of the solution (see Chap. 6).

Indirect evidence of the regularity of the developed method is also many years of experience with *ASTER* by a number of researchers, including those who did not have a special training, and a large amount of parametric studies of many various STS types, as well as some qualitatively new results that received physical justification only after some time (see Chap. 3).

1.12 *ASTER* Automated Software Package

The method described above for rigorously solving optimization problems based on PMP, taking into account actual constraints on control and/or phase variables, is implemented in the *ASTER* automated software package [62–65, 116]. The brief description of the structure and capabilities of *ASTER* given in the next section goes significantly beyond the framework of the principal formulation of the optimization problem given in this chapter, and takes into account the methodological developments described in subsequent chapters. In general, an injection into orbit is considered as a branching process, which includes: the main branch of the trajectory from take-off to insertion into a target orbit and side branches of SP return flight in regular and emergency situations.

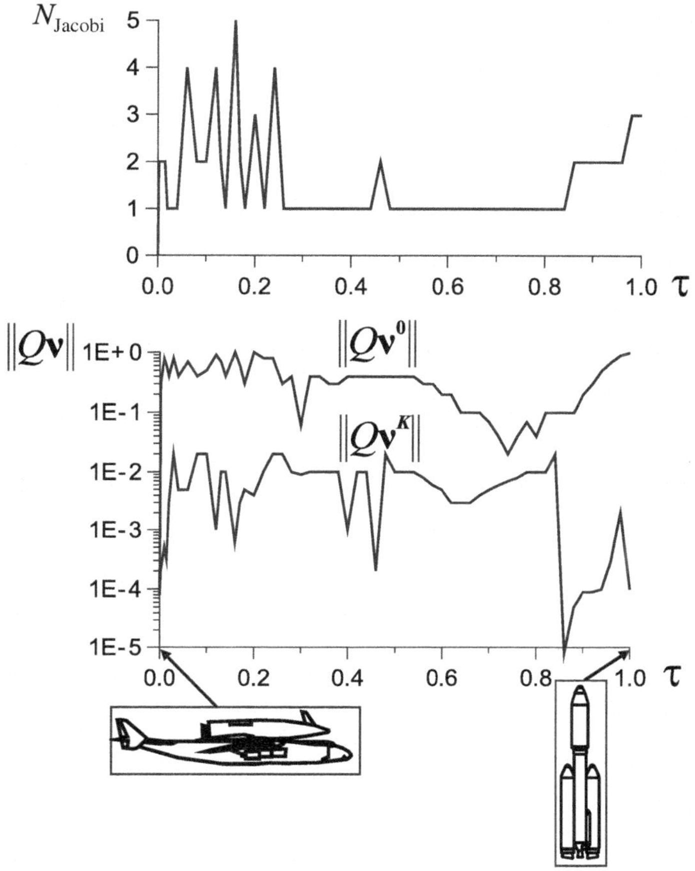

Fig. 1.11 Number N_{Jacobi} of iterations for solving TPBVP and the residual norm $\|Q\nu\|$ in dependence of parameter τ for the solution continuation from the ASTS with carrier aircraft ($\tau = 0$) to the LV with a vertical take-off ($\tau = 1$) when optimizing the injection trajectories using *ASTER* complex

When creating the *ASTER* package, the authors proceeded from the fact that the potential possibilities of PMP are unlimited, but the real conditions of the authors' work, including taking into account the widespread skepticism regarding the practical applicability of the results obtained using rigorous optimization methods, can significantly affect the sequence of development of its elements and limit the time allotted for this. Therefore, from the very beginning of work, the structure of the package was formed in such a way as to ensure the maximum flexibility and possibility of a further expansion. Another strategic criterion that guided the authors in the development of the package, especially in view of the skepticism noted above, was the possibility of its widest application in solving actual problems. In accordance with this, requirements for the package development were set in the following areas:

- the possibility of gradually increasing the level of detail and depth of study of the STS mathematical model and conditions of its motion in order to fully take into account actual constraints, including those adopted in design firms, without prejudice to the rigor of fulfilling PMP requirements;
- "maximum" automation (regularization) of the numerical procedure for finding the optimal solution;
- development of an interface convenient for the widest range of users.

1.12.1 The Main Types of Problems Solved by ASTER

The *ASTER* software package is primarily designed and passed the appropriate test trials to solve the following scientific and applied problems:

- The through optimization of control laws and spatial LV trajectories according to the criterion of maximum injected mass, taking into account constraints on control and phase vector;
- The through optimization of control laws and spatial LV trajectories according to the criterion of the maximum injected mass, taking into account the branching of trajectories and constraints at all stages of flight, including the requirements for separation safety, fail-safety ascent, return of SPs (boosters, head fairings, etc.) to specified areas, etc.;
- The study of the influence of the problem parameters (in boundary conditions, mathematical model, environment, etc.) on the maximum injected mass, optimal control laws and trajectories, and sensitivity functions;
- The study of the influence of disturbances (in aerodynamics, propulsion, atmosphere, wind, etc.), including random ones, distributed along the branching trajectory, on the maximum injected mass, optimal control laws and trajectories.

The *ASTER* package allows you to vary the objects of study by:

- LV models:
  - number of stages,
  - take-off type,
  - propulsion types (liquid and solid propellant rocket engines, turbojet, etc.);
- mathematical models of:
  - Earth's shape,
  - gravitation field,
  - atmosphere,
  - disturbances;
- types of constraints on:
  - control (angle of attack, thrust, overloads, forces, etc.),
  - phase vector (Mach number, speed, dynamic pressure, heat fluxes, etc.);

- boundary conditions:
  - initial,
  - in target orbit,
  - at separation;
- number of branches and constraints at branching points, at internal and final points of branches (including on areas of SP drop or landing);
- the functional: in addition to the injected mass, other types of functionals can be set depending on the initial and final phase variables, e.g., the initial mass with a fixed final mass, the payload mass (taking into account the dependence of the structure mass on loads during the ascent), flight range, etc.

The *ASTER* is most effective for studying the extremely flight performance, optimal control programs, the effect of variations in the problem parameters and distributed disturbances on the studied functional under the optimal control, as well as the effect of the control program deviations from the optimal one.

1.12.2 ASTER *Package Structure*

ASTER package consists of three main parts (Fig. 1.12): the solver, interface module, and database. The initial data for calculating the optimal injection trajectories are set in the dialog mode. After all the necessary parameters are set, the solution of the problem is implemented automatically with a required accuracy.

As a result of solving the optimization problem, *ASTER* provides researchers with following information:

1. The optimal control programs at all stages of flight, including SP return branches;
2. The optimal trajectory including phase and conjugate variables, dynamic pressure, overloads, heat fluxes, etc.;
3. The maximum values of dynamic pressure, aerodynamic forces, overloads, heat fluxes and Mach number at all branches;
4. Assessment of proximity to optimal values of parameters of the engine cyclogram (if it is not optimized), that is, the possibility of improving the functional by optimizing the engine operation;
5. Sensitivity coefficients of the functional to changes in any problem parameters;
6. Sensitivity functions of the functional to disturbances distributed along the trajectory;
7. The optimal take-off conditions (the azimuth, path angle, altitude, velocity);
8. Dependences of the functions of phase coordinates at the selected point of the trajectory (initial, final, staging points, etc.) on any parameters (target orbit, launch point, engine cyclogram, constraints, etc.).

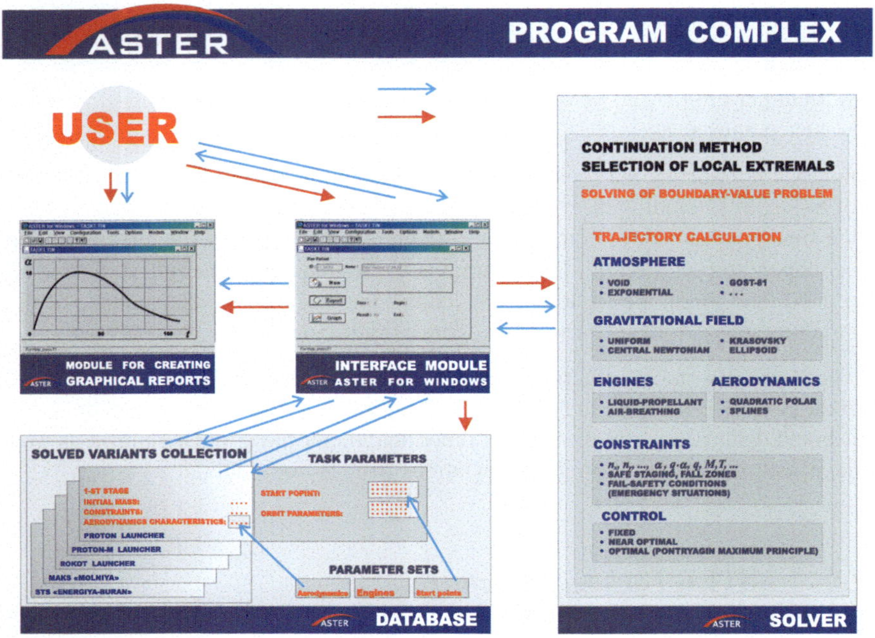

Fig. 1.12 The structure of *ASTER* complex

Due to the indirect optimization methods used, all information listed above can be issued simultaneously with the optimization problem solution. In order to implement the possibilities noted in items 4–8, a numerical differentiation is not used that increases the accuracy of the estimates and significantly reduces the research time, because it does not require additional calculations with solving TPBVP.

1.12.3 ASTER *Solver*

The *ASTER* solver implements the optimization technique for various models of the atmosphere, shape and gravity field of the Earth, method of setting aerodynamic data and propulsion characteristics, etc. The algorithm for solving the optimization problem in *ASTER* is based on the use of PMP and was first published in [116]. The LV injection is considered as a branching process, which includes: the main branch from a launch site to a target orbit and side branches of SP motion.

ASTER solver implements:

- MNM for solving TPBVP;
- The solution continuation and extremal selection methods, which regularize the procedure by using solutions of the reference (basic) variants of the problem;

- The method of reduction of a multipoint BVP for branching trajectories by transferring (all or part of) the boundary conditions from side branches to the branching points;
- The fourth-order Runge–Kutta method of numerical integration of phase and conjugate systems.

The input data of the module is set in the dialog mode using the interface module. These include: parameters of LV, launch site, atmosphere, shape and gravitational field of the Earth, etc.; parameters of the numerical procedure that determine the number of steps in the continuation method, the required accuracy of BVP solution, the trajectory integration step, etc. After all the necessary parameters are set, the solution of the problem is implemented automatically.

The block diagram of the operation of *ASTER* is reduced to the following main operations (Fig. 1.13). Based on the initial data, the database finds the closest of the previously calculated optimal trajectory options, which determines the initial estimate for varied parameters of TPBVP. The TPBVP solution for phase and conjugate systems for all flight segments, by virtue of PMP that "generated" it, ensures the fulfillment of all specified flight conditions at the end of the trajectory, at the SP landing/drop points and during the flight reaching the best value of the functional. Based on PMP, selection of extremals and MNM, by iterations over the initial estimate, the solution of TPBVP (values of varied parameters) is found for the current values of the phase vector and LV parameters, the mission target and the environment (atmosphere). At each iteration, based on the generated initial conditions, the prediction of the optimal branching trajectory is performed, the residuals are calculated in the fulfillment of the specified conditions at the right ends and at some internal points of the main and side branches, and the value of the residual norm is estimated. If the residual norm is greater than the allowable one, the Jacobi finite-difference matrix is calculated for the dependence of the residuals on the varied parameters. With the help of the Jacobi matrix, this operator dependence is linearized in the vicinity of the known (reference) solution. From the solution of the corresponding system of linear algebraic equations, the direction of the necessary correction is determined in the space of varied parameters, and the value of the step of such a correction is estimated. If in the course of iterations the residual norm becomes less than the set value, then TPBVP is considered solved and with the found varied parameters values, the prediction block is re-applied to form a numerical array with the solution of the problem (the optimal thrust-vector program, trajectory, sensitivity functions, etc.).

The *ASTER* solver is designed as a multi-level system, where each level is a closed sequence of blocks of subprograms, and each subsequent level is nested into the previous one, representing a solution to a "narrower" and more specific problem. At the first level, the initial data are read and processed, and the calculation results are stored (*DAN* block). In the block of the second level, the continuation method is implemented, with the help of which automatic selection of initial estimates for varied parameters is performed (*CYCLE* block). The third level is the block for solving TPBVP using MNM (*ANEW* block). In the process of solving TPBVP, the trajectories are repeatedly integrated by the fourth-order Runge–Kutta method using

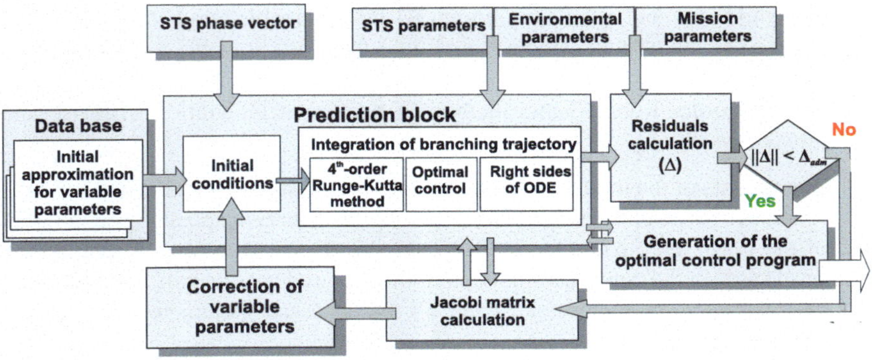

Fig. 1.13 The block diagram of *ASTER* solver

subroutines of the block of the fourth level (*INTEGR* block). When integrating the trajectory, it is necessary at each point to set the value of the control parameters that satisfy the optimality conditions. These calculations are performed at the fifth level (*UPR* block).

1.12.4 ASTER *Interface Module*

The user interface support program (interface module) uses the capabilities of the operating system to provide maximum convenience in work, provides convenient input and control of task parameters and graphical representation of the results using a special *Grapher-for-Windows* program. The interface program is implemented as *ASTER-for-Windows* module. The program provides convenient input of task parameters and control of *ASTER* module operation. The program can be configured to use additional output processing packages in a Microsoft Windows environment.

1.12.5 ASTER *Database*

The database (DB) contains the initial data (parameters) and optimal solutions for previously calculated problems. On the one hand, it allows to significantly reducing the time of obtaining the optimal solution, on the other hand, due to the selection of local extremals, it makes it easy to find the global extremum even in the case when PMP does not give the only solution to the problem.

References

1. Pontryagin, L.S., Boltyansky, V.G., Gamkrelidze, R.V., Mischenko, E.F.: The mathematical theory of optimal processes. Interscience Publishers, New York (1962)
2. Betts, J.T.: Survey of numerical methods for trajectory optimization. J. Guid. Control. Dyn. **21**(2), 193–207 (1998)
3. Rao, S.S.: Engineering Optimization: Theory and Practice. John Wiley & Sons, Hoboken, NJ, USA (2009)
4. Morante, D., Rivo, M.S., Soler, M.: A survey on low-thrust trajectory optimization approaches. Aerospace **8**, 88 (2021). https://doi.org/10.3390/aerospace8030088
5. Conway, B.A.: A survey of methods available for the numerical optimization of continuous dynamic systems. J. Optim. Theory Appl. **152**, 271–306 (2012). https://doi.org/10.1007/s10957-011-9918-z
6. Bryson, A.E., Ho, Y.C.: Applied Optimal Control: Optimization, Estimation, and Control, p. 481. Blaisdell Publishing Company, Waltham, Mass (1969)
7. Bellman, R.E.: Dynamic Programming. Princeton University Press, Princeton, N. Y. (1957)
8. Gabasov, R. F., Kirillova, F. M. (1975) Fundamentals of Dynamic Programming. Publishing House of BSU, Minsk, p. 262 (in Russian)
9. Krotov, V.F.: Global Methods in Optimal Control Theory, p. 408. Marcel Dekker, NY (1996)
10. Himmelblau, D.M.: Applied Nonlinear Programming. McGraw-Hill Book Company, (1972)
11. Aoki, M.: Introduction to Optimization Techniques, Fundamentals and Applications of Nonlinear Programming. Macmillan Series in Applied Computer Sciences. The Macmillan Company, New York (1971)
12. Rastrigin, L.A.: Extreme Regulation by the Random Search Method, No. 9. Automation and Telemechanics, Moscow, (1960) (in Russian).
13. Beasley, D., Bull, D.R., Martin, R.R.: An overview of genetic algorithms. Univercity Comput. **15**(2), 58–69 (1993)
14. Letov, A.M.: Flight Dynamics and Control. Nauka, Moscow (1969). (in Russian)
15. Rozonoer, L.I.: The Pontryagin Maximum Principle in the Theory of Optimal Systems. Automation and Telemechanics **20**, 10–12 (1959) (in Russian)
16. Isayev, V.K.: The Pontryagin maximum principle and optimal programming of rocket thrust. Automation and Telemechanics, **22**(8), 1961, and Vol. 23, No. 1, 1962 (in Russian).
17. Isayev, V.K.: On the theory of optimal programming of rocket thrust. Proceedings of TsAGI, No. 853 (1962) (in Russian)
18. Isayev, V.K., Sonin, V.V.: On a modification of Newton's method for the numerical solution of boundary value problems. J. of CM MPh. **3**, 1114–1116 (1963) (in Russian)
19. Isayev, V.K., Kurianov, A.I., Sonin, V.V.: On the application of the maximum principle to rocket flight problems. In: The Proceedings of the 14th International Astronautical Congres, Warszawa (1963)
20. Isayev, V.K.: On some features of Mayer's variational problem in rocket dynamics. Automation and Telemechanics, **26**(7) (1965) (in Russian)
21. Grigoriev, K.G., Kuzmich, V.A.: On special control of thrust vector orientation in optimization problems of rocket dynamics. Space Res. **3**(2) (1997) (in Russian)
22. Pyatakov, V.I.: On singular optimal control in the problem of launching aircraft. In the book: Complex control systems, Kyiv, pp. 32–46 (1977) (in Russian)
23. Marec, J.-P.: Optimal Space Trajectories. Elsevier Science, Amsterdam (1979)
24. Marchal, C.: Chattering arcs and chattering controls. J. Opt. Theo. Appl. **11** (1973)
25. Krotov, V.F., Bukreev, V.Z., Gurman, V.I.: New Methods of Calculus of Variations in Flight Dynamics. Mashinostroenie, Moscow (1969). (in Russian)
26. Krotov, V.F., Gurman, V.I.: Methods and Problems of Optimal Control. Nauka, Moscow (1973). (in Russian)
27. Gurman, V.I.: Degenerate Optimal Control Problems. Nauka, Moscow (1977). (in Russian)
28. Gabasov, R., Kirillova, F.M.: Singular Optimal Controls. Nauka, Moscow (1973). (in Russian)

29. Utkin, V.I.: Sliding Modes in Optimization and Control Problems. Nauka, Moscow (1981). (in Russian)
30. Grodzovsky, G.L., Kiforenko, B.N., Kuzmenko, V.V.: Influence of weight restrictions on maximum climb altitude while moving in the atmosphere. Uchenye Zapiski of TsAGI, **5**(1), 1974 (in Russian)
31. Sokolov, N.L., Tsybulsky, G.A.: Research on optimizing the design and ballistic parameters of an ambulance complex based on intercontinental ballistic missiles. Cosm. Res. **34**(3), 325–331 (1996)
32. Brusch, R.G.: Trajectory optimization for the Atlas/Centaur launch vehicle. Journal of Spacecraft and Rockets **14** (1977)
33. Nguyen, H.N.: Trajectory characteristics of horizontal takeoff single stage to orbit vehicle. AIAA Paper 87-0448 (1987)
34. Lozino-Lozinsky, G.E., Dudar, E.N., Plokhikh, V.P. Russian sub-orbital commercial passenger transport system. IPAC'91 Conference, October 1991, Gufu, Japan. SAE Paper, 1992, preprint Number 920aaa.
35. Lozino-Lozinsky, G.E., Dudar, E.N., Chelkis, F.Y.: The MAKS project and perspectives of development. In: AIAA 7th International Space Planes and Hypersonic Systems and Technologies Conference, 18-22 November 1996, Norfolk, Virginia, USA.
36. Volodin, S.V., Buzuluk, V.I.: Interrelation of trajectory control parameters and aerodynamic configuration of an aerospace vehicle with a hypersonic air-jet engine. Proceedings of the 1st International Aerospace Conference, Moscow, September 28 - October 2, 1992. Vol. 3, Aerodynamics, flight dynamics and control of aerospace systems. Moscow: (1995) (in Russian)
37. Schleich, W.T.: The Space Shuttle ascent guidance and control. AIAA P. 82-1497
38. Schleich, W.T.: Shuttle vehicle configuration impact on ascent guidance and control. AIAA Paper 82-1552
39. Nagorski, R.P. Application of the direct injection method at the start from the Vanderberg launch site. AIAA P. 84-2040
40. Noton, M., Swinerd, S.G.: Optimization of the rocket-powered ascent of HOTOL. IAF-87-329
41. Well, K.H., Markl, A., Mehlem, K.: ALTOS—A Trajectory Analysis and Optimization Software for Launch- and Reentry Vehicles. In: 48th Congress of the International Astronautical Federation, October 6–10, 1997, Turin, Italy, IAF-97-V4.04.
42. Roenneke, A. J., Jansch, C., Marki, A.: Advanced Launch and Reentry Trajectory Optimization Software documentation (ALTOS). European Space Science and Technology Center, TR, Software User Manual, Draft 1, Noordwijk, The Netherlands (1995)
43. Brauer, G.L., Comick, D.E., Stevenson, R.: Capabilities and applications of the Program to Optimize Simulated Trajectories (POST). NASA CR-2770 (1977)
44. Meder, D.S., & Searcy, J.L.: Generalized Trajectory Simulation (GTS), Volumes I-V. The Aerospace Corp., TR SAMSO-TR-75-255, El Segundo, CA (1975)
45. Betts, J.T., & Hallman, W.P.: NLP2 optimization algorithm documentation. The Aerospace Corp., Technical Operating Rept. TOR-0089(4464-06)-1, reissue A, El Segundo, CA (1997)
46. Balkanyi, L.R., & Spurlock, O.F.: DUKSUP—A High Thrust Trajectory Optimization Code. Proceedings of the AIAA/AHS/ASEE Aerospace Design Conference, AIAA, Washington, DC, 1993 (AIAA Paper 93-1127)
47. Oberle, H.J., & Grimm, W.: BNDSCO—A Program for the Numerical Solution of Optimal Control Problems, User Guide. Institut flir Dynamik der Flugsysteme, Deutsche Forschungsanstalt fur Luft und Raumfahrt DLR, fe Internal Rept. DLR IB/515-89/22, Oberpfaffenhofen, Germany (1989)
48. Betts, J.T., Huffman, W.P.: Sparse optimal control software SOCS. Boeing Information and Support Services, Mathematics and Engineering Analysis Technical Document MEA-LR-085, The Boeing Co., Seattle, WA (1997)
49. Moiseev, N.N.: Numerical Methods in the Theory of Optimal Systems. Nauka, Moscow (1971). (in Russian)

50. Fedorenko, R.P.: Approximate Solution of Optimal Control Problems. Nauka, Moscow (1978). (in Russian)
51. Canon, M.D., Cullum, C.D., Polak, E.: Theory of Optimal Control and Mathematical Programming. McGraw-Hill, New York (1970)
52. Polak, E.: Computational Methods in Optimization. Academic, New York (1971)
53. Tabak, D., & Kuo, B.C.: Optimal Control by Mathematical Programming. Prentice-Hall, Englewood Cliffs, NJ (1971)
54. Enright, P.J., Conway, B.A.: Discrete approximations to optimal trajectories using direct transcription and nonlinear programming. J. Guid. Control. Dyn. **15**, 994–1002 (1992)
55. Herman, A.L., Conway, B.A.: Direct optimization using collocation based on high-order Gauss-Lobatto quadrature rules. J. Guid. Control. Dyn. **19**(3), 592–599 (1996)
56. Von Stryk, O., Bulirsch, R.: Direct and indirect methods for trajectory optimization. Ann. Oper. Res. **37**, 357–373 (1992)
57. Von Stryk, O.: Numerical solution of optimal control problems by direct collocation. International Series of Numerical Mathematics, Vol. III, pp. 129–143. Birkhauser Verlag, Basel, Switzerland (1993)
58. Nesterov, Y.E.: Methods for minimizing non-smooth and quasi-convex functions. Econom. Math Meth 3(11), 519–531 (1983) (in Russian)
59. Benedikter, B., Zavoli, A., Colasurdo, G.: A convex approach to rocket ascent trajectory optimization. https://doi.org/10.13009/EUCASS2019-430
60. Liu, X., Lu, P., Pan, B.: Survey of convex optimization for aerospace applications. Astrodynamics, **1**(1), 23–40 (2017)
61. Yang, R., & Liu, X.: Comparison of convex optimization-based approaches to solve nonconvex optimal control problems. In: AIAA Scitech 2019 Forum, p. 1666 (2019)
62. Filatyev, A.S.: The through optimization of branching launch vehicle injection trajectories in the atmosphere based on the Pontryagin maximum principle. Dissertation for the degree of Doctor of Engineering Sciences, p. 395 (2001) (in Russian)
63. Filatyev, A., Yanova, O.: ASTER program package for the thorough trajectory optimization. AIAA Guidance, Navigation, and Control Conference and Exhibit, Montreal, Canada, Aug. 6–9, 2001, AIAA-2001-4391. https://doi.org/10.2514/6.2001-4391
64. Filatyev, A.S., Golikov, A.A., Yanova, O.V.: Indirect optimization of branched injection trajectories of aerospace vehicles. In: The 1st ESA Workshop on Astrodynamics Tools and Techniques, 17–18 July 2001, ESTEC, Noordwijk, The Netherlands
65. Filatyev, A.S., Golikov, A.A., Yanova, O.V.: ASTER program package capabilities for the through trajectory optimization of branched processes. In: Proceedings of the the 5th International Conference on Astrodynamics Tools and Techniques (ICATT), ESA/ESTEC, Noordwijk, the Netherlands, 29 May – 1 June 2012
66. Berkmann, P., Pesch, H.J.: Abort landing in windshear: optimal control problem with third-order state constraint and varied switching structure. J. Opt. Theo. Appl. **85** (1995)
67. Pesch, H.J.: A practical guide to the solution of real-life optimal control problems. Control. Cybern. **23**, 7–60 (1994)
68. Pesch, H.J.: Real-Time computation of feedback controls for constrained optimal control problems. Optimal Control Applications and Methods **10**, 147–171 (1989)
69. Lysenko, A.I.: The jump condition in the problem of optimal control of a composite object. In: Adaptive systems of automatic control. Kyiv: Technique, No. 16 (1988) (in Russian)
70. Lecohier, G., Mehlem, K.: A new trajectory optimization tool (ALTOS) applied to conventional launchers. IFAC Automatic Control in Aerospace, Ottobrunn, Germany (1992)
71. Well, K., Gath, P.F., Fischer, D.: ASTOS—Aerospace Trajectory Optimization Software, 1st ESA Workshop on Astrodynamics tools and Techniques, 17–18 July 2001, ESTEC, Noordwijk, The Netherlands
72. Okhotsimsky, D.E., Eneev, T.M.: Some variation problems associated with the launch of an artificial Earth satellite. UFN, **63**(1a) (1957) (in Russian)
73. Lowden, D.F.: Optimal Trajectories for Space Navigation, p. 126. Butterworths (1963)

74. Sokolov, N.L., Suvorov, E.A.: About one control scheme of spacecraft on the active stage of launching an artificial Earth satellite into orbit. Cosmic Res. **34**(5) (1996)
75. Fedorov, A.I.: Using the method of neighboring extremals to control the trajectory motion of spacecraft during launch into orbit. Uchenye Zapiski TsAGI **21**(4) (1990) (in Russian)
76. Jaggers, R.F.: An explicit solution to the exoatmospheric powered flight guidance and trajectory optimization problem for rocket propelled vehicles. Guidance and control conference, Hollywood, Fla, Technical paper, A77-42751 20-35, Aug 1977
77. Singh, S.K. et al.: Optimal control on active sections of extra-atmospheric flight. Astronautics and Rocket Dynamics (9) (1990)
78. Anorov, V.P.: The maximum principle for processes with general constraints. Automation and Telemech, **3**(4) (1967) (in Russian)
79. Cheng, S.S.L.: Modified maximum principle for optimal systems with constrained phase coordinates. In: Proceedings of the 2nd International Congress of IFAC, vol. 2, Moscow: Nauka (1965) (in Russian)
80. Velichenko, V.V.: The Optimal Control of Composite Systems. Doklady AN USSR **176**(4), 754–756 (1967). (in Russian)
81. Arutyunov, A.V.: On necessary optimality conditions in a problem with phase constraints. Dokl. Akad. Nauk SSSR, **280**(5) (1985) (in Russian)
82. Dreyfus, S.E.: Dynamic Programming and the Calculus of Variations. Academic Press, N.Y. (1965)
83. Speyer, J.L., Bryson, A.E.: Optimal programming problems with a bounded state space. AIAA J **6**(8) (1968)
84. Fraeijs de Veubeke, B.: Advanced Problems and Methods for Space Flight Optimization. Pergamon Press, Oxford (1969)
85. Pyatakov, V.I.: On the optimization of aircraft control taking into account constraints on the control function and phase coordinates. In: Control of Moving Objects, Moscow, Nauka, pp. 187–200 (1972) (in Russian)
86. Bryson, A.E., Denham, V.F., Dreyfus, S.E.: Optimal programming problems with inequality constraints. AIAA J. **1**(11), 2544–2550 (1963)
87. Ashchepkov, L.T.: The Optimal Control of Discontinuous Systems. Nauka, Novosibirsk (1987). (in Russian)
88. Vincent, T.L., Mason, J.D.: Disconnected optimal trajectories. JOTA **3**(4), 263–281 (1969)
89. Morozov, S.F., Sumin, M.I.: On a class of control problems for dynamical systems with a discontinuous right-hand side. Cybernetics **3** (1985) (in Russian)
90. Ortega, J., Rheinboldt, W.: Iterative Solution of Nonlinear Equations in Several Variables. Academic Press, New York (1970)
91. Asher, U.M., Mattheij, R.M.M., Russell, P.D.: Numerical solution of boundary value problems for ordinary differential equations. SIAM, p. 620 (1994)
92. Bartish, M.Y.: On a class of Newton-type methods. Bulletin of Moscow State University, Ser. Calculate. Math. Cybern **2** (1987)
93. Stechkin, S.B., Subbotin, Yu.N.: Splines in Computational Mathematics. Nauka, Moscow (1976). (in Russian)
94. Dennis, J., Schnabel, R.: Numerical Methods for Unconstrained Optimization and Nonlinear Equations. Prentice-Hall Inc., Englewood Cliffs, NY (1983)
95. Davidenko, D.F.: On a new numerical method for solving systems of nonlinear equations. Dokl. Akad. Nauk SSSR. **88**(4), 601–602 (1953). (in Russian)
96. Lidov, M.L., Teslenko, N.M.: Optimization of the solution of some problems of spacecraft flight control by the parameter descent method. Mathematical support of space experiments, Moscow, Nauka, pp. 112–141 (1978) (in Russian)
97. Goman, M.G.: Differential method for continuing solutions of systems of finite nonlinear equations depending on a parameter. Uchenye Zapiski TsAGI **17**(5) (in Russian)
98. Roberts, S., Shipman, J.: Continuation in shooting methods for two-point boundary value problems J. Math. Anal. Appl. **18**, 45–58 (1967)

99. Sagalakov, A.E., Filatyev, A.S.: New periodic solutions in the aircraft maximum range problem. Proceedings of the 29th Congress of the International Council of the Aeronautical Sciences. September 7–12, 2014, St. Petersburg, Russia, ICAS2014-8.10.4-2014_0562
100. Sagalakov, A.E., Filatyev, A.S.: New approach for investigation of maximum-range aircraft trajectories. In: The 31st Congress of the International Council of the Aeronautical Sciences, Belo Horizonte, Brazil, 9–14 Sept. 2018
101. Vinh, N.X., Busemann, A., and Culp, R.D.: Hypersonic and Planetary Entry Flight Mechanics. The University of Michigan Press (1980)
102. Yaroshevsky, V.A.: An Uncontrolled Body Motion in the Atmosphere. Mashinostroenie, Moscow (1978). (in Russian)
103. Yaroshevsky, V.A.: Spacecraft Entry Into Atmosphere. Nauka, Moscow (1988). (in Russian)
104. Yaroshevsky, V.A., Vinh, N., de Olive, F.L.: Analytical and semi-analytical methods for studying atmospheric entry trajectories. Uchenye Zapiski TsAGI **28**(3–4) (1997) (in Russian)
105. Busemann, A., Vinh, N.X., and Kelley, G.F.: Optimum Maneuvers of Hypervelocity Vehicles. NASA Report CR-1078, 1968.
106. Shkadov, L.M., Bukhanova, R.S., Illarionov, V.F., Plokhikh, V.P.: Mechanics of Optimal Aircraft Spatial Motion in the Atmosphere. Mashinostroenie, Moscow (1972). (in Russian)
107. Bliss, G.A.: Lectures on the Calculus of Variations. University of Chicago Press, pp. 292 (1963)
108. Bliss, G.A.: Mathematics for Exterior Ballistics, N.Y. (1944)
109. Hardtla, J.W., oth. Guidance requerements for future launch vehicles. AIAA P. 87-2462
110. Bakhvalov, N.S., Zhidkov, N.P., Kobelkov, G.M.: Numerical Methods. Nauka, Moscow (1987). (in Russian)
111. Hairer, E., Nursett, S., Wanner, G.: Solving Ordinary Differential Equations I: Nonstiff problems. Springer (1993)
112. Demidovich, B.P., Maron, I.A.: Fundamentals of Computational Mathematics. Nauka, Moscow (1970). (in Russian)
113. Korn, G., Korn, T.: Handbook of Mathematics. The 4th edition, Nauka, p. 696 (1977) (in Russian)
114. Filatyev, A.S.: Paradoxes of optimal solutions in problems of space vehicle injection and reentry. Acta Astronaut. **47**(1), 11–18 (2000)
115. Filatyev, A.S.: Analysis of the optimal control law for hypersonic aircraft reentry. Uchenye Zapiski TsAGI **7**(5) (1976) (in Russian)
116. Filatyev, A.S.: The optimal launch of an artificial Earth satellite using aerodynamic forces. Cosm. Res. **29**(2), 225–271 (1991)

Chapter 2
Optimization of Aerospace Vehicle Trajectories in the Atmosphere by the Pontryagin Maximum Principle

2.1 Introduction

The first artificial Earth satellite was launched by the USSR on October 4, 1957. This event was preceded by a huge theoretical work, in particular, in the field of ballistics to determine the best trajectory and control program for the launch vehicle (LV). The fundamental results of these studies were published in the paper of Okhotsimsky and Eneev [1], whose release was timed to coincide with this epochal event of humanity's entry into the third dimension. A little later, the work with the solution of a similar problem, but in a simpler formulation, was independently published by Lawden [2]. In both papers, the problem of the optimal LV injection for delivering spacecraft (SC) to an orbit was extremely simplified in the first study phase. The Earth's sphericity and atmosphere were neglected, and the gravitational field was assumed plane-parallel uniform. This made it possible to obtain and study in detail the analytical solution, in particular, the optimal control law for the thrust vector. Despite the fact that under such assumptions, the error in modeling non-inertial forces reached 100% (in a vicinity of entry into an orbit), the resulting approximate optimal control program turned out to describe the structure of a strictly optimal solution for ballistic LVs quite well and therefore became classical and has been used in almost all LV trajectory control systems in the world up to the current time. This once again confirms the wisdom of the well-known aphorism (attributed to Immanuel Kant): "*There is nothing more practical than a good theory*."

In Sect. 2.2, the Okhotsimsky-Eneev-Lawden (OEL) classical solution is presented as a reference for comparison with more rigorous solutions for refined mathematical models. However, only the problem statement is borrowed from OEL works. The solution here is given on the basis of PMP, while the original OEL works were based on the methods of the calculus of variations.

In Chaps. 3 to 6, attention is focused on the conditions under which more rigorous solutions no longer even qualitatively correspond to the classical OEL solutions.

A. S. Filatyev and V. G. Petukhov, *Through Optimization of Aerospace Vehicle Trajectories by the Pontryagin Maximum Principle*, Studies in Systems, Decision and Control 579, https://doi.org/10.1007/978-3-031-80756-5_2

2.2 Approximate Solution of the Problem of Ascent-To-Orbit Optimization

Below, we describe two cases of approximate optimal control for LV injection into an orbit. The first case corresponds to the classical solution for the motion of a material point with thrust in a uniform gravitational field. In the second case, it is only conditionally possible to speak of optimality, since we are talking about a deterministic motion, though in the field of arbitrary, gravitational and aerodynamic, forces. Since the motion is deterministic, that is, the only possible one due to a fixed control program, then it is "optimal" in a certain sense. Freedom there remains in setting part of the conditions at initial and/or intermediate points.

2.2.1 *The Okhotsimsky-Eneev-Lawden Classical Solution*

Consider the motion of a rocket in a uniform gravitational field in a vertical plane in the inertial Cartesian coordinate system (see Fig. 2.1):

$$\begin{cases} \frac{dx_1}{dt} = x_3, \\ \frac{dx_2}{dt} = x_4, \\ \frac{dx_3}{dt} = \frac{T}{x_5}\cos\vartheta, \\ \frac{dx_4}{dt} = \frac{T}{x_5}\sin\vartheta - g, \\ \frac{dx_5}{dt} = -\mu = -\frac{T}{c}, \end{cases} \tag{2.1}$$

where $\mathbf{x} = (x_1,\ldots,x_5)^{\mathrm{T}}$ is the phase vector of dimension $n = 5$; x_1 and x_2 are the longitudinal range and altitude; x_3 and x_4 are the horizontal and vertical velocities; x_5 is the rocket mass, t is the time, T is the thrust, ϑ is the pitch angle (angle of

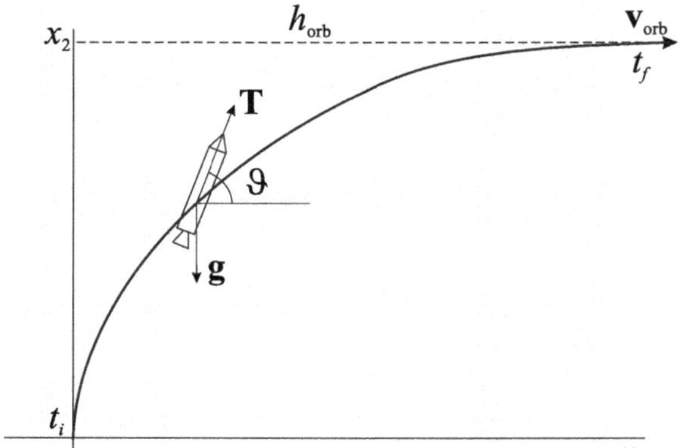

Fig. 2.1 The coordinate system for the motion model with the uniform gravitational field in a vertical plane

inclination of the thrust vector to the horizontal axis), g is the constant gravitational acceleration, μ is the fuel mass consumption rate, c is the engine exhaust velocity.

Control vector $\mathbf{u}$ consists of the thrust magnitude and pitch angle, which are subject to constraints:

$$\mathbf{u} = (T, \vartheta)^{\mathrm{T}}, T \in [0, T_{max}], \vartheta \in (-\pi, \pi]. \tag{2.2}$$

At given initial time $t_i = 0$, admissible phase vector $\mathbf{x}^i = \mathbf{x}(t_i)$ is completely specified (see (1.2)):

$$t = t_i = 0 : g_1^i = x_1^i = 0, g_{2_i}^i = x_2^i = 0, g_3^i = x_3^i = 0, g_4^i = x_4^i = 0, g_5^i = x_5^i - m_i = 0, \tag{2.3}$$

where m_i is the initial mass.

At the final time moment t_f, the following two constraint sets can be imposed on admissible phase vector $\mathbf{x}^f = \mathbf{x}(t_f)$ (see (1.3)):

$$t = t_f = \text{free} : g_1^f = x_2^f - h_{\text{orb}} = 0, g_2^f = x_3^f - v_{\text{orb}} = 0, g_3^f = x_4^f = 0, x_1^f = \text{free} \tag{2.4a}$$

where h_{orb} is the given final altitude, v_{orb} is the given final horizontal speed,
or

$$\begin{aligned} t = t_f = \text{free} : g_1^f &= x_2^f - h_{\text{orb}} = 0, g_2^f = x_3^f - v_{\text{orb}} \\ &= 0, g_3^f = x_4^f = 0, g_4^f = x_1^f - L = 0, \end{aligned} \tag{2.4b}$$

where L is the given final range.

The functional is the final mass to be maximized:

$$J = x_5^f \Rightarrow \max_{\mathbf{u}}. \tag{2.5}$$

Thus, we will solve the problem of the optimal ascent of a rocket into the orbit in the following simplified formulation. It is required to find such dependences of thrust vector (2.2) on time so that, obeying motion (2.1), the rocket from the initial (zero) altitude reaches the given final altitude h_{orb}, horizontal velocity v_{orb} and zero vertical velocity, while consuming the minimum fuel mass (maximizing the final mass). In case (2.4a) the final range is free, in case (2.4b) it is fixed.

We will solve the problem using PMP, keeping the sequence and notations from Chap. 1.

Let us introduce the Hamiltonian and conjugate system (1.6) into consideration:

$$H(\mathbf{p}, \mathbf{x}, \mathbf{u}) = \mathbf{p}^{\mathrm{T}}\mathbf{f}(\mathbf{x}, \mathbf{u}) = p_1 x_3 + p_2 x_4 + p_3 \frac{T}{x_5} \cos\vartheta + p_4\left(\frac{T}{x_5} \sin\vartheta - g\right) + p_5\left(-\frac{T}{c}\right), \tag{2.6}$$

$$\begin{cases} \frac{dp_1}{dt} = -\frac{\partial H}{\partial x_1} = 0 \\ \frac{dp_2}{dt} = -\frac{\partial H}{\partial x_2} = 0 \\ \frac{dp_3}{dt} = -\frac{\partial H}{\partial x_3} = -p_1 \\ \frac{dp_4}{dt} = -\frac{\partial H}{\partial x_4} = -p_2 \\ \frac{dp_5}{dt} = -\frac{\partial H}{\partial x_5} = \frac{T}{x_5^2}(p_3 \cos\vartheta + p_4 \sin\vartheta) \end{cases} \tag{2.7}$$

In accordance with (1.8–1.10) let us write the boundary conditions (transversality conditions) for the conjugate vector at the initial and final points:

$$\mathbf{p}^i = \mathbf{p}(t_i) = -\left(\frac{\partial \mathbf{g}^i(\mathbf{x}^i)}{\partial \mathbf{x}^i}\right)^{\mathrm{T}} \lambda^i = \text{free}; \tag{2.8}$$

$$\mathbf{p}^f = \left(\frac{\partial\left(\lambda_1^f g_1^f + \lambda_2^f g_2^f + \lambda_3^f g_3^f + J\right)}{\partial \mathbf{x}^f}\right)^{\mathrm{T}}; \; p_1^f = 0; p_{2,3,4}^f = \text{free}; \; H^f = 0; \tag{2.9a}$$

$$\mathbf{p}^f = \left(\frac{\partial\left(\lambda_1^f g_1^f + \lambda_2^f g_2^f + \lambda_3^f g_3^f + \lambda_4^f g_4^f + J\right)}{\partial \mathbf{x}^f}\right)^{\mathrm{T}}; \; \mathbf{p}^f = \text{free}; \; H^f = 0. \tag{2.9b}$$

At each point, the optimal control is determined from condition (1.7):

$$\mathbf{u}_{\mathrm{opt}}(t) = \left(\vartheta_{\mathrm{opt}}(t), T_{\mathrm{opt}}(t)\right)^{\mathrm{T}} = \arg\max_{\mathbf{u}\in\mathbf{U}} H = \arg \max_{\substack{-\pi<\vartheta\le\pi \\ 0\le T\le T_{\max}}} \left[T\left(\frac{p_3\cos\vartheta + p_4\sin\vartheta}{x_5} - \frac{p_5}{c}\right)\right]. \tag{2.10}$$

From (2.10) we find the optimal pitch angle law. The necessary optimality condition is.

$$\frac{\partial H}{\partial \vartheta} = \frac{T}{m}(-p_3 \sin\vartheta + p_4\cos\vartheta) = 0 \Rightarrow \tan\vartheta_{\mathrm{opt}} = \frac{p_4}{p_3}, \; \sin\vartheta = \frac{\pm p_4}{\sqrt{p_3^2+p_4^2}}, \; \cos\vartheta = \frac{\pm p_3}{\sqrt{p_3^2+p_4^2}} \tag{2.11}$$

From the analysis of the second derivative:

$$\frac{\partial^2 H}{\partial \vartheta^2} = \mp\frac{T}{m}\sqrt{p_3^2+p_4^2}, \tag{2.12}$$

we choose the upper sign, since we are only interested in the Hamiltonian maximum. Then

$$\tan \vartheta_{\text{opt}} = \frac{p_4}{p_3},\ \sin \vartheta_{\text{opt}} = \frac{p_4}{\sqrt{p_3^2 + p_4^2}},\ \cos \vartheta_{\text{opt}} = \frac{p_3}{\sqrt{p_3^2 + p_4^2}}. \tag{2.13}$$

From (2.10), taking into account (2.13), we find the optimal thrust magnitude control:

$$T_{\text{opt}} = \begin{cases} T_{\text{max}},\ s > 0, \\ 0,\ s < 0, \\ \tilde{T},\ s \equiv 0, \end{cases} \tag{2.14}$$

where $s = \frac{\sqrt{p_3^2+p_4^2}}{x_5} - \frac{p_5}{c}$ is the switching function, $\tilde{T} \in [0, T_{max}]$ is the singular thrust control.

Conjugate system (2.7) with final conditions (2.9), taking into account (1.54), has the following obvious first integrals:

$$\begin{aligned} p_1 &= a;\ p_2 = c;\ p_3 = b - at;\ p_4 = d - ct;\ \ a, b, c, d \\ &= \text{const};\ H = p_1 x_3 + p_2 x_4 - p_4 g + Ts = 0. \end{aligned} \tag{2.15}$$

Substituting the expressions for conjugate variables (2.15) into optimality condition (2.13), we obtain the well-known bi-linear law for the optimal pitch angle tangent:

$$\tan \vartheta_{\text{opt}} = \frac{d - ct}{b - at}. \tag{2.16a}$$

If the final range is not specified (see (2.4a)), then by virtue of (2.9a) $p_1 = a = 0$, and we obtain a linear optimal program for the pitch angle tangent:

$$\tan \vartheta_{\text{opt}} = \tilde{d} - \tilde{c}t;\quad \tilde{c} = c/b, \tilde{d} = d/b. \tag{2.16b}$$

2.2.2 The Gravity Turn

The gravity turn (g-turn) is the LV control program in which zero angle of attack is maintained:

$$\alpha = 0. \tag{2.17}$$

Control program (2.17) is traditionally used as nominal for LV injection segment in dense atmosphere layers. This allows a symmetrical LV to keep zero nominal (in the absence of wind) normal aerodynamic loads.

The fact that this is actually not always justified, either from the point of view of proximity to the optimal control program (maximization of the final mass) or from the point of view of reducing normal loads in a real atmosphere, will be shown in Chaps. 3 and 6. In the sequel, we will use a combination of the initial ascent leg with a zero angle of attack (2.17) and the subsequent flight in rarefied layers of the atmosphere with OEL suboptimal law (2.16), which qualitatively corresponds to the typical control programs for LVs with a vertical take-off, for comparison with the rigorous optimal control programs, including constraints on aerodynamic loads.

In case (2.17), the trajectory can be controlled either by changing initial flight path angle θ_i or by introducing another deterministic program with a fixed angle of attack after a short initial leg of the vertical ascent. The first mode of the g-turn is conveniently used in methodological studies of optimal injection trajectories (see Chap. 3). The second one corresponds to actual LV programs (see Chap. 6).

2.3 The General Problem Statement

2.3.1 Equations of Motion

Let us consider ASV motion in the non-inertial Cartesian coordinate system associated with the start point (Fig. 2.2).

Let us define ASV phase vector $\mathbf{x} = \{\mathbf{r}, \mathbf{v}, m\}^{\mathrm{T}} \in R^n$ by ODE system

$$\frac{d\mathbf{x}}{dt} = \mathbf{f}(\mathbf{x}, \mathbf{u}), \mathbf{f} = \left(\mathbf{v}, \frac{\mathbf{T} + \mathbf{A}}{m} + \mathbf{g} + \mathbf{\Omega}, -\mu\right)^{\mathrm{T}}, \tag{2.18}$$

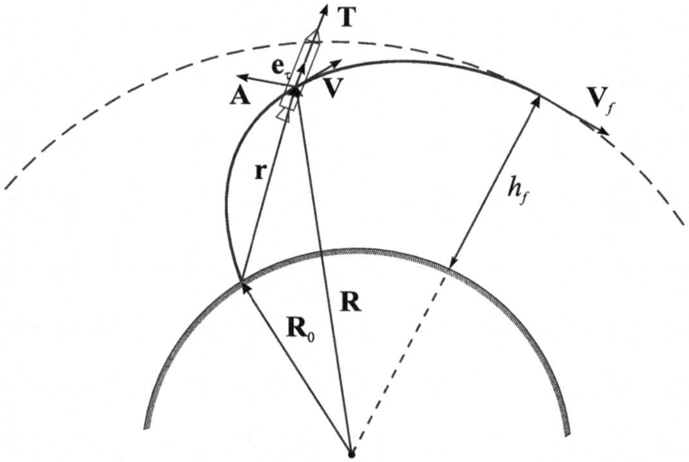

Fig. 2.2 The noninertial start coordinate system for the motion model with the central gravitational field

where t is the time, m is the ASV mass, $\mathbf{T} = T \cdot \mathbf{e}_T$ is the thrust vector, $\mathbf{e}_T$ is the unit thrust vector, $\mathbf{A}$ is the aerodynamic force, $\mathbf{g}$ is the gravitational acceleration, $\mathbf{\Omega} = (\boldsymbol{\omega} \times \mathbf{R} + 2\mathbf{v}) \times \boldsymbol{\omega}$ is the acceleration vector due to non-inertial reference frame, $\boldsymbol{\omega}$ is the angular velocity of the daily Earth rotation, $\mathbf{R} = \mathbf{r} + \mathbf{R}_o$, $\mathbf{R}_o$ is the radius vector from the Earth's center to the origin of coordinates, $R{-}R_o = h$ is the flight altitude, μ is the fuel mass consumption rate, $\eta = T/T_{\max}$ is the thrust throttling factor, $T_{\max}(\mathbf{x})$ is the maximum thrust. For the model of the central Newtonian gravitational field

$$\mathbf{g} = -\frac{g_0 R_0^2}{R^3}\mathbf{R}, \tag{2.19}$$

where g_0 is the gravitational acceleration value on the Earth's surface, R_0 is the average Earth radius.

In what follows, the variables in (2.18) will be assumed to be dimensionless

$$r = (r/R_0)_{\text{dim}},\ v = \left(v/\sqrt{g_0 R_0}\right)_{\text{dim}},\ m = (m/m_i)_{\text{dim}},\ t = \left(t\sqrt{g_0/R_0}\right)_{\text{dim}},$$
$$g = (g/g_0)_{\text{dim}},\ \omega = \left(\omega\sqrt{R_0/g_0}\right)_{\text{dim}},\ T = (T/(m_i g_0))_{\text{dim}},\ A = (A/m_i g_0)_{\text{dim}}, \tag{2.20}$$

where subscript "$_{\text{dim}}$" denotes dimensional quantities.

When flying without sliding (the plane of symmetry of the vehicle passes through the velocity vector) [3, 4],

$$\mathbf{A} = a_r \rho v^2\left(-C_D \mathbf{e}_v + C_L \frac{\mathbf{e}_v \times \mathbf{e}_\tau \times \mathbf{e}_v}{|\mathbf{e}_v \times \mathbf{e}_\tau|}\right), \tag{2.21}$$

where $a_r = \frac{\rho_0 R_0 F_{\text{ref}}}{2m_i}$, ρ is the atmosphere density referred to the one at the Earth's surface ρ_0, $\mathbf{e}_\tau$ is the unit vector directed along the longitudinal axis of the vehicle from tail to nose, α is the angle of attack, $\cos\alpha = (\mathbf{e}_\tau, \mathbf{e}_v)$, $\mathbf{e}_v = \mathbf{v}/v$, $v = |\mathbf{v}|$, F_{ref} is the reference area.

Let us represent the coefficients of aerodynamic lift C_L and drag C_D in the following form [3, 4]:

$$C_L = C_L^\alpha \sin\alpha,\ C_D = D_0 + D_\alpha \cos\alpha, \tag{2.22}$$

where C_L^α, D_0, D_α in the general case depend on phase vector $\mathbf{x}$ (flight modes). For small α relations (2.22) correspond to the quadratic polar:

$$C_L \cong C_L^\alpha \alpha,\ C_D \cong C_{D0} + A\alpha^2, \tag{2.23}$$

where $A = -D_\alpha/2$, $C_{D0} = D_0 + D_\alpha$.

According to (2.22), the maximum lift-to-drag ratio ("aerodynamic quality")

$$K_{\max}(\mathbf{x}) = \max_{\alpha} \frac{C_L(\alpha, \mathbf{x})}{C_D(\alpha, \mathbf{x})} = \frac{C_L^{\alpha}}{\sqrt{D_0^2 - D_{\alpha}^2}} \tag{2.24}$$

is reached at

$$\alpha_{K_{\max}} = \arg\max_{\alpha}(C_L/C_D) = \arcsin\sqrt{1 - D_{\alpha}^2/D_0^2}. \tag{2.25}$$

In case (2.22), expression (2.21) takes the form

$$\mathbf{A} = \tilde{C}_y^{\alpha}\mathbf{e}_{\tau} - \left[\tilde{D}_0 + \left(\tilde{C}_y^{\alpha} + \tilde{D}_{\alpha}\right)(\mathbf{e}_{\tau}, \mathbf{e}_v)\right]\mathbf{e}_v, \tag{2.26}$$

where $\tilde{D}_0 = a_r \rho v^2 D_0$, $\tilde{D}_{\alpha} = a_r \rho v^2 D_{\alpha}$, $\tilde{C}_y^{\alpha} = a_r \rho v^2 C_L^{\alpha}$.

2.3.2 *The Control Vector*

Let the thrust be directed along the vehicle longitudinal axis given by unit vector $\mathbf{e}_{\tau}$:

$$\mathbf{e}_T = \mathbf{e}_{\tau}. \tag{2.27}$$

The case when the thrust vector orientation relative to the vehicle longitudinal axis is given by a more general condition than (2.27) will be considered below in Sect. 2.8.2. The case of free orientation of the thrust vector relative to the longitudinal axis was considered, for example, in [5].

Taking into account (2.18–2.20), (2.22), (2.23), (2.26), (2.27) the control vector has form

$$\mathbf{u} = (\mathbf{e}_{\tau}, \eta)^{\mathrm{T}},\ \eta \in [\eta_{\min}, \eta_{\max}]. \tag{2.28}$$

2.3.3 *Constraints*

The admissible control, as a rule, is associated with constraints on the angle of attack, normal and longitudinal accelerations and forces. Taking into account the dependence of these constraints on the flight modes and mass, they can be approximately represented in general form [3, 4, 6]:

$$G \equiv f_{\tau}(\mathbf{x}) - (\mathbf{e}_v, \mathbf{e}_{\tau}) \le 0, \tag{2.29}$$

$$\eta T_{\max}(\mathbf{x}) - f_T(\mathbf{x}) \le 0. \tag{2.30}$$

For example, to limit the angle of attack: $\alpha_{\min} \le \alpha \le \alpha_{\max}$, $\alpha_{\min} < 0$, we can use (2.29), where

$$f_\tau(\mathbf{x}) = \begin{cases} \cos\alpha_{\max}(M), \alpha > 0, \\ \cos\alpha_{\min}(M), \alpha < 0, \end{cases} \tag{2.31}$$

$M = v/a_s$ is Mach number, a_s is the sound speed.

If it is necessary to take into account constraint $q|\alpha| \le Q_{\text{adm}}(M)$, which is widely used in practical rocket dynamics to limit local aerodynamic loads on LV structure, then in (2.29) one should put

$$f_\tau(\mathbf{x}) = \cos\left(\frac{\overline{Q}}{a_r \rho v^2}\right), \overline{Q} = \frac{Q_{\text{adm}}}{\rho_0 g_0 R_0}. \tag{2.32}$$

To limit the normal aerodynamic load factor $n_{z\,\min} \le n_z \le n_{z\,\max}$, in (2.29) it can be taken approximately

$$f_\tau(\mathbf{x}) = \begin{cases} 1 - \frac{1}{2}\left[\frac{n_{z\,\max}}{a_r \rho v^2 (C_L^\alpha(M) + C_{D0}(M))}\right]^2, n_z > 0, \\ 1 - \frac{1}{2}\left[\frac{n_{z\,\min}}{a_r \rho v^2 (C_L^\alpha(M) + C_{D0}(M))}\right]^2, n_z < 0. \end{cases} \tag{2.33}$$

The constraint on the normal force is taken into account similarly.

To limit the longitudinal load factor $n_x \le n_{x\,\text{adm}}$ when flying with a low dynamic pressure, it is sufficient to put

$$f_T(\mathbf{x}) = n_{x\text{adm}} \cdot m \tag{2.34}$$

into (2.30).

Along with constraints (2.29) containing the control explicitly, constraints on the dynamic pressure, Mach number, heat flux, and etc. are taken into account. These are the first-order constraints on phase variables (see Sect. 1.5), because they contain the control explicitly in their first time derivative.

Constraints on the dynamic pressure, heat flux and Mach number are considered in the following form [7–9]:

$$G(\mathbf{x}) \equiv f_0(\mathbf{x}) - f_{0\,\text{adm}} \le 0. \tag{2.35}$$

If in (2.35) we take

$$f_0 = q_\kappa = \rho v^{2\kappa}, \tag{2.36}$$

then under $\kappa = 1$ condition (2.35) allows limiting the dynamic pressure, and, for example, at $\kappa = 3.25$ we have the constraint on the specific heat flow, estimated by the Kemp-Riddell formula [10, 11]:

$$G \equiv q_{\kappa} - q_{\kappa\,\mathrm{adm}} \leq 0. \tag{2.37}$$

The Mach number constraint is typical, e.g., for the flight segments of a carrier aircraft (CA) of an aerospace transport system (ASTS):

$$G \equiv M - M_{\mathrm{adm}} \leq 0. \tag{2.38}$$

If we neglect the change in the sound speed with altitude on these short legs of CA flight, then constraint (2.38) is actually replaced by the restriction on the admissible speed:

$$G \equiv v - v_{\mathrm{adm}} \leq 0. \tag{2.39}$$

Taking into account (2.21–2.28) the orientation of the thrust vector, starting from the moment of reaching the boundaries Γ of the admissible areas (2.35–2.39), is bound by condition

$$G \equiv \left(T - \tilde{D}_{\alpha}\right)(\mathbf{e}_v, \mathbf{e}_{\tau}) - f_q(\mathbf{x}) \leq 0, \tag{2.40}$$

where the form of function f_q depends on the constraint physical meaning.

In particular, for the constraints on the dynamic pressure and heat flow (for the Newtonian gravitational field, $\beta(h) = -\frac{1}{\rho}\frac{d\rho}{dh}$, $\mathbf{e}_R = \mathbf{R}/R$, $\mathbf{e}_{\omega} = \boldsymbol{\omega}/\omega$,)

$$f_q = \tilde{D}_0 + m\left(1/R^2 + \beta(h)v^2/\kappa\right)\mathbf{e}_R^{\mathrm{T}}\mathbf{e}_v + m\omega^2 R[(\mathbf{e}_R, \mathbf{e}_{\omega})(\mathbf{e}_v, \mathbf{e}_{\omega}) - (\mathbf{e}_R, \mathbf{e}_v)]; \tag{2.41}$$

for the constraint on the allowable Mach number (2.38)

$$f_q = \tilde{D}_0 + m\left(\frac{1}{R^2} - Mv\frac{da_s}{dh}\right)(\mathbf{e}_R, \mathbf{e}_v) + m\omega^2 R[(\mathbf{e}_R, \mathbf{e}_{\omega})(\mathbf{e}_v, \mathbf{e}_{\omega}) - (\mathbf{e}_R, \mathbf{e}_v)]; \tag{2.42}$$

for the constraint on the admissible flight speed (2.39)

$$f_q = \tilde{D}_o + \frac{m}{R^2}(\mathbf{e}_R, \mathbf{e}_v) + m\omega^2 R[(\mathbf{e}_R, \mathbf{e}_{\omega})(\mathbf{e}_v, \mathbf{e}_{\omega}) - (\mathbf{e}_R, \mathbf{e}_v)]. \tag{2.43}$$

Note that (2.43) is obtained from (2.41) by the formal substitution $1/\kappa = 0$.

If the thrust throttling program is given (not optimized), then constraints (2.40–2.43) up to the sign of the left side of the inequality are reduced to constraint on angle of attack (2.29) if we set $f_{\tau} = f_q/(T - \tilde{D}_{\alpha})$.

The PMP admits discontinuities in the control [12]. In particular, the optimal control may undergo discontinuities at predetermined moments, when, e.g., the ASV characteristics, mass, etc., change abruptly (for example, at STS staging). If at these moments t_* the optimal ASV orientation changes from $\mathbf{e}_{\tau}^{-}(t_*)$ to $\mathbf{e}_{\tau}^{+}(t_*)$, it may be necessary to take into account the constraint on the admissible angular rate of the ASV turn relative to the center of mass:

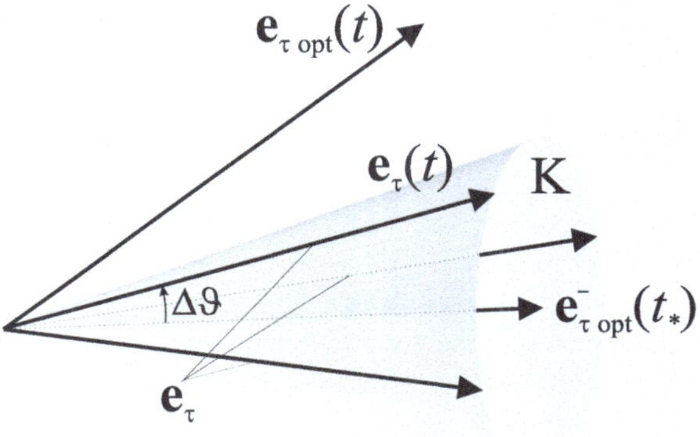

Fig. 2.3 Geometric interpretation of the constraint on the pitch angular velocity for regularization of the optimization problem statement in the PMP framework

$$|\omega_v| \le \omega_{v\ \mathrm{adm}}. \tag{2.44}$$

System (2.18) does not contain equations of angular motion, and condition (2.44) limits the rate of control change, which goes beyond the PMP formalism and requires the development of a special regularization technique. To do this, instead of (2.44) we require the fulfillment of condition

$$\cos(\omega_{v\ \mathrm{adm}} \cdot (t - t_*)) - \left(\mathbf{e}_\tau^-(t_*), \mathbf{e}_\tau\right) \le 0. \tag{2.45}$$

Inequality (2.45) coincides in form with control constraints (2.29). The geometric meaning of this inequality is that on interval $t > t_*$ control vector $\mathbf{e}_\tau$ must not go beyond the circular cone $\boldsymbol{K}$ with the symmetry axis passing through vector $\mathbf{e}^-_{\tau\mathrm{opt}}(t_*)$ and half-angle $\Delta\vartheta = \omega_{v\ \mathrm{adm}} \cdot (t - t_*)$ (Fig. 2.3). Since the optimal control vector practically remains in the same plane (the control plane in terms of [1, 2]) in the out-of-atmospheric flight leg, the proposed approach ensures that the specified pitch rate limitation is maintained with a sufficient accuracy.

2.3.4 Boundary Conditions

Consider the following types of initial conditions ($t = t_i$):

(a) phase vector is fixed:

$$\mathbf{x}(t_i) - \mathbf{x}^i = 0; \tag{2.46}$$

(b) the orientation of the velocity vector is free, and other components of the phase vector are fixed:

$$\mathbf{r}(t_i) - \mathbf{r}^i = 0;\ v(t_i) - v_i = 0;\ \ m(t_i) - m_i = 0,\ \mathbf{e}_v = \text{free}, \tag{2.47}$$

which allows to determine the optimal initial spatial orientation of the velocity vector, i.e. initial azimuth $\eta_{i\text{opt}}$ and flight path angle $\theta_{i\text{opt}}$;

(c) a more general form of the initial phase vector constraint:

$$\mathbf{x}(t_i) \in D_i \subset R^n, \tag{2.48}$$

is implemented, e.g., for ASTS with an air launch [3, 4, 6, 13, 14], and allows us to determine, together with the optimal control, the best take-off conditions from a given manifold.

The right trajectory end is determined by the conditions of motion along the target orbit. The optimal motion of a high-thrust ASV outside the atmosphere is characterized by the presence of a long passive leg [15, 16]. Therefore, this leg beyond the conditional boundary of the atmosphere (where the influence of aerodynamic forces is small) can be approximately replaced by the transition ellipse tangent to the target orbit, and a short active leg in the vicinity of the insertion into the orbit can be replaced by a velocity impulse at the apogee of the transition ellipse. In this case, the boundary conditions from the right end of the ASV injection trajectory can be transferred approximately, using Kepler's laws, to the right end of the active ASV flight leg:

$$t = t_f\ :\ \begin{cases} Q_0 \equiv v_a - v_{af} = 0, Q_1 \equiv R - R_f = 0, \\ Q_2 \equiv \frac{(\mathbf{R},\mathbf{v}_a)}{R_f v_{af}} - \sin\theta_{af} = 0, \\ Q_6 \equiv \frac{\omega \mathbf{R} \mathbf{v}_a}{|\omega||\mathbf{R}\times\mathbf{v}_a|} - \cos i_{\text{orb}} = 0, \end{cases} \tag{2.49}$$

where subscript "a" marks the variables in the inertial coordinate system, $\mathbf{v}_a = \mathbf{v} + \boldsymbol{\omega} \times \mathbf{R}$, θ_a is the flight path angle to the local horizontal plane, i_{orb} is the orbit inclination. If the target orbit is circular with radius R_{orb}, then, taking into account the impulse on the target orbit, velocity v_{af} and injected mass m_{orb} are equal to

$$v_{af} = \sqrt{\frac{2}{R_f} \cdot \frac{1 - R_f/R_{\text{orb}}}{1 - \left(\cos\theta_{af} R_f/R_{\text{orb}}\right)^2}}, \tag{2.50}$$

$$m_{\text{orb}} = m_f \exp\left\{-\frac{1}{c}\left(\frac{1}{\sqrt{R_{\text{orb}}}} - v_{af}\frac{R_f}{R_{\text{orb}}}\cos\theta_{af}\right)\right\}. \tag{2.51}$$

If the ASV motion is studied in the vertical plane, then the last condition in (2.49) is absent. One of conditions (2.49) is the attribute of the right end of the trajectory (implicit definition of t_f), e.g.,

$$t_f\ :\ v_a\left(t_f\right) - v_{af} = 0. \tag{2.52}$$

Note that if an approximate recalculation of the conditions from the end of the active leg to the target orbit is used, then final time (2.52) corresponds exactly to the end of the active (beginning of the passive) launch leg.

The above simplifications are of a non-principled nature and are usually used in methodical, qualitative studies, in which a high-precision definition of the maximum injected mass is not required. When calculating the optimal trajectories of specific LVs, such simplifications are inappropriate, and more comprehensive motion model is used for optimization.

Slightly ahead of the presentation, we note that discontinuities in the phase vector and right parts of the phase system (2.18) are allowed at the internal trajectory points. Abrupt changes in mass are due to a staging, a release of cargo or structure elements. The jumps in the components of the velocity vector and radius vector can be associated, e.g., with the desire to introduce a simplified simulation for changes in the phase vector in STS staging process. Time moments t_*, at which there are jumps of the phase vector or discontinuities of the right-hand sides of the motion equations, in the general case depend on the phase vector and are determined implicitly from relations of the form

$$t_* : Q(\mathbf{x}(t_*), t_*) = 0. \tag{2.53}$$

2.3.5 *The Functional*

The purpose of the control is to minimize the ASV fuel mass consumption:

$$J \equiv m_i - m_{\text{orb}} \Rightarrow \min_{\{\mathbf{e}_\tau, \eta\}}, \tag{2.54}$$

for given constraints (2.18), (2.28–2.53). Hereinafter, the ASV initial mass is assumed to be fixed, unless otherwise specified. Then, the problem (2.54) is equivalent to the problem for the maximum injected mass:

$$J \equiv m_{\text{orb}} \Rightarrow \max_{\{\mathbf{e}_\tau, \eta\}}. \tag{2.55}$$

We will solve this problem with the help of PMP, using the main relations and properties of optimal solutions from Chap. 1.

2.4 The Conjugate System

In accordance with (1.6), (1.21), (1.29), the Hamiltonian of system (2.18), taking into account constraints (1.24), (2.29–2.45), has the form:

$$H = \mathbf{p}^{\mathrm{T}}\dot{\mathbf{x}} + \lambda G = \mathbf{p}_r^{\mathrm{T}}\mathbf{v} + \mathbf{p}_v^{\mathrm{T}}\left(\frac{\mathbf{T}+\mathbf{A}}{m} + \mathbf{g} + \boldsymbol{\Omega}\right) - p_m\mu + \lambda\left(f_\tau(\mathbf{x}) - \mathbf{e}_\tau^{\mathrm{T}}\mathbf{e}_v\right), \tag{2.56}$$

where vector of conjugate variables $\mathbf{p} = (\mathbf{p}_r, \mathbf{p}_v, p_m)^{\mathrm{T}}$ satisfies the system of equations

$$\begin{cases} \dot{\mathbf{p}}_r = -H_R\mathbf{e}_R + \left(1/R^3 - \omega^2\right)\mathbf{p}_v + (\mathbf{p}_v, \boldsymbol{\omega})\boldsymbol{\omega} - \lambda\left(\dfrac{\partial f_\tau}{\partial \mathbf{r}}\right)^{\mathrm{T}}, \\ \dot{\mathbf{p}}_v = -\mathbf{p}_r - H_v\mathbf{e}_v + \dfrac{a_r\rho}{m}(C_{Dv}\mathbf{p}_v + C_{LD}(\mathbf{p}_v, \mathbf{v})\mathbf{e}_\tau) + 2\mathbf{p}_v \times \boldsymbol{\omega} - \lambda\left[\left(\dfrac{\partial f_\tau}{\partial \mathbf{v}}\right)^{\mathrm{T}} + \dfrac{f_\tau\mathbf{e}_v - \mathbf{e}_\tau}{v}\right], \\ \dot{p}_m = \dfrac{1}{m}\left\{(\mathbf{p}_v, \mathbf{e}_\tau)\left(\dfrac{T}{m} - \dfrac{\partial T}{\partial m}\right) + \dfrac{a_r\rho}{m}\left[-C_{Dv}(\mathbf{p}_v, \mathbf{v}) + C_L^\alpha v^2(\mathbf{p}_v, \mathbf{e}_\tau)\right]\right\} + p_m\dfrac{\partial\mu}{\partial m} - \lambda\dfrac{\partial f_\tau}{\partial m}, \end{cases} \tag{2.57}$$

where

$$H_R = \frac{(\mathbf{p}_v, \mathbf{e}_\tau)}{m}\frac{\partial T}{\partial h} + \frac{a_r}{m}\left[\begin{matrix} v^2(\mathbf{p}_v, \mathbf{e}_\tau)\dfrac{\partial\left(\rho C_L^\alpha\right)}{\partial h} - (\mathbf{p}_v, \mathbf{v})\left(\begin{matrix} v\dfrac{d(\rho D_0)}{dh} \\ +(\mathbf{v}, \mathbf{e}_\tau)\dfrac{d(\rho C_{LD})}{dh}\end{matrix}\right) \\ +\dfrac{3(\mathbf{p}_v, \mathbf{e}_R)}{R^3} - p_m\dfrac{\partial\mu}{\partial h},\end{matrix}\right]$$

$$H_v = \frac{(\mathbf{p}_v, \mathbf{e}_\tau)}{m}\frac{\partial T}{\partial v} + \frac{a_r\rho}{m}\left[\begin{matrix}(\mathbf{p}_v, \mathbf{e}_\tau)v\left(2C_L^\alpha + M\dfrac{\partial C_L^\alpha}{\partial M}\right) \\ -(\mathbf{p}_v, \mathbf{v})\left(\dfrac{\partial(D_0M)}{\partial M} + \dfrac{\partial C_{LD}}{\partial M}\dfrac{(\mathbf{v}, \mathbf{e}_\tau)}{a_s}\right)\end{matrix}\right] - p_m\frac{\partial\mu}{\partial v},$$

$$C_{Dv} = D_0v + C_{LD}(\mathbf{v}, \mathbf{e}_\tau),\ C_{LD} = C_L^\alpha + D_\alpha,\quad \frac{d}{dh} = \left.\frac{\partial}{\partial h}\right|_{M=fix} - \frac{M}{a_s}\frac{da_s}{dh}\left.\frac{\partial}{\partial M}\right|_{h=fix};$$

$\mathbf{p}_r$, $\mathbf{p}_v$ and p_m are components of the conjugate vector corresponding to radius vector $\mathbf{r}$, velocity vector $\mathbf{v}$ and mass m; $h = R$-1 is the altitude; λ is the Lagrange function

$$\lambda = \begin{cases} 0, & G < 0, \\ \frac{1}{m}\left[\left(T - \tilde{D}_\alpha\right)(\mathbf{p}_v, \mathbf{e}_v) - \tau_* f_\tau\right] \le 0, & G = 0, \end{cases}$$

$$\tau_* = \frac{\left(T + \tilde{C}_L^\alpha\right)\sqrt{|\mathbf{p}_v|^2 - (\mathbf{p}_v, \mathbf{e}_v)^2}}{\sqrt{1 - f_\tau^2}}. \tag{2.58}$$

Taking into account remarks to (2.40–2.43), expressions (2.56–2.58) can also be used in the case of first-order phase constraints (2.35–2.43). In this case, one should keep in mind the change in the Lagrange multiplier when moving along the boundary ($\lambda \geq 0$ for $G = 0$).

When moving along the boundary (2.30), one should set $\frac{\partial T}{\partial \mathbf{x}} = \frac{\partial f_T}{\partial \mathbf{x}}$ in (2.57), (2.58).

Let us consider expressions for partial derivatives of f_τ for different constraints.

For the constraint on the angle of attack $\alpha_{\min}(M, h) \leq \alpha \leq \alpha_{\max}(M, h)$, $\alpha_{\min} < 0$:

$$\frac{\partial f_\tau}{\partial \mathbf{r}} = -\sin \alpha_{\max(\min)} \frac{d\alpha_{\max(\min)}}{dh} \mathbf{e}_R^{\mathrm{T}}, \frac{\partial f_\tau}{\partial \mathbf{v}} = \frac{\partial \alpha_{\max(\min)}}{\partial M} \frac{\mathbf{e}_v^{\mathrm{T}}}{a_s}, \frac{\partial f_\tau}{\partial m} = 0,$$

where $\frac{d}{dh} = \frac{\partial}{\partial h} - \frac{M}{a_s} \frac{\partial a_s}{\partial h} \frac{\partial}{\partial M}$.

For constraint $q|\alpha| \leq Q_{\text{adm}}(M)$:

$$\frac{\partial f_\tau}{\partial \mathbf{r}} = -f_\tau \cdot \sin\left(\frac{\bar{Q}}{a_r \rho v^2}\right)\left[\frac{d\bar{Q}}{dh} + \beta(h)\right]\mathbf{e}_R^{\mathrm{T}},$$

$$\frac{\partial f_\tau}{\partial \mathbf{v}} = \frac{f_\tau}{v} \cdot \sin\left(\frac{\bar{Q}}{a_r \rho v^2}\right)\left[2 - \frac{\partial \bar{Q}}{\partial M}\frac{1}{\bar{Q}}\right]\mathbf{e}_v^{\mathrm{T}}, \frac{\partial f_\tau}{\partial m} = 0,$$

$$\overline{Q} = \frac{Q_{\text{adm}}}{\rho_0 g_0 R_0}, \beta(h) = -\frac{1}{\rho}\frac{\partial \rho}{\partial h}.$$

For the constraint on the normal load factor n_z:

$$\frac{\partial f_\tau}{\partial \mathbf{r}} = 2(f_\tau - 1)\left[\frac{1}{\left(C_L^\alpha + C_{D0}\right)} \frac{d\left(C_L^\alpha + C_{D0}\right)}{dh} + \beta(h)\right]\mathbf{e}_R^{\mathrm{T}},$$

$$\frac{\partial f_\tau}{\partial \mathbf{v}} = 2(f_\tau - 1)\left(2 + \frac{M}{C_L^\alpha + C_{D0}} \frac{\partial\left(C_L^\alpha + C_{D0}\right)}{\partial M}\right)\frac{\mathbf{e}_v^{\mathrm{T}}}{v},$$

$$\frac{\partial f_\tau}{\partial m} = 1 - \frac{2f_\tau}{m}.$$

For the constraints on the dynamic pressure ($\kappa = 1$), heat flux ($\kappa = 3.25$) and speed ($\kappa^{-1} = 0$):

$$\frac{\partial\left(T - \tilde{D}_\alpha\right)}{\partial \mathbf{r}} = \left[\frac{\partial T}{\partial h} + \tilde{D}_\alpha \beta(h) - \frac{\tilde{D}_\alpha}{D_\alpha}\frac{dD_\alpha}{dh}\right]\mathbf{e}_R^{\mathrm{T}},$$

$$\frac{\partial\left(T - \tilde{D}_\alpha\right)}{\partial \mathbf{v}} = \left[\frac{\partial T}{\partial v} - \frac{\tilde{D}_\alpha}{v}\left(2 + v^2 \frac{M}{D_\alpha}\frac{\partial D_\alpha}{\partial M}\right)\right]\mathbf{e}_v^{\mathrm{T}},$$

$$\frac{\partial f_q}{\partial \mathbf{r}} = \left[\frac{\tilde{D}_0}{D_0}\frac{dD_0}{dh} - \tilde{D}_0\beta(h) + \frac{mv^2}{\kappa}(\mathbf{e}_R, \mathbf{e}_v)\left(\frac{\partial\beta(h)}{\partial h} - \frac{\beta(h)}{R}\right) - \frac{3m}{R^3}(\mathbf{e}_R, \mathbf{e}_v)\right]\mathbf{e}_R^{\mathrm{T}} + \\ + \left[\frac{1}{R^3} + \frac{v^2\beta(h)}{R\cdot\kappa} - \omega^2\right]m\mathbf{e}_v^{\mathrm{T}} + m\omega^2(\mathbf{e}_\omega, \mathbf{e}_v)\mathbf{e}_\omega^{\mathrm{T}},$$

$$\frac{\partial f_q}{\partial \mathbf{v}} = \left[\left(2 + \frac{M}{D_0}\frac{\partial D_0}{\partial M}\right)\frac{\tilde{D}_0}{v} + \frac{2mv}{\kappa}\beta(h)(\mathbf{e}_R, \mathbf{e}_v) - \frac{m}{v}\left(\frac{1}{R^2} + \frac{v^2\beta(h)}{\kappa}\right)(\mathbf{e}_R, \mathbf{e}_v) + \right. \\ \left. + \frac{m\omega^2 R}{v}((\mathbf{e}_R, \mathbf{e}_v) - (\mathbf{e}_\omega, \mathbf{e}_R)(\mathbf{e}_\omega, \mathbf{e}_v))\right]\mathbf{e}_v^{\mathrm{T}} - \frac{m\omega^2 R}{v}(\mathbf{e}_\omega, \mathbf{e}_R)\mathbf{e}_\omega^{\mathrm{T}} \\ + \frac{m}{v}\left[\left(\frac{1}{R^2} + \frac{v^2\beta(h)}{\kappa}\right) - \omega^2 R\right]\mathbf{e}_R^{\mathrm{T}},$$

$$\frac{\partial f_q}{\partial m} = \left(\frac{1}{R^2} + \frac{v^2\beta(h)}{\kappa}\right)(\mathbf{e}_R, \mathbf{e}_v) - \omega^2 R[(\mathbf{e}_R, \mathbf{e}_v) - (\mathbf{e}_\omega, \mathbf{e}_R)(\mathbf{e}_\omega, \mathbf{e}_v)],\ \mathbf{e}_\omega = \frac{\boldsymbol{\omega}}{|\boldsymbol{\omega}|}.$$

2.5 Transversality Conditions

According to (1.8), the transversality conditions at the left end of the trajectory ($t = t_i$) have the form:

(a) for problem (2.46)

$$\mathbf{p}(t_i) = \text{free}, \tag{2.59}$$

b) for problem (2.47)

$$\mathbf{p}_r(t_i) = \text{free},\ p_m(t_i) = \text{free},\ \mathbf{p}_v(t_i) \times \mathbf{v}(t_i) = 0. \tag{2.60}$$

At the right end, the conjugate variables must satisfy the transversality condition (see (1.9), (1.10), (1.21)):

$$\left(p_{mf} - 1\right)\delta m_f - H_f\delta t_f + \left(\mathbf{p}_{rf}, \delta\mathbf{r}_f\right) + \left(\mathbf{p}_{vf}, \delta\mathbf{v}_f\right) = 0, \tag{2.61}$$

on admissible variations of the phase vector corresponding to the constraint (2.49):

$$\begin{gathered}\left(\mathbf{v}_{af}, \delta\mathbf{v}_{af}\right) = 0,\ \left(\mathbf{R}_f, \delta\mathbf{r}_f\right) = 0,\ \left(\mathbf{v}_{af}, \delta\mathbf{r}_f\right) + \left(\mathbf{R}_f, \delta\mathbf{v}_{af}\right) = 0, \\ \left(\mathbf{v}_{af} \times \boldsymbol{\omega}, \delta\mathbf{r}_f\right) + \left(\boldsymbol{\omega} \times \mathbf{R}_f, \delta\mathbf{v}_{af}\right) = 0.\end{gathered} \tag{2.62}$$

For free variations $\delta m_f, \delta t_f$ the following conditions must be met:

$$Q_3 = p_{mf} - 1 = 0,\ H_f = 0. \tag{2.63}$$

From (2.61), (2.62) and taking into account that $\delta \mathbf{v} = \delta \mathbf{v}_a - \boldsymbol{\omega} \times \delta \mathbf{R}$, it follows that the vectors $\mathbf{p}_{rf} - \mathbf{p}_{vf} \times \boldsymbol{\omega}$ and $\mathbf{p}_{vf}$ can be represented as the following linear combinations:

$$\begin{aligned} \mathbf{p}_{arf} &= \mathbf{p}_{rf} - \mathbf{p}_{vf} \times \boldsymbol{\omega} = \zeta_1 \mathbf{R}_f + \zeta_3 \mathbf{v}_{af} + \zeta_4 \left(\mathbf{v}_{af} \times \boldsymbol{\omega} \right), \\ \mathbf{p}_{vf} &= \zeta_2 \mathbf{v}_{af} + \zeta_3 \mathbf{R}_f + \zeta_4 \left(\boldsymbol{\omega} \times \mathbf{R}_f \right). \end{aligned} \tag{2.64}$$

where $\zeta_{1,\ldots,4}$ are arbitrary multipliers.

Making cross product of the first equality by $\mathbf{R}_f$, and of the second one by $\mathbf{v}_{af}$ and adding them, we eliminate ζ_j ($j = 1, 2, 3$):

$$\mathbf{p}_{arf} \times \mathbf{R}_f + \mathbf{p}_{vf} \times \mathbf{v}_{af} = \zeta_4 \left(\mathbf{v}_{af} \times \mathbf{R}_f \times \boldsymbol{\omega} \right).$$

Whence we finally obtain that, in the general case, for $\mathbf{v}_{af} \times \mathbf{R}_f \times \boldsymbol{\omega} \neq 0$:

$$\begin{aligned} Q_4 &\equiv \left(\mathbf{p}_{arf}, \mathbf{R}_f \times \mathbf{v}_{af} \times \mathbf{R}_f \right) - \left(\mathbf{p}_{vf}, \mathbf{v}_{af} \times \mathbf{R}_f \times \mathbf{v}_{af} \right) = 0, \\ Q_5 &\equiv \left(\mathbf{p}_{arf}, \mathbf{R}_f \times \boldsymbol{\omega} \right) - \left(\mathbf{p}_{vf}, \mathbf{v}_{af} \times \boldsymbol{\omega} \right) = 0. \end{aligned} \tag{2.65}$$

In case $\omega = 0$, we have

$$\mathbf{P}_{af} \times \mathbf{R}_f + \mathbf{S}_f \times \mathbf{v}_{af} = 0. \tag{2.66}$$

At points of discontinuity of the phase vector and/or the right-hand sides of the motion equations (see (2.53)), the conjugate vector suffers a discontinuity [17]. If break moment t_* is determined by reaching a given mass:

$$Q(m(t_*)) = 0, \tag{2.67}$$

then at this moment the corresponding conjugate variable p_m breaks:

$$\mathbf{p}_r^+ = \mathbf{p}_r^-, \mathbf{p}_v^+ = \mathbf{p}_v^-, H^+ = H^-, p_m^+ = p_m^- + \lambda, \tag{2.68}$$

where $()^+$, $()^-$ are limits at $t = t_* + 0$ and $t = t_* - 0$ respectively.

The jump value can be found using the Hamiltonian continuity condition in (2.68):

$$\mu^+ p_m^+ = \mu^- p_m^- + \left(\frac{\boldsymbol{\Sigma}^-}{m^-} - \frac{\boldsymbol{\Sigma}^+}{m^+}, \mathbf{p}_v \right), \tag{2.69}$$

where $\boldsymbol{\Sigma} = \mathbf{T} + \mathbf{A}$.

In particular, when changing the composition of the propulsion system without mass dropping

$$p_m^+ = \frac{1}{\mu^+} \left(\mu^- p_m^- + \frac{\left(\boldsymbol{\Sigma}^- - \boldsymbol{\Sigma}^+, \mathbf{p}_v \right)}{m} \right). \tag{2.70}$$

At the corner points, e.g., when entering the constant overload segment, the conjugate vector is continuous:

$$\mathbf{p}^{+} = \mathbf{p}^{-}. \tag{2.71}$$

However, the right-hand sides of the conjugate system suffer a discontinuity in this case. Therefore, to ensure the required accuracy of numerical integration, the corner points must be determined from the solution of the nonlinear Eq. (2.67), for example, by the chord method, with sufficient accuracy (for "sufficient" accuracy, see Sect. 1.10.1).

2.6 Integrals of Motion

On sections of the trajectory where phase system (2.18) is autonomous, under optimal control the first integral (1.53) takes place:

$$H|_{\mathbf{u}_{\text{opt}}} = \text{const}. \tag{2.72}$$

If the autonomy of the system is preserved throughout the trajectory and the boundary conditions do not explicitly depend on time, then according to (1.54):

$$H \equiv 0, t \in \left[t_i, t_f\right]. \tag{2.73}$$

2.7 Control Optimality Conditions

In accordance with PMP, the optimal control is determined from conditions (1.7) and (2.28):

$$\{\mathbf{e}_\tau,\ \eta\}_{\text{opt}} = \arg \max_{\{\mathbf{e}_\tau, \eta\}} H\ . \tag{2.74}$$

Taking into account (2.18), (2.26–2.28), (2.56) and (2.74) we have (if $\mathbf{p}_\mathbf{v} \neq 0$):

$$\mathbf{e}_{\tau\,\text{opt}} = \arg \max_{\{\mathbf{e}_\tau\}} \left\{ \left[\left(T + \tilde{C}_L^\alpha\right)\mathbf{e}_s^{\text{T}} - \left(\tilde{C}_L^\alpha + \tilde{D}_\alpha\right)\mathbf{e}_v^{\text{T}}\mathbf{e}_s\mathbf{e}_v^{\text{T}} \right]\mathbf{e}_\tau \right\},$$

where $\tilde{C}_L^\alpha = a_r C_L^\alpha \rho v^2, \tilde{D}_\alpha = a_r D_\alpha \rho v^2, \mathbf{e}_s = \frac{\mathbf{p}_v}{|\mathbf{p}_v|}$,
from which

$$\mathbf{e}_{\tau\,\text{opt}} = \boldsymbol{\tau}/\tau, \boldsymbol{\tau} = \left(T + \tilde{C}_L^\alpha\right)\mathbf{e}_s - \left(\tilde{C}_L^\alpha + \tilde{D}_\alpha\right)(\mathbf{e}_v, \mathbf{e}_s)\mathbf{e}_v,$$

$$\tau = \sqrt{\left(T + \tilde{C}_L^{\alpha}\right)^2 \sin^2\zeta + \left(T - \tilde{D}_{\alpha}\right)^2 \cos^2\zeta}, \quad \cos\zeta = (\mathbf{e}_v, \mathbf{e}_s) \tag{2.75}$$

According to (2.75), in the case of the optimal flight without sliding, as well as in the case of independent optimal control by the thrust and aerodynamic force [18], the ASV symmetry plane passes through vectors $\mathbf{v}$ and $\mathbf{p}_{\mathbf{v}}$.

This result admits a simple geometric interpretation. Vector $\boldsymbol{\Sigma} = \mathbf{T} + \mathbf{A}$ is determined by current velocity vector $\mathbf{v}$, angle of attack α and orientation of the ASV symmetry plane passing through $\mathbf{v}$. Because $|\boldsymbol{\Sigma}|$ depends only on α, then at $\alpha = \text{fix}$ and changing the orientation of the unit vector of the normal to the ASV symmetry plane $\mathbf{e}_k$ vector $\boldsymbol{\Sigma}$ will describe the surface of a circular cone $\boldsymbol{K}$, whose axis passes through velocity vector $\mathbf{v}$ (Fig. 2.4). By virtue of (2.74), the optimal control corresponds to such position of vector $\boldsymbol{\Sigma}$ on surface $\boldsymbol{K}$, at which its projection onto vector $\mathbf{p}_v$ is maximum. This position of $\boldsymbol{\Sigma}_{\text{opt}}$ is determined by the intersection of plane $\boldsymbol{P}$, containing vectors $\mathbf{v}$ and $\mathbf{p}_v$, with $\boldsymbol{K}$. This means that under optimal control vector $\boldsymbol{\Sigma}$ must lie in plane $\boldsymbol{P}$ and, consequently, the ASV symmetry plane must pass through vectors $\mathbf{v}$ and $\mathbf{p}_v$.

Note that under initial conditions (2.47) due to (2.60) and (2.75),

$$\sin \alpha_{\text{opt}}(t_i) = |\mathbf{e}_v \times \mathbf{e}_\tau|_{t_i} = 0, \tag{2.76}$$

i.e., with the optimal orientation of the initial velocity at ASV take-off, the optimal control program begins with a zero angle of attack. Therefore, if condition (2.76) is not satisfied in the calculations of optimal trajectories, then this is one of the clear signs of an error in the program or in the problem statement.

Fig. 2.4 Geometric interpretation of the control optimality conditions

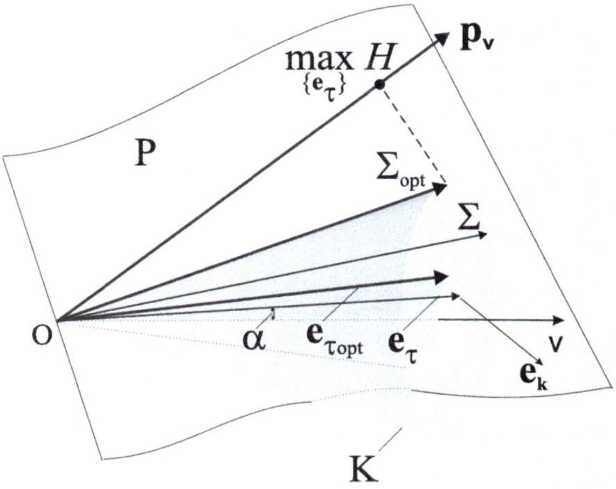

2.7.1 The Singular Attitude Control

The necessary condition for the existence of a singular ASV attitude control has the form of identity

$$\mathbf{p}_v(t) \equiv 0, t \in \mathbf{I} \subset \Re. \tag{2.77}$$

However, if the rate of mass change μ does not depend on the phase vector, which is typical for rocket engines, then for autonomous systems (2.18) the fulfillment of such a condition would lead to the trivial conjugate vector, which contradicts the PMP conditions. The proof of this statement and the analysis of the necessary conditions for the existence of the singular control of ASV attitude in the general case $\mu = \mu(\mathbf{x})$ are given below.

For existence of the singular control of $\mathbf{e}_\tau$, it is necessary that identity (2.77) takes place on some finite time interval $\mathbf{I}$. It follows from (2.77) that the following condition must also be satisfied on this interval:

$$\frac{d\mathbf{p}_v}{dt} \equiv 0, t \in \mathbf{I} \subset \Re.$$

According to conjugate system (2.57) this identity takes form

$$\frac{d\mathbf{p}_v}{dt} = -\mathbf{p}_r + p_m \frac{\partial \mu}{\partial v}\mathbf{e}_v \equiv 0.$$

Since this relation does not explicitly contain the control, we differentiate it with respect to time, taking into account (2.57) and (2.77):

$$\frac{d^2\mathbf{p}_v}{dt^2} = p_m \left\{ \begin{array}{l} -\dfrac{\partial \mu}{\partial h}\mathbf{e}_R + \left[\begin{array}{l} \dfrac{\partial \mu}{\partial m}\dfrac{\partial \mu}{\partial v} + \left(\dfrac{\partial^2 \mu}{\partial v^2} - \dfrac{1}{v}\dfrac{\partial \mu}{\partial v} \right)(\mathbf{g} + \boldsymbol{\Omega}, \mathbf{e}_v) \\ -\dfrac{\tilde{D}_0}{m}\dfrac{\partial^2 \mu}{\partial v^2} \end{array} \right] \mathbf{e}_v + \dfrac{1}{v}\dfrac{\partial \mu}{\partial v}(\mathbf{g} + \boldsymbol{\Omega}) + \\ + \left[\begin{array}{l} \left(\dfrac{\partial^2 \mu}{\partial v^2} - \dfrac{1}{v}\dfrac{\partial \mu}{\partial v} \right)(T + \tilde{C}_L^\alpha) - \\ \dfrac{\partial^2 \mu}{\partial v^2}(\tilde{C}_L^\alpha + \tilde{D}_\alpha) \end{array} \right] \dfrac{(\mathbf{e}_v, \mathbf{e}_\tau)}{m}\mathbf{e}_v + \dfrac{T + \tilde{C}_L^\alpha}{m}\dfrac{1}{v}\dfrac{\partial \mu}{\partial v}\mathbf{e}_\tau \end{array} \right\} \equiv 0$$

where $\tilde{C}_L^\alpha = a_r \rho v^2 C_L^\alpha$, $\tilde{D}_0 = a_r \rho v^2 D_0$, $\tilde{D}_\alpha = a_r \rho v^2 D_\alpha$.

The obtained identity is possible in two cases only, either.

(a) $p_m \equiv 0$,

or

(b)

$$-\frac{\partial\mu}{\partial h}\mathbf{e}_R+\begin{bmatrix}\dfrac{\partial\mu}{\partial m}\dfrac{\partial\mu}{\partial v}+\left(\dfrac{\partial^2\mu}{\partial v^2}-\dfrac{1}{v}\dfrac{\partial\mu}{\partial v}\right)(\mathbf{g}+\boldsymbol{\Omega},\mathbf{e}_v)\\-\dfrac{\tilde{D}_0}{m}\dfrac{\partial^2\mu}{\partial v^2}\end{bmatrix}\mathbf{e}_v+\frac{1}{v}\frac{\partial\mu}{\partial v}(\mathbf{g}+\boldsymbol{\Omega})+$$

$$+\begin{bmatrix}\left(\dfrac{\partial^2\mu}{\partial v^2}-\dfrac{1}{v}\dfrac{\partial\mu}{\partial v}\right)(T+\tilde{C}_L^\alpha)\\-\dfrac{\partial^2\mu}{\partial v^2}(\tilde{C}_L^\alpha+\tilde{D}_\alpha)\end{bmatrix}\frac{(\mathbf{e}_v,\mathbf{e}_\tau)}{m}\mathbf{e}_v+\frac{T+\tilde{C}_L^\alpha}{m}\frac{1}{v}\frac{\partial\mu}{\partial v}\mathbf{e}_\tau\equiv 0$$

Substituting a) into the expression for $\frac{d\mathbf{p}_v}{dt}$, we obtain $\mathbf{p}_r\equiv 0$. In combination with (2.77), this leads to the identically trivial conjugate vector, which contradicts the PMP. Thus, $\frac{d^2\mathbf{p}_v}{dt^2}$ can be identically zero only under condition b), that contains explicitly the control $\mathbf{e}_\tau$ (if $\frac{\partial\mu}{\partial v}\neq 0$), which can be found from the following linear matrix equation:

$$\mathbf{B}\mathbf{e}_\tau=\mathbf{b},$$

where

$$\mathbf{B}=\frac{T+\tilde{C}_L^\alpha}{m}\frac{1}{v}\frac{\partial\mu}{\partial v}\mathbf{E}+\left[\left(\frac{\partial^2\mu}{\partial v^2}-\frac{1}{v}\frac{\partial\mu}{\partial v}\right)\frac{(T+\tilde{C}_L^\alpha)}{m}-\frac{\partial^2\mu}{\partial v^2}\frac{(\tilde{C}_L^\alpha+\tilde{D}_\alpha)}{m}\right]\mathbf{e}_v\mathbf{e}_v^{\mathrm{T}};$$

$$\mathbf{b}=\frac{\partial\mu}{\partial h}\mathbf{e}_R+\left[\frac{\tilde{D}_0}{m}\frac{\partial^2\mu}{\partial v^2}-\frac{\partial\mu}{\partial m}\frac{\partial\mu}{\partial v}-\left(\frac{\partial^2\mu}{\partial v^2}-\frac{1}{v}\frac{\partial\mu}{\partial v}\right)(\mathbf{g}+\boldsymbol{\Omega},\mathbf{e}_v)\right]\mathbf{e}_v-\frac{1}{v}\frac{\partial\mu}{\partial v}(\mathbf{g}+\boldsymbol{\Omega});$$

$\mathbf{E}$ is the identity matrix.

As $\det\mathbf{B}\neq 0$, the singular control is

$$\mathbf{e}_\tau=\mathbf{B}^{-1}\mathbf{b}.$$

To invert matrix $\mathbf{B}$, we use the Sherman–Morrison–Woodbury formula [19]:

$$\mathbf{B}^{-1}=\frac{1}{k_1}\left[\mathbf{E}-\frac{k_2}{k_1+k_2}\mathbf{e}_v\mathbf{e}_v^{\mathrm{T}}\right].$$

Then

$$\mathbf{B}=k_1\mathbf{E}+k_2\mathbf{e}_v\mathbf{e}_v^{\mathrm{T}},$$

where

$$k_1=\frac{(T+\tilde{C}_L^\alpha)}{mv}\frac{\partial\mu}{\partial v},$$
$$k_2=\left(\frac{\partial^2\mu}{\partial v^2}-\frac{1}{v}\frac{\partial\mu}{\partial v}\right)\frac{(T+\tilde{C}_L^\alpha)}{m}-\frac{\partial^2\mu}{\partial v^2}\frac{(\tilde{C}_L^\alpha+\tilde{D}_\alpha)}{m},$$
$$k_1+k_2=\frac{\partial^2\mu}{\partial v^2}\frac{(T-\tilde{D}_\alpha)}{m}.$$

After that we can explicitly write the singular control as a linear combination of vectors $\mathbf{e}_R$, $\mathbf{e}_v$, $\mathbf{g}$, and $\boldsymbol{\Omega}$:

$$\mathbf{e}_\tau = k_R\mathbf{e}_R + k_v\mathbf{e}_v + k_g\mathbf{g} + k_\Omega\boldsymbol{\Omega},$$

where

$$\begin{aligned}
k_R &= \frac{mv\frac{\partial\mu}{\partial h}}{(T+\tilde{C}_L^\alpha)\frac{\partial\mu}{\partial v}};\\
k_v &= \frac{mv}{(T+\tilde{C}_L^\alpha)(T-\tilde{D}_\alpha)\frac{\partial\mu}{\partial v}\frac{\partial^2\mu}{\partial v^2}}\left\{-(\tilde{C}_L^\alpha+\tilde{D}_\alpha)\frac{\partial\mu}{\partial v}\frac{\partial^2\mu}{\partial v^2}\frac{(\mathbf{g}+\boldsymbol{\Omega},\mathbf{e}_v)}{v}+\right.\\
&\left.\frac{\partial\mu}{\partial h}(\mathbf{e}_v,\mathbf{e}_R)\left[\frac{T+\tilde{C}_L^\alpha}{v}\frac{\partial\mu}{\partial v}-(T+\tilde{D}_\alpha)\frac{\partial^2\mu}{\partial v^2}\right]+\frac{T+\tilde{C}_L^\alpha}{v}\frac{\partial\mu}{\partial v}\left(\frac{\tilde{D}_0}{m}\frac{\partial^2\mu}{\partial v^2}-\frac{\partial\mu}{\partial m}\frac{\partial\mu}{\partial v}\right)\right\};\\
k_g &= k_\Omega = -\frac{m}{(T+\tilde{C}_L^\alpha)},
\end{aligned}$$

must satisfy the normalization condition:

$$\begin{aligned}
&k_R^2+k_v^2+k_g^2g^2+k_\Omega^2\Omega^2+2\big[k_Rk_v(\mathbf{e}_v,\mathbf{e}_R)+k_Rk_g(\mathbf{e}_R,\mathbf{g})+\\
&+k_Rk_\Omega(\mathbf{e}_R,\boldsymbol{\Omega})+k_vk_g(\mathbf{e}_v,\mathbf{g})+k_vk_\Omega(\mathbf{e}_v,\boldsymbol{\Omega})+k_gk_\Omega(\mathbf{g},\Omega)\big]-1\equiv 0.
\end{aligned}$$

This condition does not contain the conjugate variables and defines the singular surface in the phase space.

Thus, for the existence of the singular control legs, it is necessary to fulfill simultaneously the following three vector conditions on phase and conjugate variables:

$$\begin{cases}
\mathbf{p}_v = 0,\\
-\mathbf{p}_r + p_m\frac{\partial\mu}{\partial v}\mathbf{e}_v = 0,\\
k_R^2+k_v^2+k_g^2g^2+k_\Omega^2\Omega^2+2\big[k_Rk_v(\mathbf{e}_v,\mathbf{e}_R)+k_Rk_g(\mathbf{e}_R,\mathbf{g})+\\
+k_Rk_\Omega(\mathbf{e}_R,\boldsymbol{\Omega})+k_vk_g(\mathbf{e}_v,\mathbf{g})+k_vk_\Omega(\mathbf{e}_v,\boldsymbol{\Omega})+k_gk_\Omega(\mathbf{g},\boldsymbol{\Omega})\big]-1\equiv 0.
\end{cases}$$

If we add the first integral (2.72)

$$H = \mathbf{p}_\mathbf{r}^\mathrm{T}\mathbf{v} + \mathbf{p}_\mathbf{v}^\mathrm{T}\mathbf{v} - p_m\mu \equiv \text{const} = c_H,$$

then we get that on singular legs of the extremal

$$\left(\frac{\partial\mu}{\partial v}v-\mu\right)p_m = c_H.$$

If $\frac{\partial\mu}{\partial v}v-\mu\neq 0$, i.e. the fuel mass consumption rate is not proportional to the flight speed, then the conjugate variables are completely specified as functions of the phase variables:

$$\mathbf{p}_v \equiv 0, \quad \mathbf{p}_r = \frac{c_H}{\frac{\partial \mu}{\partial v} v - \mu} \cdot \frac{\partial \mu}{\partial v} \mathbf{e}_v, \quad p_m = \frac{c_H}{\frac{\partial \mu}{\partial v} v - \mu}.$$

If system (2.18) is autonomous and the boundary conditions (including those at internal points) do not explicitly depend on time, then $c_H = 0$ and the conjugate vector is identically trivial, which contradicts the PMP. Consequently, for autonomous systems in the general case the optimal trajectory contains no singular attitude control legs.

Nevertheless, let us consider the physical meaning of the trivial conjugate variables. The physical meaning of condition $\mathbf{p}_v \equiv 0$ is that small variations of the velocity vector in the singular control segment do not affect the functional (the injected mass): $\mathbf{p}_v = \left[\frac{\partial m_f}{\partial \mathbf{v}}\right]^{\mathrm{T}} \equiv 0$

Also, this relation can be interpreted as a necessary condition for the optimality of the current velocity vector value.

Similarly, the second condition $\mathbf{p}_r \equiv 0$ means the independence (in the small) of the functional from the current flight altitude and geographic coordinate values:

$$\mathbf{p}_r = \left[\frac{\partial m_f}{\partial \mathbf{r}}\right]^{\mathrm{T}} \equiv 0.$$

Also, it means the probable existence of regions of the phase space with the optimal altitude and geographic coordinates.

Condition $p_m \equiv 0$ means the independence (in the small) of the functional from the current mass:

$$p_m = \frac{\partial m_f}{\partial m} \equiv 0.$$

Based on the physical interpretation of the necessary conditions for the existence of a singular control segments on the extremal, it may seem that such flight modes on the optimal trajectory are not realizable. However, in fact, this conclusion is not so evident. As calculations show, there are locally optimal values of velocity vector orientation $\mathbf{e}_{v\ \mathrm{opt}}$, flight altitude h_{opt} and mass m_{opt}.

In particular, the existence of $\mathbf{e}_{v\ \mathrm{opt}}$ is demonstrated in the next Chapters when the optimization problem statement with free (optimized) initial unit vector $\mathbf{e}_v(t_i)$ is considered.

The existence of optimal altitude h_{opt} and mass m_{opt} follows from Fig. 2.5, which shows the dependences of the dimensionless sensitivity coefficient $p_h = \frac{\partial m_f}{\partial h} = (\mathbf{p}_r, \mathbf{e}_R)$ (for air launch from the initial altitude of $h_i = 17.22$ km) and the sensitivity coefficient $p_m = \frac{\partial m_f}{\partial m}$ (for the air launch with initial thrust-to-weight ratio $n_i = 0.9$) on the maximum lift-to-drag ratio at subsonic speeds K_{sub} (maximum lift-to-drag ratio at supersonic speeds $K_{\mathrm{sup}} = 1$). It can be seen that there are such values K^*_{sub} depending on the ASV parameters and initial conditions under which $\frac{\partial m_f}{\partial h} = 0$ or $\frac{\partial m_f}{\partial m} = 0$. Assuming the continuity of the dependence of conjugate variables on a parameter

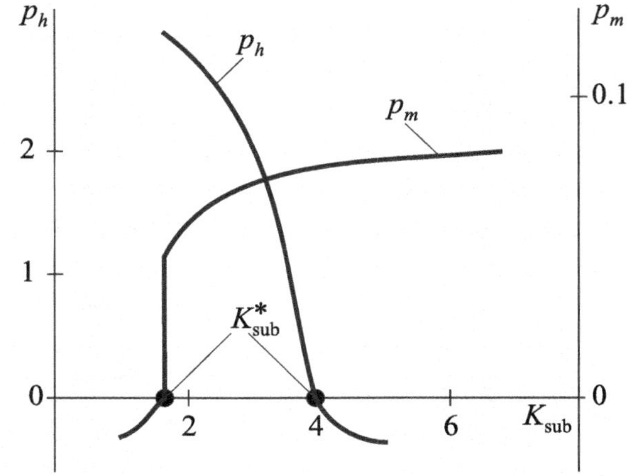

Fig. 2.5 Sensitivity coefficients $p_h = \frac{\partial m_f}{\partial h}$ and $p_m = \frac{\partial m_f}{\partial m}$ vs the ASV maximum lift-to-drag ratio K_{sub} at subsonic speeds

(initial conditions), it is obvious that in the vicinity of the found "singular" points there can be a singular surface in the phase space, which are a generalization of singular extremals studied by Miele [20].

2.7.2 Control and Phase Constraints

When moving along the boundary (2.29), (2.40), the optimal attitude control is determined from condition (2.74) for extended Hamiltonian (2.56), (2.57):

$$\mathbf{e}_{\tau\,\text{opt}} = (\boldsymbol{\tau} - \lambda m\mathbf{e}_v)/|\boldsymbol{\tau} - \lambda m\mathbf{e}_v|$$
$$= \frac{T + \tilde{C}_L^\alpha}{\tau_*}\left[\mathbf{e}_s - \left(\cos\varsigma - \frac{f_\tau}{\sqrt{1 - f_\tau^2}}\sin\varsigma\right)\mathbf{e}_v\right],$$
$$= \frac{\sin\bar{\alpha}\mathbf{e}_s + \sin(\varsigma - \bar{\alpha})\mathbf{e}_v}{\sin\varsigma}, \quad \mathbf{e}_s = \frac{\mathbf{p}_v}{|\mathbf{p}_v|}$$
$$\lambda = \frac{|\mathbf{p}_v|}{m}\left[(T - \tilde{D}_\alpha)\cos\varsigma - (T + \tilde{C}_L^\alpha)\cot\bar{\alpha}\sin\varsigma\right], \cos\bar{\alpha} = f_\tau, \sin\bar{\alpha} = \sqrt{1 - f_\tau^2}, \tag{2.78}$$

where in case (2.40–2.43) one should set $f_\tau = f_q/(T - \tilde{D}_\alpha)$.

Thus, the optimal ASV orientation when moving along boundary (2.29), (2.40) is determined by maintaining the maximum admissible angle of attack $\bar{\alpha}$, while the ASV symmetry plane passes through vectors $\mathbf{v}$ and $\mathbf{p}_v$.

Note that the optimality condition for the ASV orientation on boundary (2.29) does not contain a singularity, since in this case $\varsigma > \bar{\alpha} > 0$.

2.7.3 The Optimal Thrust Control

To determine the optimal control of the thrust magnitude, consider the part of the Hamiltonian H_u, which depends on the control $\mathbf{u}$:

$$H_u = \frac{|\mathbf{p}_v|}{m}\left[(T + \tilde{C}_L^\alpha)\mathbf{e}_s^T - (\tilde{C}_L^\alpha + \tilde{D}_\alpha)\mathbf{e}_v^T\mathbf{e}_s\mathbf{e}_v^T\right]\mathbf{e}_\tau - p_m\mu. \tag{2.79}$$

For $\mathbf{e}_\tau = \mathbf{e}_{\tau\,\text{opt}}$, $\mu = T/c$ we have

$$\begin{aligned} H_u &= \frac{|\mathbf{p}_v|\tau}{m} - \frac{p_m T}{c}, \\ \frac{\partial H_u}{\partial T} &= \frac{|\mathbf{p}_v|\left(T + \tilde{C}_L^\alpha \sin^2\varsigma - \tilde{D}_\alpha \cos^2\varsigma\right)}{m\tau} - \frac{p_m}{c}, \\ \frac{\partial^2 H_u}{\partial T^2} &= \frac{|\mathbf{p}_v|\left(\tilde{C}_L^\alpha + \tilde{D}_\alpha\right)^2 \sin^2 2\varsigma}{4m\tau^3} \geq 0. \end{aligned} \tag{2.80}$$

Likewise, when moving along boundary (2.40):

$$H_u = \frac{\tau_*(T)}{m} - p_m\frac{T}{c},\ \tau_*(T) = \sqrt{(T + \tilde{C}_L^\alpha)^2|\mathbf{p}_v|^2\sin^2\varsigma + \left(T - \tilde{D}_\alpha\right)^2(|\mathbf{p}_v|\cos\varsigma + \lambda m)^2}\ ,$$

$$\frac{\partial H_u}{\partial T} = \frac{(T + \tilde{C}_L^\alpha)|\mathbf{p}_v|^2\sin^2\varsigma + \left(T - \tilde{D}_\alpha\right)(|\mathbf{p}_v|\cos\varsigma + \lambda m)^2}{m\tau_*} - \frac{p_m}{c},$$

$$\frac{\partial^2 H_u}{\partial T^2} = \frac{|\mathbf{p}_v|^2\sin^2\varsigma(|\mathbf{p}_v|\cos\varsigma + \lambda m)^2(\tilde{C}_L^\alpha + \tilde{D}_\alpha)^2}{m\tau_*^3} \geq 0. \tag{2.81}$$

Thus, if

$$(\mathbf{p}_v \nparallel \mathbf{e}_v) \wedge (\mathbf{p}_v \perp \mathbf{e}_v) \wedge (\rho v \neq 0) \tag{2.82}$$

then $H_u(T)$ is a convex function and can reach its maximum only at limit points of the admissible interval:

$$T_{\text{opt}} = \begin{cases} T_{\min}, & s < 0, \\ T_{\max}, & s > 0, \end{cases} \tag{2.83}$$

where switching function

$$\begin{aligned} s = H_u(T_{\max}) - H_u(T_{\min}) &= \frac{2|\mathbf{p}_v|}{m}\frac{\bar{T} + \tilde{C}_L^\alpha\sin^2\varsigma - \tilde{D}_\alpha\cos^2\varsigma}{\tau(T_{\max}) + \tau(T_{\min})} - \frac{p_m}{c}, \\ \bar{T} &= (T_{\max} + T_{\min})/2. \end{aligned}$$

If

$$(\mathbf{p_v}||\mathbf{e}_v) \vee (\mathbf{p_v}\perp\mathbf{e}_v) \vee (\rho v = 0), \tag{2.84}$$

then

$$H_u = T \cdot s_0,\, s_0 = \frac{|\mathbf{p}_v|}{m} - \frac{p_m}{c}, \tag{2.85}$$

where the form of s_0 coincides with the switching function for the case of motion in a vacuum (see, e.g., [1, 15] and compare with (2.14) in the OEL solution). In this case:

$$T_{\text{opt}} = \begin{cases} T_{\min}, & s_0 < 0, \\ T_{\max}, & s_0 > 0, \\ \tilde{T} \in [T_{\min}, T_{\max}], & s_0 \equiv 0, \end{cases} \tag{2.86}$$

where $\tilde{T}$ is the singular control of the thrust magnitude.

2.7.4 The Singular Thrust Control

The singular control of the thrust magnitude $T_{\text{opt}} = \tilde{T}$ in the atmosphere is possible, according to (2.84–2.86), if one of the groups of conditions is met either

$$\begin{cases} (\mathbf{e}_s, \mathbf{e}_v) = 1, \\ s_0 \equiv 0, \end{cases} \tag{2.87}$$

or

$$\begin{cases} (\mathbf{e}_s, \mathbf{e}_v) = -1, \\ s_0 \equiv 0, \end{cases} \tag{2.88}$$

or

$$\begin{cases} (\mathbf{e}_s, \mathbf{e}_v) = 0, \\ s_0 \equiv 0. \end{cases} \tag{2.89}$$

Together with (2.84), we should consider the identity:

$$\frac{ds_0}{dt} = \frac{(\mathbf{e}_s, \dot{\mathbf{p}}_v)}{m} + \frac{|\mathbf{p}_v|T}{m^2 c} - \frac{\dot{p}_m}{c} + \frac{p_m}{c^2}\left[\frac{\partial c}{\partial h}(\mathbf{v}, \mathbf{e}_R) + \frac{\partial c}{\partial v}(\dot{\mathbf{v}}, \mathbf{e}_v)\right] \equiv 0. \tag{2.90}$$

Since, after the transformations, thrust does not explicitly enter into condition (2.90), to determine the singular control $\tilde{T}$, it is necessary to consider the second derivative of s_0 with respect to time:

$$\frac{d^2 s_0(\tilde{T})}{dt^2} = 0,\ \tilde{T} \in [T_{\min}, T_{\max}.]$$

In case (2.87), we have $\mathbf{p}_v \uparrow\uparrow \mathbf{e}_v$ and taking into account (2.75):

$$\mathbf{e}_{\tau\ \mathrm{opt}} \equiv \mathbf{e}_v. \tag{2.91}$$

Condition (2.91) means optimality of the flight with zero angle of attack.

Case (2.88), $\mathbf{p}_v \uparrow\downarrow \mathbf{e}_v$, corresponds to the optimal flight with angle of attack of 180°:

$$\mathbf{e}_{\tau\ \mathrm{opt}} \equiv -\mathbf{e}_v. \tag{2.92}$$

Case (2.89) corresponds to the optimal flight with angle of attack of 90°.

Let us consider case (2.87). Taking into account conditions for optimal attitude (2.75), we have

$$\mathbf{e}_{\tau\mathrm{opt}} \equiv \mathbf{e}_v,\ \tau = \left(T - \tilde{D}_\alpha\right)|\mathbf{p}_v|.$$

Then, using conjugate system (2.57) we find

$$\begin{aligned}
&\dot{\mathbf{p}}_v = -\mathbf{p}_r - \mathbf{e}_v\left\{\tfrac{|\mathbf{p}_v|}{m}\left(\tfrac{\partial T}{\partial v} - 2\tfrac{\tilde{C}_{D0}}{v} - \tfrac{a_r \rho v^2}{a_s}\tfrac{\partial \tilde{C}_{D0}}{\partial M}\right) - p_m \tfrac{\partial \mu}{\partial v}\right\} - \tfrac{\partial \boldsymbol{\Omega}}{\partial \mathbf{v}}\mathbf{p}_v,\\
&\dot{p}_m = \tfrac{|\mathbf{p}_v|}{m^2}\left[\left(T - \tilde{C}_{D0}\right) - m\tfrac{\partial T}{\partial m}\right] + p_m \tfrac{\partial \mu}{\partial m},\\
&\tilde{C}_{D0} = \tilde{D}_0 + \tilde{D}_\alpha.
\end{aligned}$$

Substituting these equations and (2.87) into (2.90) we get

$$\begin{aligned}
&\frac{ds_0}{dt} = -\frac{1}{m}(\mathbf{p}_r, \mathbf{e}_v) + \frac{|\mathbf{p}_v|}{m^2}\left[\frac{\tilde{C}_{D0}}{v}\left(2 + \frac{v}{c}\right) + \frac{a_r \rho v^2}{a_s}\frac{\partial \tilde{C}_{D0}}{\partial M}\right] +\\
&+\frac{p_m}{c^2}\left[\left[(\mathbf{g}, \mathbf{e}_v) + (\boldsymbol{\omega} \times \mathbf{R} \times \boldsymbol{\omega}, \mathbf{e}_v) - \frac{\tilde{C}_{D0}}{m}\right]\frac{\partial c}{\partial v} + (\mathbf{v}, \mathbf{e}_R)\frac{\partial c}{\partial h} + \mu\frac{\partial c}{\partial m}\right] \equiv 0.
\end{aligned}$$

Let $\frac{\partial c}{\partial m} \equiv 0$ then

$$\begin{aligned}
&\frac{ds_0}{dt} = -\frac{1}{m}(\mathbf{p}_r, \mathbf{e}_v) + \frac{|\mathbf{p}_v|}{m^2}\left[\frac{\tilde{C}_{D0}}{v}\left(2 + \frac{v}{c}\right) + \frac{a_r \rho v^2}{a_s}\frac{\partial \tilde{C}_{D0}}{\partial M}\right] +\\
&+\frac{p_m}{c^2}\left[\left[(\mathbf{g}, \mathbf{e}_v) + (\boldsymbol{\omega} \times \mathbf{R} \times \boldsymbol{\omega}, \mathbf{e}_v) - \frac{\tilde{C}_{D0}}{m}\right]\frac{\partial c}{\partial v} + (\mathbf{v}, \mathbf{e}_R)\frac{\partial c}{\partial h}\right] \equiv 0.
\end{aligned}$$

If the system is autonomous

$$H \equiv (\mathbf{p}_r, \mathbf{v}) + \left[\frac{(T - \tilde{C}_{D0})}{m} + (\mathbf{g} + \boldsymbol{\Omega}, \mathbf{e}_v)\right]|\mathbf{p}_v| - p_m \frac{T}{c} \equiv 0,$$

then

$$(\mathbf{p}_r, \mathbf{v}) + |\mathbf{p}_v|\left[-\frac{\tilde{C}_{D0}}{m} + (\mathbf{g} + \boldsymbol{\Omega}, \mathbf{e}_v)\right] \equiv 0.$$

Thus, we obtain the necessary condition for existence of the singular thrust control in the form of the following system of linear homogeneous algebraic equations with respect to $(\mathbf{p}_r, \mathbf{v})$, $\frac{|\mathbf{p}_v|}{m}$ and $\frac{p_m}{c}$:

$$\begin{cases} \frac{|\mathbf{p}_v|}{m} - \frac{p_m}{c} = 0, \\ -(\mathbf{p}_r, \mathbf{v}) + \left[\tilde{C}_{D0}\left(2 + \frac{v}{c}\right) + a_r \rho v^2 M \frac{\partial \tilde{C}_{D0}}{\partial M}\right]\frac{|\mathbf{p}_v|}{m} + \\ +\left[\frac{m}{c}(\mathbf{g} + \boldsymbol{\omega} \times \mathbf{R} \times \boldsymbol{\omega}, \mathbf{v})\frac{\partial c}{\partial v} + \frac{mv}{c}(\mathbf{v}, \mathbf{e}_R)\frac{\partial c}{\partial h} - \tilde{C}_{D0}\frac{v}{c}\frac{\partial c}{\partial v}\right]\frac{p_m}{c} = 0, \\ (\mathbf{p}_r, \mathbf{v}) + \left[-\tilde{C}_{D0} + m(\mathbf{g} + \boldsymbol{\Omega}, \mathbf{e}_v)\right]\frac{|\mathbf{p}_v|}{m} = 0. \end{cases}$$

Suppose that system determinant

$$\Delta = \begin{vmatrix} 0 & 1 & -1 \\ -1 & \tilde{C}_{D0}\left(2 + \frac{v}{c}\right) + a_r \rho v^2 M \frac{\partial \tilde{C}_{D0}}{\partial M} & \frac{m}{c}(\mathbf{g} + \boldsymbol{\omega} \times \mathbf{R} \times \boldsymbol{\omega}, \mathbf{v})\frac{\partial c}{\partial v} + \frac{mv}{c}(\mathbf{v}, \mathbf{e}_R)\frac{\partial c}{\partial h} - \tilde{C}_{D0}\frac{v}{c}\frac{\partial c}{\partial v} \\ 1 & -\tilde{C}_{D0} + m(\mathbf{g} + \boldsymbol{\Omega}, \mathbf{e}_v) & 0 \end{vmatrix} \neq 0$$

then the solution is

$$|\mathbf{p}_v| \equiv 0, \quad P_m \equiv 0, \quad (\mathbf{p}_r, \mathbf{v}) \equiv 0,$$

and taking into account conjugate system (2.57), we obtain that the conjugate vector is identically trivial, which contradicts the PMP. Thus, if the singular control segment exists, the following identity must hold:

$$\Delta = \tilde{C}_{D0}\left(1+\frac{v}{c}\right) + m(\mathbf{g}+\boldsymbol{\omega}\times\mathbf{R}\times\boldsymbol{\omega},\mathbf{e}_v)\left(1+\frac{v}{c}\frac{\partial c}{\partial v}\right) +$$
$$+\, a_r\rho v^2 M\frac{\partial \tilde{C}_{D0}}{\partial M} + \frac{mv}{c}(\mathbf{v},\mathbf{e}_R)\frac{\partial c}{\partial h} - \tilde{C}_{D0}\frac{v}{c}\frac{\partial c}{\partial v} \equiv 0. \tag{2.93}$$

The condition (2.93) defines a hypersurface in the phase space of dimension seven.

In a particular case of the Goddard problem on the vertical ascent of a rocket, the singular surface (2.93) coincides with the well-known Tsien-Evans solution [21].

Outside the atmosphere, the singular surface is described by equation $\Delta = m(\mathbf{g}+\boldsymbol{\Omega},\mathbf{e}_v) = 0$, which corresponds to motion along an "equipotential" surface in a gravitational field: $(\dot{\mathbf{v}}|_{T=0},\mathbf{v}) = 0$.

For the central gravitational field without Earth rotation

$$\Delta = \tilde{c}_{x0}\left[1+\frac{v}{c}\left(1-\frac{\partial c}{\partial v}\right)\right] - mg\sin\theta\left[1+\frac{v}{c}\left(\frac{\partial c}{\partial v}-\frac{v}{g}\frac{\partial c}{\partial h}\right)\right] + a_r\rho v^2 M\frac{\partial c_{x0}}{\partial M} = 0.$$

For rocket engines in the isothermal atmosphere

$$c = c_\infty - \Delta c\cdot\rho,$$

where $c_\infty = \text{const}$, $\Delta c = \text{const}$, $\rho = e^{-\beta h}$, $\beta = \text{const}$.

Neglecting the dependence of C_{D0} on the Mach number and the change of g, we obtain

$$\Delta = a_r C_{D0}\rho v^2\left(1+\frac{v}{c}\right) - m\sin\theta\left(1-v^2\beta\frac{\Delta c}{c}\rho\right) = 0.$$

It can be seen that in the exoatmospheric flight segment ($\rho = 0$), the singular thrust control can only be possible when moving with $\theta = 0$. Indeed, when moving out of the atmosphere in a direction orthogonal to the gravitational field, there are no losses in the characteristic velocity, and the injected mass does not depend on duration of the flight, but only on the relative mass consumption and engine specific impulse.

The last equation can be represented as a quadratic equation for density

$$A\rho^2 - B\rho + C = 0,\ A = \sigma_{x0}v^2\Delta c,$$
$$B = \sigma_{x0}v^2(c_\infty + v) + \sin\theta\cdot\Delta c\left(1+v^2\beta\right),\ C = \sin\theta\cdot c_\infty,$$

where $\sigma_{x0} = \frac{a_r C_{D0}}{m}$.

From where we find two solutions:

$$\rho_{1,2} = \begin{cases} \frac{2C}{B+\sqrt{B^2-4AC}}, \\ \frac{B+\sqrt{B^2-4AC}}{2A}. \end{cases}$$

The projections of the singular surface (2.93) onto the altitude-velocity phase plane are shown in Fig. 2.6 for various flight path angles θ. The following values of dimensionless parameters are used here: $\sigma_{x0} = 50$, $c_\infty = 0.45$, $\Delta c = 0.1$, $\beta = 888.56$. These sections of the singular surface are analogous to the Miele degenerate extremals [5].

At low speed $\frac{v}{c} << \left|1 + \frac{\beta\Delta\bar{c}}{\sigma_{x0}}\sin\theta\right|$ and density $\rho << \left(\frac{\sigma_{x0}}{\sin\theta\cdot\beta\Delta\bar{c}} + 1\right)\frac{1}{\Delta\bar{c}}$ singular surface (2.93) degenerates into a surface with a constant dynamic pressure:

$q = \rho v^2 \approx const = \tilde{q} = \frac{1}{\frac{\sigma_{x0}}{\sin\theta} + \beta\Delta\bar{c}}$, $\Delta\bar{c} = \frac{\Delta c}{c_\infty}$.

It can be seen that for $\sin\theta > 0$, the singular surface (2.93) lies in the area of the dynamic pressure, limited above by the inequality $\rho v^2 \le \tilde{q}_{\max} = \max\limits_{\sin\theta > 0} \tilde{q} = \frac{1}{\sigma_{x0} + \beta\Delta\bar{c}}$.

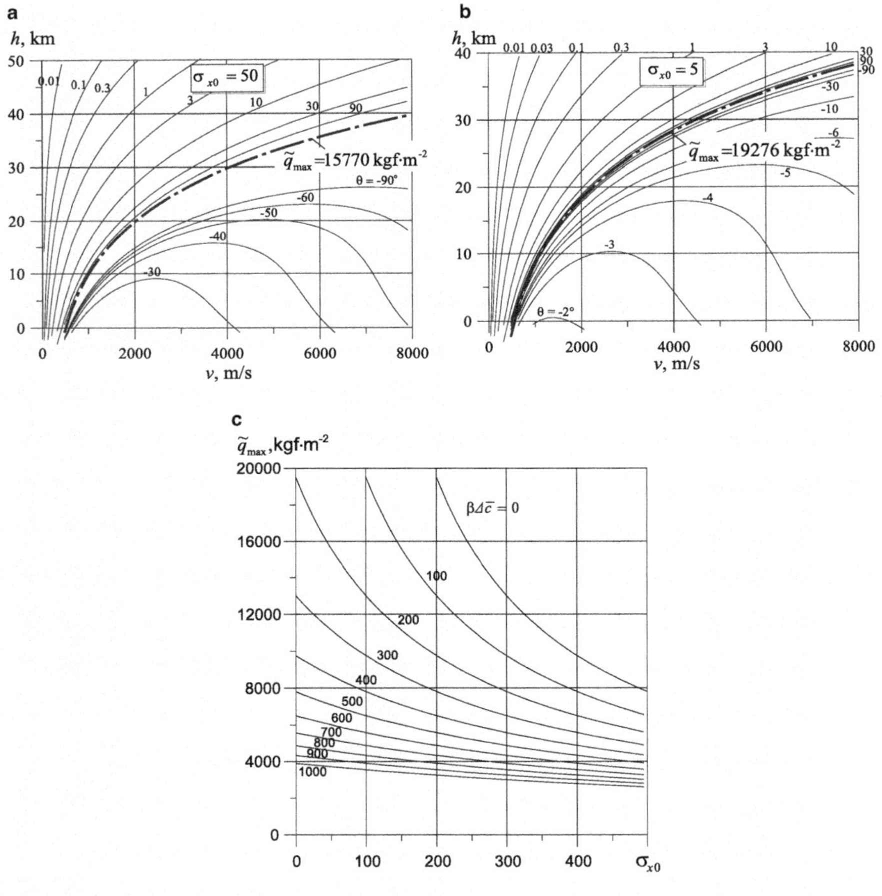

Fig. 2.6 Projections of singular surface (2.93) onto the phase plane {velocity-altitude} for different flight path angles θ; and upper limits $\tilde{q}_{\max}$ for dynamic pressure at the singular extremals vs ballistic coefficient σ_{x0}

Relationship $\tilde{q}_{\max}(\sigma_{x0}, \beta\Delta\overline{c})$ is shown in Fig. 2.6. For $\sin\theta = 0$, from (2.93) we obtain the limiting case $v = 0$.

Thus, the degenerate extremals in the plane $\{v, h\}$ with the variation θ in the range $0 \leq \theta \leq \pi/2$ fill densely the velocity range $0 \leq v \leq \sqrt{\frac{\tilde{q}_{max}}{\rho(h)}}$ (see Fig. 2.6). The singular surfaces corresponding to the second solution of the quadratic equation lie in the region of high dynamic pressure, which is out the assumption above.

2.8 Non-classical Constraints

2.8.1 Constraint on Attitude Angular Rate

Taking into account the constraint on the angular rate of ASV attitude change (2.44) for the considered model of trajectory motion is beyond the scope of PMP, since it constraints the rate of control change, while PMP is based on the use of needle control variations. Therefore, it is possible to limit the angular rate only by preliminary regularization of the statement.

In particular, suppose that the need for such the constraint arises only in a vicinity of moments of take-off and staging, when there is a jump in the ASV parameters included into the optimality condition for unit vector $\mathbf{e}_{\tau\text{opt}}$ (2.75). Moreover, according to [1, 2], in rarefied layers of the atmosphere (as a rule, the staging occurs there) vector $\boldsymbol{\tau}$ in (2.75) changes linearly in time, and $\mathbf{e}_{\tau\text{opt}}$ remains in one plane (so-called "control plane").

In accordance with optimality conditions (2.74) on the segment where constraint (2.45) is substantial

$$\mathbf{e}_{\tau\ \text{opt}}(t) = \arg\max_{\mathbf{e}_\tau \subset \boldsymbol{K}} H = \arg\max_{\mathbf{e}_\tau \subset \boldsymbol{K}} (\mathbf{e}_\tau, \tau), \tag{2.94}$$

where $\boldsymbol{K}$ is a circular cone with axis $\mathbf{e}_\tau^{-}(t_*)$ and half angle $\Delta\vartheta = \omega_{\max}(t - t_*)$ (Fig. 2.7).

To provide the maximum projection on $\boldsymbol{\tau}$, vector $\mathbf{e}_\tau(t)$, by virtue of (2.94), lies at the intersection of the cone $\boldsymbol{K}$ surface and plane $\boldsymbol{\Pi}$, containing vectors $\mathbf{e}^{-}_{\tau\text{opt}}(t_*)$ and $\mathbf{e}_{\tau\text{opt}}(t) = \frac{\boldsymbol{\tau}}{\tau}$ (the optimal ASV orientation without the constraint on the angular rate). Thus,

$$\mathbf{e}_\tau(t) = \xi_1 \cdot \mathbf{e}^{-}_{\tau\text{opt}}(t_*) + \xi_2 \cdot \mathbf{e}_{\tau\text{opt}}(t), \tag{2.95}$$

where coordinates ξ_1, ξ_2 are determined from the conditions of normalization $\mathbf{e}_\tau(t)$ and belonging of this vector to the surface of cone $\boldsymbol{K}$:

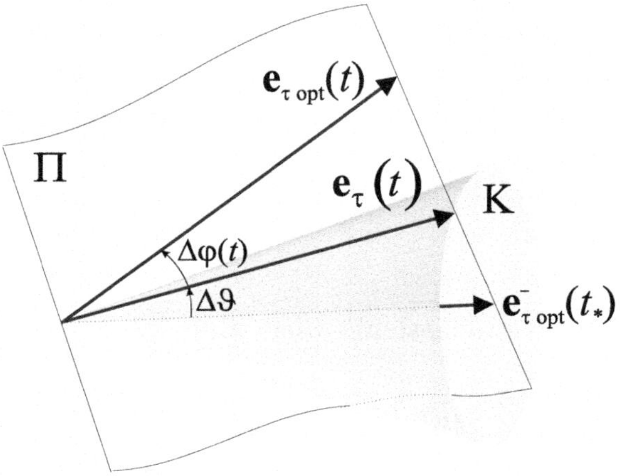

Fig. 2.7 Geometrical interpretation of the optimality condition for the ASV attitude vector with constraint on the angular rate

$$\begin{cases} \xi_1 = \dfrac{\left|\left[\mathbf{e}_\tau(t), \mathbf{e}_{\tau\text{opt}}(t)\right]\right|}{\left|\left[\mathbf{e}_\tau^-(t_*), \mathbf{e}_{\tau\text{opt}}(t)\right]\right|} = \dfrac{\sin(\Delta\varphi(t))}{\sin\varphi(t)}, \quad \Delta\varphi(t) = \varphi(t) - \Delta\vartheta, \\ \xi_2 = \dfrac{\left|\left[\mathbf{e}_\tau(t), \mathbf{e}_\tau^-(t_*)\right]\right|}{\left|\left[\mathbf{e}_\tau^-(t_*), \mathbf{e}_{\tau\text{opt}}(t)\right]\right|} = \dfrac{\sin(\Delta\vartheta)}{\sin\varphi(t)}, \end{cases}$$

where $\varphi(t)$ is the current angle between vectors $\mathbf{e}_{\tau\text{opt}}(t)$ and $\mathbf{e}^-_{\tau\text{opt}}(t_*)$.

Moment t^{**} of departure the constraint on the angular velocity is defined using an iterative procedure for solving the nonlinear operator equation for residual:

$$\varepsilon\left(t^{**}\right) = 0,$$

for example, by the chord method, in which

$$\varepsilon = \left|\left[\mathbf{e}^-_{\tau opt}\left(t^*\right), \mathbf{e}_{\tau opt}\left(t^{**}\right)\right]\right| - \sin\left(\omega_{\max} \cdot \left(t^{**} - t^*\right)\right). \tag{2.96}$$

2.8.2 Conditions of Moment Balancing

Engines can be located above or below the ASV longitudinal axis. To balance the pitching moment of the thrust vector, its orientation program, as a rule, is formed so that nominally (in the absence of other forces) the thrust vector is directed to the center of mass. As a result, the thrust vector during ASV ascent can be deviated from the longitudinal axis by an angle φ_E, which we will call the engine installation angle.

The engine installation angle can be tens of degrees, which is commensurate with the operating range of the pitch angle change on the optimal injection trajectory and have a noticeable effect on the optimal trajectory, control program and functional. Therefore, the ballistic analysis of such asymmetric layouts should be carried out taking this factor into account already at the early design stages.

Let us consider the trajectory optimization technique [3, 4] with a generalization to the case when the engine installation angle is not equal to zero and can change along the trajectory depending on the phase variables (mass, Mach number, etc.) and, possibly, the flight time.

Let us consider the model of motion in the vertical plane, which is described by Eqs. (2.18) under the assumption that $\mathbf{\Omega} = 0$, in the Cartesian coordinate system associated with the start point.

The orientation of ASV and thrust vector is determined by the vectors:

$\mathbf{e}_\tau$ is the unit vector in the direction of the longitudinal axis;

$\mathbf{e}_n$ is the unit vector in the direction of the normal axis;

$\mathbf{e}_T$ is the unit thrust vector.

Then, the angle φ_E is defined as follows:

$$\cos\varphi_E = (\mathbf{e}_\tau, \mathbf{e}_T), \sin\varphi_E = (\mathbf{e}_n, \mathbf{e}_T). \tag{2.97}$$

In accordance with (2.97), the thrust vector is

$$\mathbf{T} = T \cdot \mathbf{e}_T = T(\cos\varphi_E \cdot \mathbf{e}_\tau + \sin\varphi_E \cdot \mathbf{e}_n). \tag{2.98}$$

The coefficients of aerodynamic lift C_L and drag C_D will be represented as (compare with (2.22))

$$C_L = C_{L0} + C_L^\alpha \sin\alpha, C_D = D_0 + D_\alpha \cos\alpha, \tag{2.99}$$

where C_{L0}, C_L^α, D_o, D_α are functions of the Mach number, α is the angle of attack, determined by equalities: $\cos\alpha = (\mathbf{e}_\tau, \mathbf{e}_v)$, $\sin\alpha = -(\mathbf{e}_n, \mathbf{e}_v)$.

The vector of aerodynamic forces can be represented as follows (see (2.21), (2.26)):

$$\mathbf{A} = a_r \rho v^2 \big[C_L^\alpha \mathbf{e}_\tau - \big(D_\alpha + C_L^\alpha\big)(\mathbf{e}_\tau, \mathbf{e}_v)\mathbf{e}_v - D_0 \mathbf{e}_v + C_{L0} \cdot \mathbf{e}_L\big], \tag{2.100}$$

where $\mathbf{e}_L$ is the unit lift vector.

When considering the ASV motion in the vertical plane, it is convenient to represent the unit vectors $\mathbf{e}_n$ and $\mathbf{e}_L$ by introducing the rotation matrix of the vector by angle π/2:

$$\mathbf{\Pi} = \begin{pmatrix} 0 & -1 \\ 1 & 0 \end{pmatrix}.$$

Taking into account conditions $(\mathbf{e}_n, \mathbf{e}_\tau) = 0$ and $(\mathbf{e}_L, \mathbf{e}_v) = 0$ we will have:

$$\mathbf{e}_n = k\Pi \cdot \mathbf{e}_\tau,\, \mathbf{e}_L = k\Pi \cdot \mathbf{e}_v, \tag{2.101}$$

where k can take the values $\pm$ 1, which physically corresponds to two ASV roll angles γ allowed by the motion model considered here:

$$\gamma = \begin{cases} 0, k = 1, \\ 180^\circ, k = -1. \end{cases} \tag{2.102}$$

In solving the optimization problem, it is advisable to take the unit vector $\mathbf{e}_\tau$ and the integer variable k as independent control functions.

We single out the part Hamiltonian (2.56) of system (2.18), that depends explicitly on control vector $\mathbf{u} = (\mathbf{e}_\tau, k)$:

$$H_u = \frac{\mathbf{p}_v^T}{m}\left\{\begin{bmatrix} \left(T\cos\varphi_E + a_r\rho v^2 C_L^\alpha\right)\mathbf{E} - \\ - a_r\rho v^2\left(D_\alpha + C_L^\alpha\right)\mathbf{e}_v\mathbf{e}_v^T + \\ + T\sin\varphi_E k\boldsymbol{\Pi} \end{bmatrix}\mathbf{e}_\tau + a_r\rho v^2 C_{L0} k\boldsymbol{\Pi}\mathbf{e}_v\right\}, \tag{2.103}$$

where $\mathbf{E}$ is the 2×2 identity matrix.

For optimal control, from condition (2.74), taking into account (2.103), we have:

$$\begin{aligned} &\mathbf{e}_{\tau\,\mathrm{opt}} = \boldsymbol{\tau}_{\mathrm{opt}}(k)/\tau_{\mathrm{opt}}(k), \\ &\boldsymbol{\tau}_{opt}(k) = \left(T\cos\varphi_E + a_r\rho v^2 C_D^\alpha\right)\mathbf{p_v} - a_r\rho\left(D_\alpha + C_D^\alpha\right)(\mathbf{v}, \mathbf{p_v})\mathbf{v} - T\sin\varphi_E \cdot k\Pi\mathbf{p_v}. \end{aligned} \tag{2.104}$$

If an instantaneous change in the roll angle by 180° is allowed, then the choice of optimal control k_{opt} is based on a comparison of the Hamiltonian at $k = \pm$ 1:

$$k_{\mathrm{opt}} = \arg\max_{k=\pm 1} H_u\left(\mathbf{e}_{\tau\,\mathrm{opt}}(k), k\right),$$

or

$$k_{\mathrm{opt}} = \mathrm{sign}\left[H_u\left(\mathbf{e}_{\tau\,\mathrm{opt}}(+1), +1\right) - H_u\left(\mathbf{e}_{\tau\,\mathrm{opt}}(-1), -1\right)\right]. \tag{2.105}$$

In accordance with (2.56), the conjugate system has form

$$\begin{cases} \dot{\mathbf{p}}_r = -\frac{\partial \mathbf{T}^{\mathrm{T}}}{\partial \mathbf{r}} \cdot \frac{\mathbf{p}_v}{m} - \frac{\partial \mathbf{A}^{\mathrm{T}}}{\partial \mathbf{r}} \cdot \frac{\mathbf{p}_v}{m} - \frac{\partial \mathbf{g}^{\mathrm{T}}}{\partial \mathbf{r}} \cdot \mathbf{p}_v, \\ \dot{\mathbf{p}}_v = -\mathbf{p}_r - \frac{\partial \mathbf{T}^{\mathrm{T}}}{\partial \mathbf{v}} \cdot \frac{\mathbf{p}_v}{m} - \frac{\partial \mathbf{A}^{\mathrm{T}}}{\partial \mathbf{v}} \cdot \frac{\mathbf{p}_v}{m}, \\ \dot{p}_m = \frac{(\mathbf{T}+\mathbf{A})^{\mathrm{T}}}{m^2} \cdot \mathbf{p}_v. \end{cases} \tag{2.106}$$

We note the differences between this conjugate system and the one given in (2.57). Here, the engine installation angle and aerodynamic lift at zero angle of attack $\mathbf{L}_0 = a_r\rho v^2 C_{L0} \cdot \mathbf{e}_L$ are taken into account, which leads to a change in vectors $\mathbf{T}$, $\mathbf{A}$ and the following terms in (2.106):

$$\begin{aligned}\frac{\partial \mathbf{T}^{\mathrm{T}}}{\partial \mathbf{r}} \cdot \frac{\mathbf{p}_v}{m} &= \frac{1}{m}\frac{\partial T}{\partial h}(\mathbf{e}_T, \mathbf{p}_v)\mathbf{e}_R, \\ \frac{\partial \mathbf{T}^{\mathrm{T}}}{\partial \mathbf{v}} \cdot \frac{\mathbf{p}_v}{m} &= \frac{1}{m}\frac{\partial T}{\partial v}(\mathbf{e}_T, \mathbf{p}_v)\mathbf{e}_v,\end{aligned} \tag{2.107}$$

where, according to (2.98) and (2.101): $\mathbf{e}_T = (k\boldsymbol{\Pi} \sin \varphi_E + \mathbf{E} \cos \varphi_E)\mathbf{e}_\tau$,
as well as adding terms:

$$\begin{aligned}\frac{\partial \mathbf{L}_0^{\mathrm{T}}}{\partial \mathbf{r}} \cdot \mathbf{p}_v &= \frac{a_r v^2}{m}\frac{\partial T}{\partial h}\left(\rho \frac{\partial L_0}{\partial M} \cdot \frac{M}{a_s} \cdot \frac{\partial a_s}{\partial h}\right)(\mathbf{e}_L, \mathbf{p}_v)\mathbf{e}_R, \\ \frac{\partial \mathbf{L}_0^{\mathrm{T}}}{\partial \mathbf{v}} \cdot \mathbf{p}_v &= \frac{a_r \rho}{m}\left[L_0(\mathbf{v}, \mathbf{p}_v)\mathbf{e}_L - \left(2 \cdot L_0 + M\frac{\partial L_0}{\partial M}\right)(\mathbf{e}_L, \mathbf{p}_v)\mathbf{v} - vL_0 k\boldsymbol{\Pi}\mathbf{p}_v\right].\end{aligned} \tag{2.108}$$

2.9 Features of Boundary Value Problem Solution with Phase Constraints

Taking into account the constraint on the phase vector in form (2.35), one should pay attention to the following features to ensure the regularity of the BVP procedure solving.

1. Although additional transversality condition (1.29) is specified at the same point where varied parameter λ is introduced, this condition can only be satisfied by varying one of the (n–1) varied components of initial conjugate vector $\mathbf{p}^i$ (if the initial phase vector is fixed). Parameter λ (together with the remaining (n–2) components of $\mathbf{p}^i$) ensures the fulfillment of (n–1) boundary conditions (including transversality conditions) at the right end of the trajectory.
2. It is inconvenient to use conditions (1.29), (1.30) to determine time moment t_* of reaching the constraint. The left side in (1.30) is the difference between Hamiltonians with limit and optimal controls: $\Delta H = H(\overline{\mathbf{u}}) - H(\mathbf{u}_{\text{opt}})$. Due to optimality condition $\mathbf{u}_{\text{opt}} = \arg\max\limits_{\mathbf{u}} H$, inequality $\Delta H(t) \leq 0$ is satisfied on the entire trajectory. If in a vicinity of t_* the smoothness of the time dependence of the Hamiltonian is not violated, then $\left.\frac{dH}{dt}\right|_{t_*} = 0$, which excludes the possibility of using quasilinear methods for numerically solving equation $\Delta H(t_*) = 0$.
3. To find t_* one cannot directly use the condition of reaching the boundary of admissible area $\Gamma = \{\mathbf{x} : f_0(\mathbf{x}) - f_{0\,\text{adm}} = 0\}$, since in the iterative process of solving the BVP on some of the trial trajectories, especially in a small vicinity of the solution, this condition may not be satisfied at all at any point of the trajectory. It means that one of the residuals will be undefined, which will lead to convergence violation.
4. If the local maximum of function $f_0(t)$ is pronounced, and constraint (2.35) is essential, then for the definition t_* it is proposed to use condition $\left.\frac{df_0}{dt}\right|_{t_*-0} = 0$ (see (1.38)). In this case, a smooth exit to the constraint will be realized on the optimal trajectories. Then, the additional boundary condition (the residual) at

moment t_* is the condition for function f_0 to reach the given admissible value at the maximum point: $f_0(t_*) - f_{0adm} = 0$.

On Figs. 2.8 and 2.9 the example of taking into account the dynamic pressure limitation (2.37) on the optimal ASV injection trajectory with an optimized initial flight path angle is shown. (The ASV parameters correspond to the rocket stage of ASTS-1, described in Sect. 6.3). The time moment t_{**} of the end of the boundary section is determined from condition $G(\mathbf{u}_{opt}(t_{**})) = \frac{dq}{dt}(\mathbf{u}_{opt}(t_{**})) = 0$. Lagrange multiplier λ at boundary $q = q_{max}$ has a negative sign, because in the calculations the condition of motion along the boundary was given in form $G \leq 0$. As can be seen from the presented results (Fig. 2.8), in order to fulfill more strict phase constraint, it is advantageous not so much to increase the duration $\Delta t = t_{**} - t_*$ of the motion along the boundary as to reorganize the ascent trajectory as a whole accordingly. Changes in the time-derivative of the dynamic pressure $\frac{dq}{dt}(\mathbf{u}_{opt}, t)$ for the unconstrained optimal control $\mathbf{u}_{opt}$ are shown on Fig. 2.9. Recall that under unconstrained optimal control we assume here a control that satisfies PMP with a Lagrange multiplier $\lambda = 0$. The time when this derivative goes to zero for the first time corresponds to t_*, i.e. reaching the constraint. This derivative goes to zero the second time at the moment t_{**} of leaving this constraint. Of course, the optimal control on interval (t_*, t_{**}) provides in fact motion along the boundary with constant dynamic pressure, and unconstrained optimal control is used in this interval only to accurately find time moment t_{**}. The real behaver of the Lagrange function in a vicinity of the constrained trajectory segment is shown on Fig. 2.9. The correctness of taking into account such a phase constraint can be indirectly controlled by observing the change in the first integral. In this case, the permissible local error is determined in accordance with the methodology given in Sect. 1.10, by recalculating this error to the functional increment at the right end of the optimal trajectory. According to Fig. 2.9, the relative error of conservation of zero Hamiltonian on the constrained segment of the optimal trajectory is less than 10^{-10}. Taking into account the relative duration of this segment, we get that the absolute integration error accumulated at the final point could be about 0.03 kg of the injected mass, which is significantly less than acceptable in these calculations (~ 10 kg).

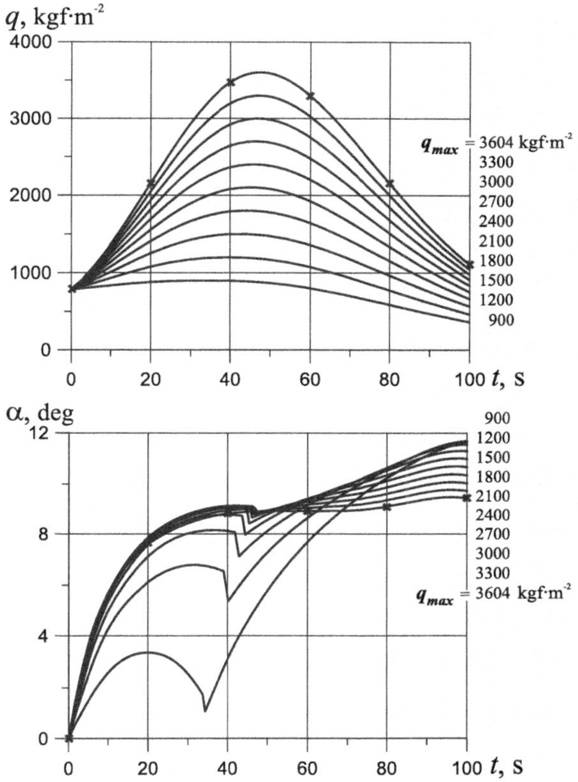

Fig. 2.8 The changes in the optimal time-dependences of dynamic pressure q and angle of attack α due to the constraint on dynamic pressure; "x" marks the optimal dependencies without the constraint

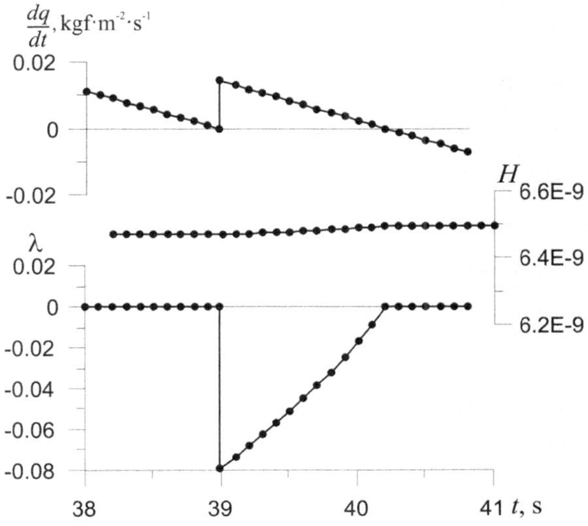

Fig. 2.9 The changes in the conventional time-derivative of the dynamic pressure for the unconstrained control; the Hamiltonian and the Lagrange function on the optimal trajectory with the constraint on the dynamic pressure

References

1. Okhotsimsky, D.E., Eneev, T.M.: Some variation problems related to launching artificial earth satellites. Usp. Fiz. Nauk **63**(1a), 5–32 (1957). (in Russian)
2. Lawden, D.F.: Optimal trajectories for space navigation. Butterworths, London (1963)
3. Filatyev, A.S.: Optimal launch of an artificial earth satellite using aerodynamic forces. Cosm. Res. **29**(2), 255–271 (1991)
4. Filatyev, A.S.: Optimization of spacecraft ascent using aerodynamic forces. IAF-92–0022, 43th IAF Congress, Aug. 28-Sep. 5, 1992, Washington, USA
5. Miele, A.: Optimum climbing technique for a rocket-powered aircraft. Jet Propulsion **25**(8), 385–391 (1955)
6. Filatyev, A.S.: Paradoxes of optimal solutions in problems of space vehicle injection and reentry. Acta Astronaut. **47**(1), 11–18 (2000)
7. Filatyev, A.S.: Reduction of maximum dynamic and thermal loads at passive reentry of hypersonic vehicles with help of small control actions. Uchenye Zapiski TsAGI. **9**(2) (1978). (in Russian)
8. Filatyev, A.S.: Approximate analytical synthesis of optimal control of hypersonic aircraft when moving in the atmosphere with sub-circular speed, parts 1 and 2. Uchenye Zapiski TsAGI. **11**(1 and 2) (1980) (in Russian)
9. Filatyev, A.S., Yanova, O.V.: Nonequilibrium reentry. ICAS-2.2–1, 21st ICAS Congress, September 7–14, Sorrento, Italy (1996)
10. Kemp, N., Riddell, F.: Heat transfer to satellite vehicles reentry atmosphere. Jet Propulsion. **21**(2) (1957)
11. Detra, R., Kemp, N., Riddell, F.: Addendum to "Heat Transfer To Satellite Vehicles Re-entering the Atmosphere. Jet Propulsion, pp. 1256 (1957)
12. Pontryagin, L.S., Boltyansky, V.G., Gamkrelidze, R.V., Mischenko, E.F.: The mathematical theory of optimal processes. Interscience Publishers, NY (1962)
13. Filatyev, A.S.: Through optimization of branching trajectories of launching aerospace vehicles in the atmosphere based on the Pontryagin maximum principle. Dissertation for the degree of Doctor of Technical Sciences, p. 395 (2001) (in Russian)
14. Filatyev, A.S.: Optimization of branched trajectories for aerospace transport systems, ICAS-94–5.2.3, 19th ICAS Congress, 18–23 September 1994, Anaheim CA, USA
15. Gobetz, F.W., Doll, J.R.: A survey of impulsive trajectories. AIAA J. **7**(8) (1969)
16. Ilyin, V.A., Kuzmak, G.E.: Optimal flights of spacecraft with high thrust engines. Nauka Publisher, Moscow, p. 744 (1976) (in Russian)
17. Bryson, A.E., Jr., Ho, Y.C.: Applied optimal control. Hemisphere Publishing Corporation, Washington, DC (1975)
18. Vinh, N.X.: General theory of optimal trajectory for rocket flight in a resisting medium. JOTA **11**, 189 (1973)
19. Golub, G.H., Van Loan, C.F.: Matrix computations. JHU Press, p. 756 (2013)
20. Cicala, P., Miele, A.: General theory of the optimum thrust programming for the level flight of a rocket powered aircraft. Jet Propulsion. **26**(6) (1956)
21. Tsien, H.S., Evans, R.C.: Optimum thrust programming for a sounding rocket. J. Am. Rocket Soc. **201**(5), 99–107 (1951)

Chapter 3
Qualitative Analysis of the Aerodynamic Forces Effect on the Optimal Ascent Trajectories and Control

3.1 Introduction

The strategy for optimal control of the ASV thrust vector is determined by the ratio of thrust **T**, aerodynamic **A** and gravitational **G** forces. If the thrust dominates, then the optimal control law and ascent trajectory qualitatively correspond to the classical OEL optimal solution [1, 2] (see Sect. 2.1) obtained in absence of aerodynamic forces. However, even if $|\mathbf{T}| \gg |\mathbf{A}|$ and $|\mathbf{G}| \gg |\mathbf{A}|$, but $|\mathbf{T} + \mathbf{G}| \approx |\mathbf{A}|$ (see Fig. 3.1), which is typical, e.g., for initial flight legs of LVs with vertical take-off, the presence of aerodynamic forces can qualitatively change the structure of optimal control laws and trajectories and, in particular, generate additional local extremals [3–5].

The first analytical solutions to the problem of optimal ASV injection in a uniform gravitational field [1, 2] were published back in the late 1950s and early 1960s. In [1, 6] there is also an approximate analysis of the influence of a gravitational field nonuniformity and aerodynamic drag. The qualitative influence of aerodynamic drag on the structure of the optimal thrust control of a rocket at a vertical ascent is shown in [7, 8]. Qualitative changes, including those of the bifurcation type, in the structure of the optimal thrust vector laws and launch trajectories with variations in the ASV aerodynamic lifting properties, apparently, were first demonstrated using rigorous solutions based on PMP in [3, 4, 9–12]. Subsequently [5], the accumulated data made it possible to introduce a classification of such types of solutions.

When solving optimal control problems with help of PMP, the necessary optimality conditions are used. Therefore, the solutions are, in the general case, only local extremals. As for the global optimum, this problem requires a special study.

Let us agree on the term "global extremal" used below. In the monograph, this term is used primarily to identify the extremal that provides the best value of the functional from several revealed local extremals. Unfortunately, this issue is rarely discussed constructively. In practical research, local extremals are often ignored (if they are obtained at all), which have an "exotic" character, do not correspond to traditional solutions and are inferior to them in the functional. However, if local extremals are

A. S. Filatyev and V. G. Petukhov, *Through Optimization of Aerospace Vehicle Trajectories by the Pontryagin Maximum Principle*, Studies in Systems, Decision and Control 579, https://doi.org/10.1007/978-3-031-80756-5_3

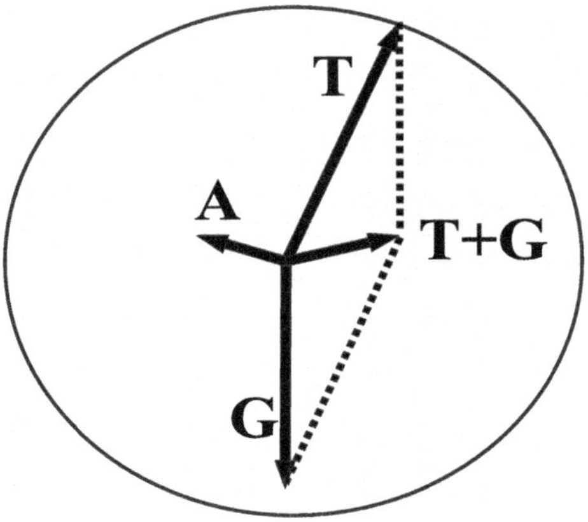

Fig. 3.1 Scheme of forces acting on an aerospace vehicle in presence of atmosphere that illustrate the possibility of small aerodynamic force **A** in comparison with weight **G** and thrust **T** to qualitatively effect on the optimal trajectory and control law

obtained on the basis of rigorous methods, then they, including "exotic" ones, reflect the objective properties of dynamic systems and therefore, on the contrary, require increased attention, research and systematization. Different types of local extremals are characterized by qualitatively different optimal control laws. When changing the values of the problem parameters, a local extremal can become global and vice versa, which is a characteristic property of optimization problems. Thus, the identification of various locally optimal solutions makes it possible to discover fundamentally new structures of optimal trajectories and control laws, which can lead to new design solutions.

This chapter presents the main results of such studies carried out by the author and confirming the qualitative influence of aerodynamic forces on optimal control laws, ascent trajectories, and sensitivity functions of the optimal functional to variations in main parameters of the problem.

Optimization of ASV ascent trajectories under initial conditions (2.47) allows determining, simultaneously with the optimal control program, also the optimal flight path angle $\theta_{i\ \mathrm{opt}}$ and azimuth $\eta_{i\ \mathrm{opt}}$ at the start moment, and, when initial conditions are specified in the general form (2.48), to find the best point in the admissible initial set. These possibilities sometimes turn out to be useful both in practical applications and for methodological purposes. In this chapter, the optimization of the initial velocity vector makes it possible to get rid of the "side effect" of the initial conditions and, thus, "refine" the study of the qualitative influence on the optimal solutions of the aerodynamic forces themselves. Moreover, the aerodynamic forces do not directly depend on the flight path angle, which is a "fast" variable at low flight speeds. The optimal initial flight path angle is an objective parameter that characterizes an extremal as a whole and has a clear physical meaning. Therefore, as will be shown below, using $\theta_{i\ \mathrm{opt}}$ turns out to be convenient for describing qualitative transformations of extremals and understanding the nature of such changes. Note that the concept of the optimal take-off angle retains its meaning even in the case when the take-off conditions are fixed and the optimal initial path angle is not reachable! In this case, parameter $\theta_{i\,\text{opt}}$ characterizes tendencies of the optimal control in the initial atmospheric flight leg to correct the initial conditions specified not in the best way.

3.2 Aerodynamic Characteristics Influence on Optimal Ascent Trajectories and Control Laws

Aerodynamic forces (2.47) with polar (2.22), (2.23) depend on three parameters: maximum lift-to-drag ratio $K_{\max} = \max\limits_{\alpha} C_L/C_D$ (the aerodynamic quality), dimensionless ballistic coefficient $\overline{\sigma}_x = a_r C_{D0}$, and relative induced drag factor $\overline{A} = \frac{A}{C_{D0}}$. The thrust will be characterized by conditional initial thrust-to-weight ratio n_i^{∞}, equal to the ratio of the maximum vacuum thrust to the initial ASV weight. When starting in dense atmosphere layers, its pressure reduces the thrust, so the real initial thrust-to-weight ratio is always less than n_i^{∞}.

In the parametric studies described below, the calculations were performed using *ASTER* software package.

3.2.1 Classification of Extremals Depending on AeroSpace Vehicle Lift-to-Drag Ratio

The strategy for the optimal thrust vector control is determined by the ratio of forces acting on the vehicle: the thrust, aerodynamic and gravitational forces. If the thrust dominates over the aerodynamic forces, then the optimal launch trajectory and control law qualitatively correspond to the classical OEL solutions obtained without taking into account the atmosphere. However, if the sum of the thrust vector and gravitational force is much less than the thrust in absolute value, as, e.g., in the initial section of a typical LV ascent with a vertical take-off, then the aerodynamic forces can qualitatively change the structure of the optimal control law and trajectory.

Figures 3.2, 3.3 and 3.4 illustrate this effect on examples of the optimal insertion of a hypothetical LV launched from a carrier aircraft (CA) into a circular orbit with the altitude of 275 km. The initial conditions and LV parameters are as follows: altitude $h_i = 12$ km, velocity $v_i = 180$ m/s, mass $m_i = 146{,}870$ kg, thrust-to-weight ratio (in vacuum) $n_i^{\infty} = 0.9$, specific impulse $I_{\infty}/I_0 = 461/360$ s, dimensionless ballistic coefficient $\overline{\sigma}_x = 26.61$.

Figure 3.2 shows the optimal initial flight path angle depending on the LV maximum lift-to-drag ratio. It is assumed that dependence $C_D(\alpha)$ is fixed, $K_{\max}$ does not depend on a flight mode, variations of $K_{\max}$ do not affect other LV parameters. It can be seen that in the shaded area ($1.05 < K_{\max} < 1.35$) there exist simultaneously three local extremals corresponding to sections ***A***, ***B***, and ***M***, which are characterized by different ranges of the optimal initial flight path angle. The bifurcation point $K_{\max}^{bif}$

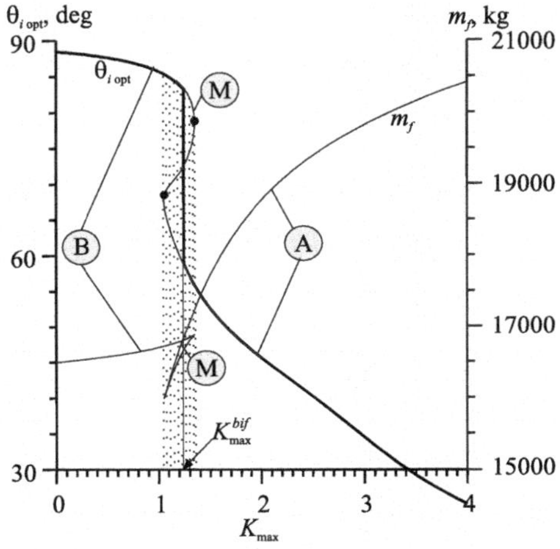

Fig. 3.2 The bifurcation of optimal path angle $\theta_{i\ opt}$ at the take-off moment with variation of maximum lift-to-drag ratio K_{max}; and the nonsmooth many-valued dependence of maximum injected mass m_f on K_{max}

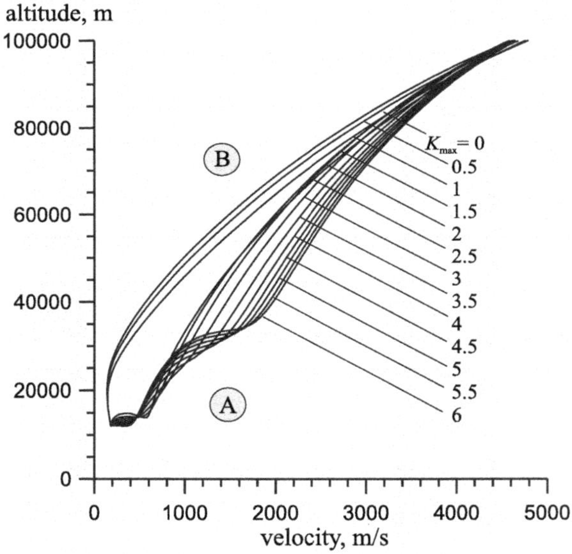

Fig. 3.3 Two qualitatively different types (**A** and **B**) of global extremals depending on maximum lift-to-drag ratio K_{max}

of the global extremals is found based on the dependence of maximum functional m_f on the maximum lift-to-drag ratio.

The given extremals have the following features, shown in Table 3.1.

Note that for global extremals of type ***A***, the variation of K_{max} leads to the effect of a "standing wave" in families of the oscillatory optimal pitch angle control programs with the nodal points that are practically independent of K_{max}

Critical value K_{max}^* is relatively low. Such the values are peculiar to traditional cylindrical layouts with a blunt nose [13, 14]. Therefore, the emergence of the new

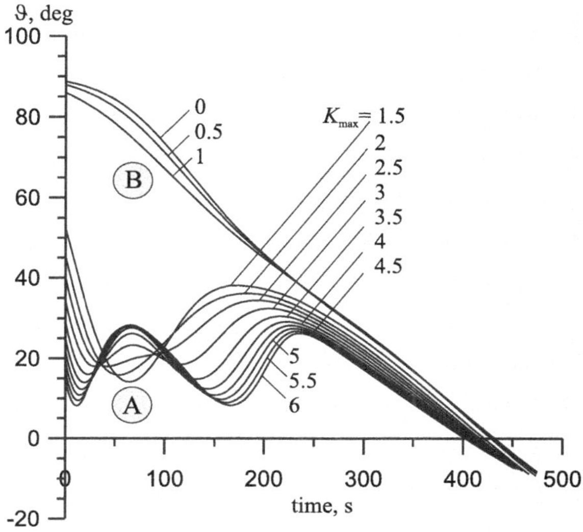

Fig. 3.4 Two qualitatively different types (**A** and **B**) of the optimal pitch angle programs depending on the maximum lift-to-drag ratio

optimal solutions can be expected not only in development of winged LVs, but within the framework of the traditional ballistic LVs, especially of a packet scheme, or when using small additional aerodynamic surfaces that increase the lifting properties. It should also be taken into account that parameter $K_{\max}$ is a characteristic of the ASV aerodynamic polar, and not the optimal control program. Therefore, the optimal angles of attack for LV with maximum lift-to-drag ratio $K_{\max}$, may be less than the angle-of-attack $\alpha_{K_{max}} = \arg\max_{\alpha} C_L/C_D$, at which $K_{\max}$ is reached.

Qualitative restructuring of the optimal solutions with a change in the ASV lifting properties does not necessarily have a bifurcation character. The restructuring of optimal control programs and dependencies on parameters can also be of a completely "evolutionary" nature, which will be demonstrated below. In this case, by the critical value $K^*_{\max}$ we mean the value of $K_{\max}$ at which the sharpest change in $\theta_{i\ \mathrm{opt}}$ is observed with variation of $K_{\max}$.

Let us continue consideration of the dependence of the optimal initial flight path angle on the maximum lift-to-drag ratio shown in Fig. 3.2. It can be seen that with growth of $K_{\max}$ in the supercritical domain, there is one more interesting point. It corresponds to such value $K^{**}_{\max}$, after which negative initial angles become optimal: $\theta_{i\ \mathrm{opt}}\left(K_{\max} \geq K^{**}_{\max}\right) \leq 0$. During one of the discussions of the presented results at the scientific seminar in TsAGI, Prof. A.A. Shilov pointed out a visible analogy between this paradoxical result and the take-off of a crow from a tree (the maximum lift-to-drag ratio of a crow in a flight configuration is, of course, much higher than in traditional LVs): a crow first glides down, and only then, having gained speed, uses own high lifting properties for climbing. (Although crows are considered one of the most intelligent representatives of the fauna, it is unlikely that they found the optimal solution strictly in accordance with PMP.) This type of ASV launch with small and

Table 3.1 Classification of extremals in the problem of the optimal launch to an orbit through the atmosphere

Type of extremals	Characteristic features	Overall score
B (“***B***allistic”)	The extremals deliver the global extremum at a low maximum lift-to-drag ratio $K_{\max} < K^*_{\max}$, where $K^*_{\max}$ is the critical (bifurcation) value (here $K^*_{\max} = K^{bif}_{\max} \approx 1.2$) The time-program for changing the tangent of the optimal pitch angle is quasi-linear, which qualitatively corresponds to the OEL classical solutions, aerodynamic forces have a little effect on the optimal control structure The atmosphere is "perceived", basically, as a medium with a drag. Therefore, the optimal trajectories pass in the area with a low level of the dynamic pressure Almost vertical take-off is optimal (here $85^\circ < \theta_{i\ \mathrm{opt}} < 90^\circ$)	***B***-type extremals are traditional for ballistic LVs
A (“***A***erodynamic”)	The extremals deliver the global optimum with a relatively high maximum lift-to-drag ratio $K_{\max} > K^*_{\max}$, The optimal pitch angle program in the atmospheric segment has a pronounced oscillatory character The atmosphere is “perceived” mainly as a medium that creates a lift. Therefore, the trajectories pass in the area with higher level of dynamic pressure compared to extremals ***B*** An inclined and quasi-horizontal take-off is optimal (here $\theta_{i\ \mathrm{opt}} < 55^\circ$)	***A***-type extremals are a qualitatively new type of the global extremals for rocket and space technology
M (“***M***edium”)	The extremals do not deliver the global extremum for the considered range of parameters The optimal trajectories and control laws occupy an intermediate position between types ***A*** and ***B***	***M***-type extremals are a qualitatively new type of the local extremals

even negative flight path angles has since then been called “crow’s take-off” among the colleagues.

When considering air launch projects, one often hears about the need for CA to perform steep climb maneuvers before LV release in order to give the largest possible flight path angles. For example, the AirLaunch project of the Polet company [15] generally refers to the “verticalization” of the LV after separation from the CA. The results presented here, which demonstrate the optimality of small launch angles for LV with sufficiently high lifting properties, provide a basis for a more detailed and, most importantly, rigorous study of the optimal control and design of LV shape in such projects. The results of complex optimization of LV layouts and control programs are demonstrated in Chap. 6.

The presence of two main types of the global extremal and possible bifurcation character of their change with a parameter variation can be explained by the following physical reasons. Losses of the characteristic velocity during launch are made up of aerodynamic, gravitational and control losses. To minimize aerodynamic losses, it would be necessary to reduce the angle of attack and increase the take-off angle (which, by the way, is incorporated in traditional launch programs). To minimize gravitational losses, it would be necessary to use lift and reduce the take-off angle. The optimal control implements a compromise between these two opposite tendencies. Since these trends turn out to be opposite from the point of view of the control strategy, they lead to qualitatively different optimal laws depending on the ratio of the lifting properties and drag. The measure of this relation is the maximum lift-to-drag ratio. Therefore, $K_{\max}$ appears as an important generalized parameter that determines extensively the optimal control structure. The opposite of the trends causes (as it is often observed in solutions of optimization problems with two separated local extrema) a high probability that the change in the optimal control with a parameter variation will be of a bifurcational nature. Indeed, in order to gain from the use of, for example, type ***A*** control laws, it is necessary to exceed a certain critical level of the lift-to-drag ratio. Otherwise, when using laws of type ***A***, an increase in aerodynamic drag losses is not compensated by a decrease in gravitational losses at the use of the lift. This means that the optimal control of LV with a subcritical lift-to-drag ratio must correspond to another type, type ***B***.

Remaining within the framework of such reasoning, one can also explain the relatively low value of $K^*_{\max}$. Indeed, taking into account that gravitational losses, as a rule, exceed the drag losses by an order of magnitude, and their main part falls on the initial (atmospheric) flight leg, it should be expected that the optimality of the “aerodynamic” type of the optimal control law affecting the dominant factor is realized already at a relatively small $K_{\max}$.

Thus, when designing advanced LVs with improved aerodynamic lifting properties, it is necessary to take into account the possibility of a qualitative, including bifurcation, change in optimal trajectories and control programs. This circumstance can also lead to a qualitative change in the optimal LV layout, since the transition through the critical lift-to-drag ratio can be accompanied by a significant, including abrupt, change in the sensitivity functions of the injected mass to variations in almost all (not only $K_{\max}$) LV parameters. At the bifurcation point, the influence functions can change by orders of magnitude and even in sign [16]. The example in Fig. 3.5 shows that derivative $\frac{\partial m_f}{\partial K_{\max}}$, characterizing the influence of the lift-to-drag ratio on the maximum injected mass changes abruptly by ~ 8 times at $K^*_{\max} = K^{bif}_{\max} \approx 1.2$. When using approximate methods based on the traditional (OEL) control program structure of type ***B***, it is impossible to notice such the significant change in the real influence of LV lifting properties. In Fig. 3.6 it is shown that the passage through the critical value of the maximum lift-to-drag ratio leads to a bifurcation change in the optimal LV ballistic coefficient.

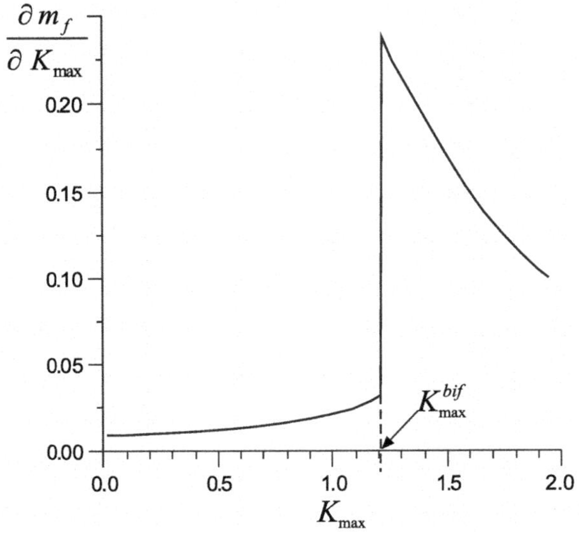

Fig. 3.5 A typical influence function of the LV maximum lift-to-drag ratio on the dimensionless injected mass vs the LV maximum lift-to-drag ratio

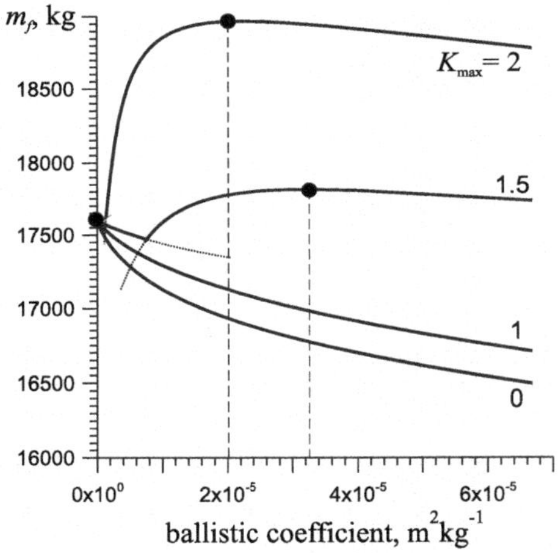

Fig. 3.6 Dependences of the maximum injection mass on the LV ballistic coefficient for various maximum lift-to-drag ratios. Markers (•) correspond to the optimal ballistic coefficient values

Consequently, the use of the rigorous optimization method is advisable not only at the final ASV design stages, but also at preliminary stages of initial development of a concept and overall appearance.

In conclusion, we note that the types of extremal, described above, single out only those classes of optimal solutions that are fundamentally different in relation to the use of aerodynamic lifting properties in the atmosphere. Additional local extremals may appear inside these classes. So, for example, Fig. 3.7 shows the results obtained by Alexander Golikov [17] using *ASTER* software package: the dependence of optimal take-off angle $\theta_{i\ \mathrm{opt}}$ on maximum lift-to-drag ratio $K_{\max}$ and relative maximum injected mass m_f of a hypothetical LV with a combined propulsion, characteristics of which depend not only on the altitude, but also on the flight velocity. It can be seen that up to five local extremals can exist simultaneously in a wide range of parameter values. Three types of extremals (***A***$_1$, ***A***$_2$ and ***B***) can provide the global optimum. The transition from one type to another with variation of the parameter in this case occurs bifurcationally. Extremals of types ***A***$_1$ and ***A***$_2$ are assigned to the same class ***A***, since they have all the features listed above that are inherent in this class.

The emergence of new types of extremals, as noted above, is due to the ratio of aerodynamic load factor and thrust-to-weight ratio. The critical level of ASV lifting properties, in a vicinity of which the existence of qualitatively different extremals is possible, will be the lower, the lower the thrust-to-weight ratio. This trend is

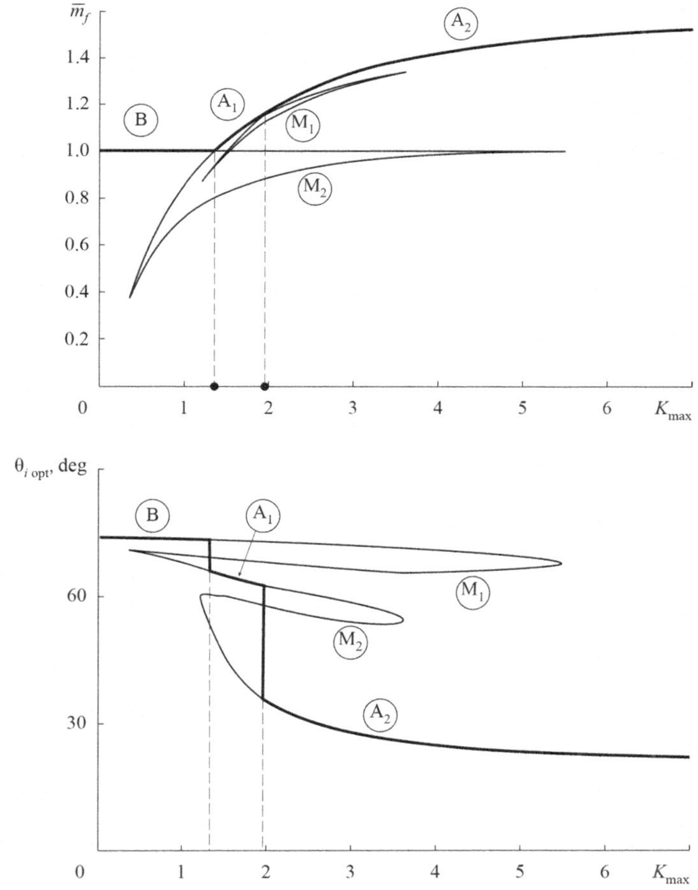

Fig. 3.7 Dependences of relative maximum injected mass $\overline{m}_f$ and optimal initial flight path angle $\theta_{i\ opt}$ on maximum lift-to-drag ratio K_{max} of a hypothetical LV with a combined propulsion

especially strong when the thrust-to-weight ratio is close to unity, corresponding to the balance of the gravity and thrust.

A physical explanation for this fact was given at the beginning of the Chapter: if the thrust is almost compensated by gravity (since the thrust and weight vectors are not necessarily collinear, the flight path angle should also be taken into account), then the aerodynamic force becomes dominant in the resulting force. As a rule, such a situation is typical for the initial segment of LV ascent, when the atmosphere density and LV mass are maximal, the thrust is minimum due to the atmosphere counterpressure, and for LV with a vertical take-off, additionally, the flight path angle is close to 90°. Due to the instability of ascent trajectories, strong perturbations of the trajectory in the initial segment, caused by aerodynamic forces, lead to significant changes in the trajectory as a whole.

Figure 3.8 shows typical dependences of relative injected mass m_{orb} and optimal take-off angle $\theta_{i\,\text{opt}}$ on $K_{\max}$ at various initial thrust-to-weight ratios n_i^{∞}. Increasing in the maximum injected mass and decreasing in $\theta_{i\,\text{opt}}$ with an increase in $K_{\max}$ is due to a decrease in gravitational losses of the characteristic velocity with an increase in LV lifting properties. It can be seen that there is critical maximum lift-to-drag ratio $K_{\max}^{*}$, in a vicinity of which an increase in $K_{\max}$ leads to the most rapid increase in the maximum injected mass and a decrease in the optimal initial flight path angle. Intensity of decreasing $\theta_{i\,\text{opt}}(K_{\max})$ in the neighborhood of $K_{\max}^{*}$ is the greater, the lower n_i^{∞}, which, as noted above, is associated with an increase in the relative influence of aerodynamic forces compared to the thrust.

Influence on maximum injected mass m_{orb} of variations of θ_i relative to $\theta_{i\,\text{opt}}$ is noticeably smaller for ASV with supercritical maximum lift-to-drag ratio: $K_{\max} > K_{\max}^{*}$, than with $K_{\max} < K_{\max}^{*}$ (Fig. 3.9). Note that the value of $K_{\max}^{*}$ is practically invariant to changes in parameters $\overline{\sigma}_x$, $\overline{A}$ and n_i^{∞} (Fig. 3.10), which confirms the validity of choosing the maximum lift-to-drag ratio as one of important generalized parameters that determine the qualitative behavior of extremals.

The structure of dependency m_{orb} on the ballistic coefficient $\overline{\sigma}_x$ is also determined by $K_{\max}$ (Fig. 3.11). For $K_{\max} < K_{\max}^{*} \approx 1.8$ these dependences are monotonically decreasing. This is explained by the fact that the decrease in the ballistic coefficient leads to a decrease in aerodynamic drag and a corresponding decrease in energy losses at the ASV ascent, which is typical for the modern class of ballistic-type LVs that do not use lifting properties during ascent to an orbit. As $K_{\max}$ increases, the structure of dependence $m_{\text{orb}}(\overline{\sigma}_x)$ changes: a local maximum appears and, upon reaching a certain critical (here, bifurcation) value $K_{\max} = K_{\max}^{*} \approx 1.8$, the optimum $\overline{\sigma}_{x\text{opt}}$ jumps from $\overline{\sigma}_{x\text{opt}} = 0$ to $\overline{\sigma}_{x\text{opt}} \approx 10^2$.

The maximum lift-to-drag ratio also determines the structure of the optimal control law (Fig. 3.12). For ASV with low lifting properties $\left(K_{\max} < K_{\max}^{*}\right)$ the optimal pitch angle program $\vartheta_{\text{opt}}(t)$ is quasilinear, and it is close in structure to the well-known OEL classical solutions, which do not take into account the effect of aerodynamic forces.

Parameters $\overline{\sigma}_x$ and $\overline{A}$ determine the relative level of aerodynamic forces. At limit $\overline{\sigma}_x \to 0$, there are no aerodynamic forces, and the optimal solutions have a structure corresponding to the OEL solutions (Fig. 3.13). With a decrease in relative induced

Fig. 3.8 Dependences of optimal initial flight path angle $\theta_{i\ \mathrm{opt}}$ and injected mass m_{orb} on maximum lift-to-drag ratio $K_{\max}$ for various thrust-to-weight ratio at $\overline{\sigma}_x = 260, \overline{A} = 8$

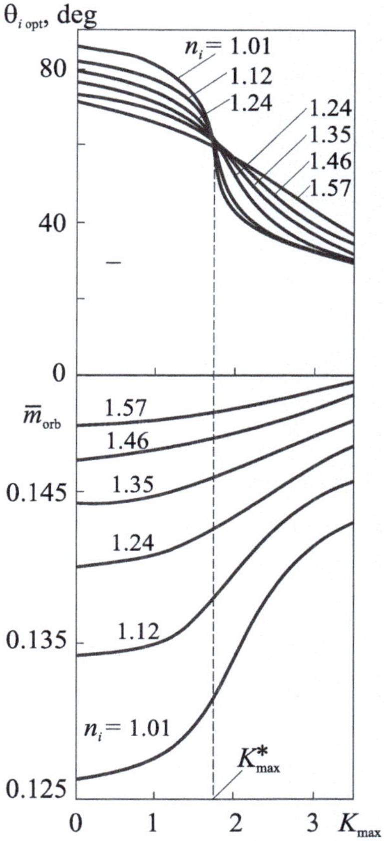

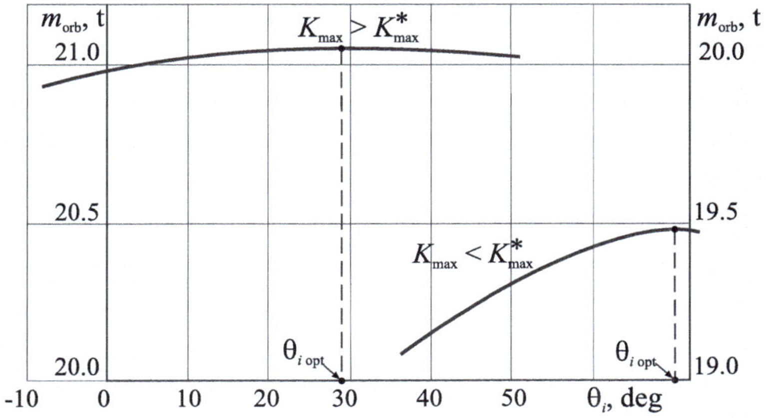

Fig. 3.9 Influence of initial flight path angle θ_i on maximum injected mass m_{orb} for sub- (lower curve) and super-critical (upper curve) maximum lift-to-drag ratios

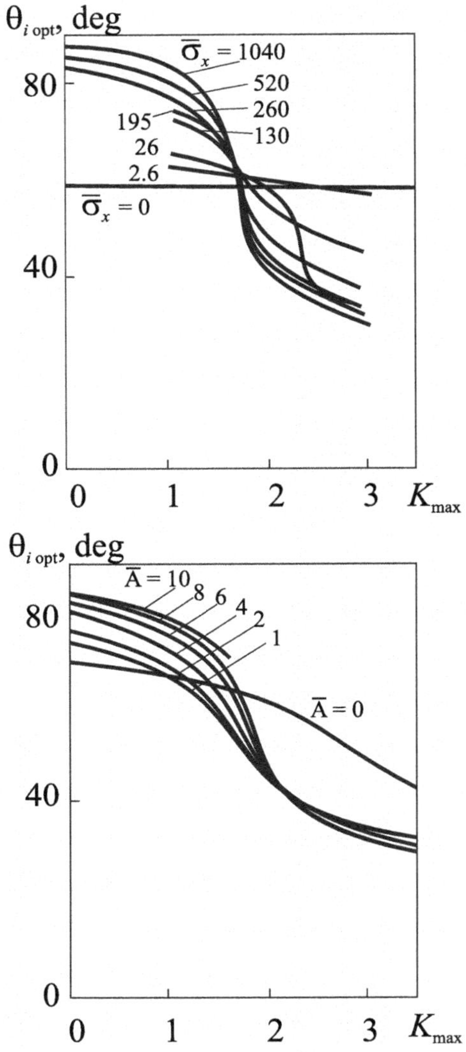

Fig. 3.10 Dependences of optimal initial flight path angle $\theta_{i\ opt}$ on maximum lift-to-drag ratio K_{max} for various relative induced drag factor $\bar{A}$ (at $\bar{\sigma}_x = 260$) and dimensionless ballistic coefficient $\bar{\sigma}_x$ (at $\bar{A} = 8$)

drag factor $\bar{A}$, angles-of-attack $\alpha_{K_{max}}$ corresponding to the maximum lift-to-drag ratio increase. Therefore, a decrease in the induced drag factor leads to an increase in the range of changing the angle of attack on the optimal ascent trajectory (Fig. 3.14).

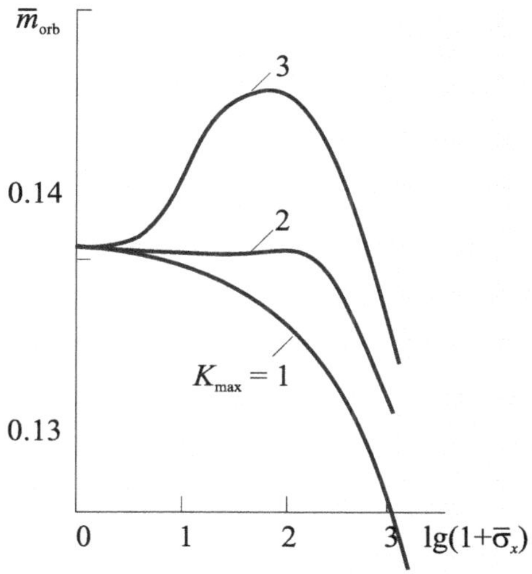

Fig. 3.11 Dependences of relative maximum injected mass on relative ballistic coefficient $\overline{\sigma}_x$ for three maximum lift-to-drag ratios K_{max} near the critical value

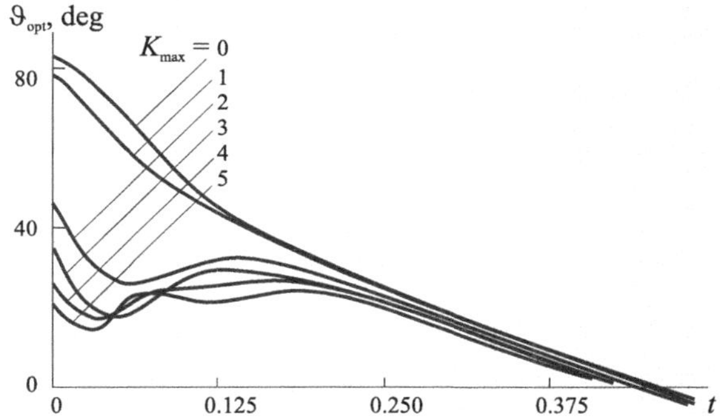

Fig. 3.12 Optimal time-programs of pitch angle ϑ_{opt} (t) for various maximum lift-to-drag ratios

3.2.2 The Influence of Aerodynamic Characteristics at Subsonic and Supersonic Speeds on the Extremals

ASV aerodynamic characteristics, as a rule, differ significantly at subsonic and high supersonic speeds [13, 14]. Subsonic legs of LV ascent trajectories are relatively short, if they exist at all [18], but, as a rule, the take-off and initial phase of LV ascent take place at subsonic modes. Therefore, the optimal take-off conditions can significantly depend not only on K_{max} averaged over the Mach number M, but also to a greater extent on the maximum lift-to-drag ratio at subsonic speeds K_{sub}.

To find out the influence of dependence $K_{max}(M)$ on the optimal injection trajectories, we approximate this dependence by a smooth function with constant values at subsonic (K_{sub}) and supersonic (K_{sup}) speeds (Fig. 3.15):

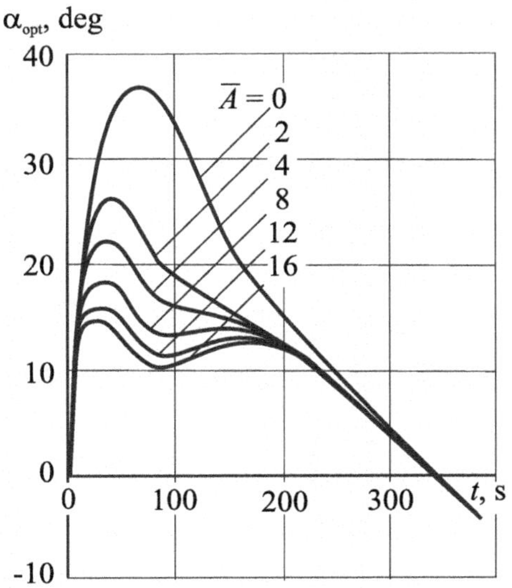

Fig. 3.13 Optimal time-programs of angle-of-attack $\alpha_{opt}(t)$ for various relative induced drag factor $\overline{A}$ at $K_{max} = 3$

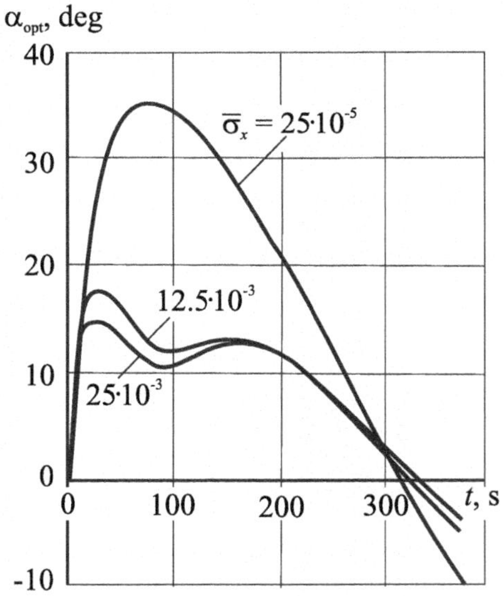

Fig. 3.14 Optimal time-programs of angle-of-attack $\alpha_{opt}(t)$ for various relative ballistic coefficient $\overline{\sigma}_x$ at $K_{max} = 3$

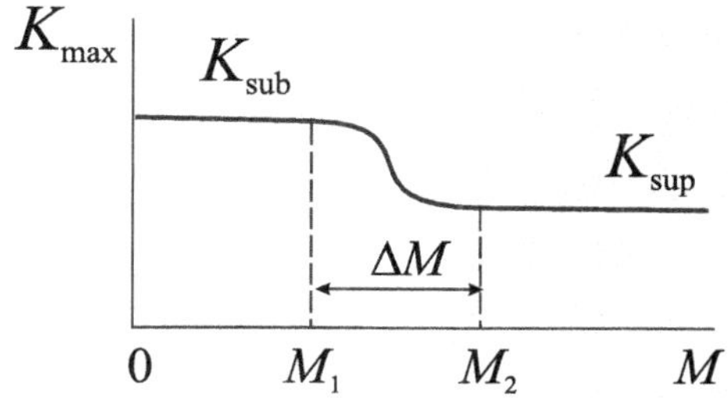

Fig. 3.15 The approximation of the dependence of the maximum lift-to-drag ratio on the Mach number

$$K(M)\begin{cases} K_{sub}, & M \le M_1, \\ \widetilde{K}(M), & M_1 < M \le M_2, \\ K_{sup}, & M_2 \le M, \end{cases}$$

$$\widetilde{K}(M) = \frac{K_{sub} + K_{sup}}{2} + 2\frac{K_{sub} - K_{sup}}{(M_2 - M_1)^3}\left(M - \frac{M_1 + M_2}{2}\right)$$
$$\left[\left(M - \frac{M_1 + M_2}{2}\right)^2 - \frac{3}{4}(M_1 - M_2)^2\right]$$

In parametric calculations $M_1 = 1$ and $M_2 = 2$ were taken as nominal values.

A typical change in the maximum injected mass m_{orb} and the optimal initial path angle $\theta_{i\ opt}$ with variations of K_{sub} and K_{sup} is shown in Fig. 3.16 [19]. An increase in K_{max} in any of the flight modes leads to an increase in injected mass m_{orb}. Moreover, the effect of subsonic and supersonic lift-to-drag ratio on m_{orb} is the stronger, the lower the ASV lifting properties in the rest part of the trajectory. The greatest impact on $\theta_{i\ opt}$ provides the maximum subsonic lift-to-drag ratio. This is explained by the fact that the process of changing the flight path angle is relatively fast, so the optimal take-off conditions are determined by flight mode in a vicinity of the start point, that is, in this case, with subsonic speeds.

By analogy with K^*_{max} one can single out the critical value of K^*_{sub}, in the vicinity of which the most intense change occurs in m_{orb} and $\theta_{i\ opt}$. The existence of critical value K^*_{sub} is especially evident at low supersonic lift-to-drag ratio $K_{sup} < K^*_{max}$. So, with $K_{sup} < 1$ an increase of K_{sub} in the neighborhood of $2.5 < K^*_{sub} < 3.5$ leads to a sharp decrease in $\theta_{i\ opt}$ from 55–60° to 18°, and in the domain of $K_{sub} > K^{**}_{sub} \approx 5$ the "crow's take-off", when LV starts under near-zero and negative path angles to the horizon, is optimal. In the domain of $K_{sub} > K^*_{sub}$, the effect of variations in the initial flight path angle θ_i with respect to $\theta_{i\ opt}$ on the injected mass m_{orb} is reduced appreciably (Fig. 3.17).

Thus, ASVs are divided into two classes according to the optimal launch type: for ASV with $K_{sub} < K^*_{sub}$, almost vertical (with large θ_i) take-off is optimal, for $K_{sub} > K^*_{sub}$, an inclined or quasi-horizontal start is optimal.

Critical value K^*_{sub} depends on initial thrust-to-weight ratio n_i^∞ due to the change in the proportions of thrust, aerodynamic and gravitational forces. Values of K^*_{sub} for various K_{sup} and n_i^∞ are given in Table 3.2.

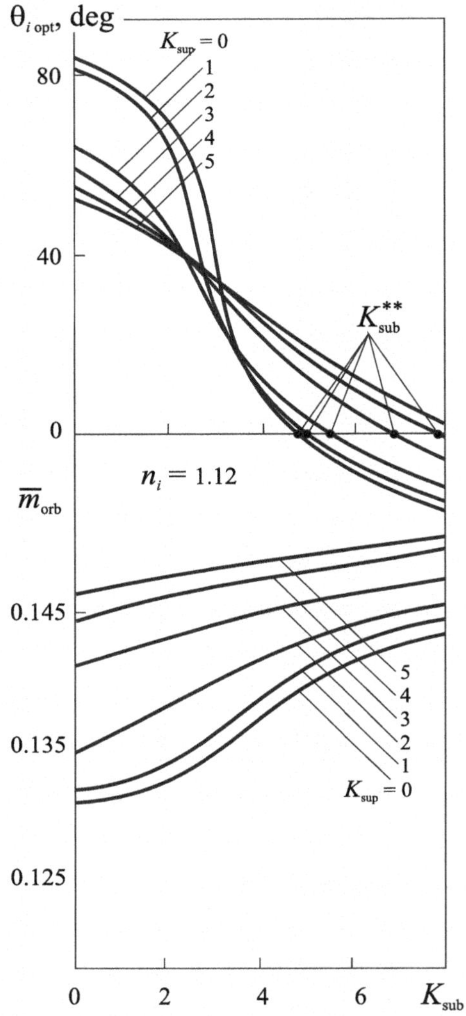

Fig. 3.16 Dependences of optimal initial flight path angle $\theta_{i\ \mathrm{opt}}$ and relative maximum injected mass m_{orb} on maximum lift-to-drag ratio at subsonic (K_{sub}) and supersonic (K_{sup}) velocities at $n_i = 1.12$

The less n_i^{∞}, the more intensive is the rearrangement of the optimal take-off conditions in a vicinity of critical value K_{sub}^{*}. At low n_i^{∞} and K_{sup}^{*} the change in $\theta_{i\,\mathrm{opt}}$ is bifurcational in nature. So, for $n_i^{\infty} = 0.9$ and $K_{\mathrm{sup}}^{*} = 1$ the optimal initial flight path angle at $K_{\mathrm{sub}} = K_{\mathrm{sub}}^{*} \approx 1.56$ jumps from $\theta_{i\,\mathrm{opt}} = 77°$ to $\theta_{i\,\mathrm{opt}} = 31°$ (Fig. 3.18). The jump is associated with the transition between local extremals that coincide in the functional at the bifurcation point, but have qualitatively different optimal control programs. One of the extremals (type ***B***) passes in the region of high altitudes and lower dynamic pressures, thereby seeking to minimize the loss of characteristic velocity to overcome the aerodynamic drag. The other one (type ***A***) uses the aerodynamic lift to reduce gravitational losses and therefore implements

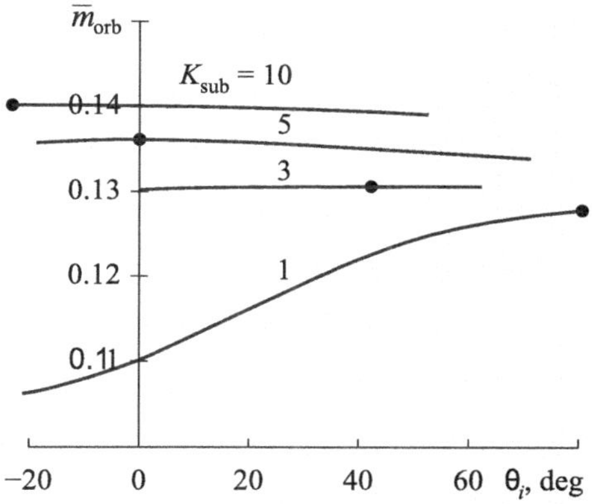

Fig. 3.17 Influence of initial flight path angle θ_i on relative maximum injected mass m_{orb} for various maximum subsonic lift-to-drag ratios K_{sub} at $K_{sup} = 0$, $n_i = 1.12$; marker "•" corresponds to the optimal initial flight path angle

Table 3.2 Critical values of subsonic maximum lift-to-drag ratio K^*_{sub} depending on the supersonic maximum lift-to-drag ratio K_{sup} and initial thrust-to-weight ratio n_i^{∞}

K_{sup}	n_i^{∞}		
	0.9	1.12	1.35
1	1.56	2.7–2.9	3.6–4.3
2	1.5–2.0	2.6–3.0	3.0–3.6
3	1.5–2.0	2.6–3.3	3.5–4.0

The uncertainty in the estimates of K^*_{sub} is due to the large step of K_{sub} variation taken in the parametric calculations.

motion with a smaller initial flight path angle and has a longer flight leg in dense atmosphere layers (Fig. 3.19).

The study of extremals with variations in the lift-to-drag ratio and initial thrust-to-weight ratio was carried out using the continuation method with respect to the corresponding parameter (see Sect. 1.10.2). In the case of the occurrence of a singularity, as in the case of $n_i = 0.9$, $K_{sub} = 1$ (Fig. 3.18), to continue the solution through the bifurcation point at $K^*_{sub} \approx 1.56$ the fold of the surface $\theta_{i\,opt}(K_{sub}, n_i^{\infty})$ is bypassed along contour $\boldsymbol{\Gamma}$, which does not contain singularities in the extended space of variable parameters $\tilde{\nu} = \left\{\nu, K_{sub}, n_i^{\infty}\right\}$ (Fig. 3.20). To build a section with $\frac{\partial \theta_{iopt}}{\partial K_{sub}} > 0$, $\theta_{i\,opt}$ was assigned as the continuation parameter, since dependency $K_{sub}\left(\theta_{i\,opt}\right)$ does not contain a singularity in this case.

When K_{sub} and K_{sup} are changed in a vicinity of the critical values K^*_{sub} and K^*_{max} the structure of the optimal control program is rearranged (Fig. 3.21). At $K_{sub} < K^*_{sub}$ and $K_{sup} < K^*_{max}$ the optimal pitch angle program is close to linear. At $K_{sub} > K^*_{sub}$ or $K_{sup} > K^*_{max}$, it has a pronounced oscillatory character in dense atmosphere layers. For high lift-to-drag ratio, the variations of K_{sub} result to the standing wave effect with nodal points practically independent of K_{sub} in families of optimal programs $\{\vartheta_{opt}(t), K_{sub} = \text{var}\}$.

Note, under the optimal control, the aerodynamic forces are small relative to gravitational ones even when trajectories of type $\boldsymbol{A}$ are optimal.

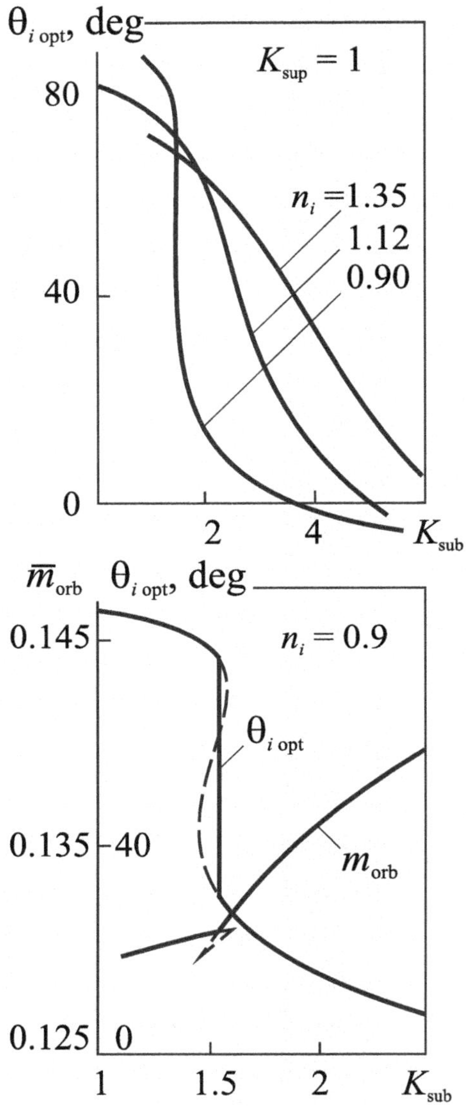

Fig. 3.18 Dependences of optimal initial flight path angle $\theta_{i\ opt}$ and maximum injected mass m_{orb} on maximum subsonic lift-to-drag ratio K_{sub} for various initial thrust-to-weight ratios at $K_{sup} = 0$

The results obtained for the refined model of the dependence of the maximum lift-to-drag ratio on the Mach number once again indicate that the influence of even small aerodynamic forces compared to the thrust can qualitatively affect the structure of the optimal trajectories and control laws during LV ascent into an orbit.

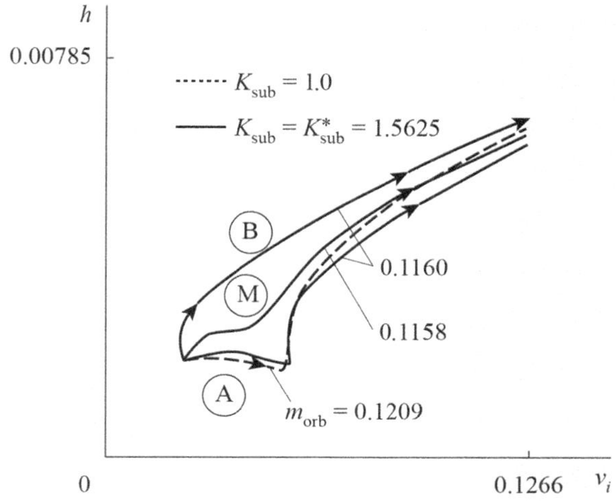

Fig. 3.19 Three local extremals (solid lines) in the dimensionless phase plane at critical LV maximum subsonic lift-to-drag ratio K^*_{sub}, $v_i = 0.02278$, $n_i = 0.9$

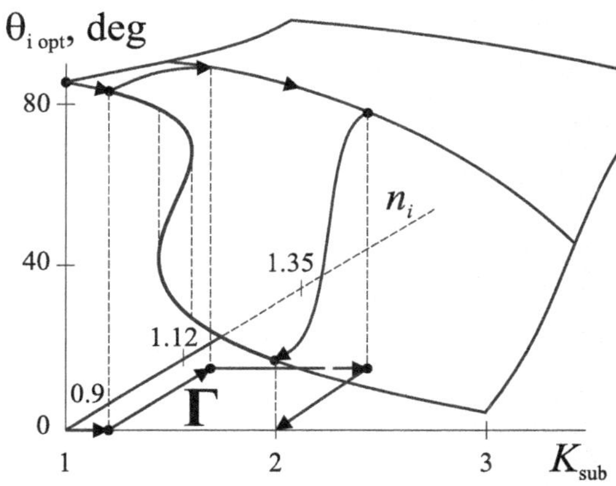

Fig. 3.20 The example of solution continuation bypassing the singular point in extended space of parameters $\theta_{i\ opt}$ (K_{sub}, n_i)

3.2.3 *The Influence of Aerodynamic Characteristics and Thrust on Parameters of Optimal Trajectories and Control Laws*

The change in flight path angle θ during ASV optimal ascent motion in dense atmosphere layers has a pronounced oscillatory character (Fig. 3.22). At sufficiently large values of K_{sub}, ($K_{sub} \geq 4$) on Fig. 3.22, negative θ_{opt} can be realized. It means that the optimal ascent trajectory can have the corresponding descent segment. The higher K_{sub}, the more profitable the flight at subsonic speeds in dense atmosphere layers. Therefore, as K_{sub} increases such the descending flight segment increases, and its beginning shifts to the initial point. For most of the trajectory passing through

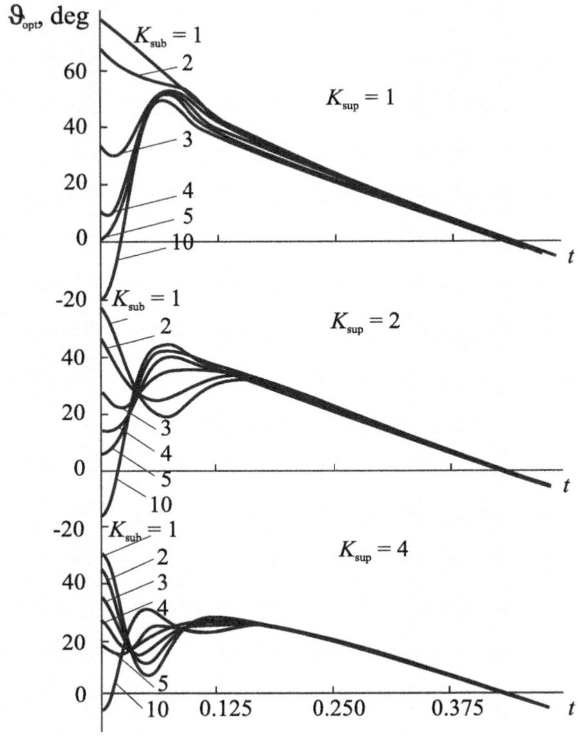

Fig. 3.21 Optimal time-programs of pitch angle $\vartheta_{opt}(t)$ for various subsonic and supersonic maximum lift-to-drag ratios at $n_i^{\infty} = 1.12$

rarefied layers of the atmosphere, dependence $\theta(t)$ corresponds qualitatively to the OEL classical solution and weakly depends on LV aerodynamic characteristics.

Since with increasing in K_{sub} the influence of the subsonic flight segment on the functional increases, the time interval at which the greatest angles of attack are realized in the optimal angle-of-attack program $\alpha_{opt}(t)$ shifts to the take-off moment (Fig. 3.23). The range and maximum angle of attack in optimal program $\alpha_{opt}(t)$ with

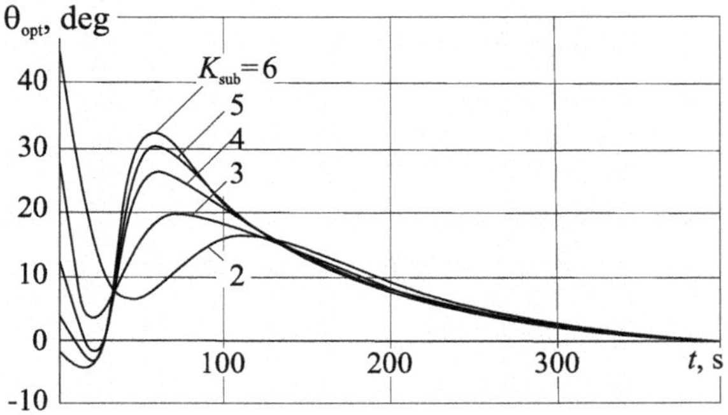

Fig. 3.22 Time-dependencies of optimal flight path angle $\theta_{opt}(t)$ for various subsonic maximum lift-to-drag ratios K_{sub} at $K_{sup} = 2$, $v_i = 180$ m/s, $h_i = 12$ km, $\Delta M = 0.47$, $n_i^{\infty} = 1.12$, $\theta_i = \theta_{i\,opt}$

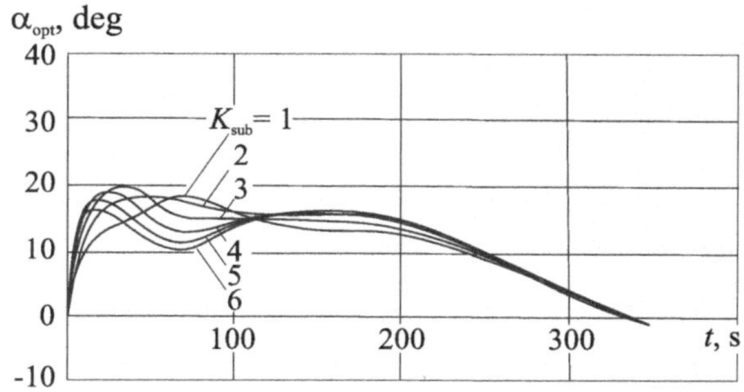

Fig. 3.23 Optimal time-programs of angle-of-attack $\alpha_{\text{opt}}(t)$ for various subsonic maximum lift-to-drag ratios K_{sub} at $K_{\text{sup}} = 2$, $v_i = 180$ m/s, $h_i = 12$ km, $\Delta M = 0.47$, $n_i^{\infty} = 1.12$, $\theta_i = \theta_{i\ \text{opt}}$

optimal initial flight path angle $\theta_i = \theta_{i\ \text{opt}}$ depends weakly on K_{sub}. This is a possible consequence of the method of varying K_{sub} by changing aerodynamic lift coefficient C_L^{α} with a fixed dependence of drag on the angle of attack. This method is adopted throughout these studies. According to (2.25), with such changes in $K_{\max}$, angle $\alpha_{K_{\max}} \approx \sqrt{2C_{D0}/D_0}$ remains constant (here, $\alpha_{K_{\max}} \approx 14°$).

The dependence of dynamic pressure $q = \rho v^2/2$ on the time has the pronounced maximum $q_{\max} = \max\limits_{t_i \le t \le t_f} q$ (Fig. 3.24). The dependence of $q_{\max}(K_{\text{sub}})$ is, as a rule, nonmonotonic: as K_{sub} increases the $q_{\max}$ first noticeably decreases, and then its growth is observed. Time t_q of reaching the maximum dynamic pressure, $t_q = \arg\max\limits_{\{t\}} q$, shifts with increasing of K_{sub} to take-off moment $t = 0$. This effect has the same physical explanation as given above in the analysis of dependences of $\theta(t)$ and $\alpha_{\text{opt}}(t)$.

The maximum dynamic pressure and maximum angle of attack are realized at different times (Figs. 3.23 and 3.24). This indicates that non-zero angles of attack are needed for optimal insertion into orbit not only to create as much lift as possible,

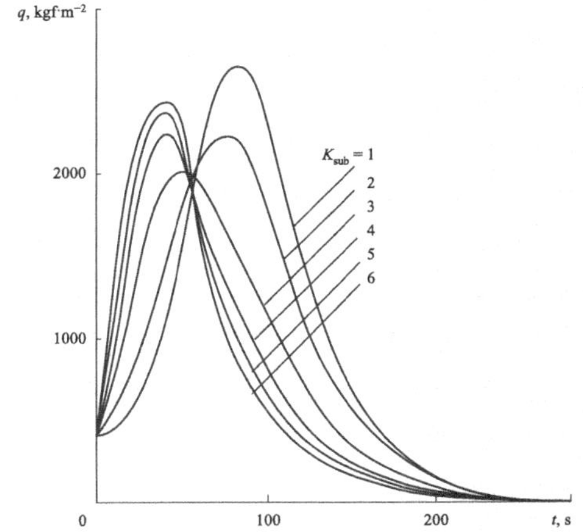

Fig. 3.24 Time-dependencies of the dynamic pressure on optimal injection trajectories for various subsonic maximum lift-to-drag ratios K_{sub} at $K_{\text{sup}} = 2$, $v_i = 180$ m/s, $h_i = 12$ km, $\Delta M = 0.47$, $n_i^{\infty} = 1.12$, $\theta_i = \theta_{i\ \text{opt}}$

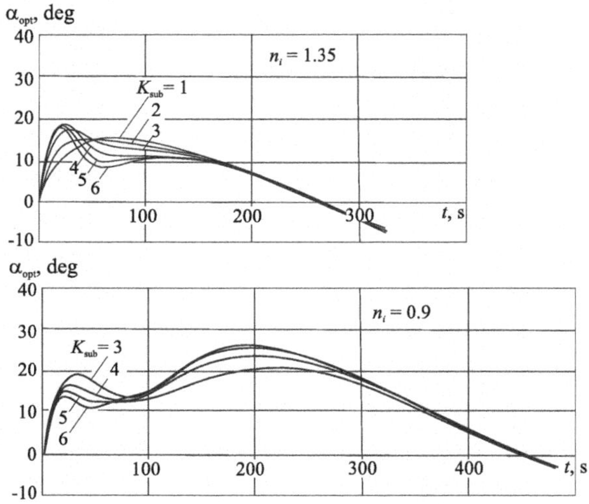

Fig. 3.25 Optimal time-programs of angle of attack $\alpha_{opt}(t)$ for various subsonic maximum lift-to-drag ratios K_{sub} and two initial thrust-to-weight ratios n_i^∞ at $K_{sup} = 2$, $\theta_i = \theta_{i\ opt}$

but also to create an optimal orientation of the thrust vector, which is directed along the LV longitudinal axis.

The optimal angle-of-attack program $\alpha_{opt}(t)$ for a moderate supersonic lift-to-drag ratio ($K_{sup} \leq 2$), as a rule, have two local maxima (Figs. 3.23 and 3.25). The first (in time) is realized at high dynamic pressures, and the second is reached in rarefied layers. Obviously, the first maximum is the most critical in terms of g-loads.

As can be seen from Fig. 3.25, when ASV starting at optimal angle $\theta_{i\ opt}$, the maximum angles of attack α_{max} have a slight excess relative to $\alpha_{K_{max}}$ (in these calculations, a constraint on the angle of attack has not yet been introduced). The influence of n_i^∞ on α_{max} is weak and is determined by K_{sub}. If $K_{sub} < K_{sub}^*$, then with decreasing in n_i^∞ the value of α_{max} increases, and if $K_{sub} > K_{sub}^*$, then α_{max} decreases. Table 3.3 shows the values of α_{max} for several values of n_i^∞ and K_{sub} at $K_{sup} = 2$.

With a larger initial thrust-to-weight ratio, a more intensive velocity increase is possible in the initial segment of motion in dense atmosphere layers, which, at positive flight path angles, also means a more intensive climb. These two factors have an opposite effect on the dynamic pressure during ASV ascent. The calculations show (Fig. 3.24) that the second factor is decisive, and the maximum dynamic pressure

Table 3.3 The maximum angle of attack (in degrees) on optimal trajectories with various n_i^∞ and K_{sub}

n_i^∞ / K_{sub}	1	2	3	4	5	6
0.90			19	17	15	14
1.12	18	18	20	19	18	16
1.35	16	15	17	18	18	18

Table 3.4 The dependence of the maximum dynamic pressure on the initial thrust-to-weight ratio

n_i^{∞}	0.90	1.12	1.35
q_{max}, kgf/m^2	2600	2000	1300

q_{max} decreases with an increase in the initial thrust-to-weight ratio. For $K_{sup} = 2$, $K_{sub} = 3$ the change in q_{max} is shown in Table 3.4.

As will be shown below, a similar trend is also observed on optimal ascent trajectories with fixed angle θ_i, which is generally not optimal.

Assessing the nature of optimal ASV trajectories and control laws with various thrust-to-weight ratio at the optimal initial flight path angle ($\theta_i = \theta_{i\ opt}$), it should be noted that they are not accompanied by excessive g-loads and angles of attack.

3.2.4 The Influence of Aerodynamic Characteristics and Engine Installation Angle on Parameters of Optimal Trajectories and Control Laws

We defined the engine installation angle as the angle between the direction of the thrust vector towards the center of mass and the longitudinal axis of the vehicle. The parametric investigation of the influence of engine installation angle φ_E on the functional, optimal take-off conditions and control programs for various ASV layout parameters with the maximum lift-to-drag ratio in range $0 \le K_{max} \le 5$ is performed. In this section, K_{max} is assumed to be constant along trajectory and averaged over the Mach numbers.

The dependences of maximum injected mass m_f on φ_E are shown in Fig. 3.26 at optimal initial flight path angle θ_{iopt}. Roll overs by 180° in this case were not allowed, so dependences $m_f(\varphi_E)$ are not symmetrical with respect to the line $\varphi_E = 0$. As can be seen, optimal engine installation angle φ_{Eopt} depends on maximum lift-to-drag ratio. For ballistic ASVs ($K_{max} = 0$), φ_{Eopt} is maximum, $\varphi_{Eopt}\big|_{K_{max}=0} \approx 22°$. As K_{max} increases, φ_{Eopt} decreases (Table 3.5).

If φ_E differs from φ_{Eopt} within $\pm$ 15°, then the losses in the injected mass can be up to ~ 2% (Fig. 3.26). The greatest loss of m_f due to the difference between φ_E and φ_{Eopt} are observed when the maximum lift-to-drag ratio is close to the critical one (K^*_{max}).

The optimal initial flight path angle changes with variation of φ_E only at $K_{max} \le K^*_{max}$ (Fig. 3.27). For larger K_{max} ($K_{max} \ge 2$) optimal angle θ_{iopt} is practically independent of φ_E. For ASV with low lifting properties, an increase in φ_E leads to an almost proportional decrease in $\theta_{i\ opt}$.

The variation of φ_E in the practical range ($|\varphi_E| \le 20°$) does not change the optimal control structure. In quantitative terms, it leads to the following effects. On the segment in rarefied atmosphere layers ($t \ge 200$ s), optimal attitude program

Fig. 3.26 Dependences of maximum injected mass m_f on engine installation angle φ_E for various maximum lift-to-drag ratios $K_{\max}$, $\theta_i = \theta_{i\ \text{opt}}$

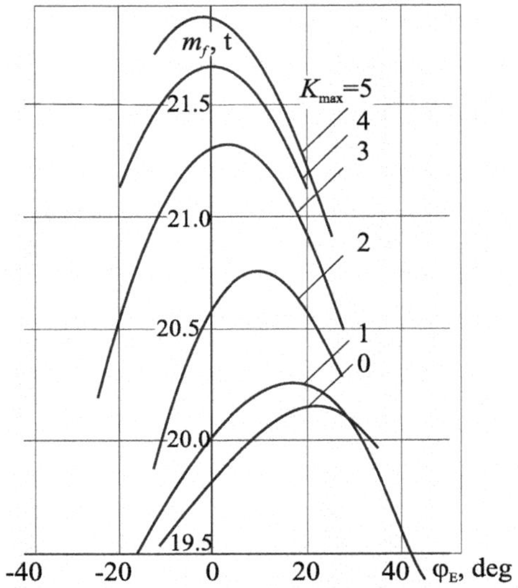

Table 3.5 The dependence of optimal engine installation angle φ_{Eopt} on maximum lift-to-drag ratio $K_{\max}$

$K_{\max}$	0	1	2	3	4	5
φ_{Eopt}, deg	22	17	10	3	0	-2

Fig. 3.27 Dependencies of the optimal initial flight path angle on the engine installation angle for various maximum lift-to-drag ratios $K_{\max}$

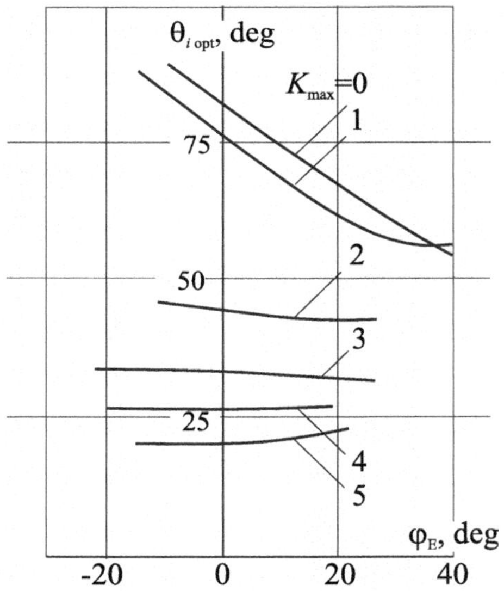

$\vartheta_{\text{opt}}(t)$ is close to quasilinear and practically preserved, regardless of φ_E. In accordance with this, the optimal angle of attack programs are equidistantly shifted relative to program $\alpha_{\text{opt}}(t)\big|_{\varphi_E=0}$ by angle $\Delta\alpha_{\text{opt}}(t) \cong -\varphi_E$. On the flight segment in dense atmosphere layers ($t \le 100$ s) the difference between these programs $\left|\alpha_{\text{opt}}(t)\big|_{\varphi_E\neq 0} - \alpha_{\text{opt}}(t)\big|_{\varphi_E=0}\right|$ is noticeably smaller than $|\varphi_E|$, which corresponds to an increase in the relative contribution of aerodynamic forces compared to the thrust.

If the initial flight path angle is optimized, $\theta_i = \theta_{i\,\text{opt}}$, then, in accordance with PMP, initial angle of attack $\alpha_{i\text{opt}}$ is determined from the condition of the maximum projection onto the velocity vector of the sum of thrust vector and aerodynamic forces:

$$\alpha_{i\,\text{opt}} = \arg\max_{\alpha}\left[(\mathbf{e}_{vi}, \mathbf{T}_i) - C_D(\alpha)q_iF_{\text{ref}}\right].$$

In case of $|\varphi_E| \ll 1$ and $\frac{\partial C_D}{\partial\alpha} \cong -D_\alpha\alpha$ we have

$$\alpha_{i\,\text{opt}} \cong \frac{-\varphi_E}{1 + \frac{D_\alpha q_i F_{\text{ref}}}{T_i}}. \tag{3.1}$$

Since φ_E does not change the optimal control structure, then (3.1) gives the estimate of the shift of both optimal initial angle of attack $\alpha_{i\,\text{opt}}$, and program $\alpha_{\text{opt}}(t)$ on the entire flight segment in dense atmosphere layers.

In rarefied atmosphere layers ($t \ge 200$ s), regardless of φ_E, the optimal pitch angle program $\vartheta_{\text{opt}}(t)$ tends to a quasilinear one, and in the region where aerodynamic forces are noticeable ($t \le 100$ s), the corresponding initial section of optimal program $\alpha_{\text{opt}}(t)$ turns out to be more stable. This effect is explained by the fact that in the extra-atmospheric section the trajectory is controlled by changing the orientation in space of the thrust vector (the characteristic control variable is the pitch angle), and in dense atmosphere layers, aerodynamic forces become decisive, which depend on ASV attitude relative to the velocity vector (the characteristic control variable is the angle of attack).

3.3 The Influence of Initial Conditions on Optimal Trajectories and Control Laws

3.3.1 Initial Velocity Influence

The change in initial velocity v_i affects the optimal ascent trajectories and functional m_f due to three reasons:

1. An increase in v_i leads to a reduction in the margin of characteristic velocity $\Delta v_c = v_{orb} - v_i$ necessary to accelerate ASV to the given orbital velocity v_{orb}, which contributes to an increase in m_f:

$$m_f = m_i \exp\left\{\frac{-\Delta v_c}{c}\right\}. \tag{3.2}$$

2. An increase in v_i leads to an increase in dynamic pressure q_i and aerodynamic forces in the initial segment of the ascent trajectory. As a result, at low ASV lifting properties ($K_{sub} < K^*_{sub}$), the fuel consumption for overcoming aerodynamic drag increases, and at sufficiently high lifting properties ($K_{sub} > K^*_{sub}$), the fuel consumption decreases due to reduction in gravitational losses.
3. An increase in v_i means an increase in initial Mach number M_i, which leads to a reduction in the flight segment $\Delta M = M_1 - M_i$ at subsonic speed. This reduces the influence of K_{sub} on the optimal trajectory and, in particular, increases critical value K^*_{sub}.

3.3.2 *Initial Altitude Influence*

An increase in the initial altitude h_i leads to an increase in the initial ASV potential energy, engine thrust and to a decrease in the dynamic pressure at the initial trajectory segment.

The first two factors cause an increase in the injected mass and a decrease in $\theta_{i\,opt}$. The third factor weakens the influence of aerodynamic forces and K_{sub} on the optimal trajectory structure and functional m_f. So an increase in h_i should lead to an increase in the critical value K^*_{sub}.

The performed computations (Figs. 3.16, 3.28 and 3.29) confirm this qualitative analysis.

3.3.3 *Initial Flight-Path-Angle Influence*

Above, the optimal injection trajectories were investigated at optimal take-off angle $\theta_i = \theta_{i\,opt}$. The study of such class of trajectories makes it possible, on the one hand, to formulate objective requirements for the best initial conditions for ASV ascent, and on the other hand, to obtain an upper estimate for injected mass m_f. Let consider the changes in the optimal solutions caused by fixing the initial flight path angle.

There are two "limiting" cases: $\theta_i = 90°$ and $\theta_i = 0°$. Ascent with a traditional vertical take-off ($\theta_i = 90°$) are characterized by low initial (after a short vertical leg) flight speeds. Changing the flight path angle occurs quickly and without additional energy costs. As a result, the "steep slope" section of the trajectory, where the flight path angle decreases to the optimal value, has practically no effect on the subsequent

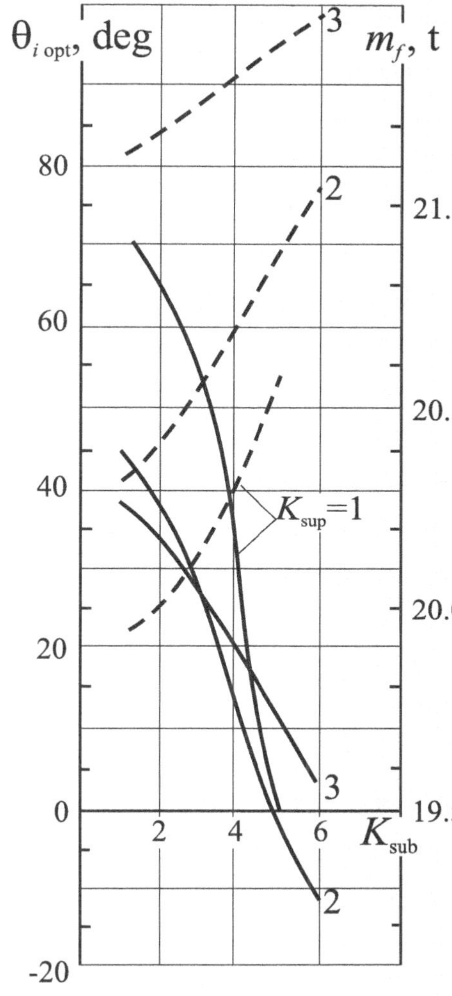

Fig. 3.28 Dependencies of optimal initial flight path angle $\theta_{i\ opt}$ and maximum injected mass m_f on subsonic K_{sub} and supersonic K_{sup} maximum lift-to-drag ratios at $h_i = 17.22$ km

optimal trajectory. (The optimal ASV ascent trajectories with a vertical take-off will be discussed in subsequent chapters.)

Below, another important class of ASV trajectories with zero initial flight path angle $\theta_i = 0$ is studied. On the one hand, the launch in the quasi-level flight mode is relatively new for space rockets and has a number of practical advantages. On the other hand, the study of such a class of trajectories makes it possible to obtain a lower estimation for the injected mass (of course, if the optimal path angle is greater than zero). Such estimations are given below in the form of losses Δm_f of the mass inserted into the target orbit with altitude $h_{orb} = 200$ km, due to the use of the horizontal take-off instead of an inclined optimally at $\theta_i = \theta_{i\ opt}$. A comparative analysis of the optimal trajectories and control programs for these two types of launch

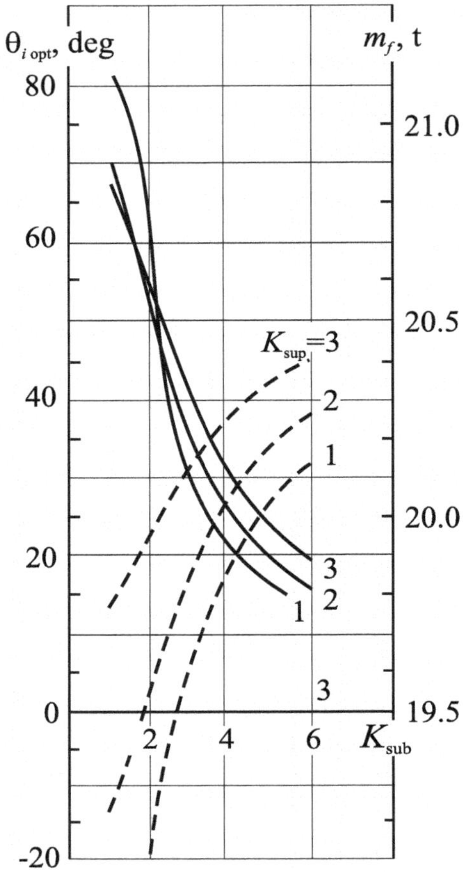

Fig. 3.29 Dependencies of optimal initial flight path angle $\theta_{i\ \mathrm{opt}}$ and maximum injected mass m_f on subsonic K_{sup} and supersonic K_{sup} maximum lift-to-drag ratios, at $h_i = 6.78$ km

was carried out in the following ranges of parameters: $1 \le K_{\mathrm{sub}} \le 6$, $0.24 \le \Delta M \le 0.47$, $0.9 \le n_i^{\infty} \le 1.35$, $180\ \mathrm{m/s} \le v_i \le 260\ \mathrm{m/s}$, $7.8\ \mathrm{km} \le h_i \le 17\ \mathrm{km}$.

The injected mass losses

As shown above, $\theta_{i\ \mathrm{opt}}$ depends essentially on K_{sub}. In a vicinity of K^*_{sub} the value of $\theta_{i\ \mathrm{opt}}$ decreases sharply with growth of K_{sub}, and for sufficiently large $K_{\mathrm{sub}} \ge K^{**}_{\mathrm{sub}}$ the optimal take-off angle $\theta_{i\ \mathrm{opt}} \le 0$. In practice, the lower limit of the range of K_{sub}, at which the launch from the quasi-horizontal flight mode of CA is preferable, corresponds to smaller K_{sub} than K^{**}_{sub}, since already at $K_{\mathrm{sub}} > K^*_{\mathrm{sub}}$ the influence of θ_i on functional m_f decreases sharply. Figure 3.30 shows the losses in injected mass Δm_f when ASV is launched with $\theta_i = 0$ instead of $\theta_{i\ \mathrm{opt}}$: $\Delta m_f = m_f\big|_{\theta_i=\theta_{i\ \mathrm{opt}}} - m_f\big|_{\theta_i=0}$. Comparing the dependences of $\Delta m_f(K_{\mathrm{sub}})$ for various K^*_{sub}, one can be convinced that the critical value of the maximum lift-to-drag ratio really highlights range $K_{\mathrm{sub}} >$

Fig. 3.30 Injected mass losses Δm_f caused by the horizontal take-off instead of the optimal inclined one; $\varepsilon_m = 100$ kg, $K_{sup} = 2$; "•" marks $K_{sub}|_{\varepsilon_m = 100\ \text{kg}}$

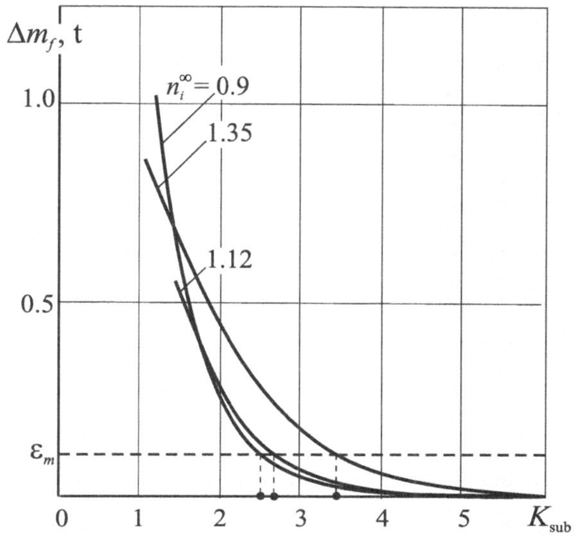

K^*_{sub}, in which the losses due to the difference between θ_i and $\theta_{i\ opt}$ are sharply reduced compared to range $K_{sub} < K^*_{sub}$.

Losses of Δm_f can vanish only for sufficiently large $K_{sub} > K^{**}_{sub} \approx 4$–6. But the dependencies of $\Delta m_f(K_{sub})$ are such that when reduction of K_{sub} compared to K^{**}_{sub} the loss $|\Delta m_f|$ increases very slowly. If we specify some allowable loss ε_m in m_f, then range $K_{sub} \geq \widetilde{K}_{sub}$, in which $|\Delta m_f| \leq \varepsilon_m$, will be in the practical range of K_{sub} for the considered ASV classes. Figure 3.31 shows the ranges of maximum subsonic lift-to-drag ratio and initial thrust-to-weight ratio, in which the take-off with $\theta_i = 0$ differs in injected mass m_f from optimal angle $\theta_{i\ opt}$ by no more than ε_m.

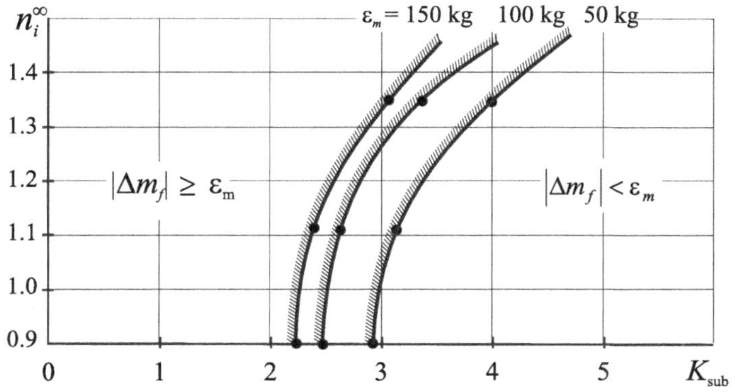

Fig. 3.31 Areas of subsonic maximum lift-to-drag ratio and initial thrust-to-weight ratio, in which the horizontal take-off differs from the optimal one in maximum injected mass m_f by no more than ε_m

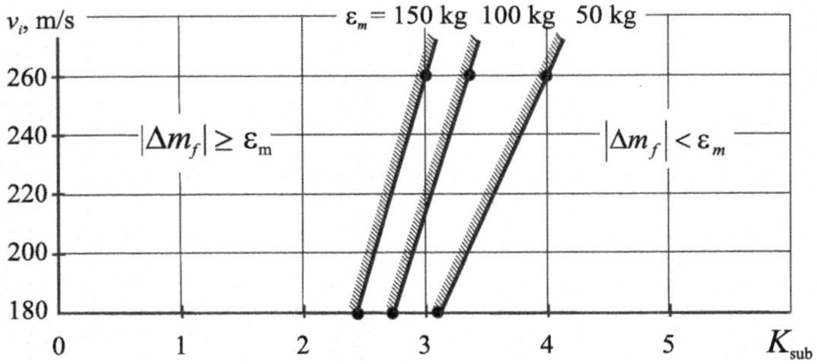

Fig. 3.32 Areas of parameters $\{K_{sub}, v_i, \varepsilon_m\}$, in which the horizontal take-off is quasi-optimal ($|\Delta m_f| < \varepsilon_m$)

Since with increasing in n_i^∞ the critical subsonic aerodynamic quality K_{sub}^* shifts towards larger values (Table 3.2), then the boundary value of K_{sub}, from which $|\Delta m_f| \leq \varepsilon_m$, increases accordingly, and the horizontal take-off does not practically differ from the optimal one.

Variation of the initial velocity in the range of 180 m/s $\leq v_i \leq$ 260 m/s shifts the boundary K_{sub} by no more than one (Fig. 3.32).

The loss of injected mass Δm_f at $K_{sub} > K_{sub}^*$ weakly depends on initial altitude h_i. The area of parameters $\{K_{sub}, h_i\}$, in which the horizontal take-off is quasi-optimal ($|\Delta m_f| < \varepsilon_m$), is shown in Fig. 3.33.

Areas of parameters $\{K_{sub}, n_i^\infty\}$, $\{K_{sub}, v_i\}$, and $\{K_{sub}, h_i\}$, in which the horizontal take-off is quasi-optimal, are calculated for case $K_{sup}^* = 2$. When K_{sup} decreases (increases) relative to this reference value, the boundaries of the indicated areas shift to the left (right), i.e. in the direction of decreasing (increasing) of K_{sub}.

These results can be interpreted for the case of air launch as follows. When a rocket stage (RS) with a high lift-to-drag ratio at subsonic speeds ($K_{sub} \geq \tilde{K}_{sub} \approx K_{sub}^*$) is

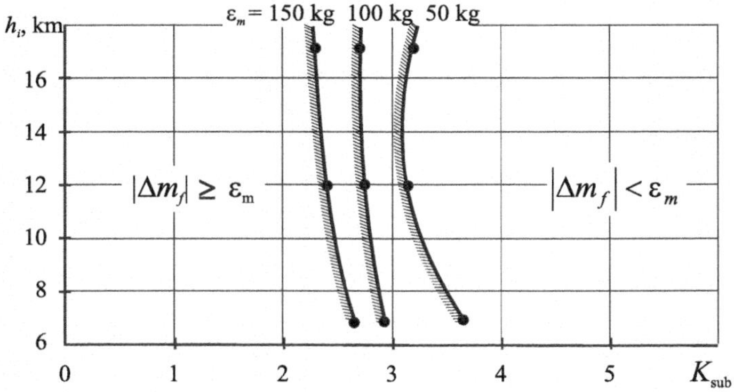

Fig. 3.33 Areas of parameters $\{K_{sub}, h_i, \varepsilon_m\}$, in which the horizontal take-off is quasi-optimal ($|\Delta m_f| < \varepsilon_m$)

used as a part of the ASTS, it is not advisable intensive CA maneuvers to create large flight path angles by the release time moment. Boundary values $\tilde{K}_{sub}$ belong to range $2 \leq \tilde{K}_{sub} \leq 4$, which corresponds to the actually reachable range of RS maximum lift-to-drag ratios at subsonic speeds.

The optimal control laws

Optimal angle-of-attack programs $\alpha_{opt}(t)$ at the horizontal take-off have a structure similar to programs with the optimal launch angle (Figs. 3.23 and 3.25). The most significant difference between them is that program $\alpha_{opt}(t)\big|_{\theta_i=\theta_{i\,opt}}$ always starts with $\alpha_{opt}(t_i) = 0$, while for ASV with a moderate subsonic maximum lift-to-drag ratio initial optimal angle-of-attack $\alpha_{opt}(t_i)\big|_{\theta_i=0} > 0$.

On Figs. 3.34, 3.35, 3.36 and 3.37 families of optimal control programs $\{\alpha_{opt}(t), 2 \leq K_{sub} \leq 6\}$ for various initial conditions and ASV parameters are shown. Figure 3.34 corresponds to the reference values of the parameters, Figs. 3.35 and 3.36 differ in the thrust-to-weight ratio ($n_i^{\infty} = 1.35$ and $n_i^{\infty} = 0.9$), Fig. 3.37 differs in initial velocity $v_i = 260$ m/s (while $\Delta M = M_1 - M_2 = 0.235$, $q_i = 778$ kgf/m^2).

An important characteristic of optimal programs $\alpha_{opt}(t)$ is the maximum angle of attack $\alpha_{max} = \max\limits_{\{t\}} \alpha_{opt}$ in dense atmosphere layers. As the calculations show, the

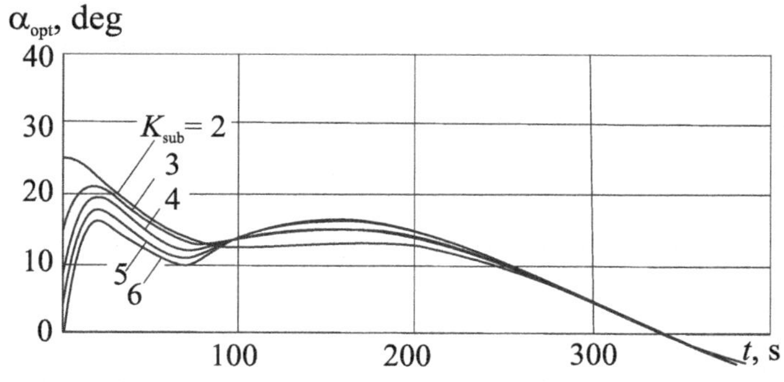

Fig. 3.34 Optimal time-programs of angle-of-attack $\alpha_{opt}(t)$ for various subsonic maximum lift-to-drag ratios K_{sub} at $K_{sup} = 2$, $v_i = 180$ m/s, $h_i = 12$ km, $\Delta M = 0.47$, $n_i^{\infty} = 1.12$, $\theta_i = 0$

Fig. 3.35 Optimal time-programs of angle-of-attack $\alpha_{opt}(t)$ for various subsonic maximum lift-to-drag ratios K_{sub} at $K_{sup} = 2$, $v_i = 180$ m/s, $h_i = 12$ km, $\Delta M = 0.47$, $n_i^{\infty} = 1.35$, $\theta_i = 0$

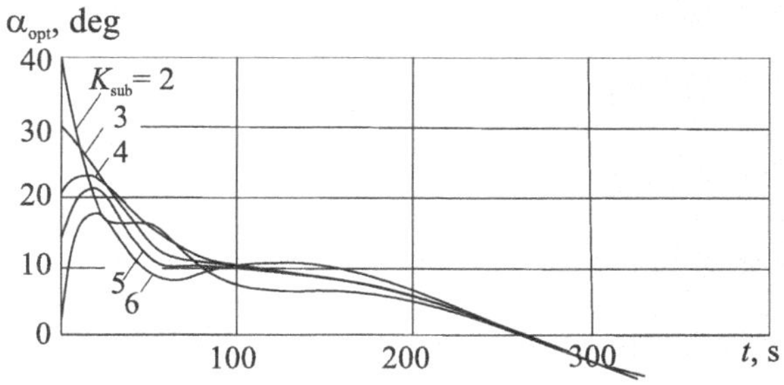

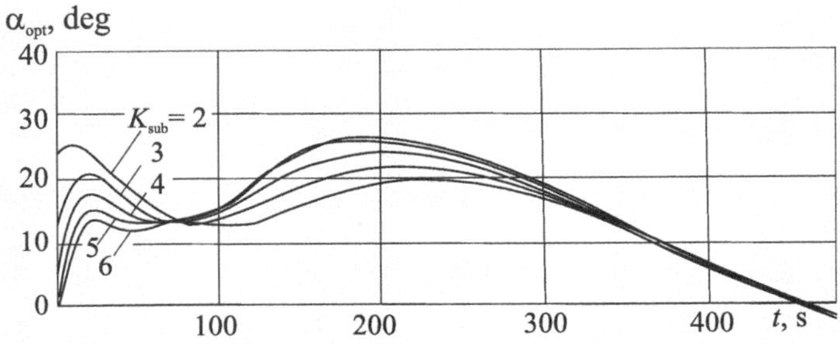

Fig. 3.36 Optimal time-programs of angle-of-attack $\alpha_{opt}(t)$ for various subsonic maximum lift-to-drag ratios K_{sub} at $K_{sup} = 2$, $v_i = 180$ m/s, $h_i = 12$ km, $\Delta M = 0.47$, $n_i^{\infty} = 0.9$, $\theta_i = 0$

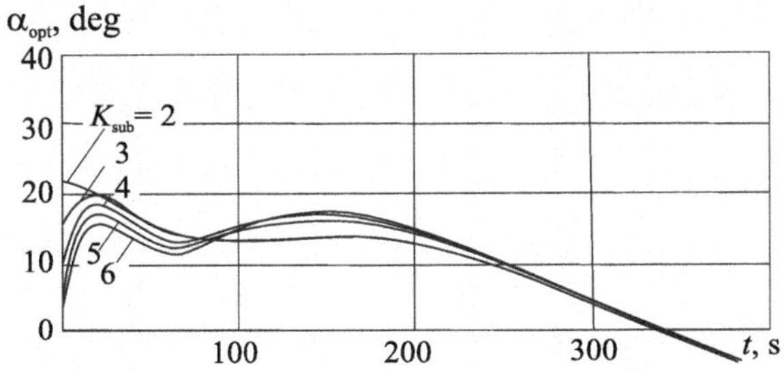

Fig. 3.37 Optimal time-programs of angle-of-attack $\alpha_{opt}(t)$ for various subsonic maximum lift-to-drag ratios K_{sub} at $K_{sup} = 2$, $v_i = 260$ m/s, $h_i = 12$ km, $\Delta M = 0.47$, $n_i^{\infty} = 0.9$, $\theta_i = 0$

setting initial path angle $\theta_i = 0$ causes the increase of α_{max} in comparison with the case of the optimal take-off angle. However, if ASV has a sufficiently high lift-to-drag ratio ($K_{sub} > K_{sub}^*$), then for the considered parameter ranges the maximum angle $\alpha_{max}\big|_{\theta_i=0}$ does not practically differ from $\alpha_{max}\big|_{\theta_i=\theta_{i\,opt}}$.

Aerodynamic loads

As a result of the decrease in initial path angle θ_i in comparison with $\theta_{i\,opt}$, the initial trajectory leg takes place in denser atmosphere layers, which, as a rule, causes an increase in the level of aerodynamic loads. With an increase in ASV lifting properties in range $0 \le K_{sub} \le 3$–4, maximum dynamic pressure q_{max} decreases sharply. A further increase of K_{sub} has practically no effect on q_{max}. Dependences of $q_{max}(K_{sub})$ for various n_i^{∞}, v_i, and h_i at $\theta_i = 0$ in comparison with similar dependences for $\theta_i = \theta_{i\,opt}$ are shown in Figs. 3.38, 3.39 and 3.40.

The growth of initial thrust-to-weight ratio n_i^{∞} is accompanied by a decrease in q_{max} on the optimal trajectory at $\theta_i = 0$. An example of such a dependence at $K_{sup} = 2$, $K_{sub} \ge 3$ is given in Table 3.6.

A decrease in h_i also leads to a noticeable increase in q_{max} (Fig. 3.40), which is due to the immersion of the initial trajectory section into denser atmosphere layers and a decrease in the thrust. An example of such a dependence for $K_{sup} = 2$, $K_{sub} \ge 3$ at $\theta_i = 0$ is given in Table 3.7.

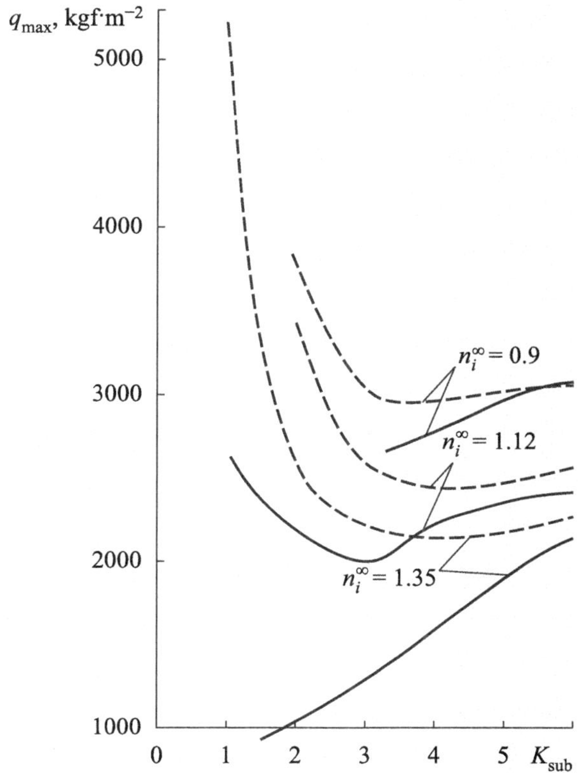

Fig. 3.38 Dependences of maximum dynamic pressure q_{max} on subsonic maximum lift-to-drag ratio K_{sub} for various initial thrust-to-weight ratios n_i^∞ at $\theta_i = 0$ (dashed lines) and $\theta_i = \theta_{i\ opt}$ (solid lines)

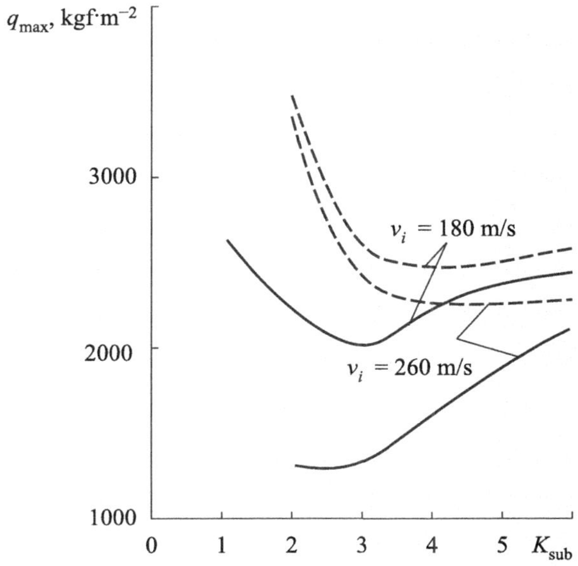

Fig. 3.39 Dependences of maximum dynamic pressure q_{max} on subsonic maximum lift-to-drag ratio K_{sub} for various initial velocities at $\theta_i = 0$ (dashed lines) and $\theta_i = \theta_{i\ opt}$ (solid lines)

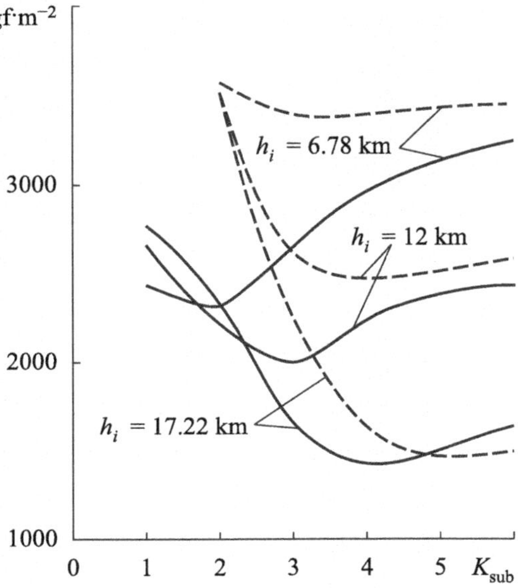

Fig. 3.40 Dependences of maximum dynamic pressure q_{max} on subsonic maximum lift-to-drag ratio K_{sub} for various initial altitudes at $\theta_i = 0$ (dashed lines) and $\theta_i = \theta_{i\ opt}$ (solid lines)

Table 3.6 The dependence of maximum dynamic pressure q_{max} on initial thrust-to-weight ratio n_i^{∞} at the optimal ascent with $\theta_i = 0$

n_i^{∞}	0.90	1.12	1.35
q_{max}, kgf m^{-2}	3000–3100	2500–2600	2100–2300

Table 3.7 The dependence of maximum dynamic pressure q_{max} on initial altitude h_i, at the optimal ascent with $\theta_i = 0$

h_i, km	17.22	12.00	6.78
q_{max}, kgf/m^2	1500–2200	2400–2600	3400

Although the increase in initial velocity v_i is accompanied by an increase in initial dynamic pressure q_i, however, q_{max} decreases in this case (Fig. 3.39). This result is explained by the fact that an increase in v_i leads to a reduction of $\Delta M = M_1 - M_i$ and, as a result, of the flight segment with a high level of maximum lift-to-drag ratio $K_{max}(M) = K_{sub}$, $M \in [M_i, M_1]$.

In the entire considered area of ASV parameters and initial conditions, the maximum normal load factor (in the body axis system) $n_{z\ max} = \max_{\{t\}} n_z$ (Fig. 3.41). Dependences $n_{z\ max}(K_{sub})$ are nonmonotonic and, as a rule, have a minimum in range $1 \leq K_{sub} \leq 3$. This is explained by the following reasons. The increase in K_{sub} is linearly related to the increase in the aerodynamic lift coefficient, and the maximum dynamic pressure depends nonmonotonically on K_{sub}: q_{max} sharply decreases with increasing in K_{sub} at low ASV lifting properties and slowly increases at high lifting properties. Since n_z is proportional to the coefficient of aerodynamic lift and dynamic pressure, then three sections can be distinguished qualitatively in dependency $n_{z\ \max}(K_{\text{sub}})$ (Fig. 3.42):

I. In a small neighborhood of $K_{\text{sub}} = 0$, we have $C_D^{\alpha} \cong 0$ and, therefore $n_{z\ max} \cong 0$. Since $n_{z\ max} \geq 0$, as K_{sub} increases the value of $n_{z\ max}$ increases from zero to some positive (maximum) value.

II. With a further increase in K_{sub} up to $K_{\text{sub}} \approx$1–3 a decrease in $q_{\max}$, as a rule, turns out to be more intense (Figs. 3.38, 3.39, 3.40 and 3.41), compared to the increase in C_D^{α}, which leads to a decrease in $n_{z\ max}$.

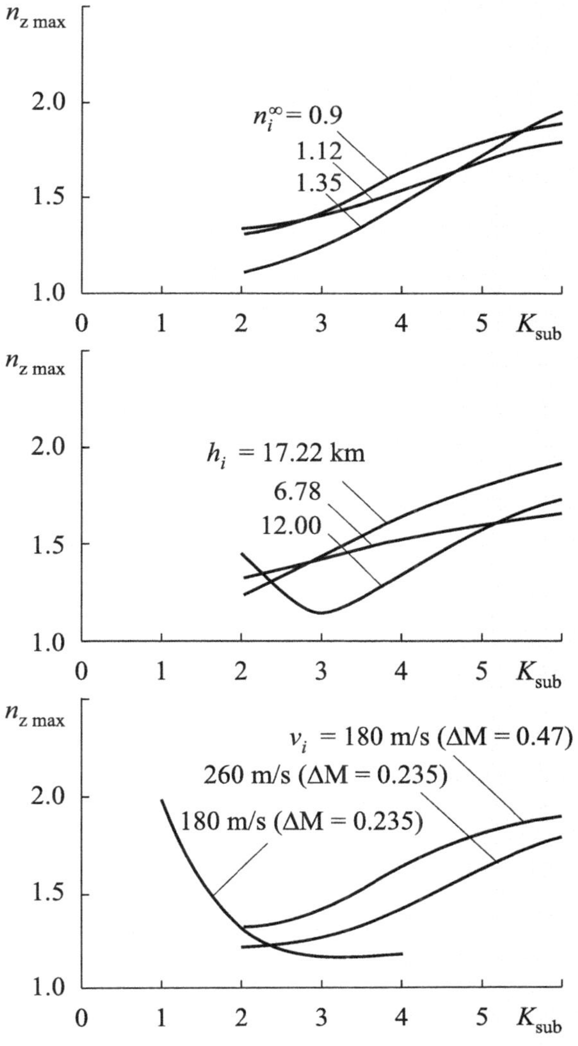

Fig. 3.41 Dependences of maximum normal load factor $n_{z\ \max}$ on subsonic maximum lift-to-drag ratio K_{sub} for various initial velocities and Mach number ranges in the aerodynamic model of $K(M)$ at the optimal ascent with $\theta_i = 0$

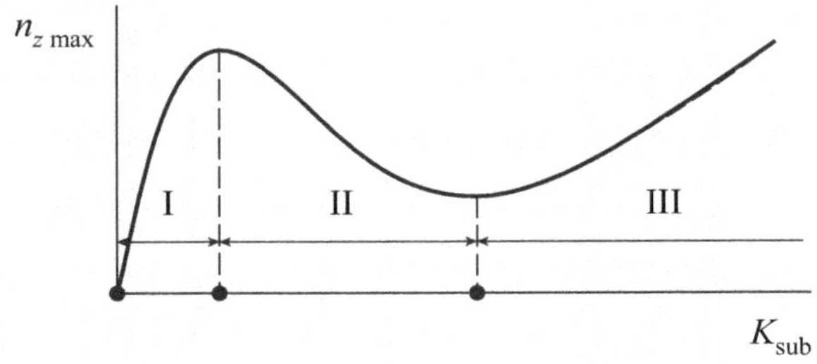

Fig. 3.42 A typical structure of the dependency of normal load factor $n_{z\,max}$ on subsonic maximum lift-to-drag ratio K_{sub}

III. In the range of large K_{sub}, both C_D^α and q_{max} increase, therefore, $n_{z\,max}$ also increases with K_{sub}.

Increasing initial path angle from $\theta_i = 0$ to $\theta_{i\,opt}$ leads to a decrease in the level of maximum dynamic loads (q_{max} and $n_{z\,max}$).

3.4 Analysis of Injected Mass Sensitivity to Perturbations of the Optimal Trajectory

When optimizing trajectories by PMP, a conjugate system solution has the physical meaning of the influence (sensitivity) functions of variation $\delta\mathbf{x}$ of current phase vector $\mathbf{x}(t)$ on maximum functional m_f (see 1.60). In the accepted start coordinate system at initial time t_i vertical component y_i of radius vector $\mathbf{r}_i$ coincides with altitude h_i. Then:

$$p_{x\,i} = \frac{\partial m_f}{\partial x_i}, p_{y\,i} = \frac{\partial m_f}{\partial h_i}, p_{v_x\,i} = \frac{\partial m_f}{\partial v_{x\,i}}, p_{v_y\,i} = \frac{\partial m_f}{\partial v_{y\,i}}, p_{m\,i} = \frac{\partial m_f}{\partial m_i}. \tag{3.3}$$

If initial velocity v_i is given, and initial path angle θ_i is free (optimized), then the initial components of the velocity vector are concerned with relations:

$$v_{x\,i} = v_i \cos\theta_i,\ v_{y\,i} = v_i \sin\theta_i. \tag{3.4}$$

Since θ_i must provide an extremum to m_f, the necessary extremum condition takes place

$$\left.\frac{\partial m_f}{\partial \theta_i}\right|_{\theta_i\,opt} = 0. \tag{3.5}$$

Taking into account (3.4), it follows from (3.5):

$$\left.\frac{\partial m_f}{\partial \theta_i}\right|_{\theta_i\,opt} = \frac{\partial m_f}{\partial v_{x\,i}}\frac{\partial v_{x\,i}}{\partial \theta_i} + \frac{\partial m_f}{\partial v_{y\,i}}\frac{\partial v_{y\,i}}{\partial \theta_i} = v_i\left(\frac{\partial m_f}{\partial v_{y\,i}}\cos\theta_{i\,opt} - \frac{\partial m_f}{\partial v_{x\,i}}\sin\theta_{i\,opt}\right) = 0,$$

whence we obtain the relation between initial components of the conjugate vector:

$$p_{v_y\,i} = \frac{\partial m_f}{\partial v_{yi}} = \frac{\partial m_f}{\partial v_{xi}} \cdot \tan\theta_{i\ \mathrm{opt}} = p_{v_x\,i} \cdot \tan\theta_{i\ \mathrm{opt}}. \tag{3.6}$$

Time-dependences of dimensionless conjugate variables $p_y, p_{v_x}, p_{v_y}, p_m$ on the optimal ascent trajectory in the vertical plane are shown in Figs. 3.43, 3.44, 3.45 and 3.46 for various K_{sub} at $K_{\mathrm{sup}} = 2$ taking into account the standard atmosphere, Earth's sphericity and central gravitational field. Dimensional and dimensionless variables are connected, according to (2.20) and (3.3), by the following approximate relations: $p_{x,y\ \mathrm{dim}} \cong p_{x,y} \cdot 0.023\mathrm{kg\ m}^{-1}$, $p_{v_x,v_y\ \mathrm{dim}} \cong p_{v_x,v_y} \cdot 0.023\mathrm{kg\ s\ m}^{-1}$, $p_{m\ \mathrm{dim}} = p_m$. Comparing the obtained exact optimal solutions with (2.15), it can be seen that they agree well in rarefied atmosphere layers (h > 40–50 km, q < 100–300 kgf/m^2). On the initial flight segment in dense atmosphere layers the influence functions differ noticeably from the OEL optimal solutions (2.15) both in orders of magnitude and in the dependences nature.

The dependencies of conjugate variables (influence functions) on $K_{\mathrm{sub}} \in [1, 6]$, $n_i^{\infty} \in [0.9, 1.35]$, $h_i \in [6.8, 17.2]$ km, $v_i \in [180, 260]$ m/s, $\Delta M \in (0.235, 0.470)$

Fig. 3.43 Time-dependences of the dimensionless conjugate variable p_y on the optimal ascent trajectory for various subsonic maximum lift-to-drag ratios K_{sub}

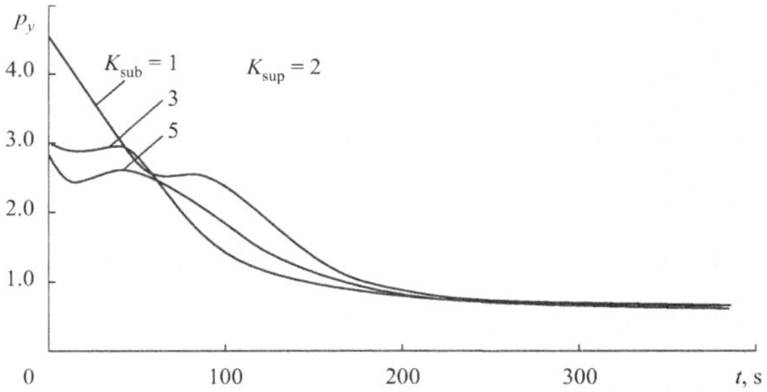

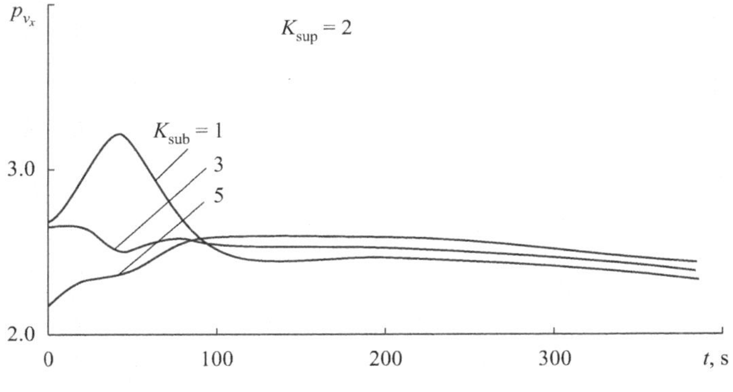

Fig. 3.44 Time-dependences of dimensionless conjugate variable p_{v_x} on the optimal ascent trajectory for various subsonic maximum lift-to-drag ratios K_{sub}

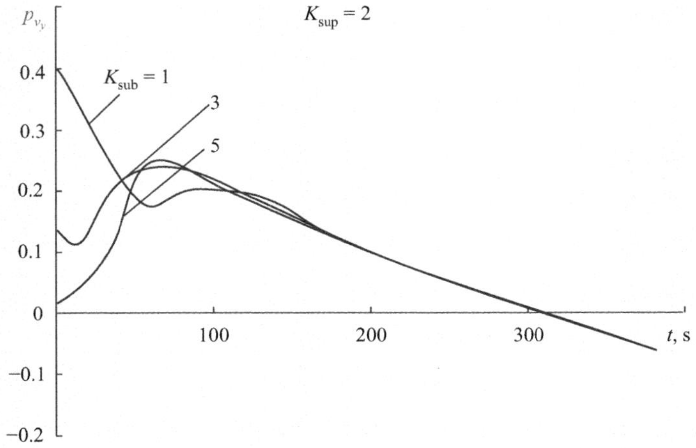

Fig. 3.45 Time-dependences of dimensionless conjugate variable p_{v_y} on the optimal ascent trajectory for various subsonic maximum lift-to-drag ratios K_{sub}

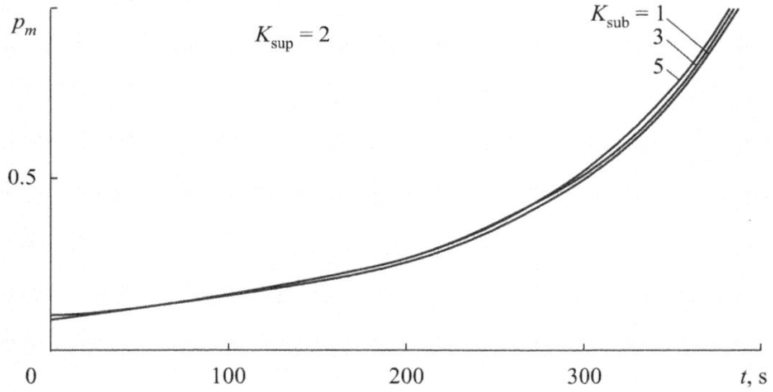

Fig. 3.46 Time-dependences of conjugate variable p_m on the optimal ascent trajectory for various subsonic maximum lift-to-drag ratios K_{sub}

are shown in Figs. 3.47, 3.48, 3.49, 3.50, 3.51, 3.52, 3.53, 3.54 and 3.55. Of the characteristic features of the dependences, we note the presence of the minimum of the initial altitude influence function $p_{yi} = \frac{\partial m_f}{\partial h_i}(K_{sub})$ (Figs. 3.47, 3.50, 3.53) and the maximum of initial horizontal velocity influence function $p_{v_x i} = \frac{\partial m_f}{\partial v_{xi}}(K_{sub})$ (Figs. 3.48, 3.51 and 3.54) near critical value K^*_{sub}. As K_{sub} moves away from the values, at which $\frac{\partial m_f}{\partial h_i}$ and $\frac{\partial m_f}{\partial v_{xi}}$ reach the indicated extremes, the influences on m_f of the initial altitude and velocity noticeably increases and decreases respectively.

With an increase in initial altitude h_i its influence on the injected mass falls (Fig. 3.50). Table 3.8 shows derivatives $\frac{\partial m_f}{\partial h_i}$ for $K_{sup} = 2$ and $K_{sub} = 4$.

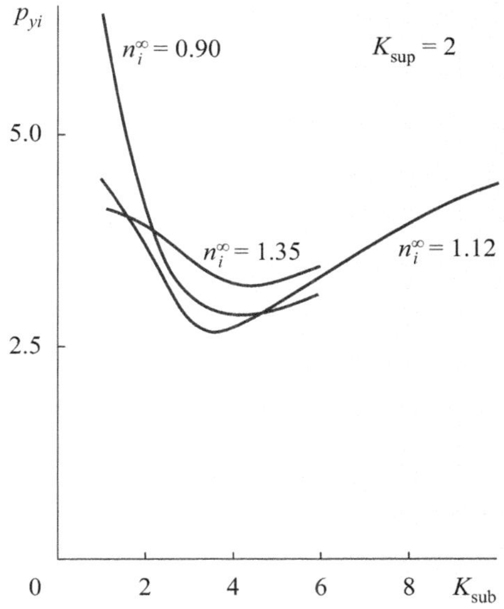

Fig. 3.47 Dependencies of optimal initial dimensionless conjugate variable $p_{y\,i}$ on subsonic maximum lift-to-drag ratio K_{sub} for various initial thrust-to-weight ratios n_i^∞

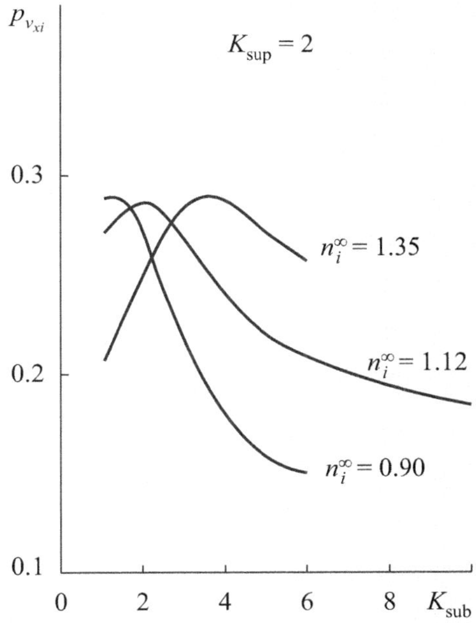

Fig. 3.48 Dependencies of optimal initial dimensionless conjugate variable $p_{v_x i}$ on subsonic maximum lift-to-drag ratio K_{sub} for various initial thrust-to-weight ratios n_i^∞

Fig. 3.49 Dependencies of optimal initial conjugate variable $p_{m\,i}$ on subsonic maximum lift-to-drag ratio K_{sub} for various initial thrust-to-weight ratios n_i^{∞} and supersonic maximum lift-to-drag ratios K_{sup}

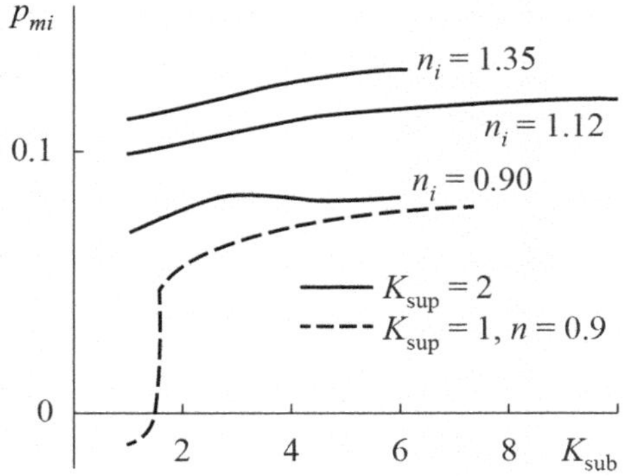

Fig. 3.50 Dependencies of optimal initial dimensionless conjugate variable $p_{y\,i}$ on subsonic maximum lift-to-drag ratio K_{sub} for various initial altitudes h_i and supersonic maximum lift-to-drag ratios K_{sup}

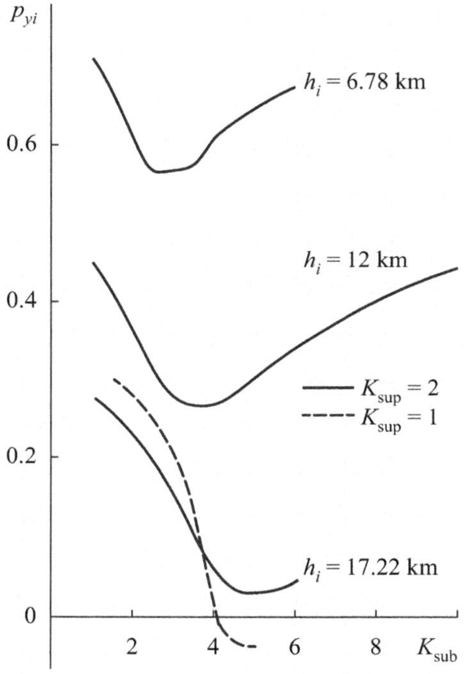

A sharp decrease in the influence of altitude h_i on the injected mass, especially with a high subsonic maximum lift-to-drag ratio is due to the fact that at high altitudes (low dynamic pressure), ASV cannot take full advantage of its lifting properties. This effect is the stronger, the lower the maximum lift-to-drag ratio at supersonic speeds, since the greater the difference between K_{sub} and K_{sup}, the more efficiently using the aerodynamic forces in the initial trajectory segment with subsonic speeds. In particular, as can be seen from Fig. 3.50, for $K_{\text{sup}} = 1$ there is range $K_{\text{sub}} \geq 4$, in

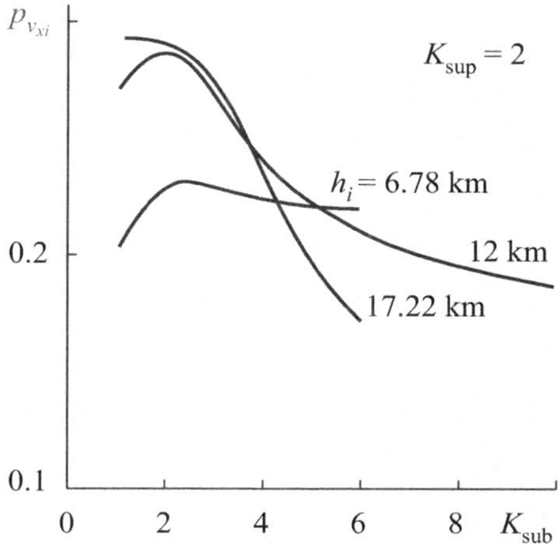

Fig. 3.51 Dependencies of the optimal initial dimensionless conjugate variable $p_{v_x i}$ on subsonic maximum lift-to-drag ratio K_{sub} for various initial altitudes h_i

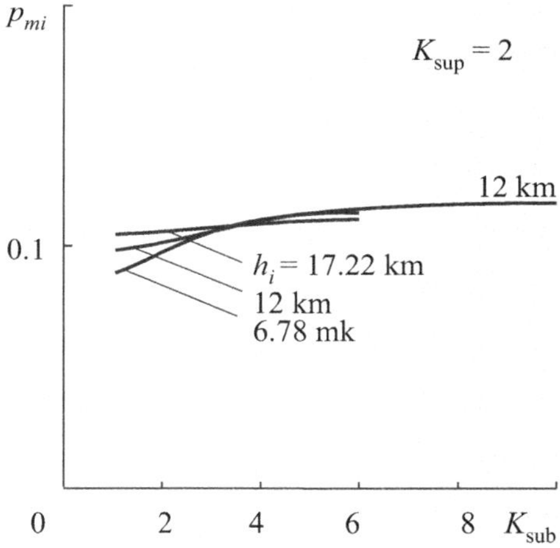

Fig. 3.52 Dependencies of optimal initial conjugate variable $p_{m\,i}$ on subsonic maximum lift-to-drag ratio K_{sub} for various initial altitudes h_i

which influence function $\frac{\partial m_f}{\partial h_i}$ even changes sign and becomes negative: $\frac{\partial m_f}{\partial h_i} < 0$. This means that for such ASV, a decrease in the initial altitude allows increasing maximum injected mass m_f. Altitude $h_{i\,\text{opt}}$, at which $\left.\frac{\partial m_f}{\partial h_i}\right|_{h_{i\,\text{opt}}} = 0$, is the locally optimal initial altitude. Since at each current time moment t phase vector $\mathbf{x}(t)$ can be interpreted as the initial one for the subsequent segment of the trajectory, then in the general case, by varying the velocity $v = v_i$ and mass $m = m_i$, one can get in the phase space set $\boldsymbol{\Gamma}_{\text{opt}}$ of optimal trajectories (trajectory sections) in a parametric form:

$$\Gamma_{\text{opt}} = \left\{\mathbf{x} : h_{\text{opt}} = h_{i\,\text{opt}}(v, m), \theta_{\text{opt}} = \theta_{i\,\text{opt}}(v, m)\right\},$$

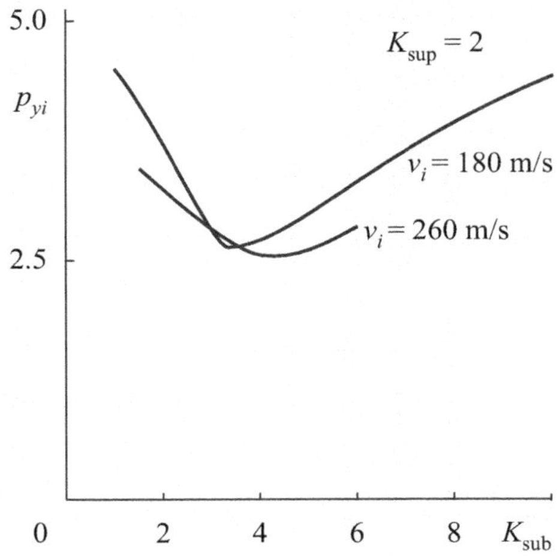

Fig. 3.53 Dependencies of optimal initial dimensionless conjugate variable $p_{y\,i}$ on subsonic maximum lift-to-drag ratio K_{sub} for two initial velocities v_i

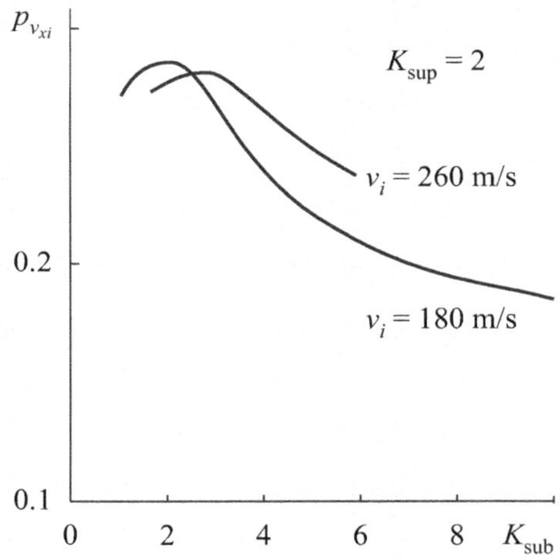

Fig. 3.54 Dependencies of optimal initial dimensionless conjugate variable $p_{v_x i}$ on subsonic maximum lift-to-drag ratio K_{sub} for two initial velocities v_i

which is determined only by ASV characteristics and parameters of the target orbit. The trajectories in $\boldsymbol{\Gamma}_{opt}$ are close in their mathematical basis to the degenerate extremals studied by Miele [20] in problems of optimal aircraft climbing. In the general case, the optimal ASV injection trajectory γ_{opt} may contain segments of $\boldsymbol{\Gamma}_{opt}$ in those cases when, firstly, γ_{opt} intersects the surface $\boldsymbol{\Gamma}_{opt}$ in the phase space, and, second, the motion along $\boldsymbol{\Gamma}_{opt}$ does not contradict the equations of motion and accepted constraints (Fig. 3.56). In the calculations performed, the identified segments $\boldsymbol{\Gamma}_{opt}$ pass at altitudes $h \geq 15$ km.

The existence of such "degenerate" segments is a new distinctive property of optimal ascent trajectories of LVs with high lift-to-drag ratio.

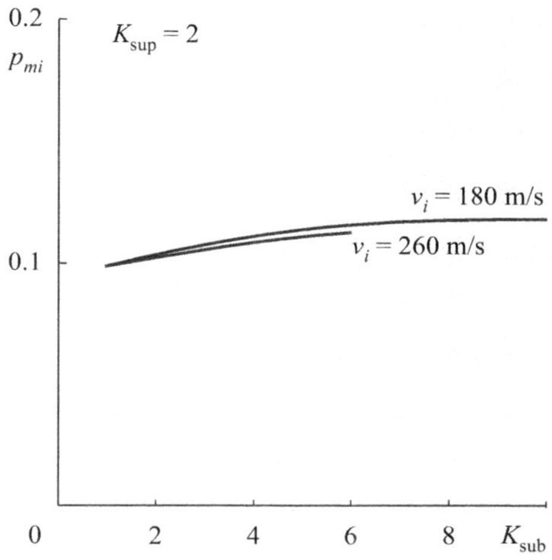

Fig. 3.55 Dependencies of optimal initial conjugate variable $p_{m\ i}$ on subsonic maximum lift-to-drag ratio K_{sub} for two initial velocities v_i

Table 3.8 The initial altitude influence function values for some initial altitudes

h_i, km	6.78	12.0	17.22
$\partial m_f/\partial h_i$, kg km^{-1}	132	61	13

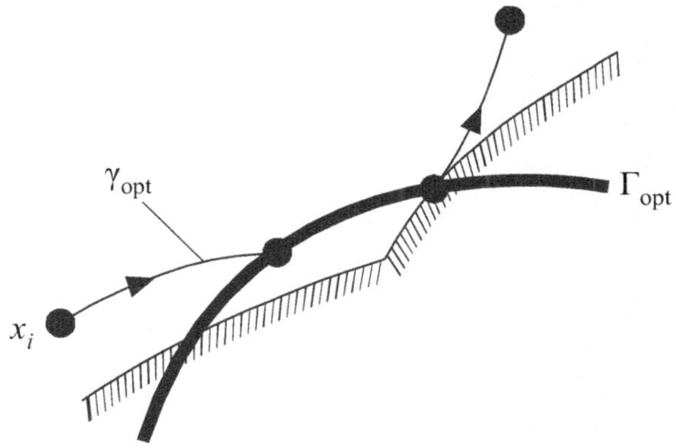

Fig. 3.56 Structure of optimal trajectory γ_{opt} with segment of the degenerate extremal Γ_{opt} in the phase space

The effect of initial (current) mass m on injected mass m_f is weakly related to subsonic aerodynamic characteristics, initial altitude, and velocity (Figs. 3.52 and 3.55). At the same time, with a decrease in initial thrust-to-weight ratio n_i^{∞}, the influence of initial ASV mass $\frac{\partial m_f}{\partial m_i}$ decreases noticeably (Fig. 3.49). With a low thrust-to-weight ratio ($n_i^{\infty} \leq 1$) the influence of other factors, in particular, of the maximum lift-to-drag ratio, on $\frac{\partial m_f}{\partial m_i}$ also increases.

When analyzing the influence of m_i on m_f it should be taken into account that an increase in m_i leads not only to an increase in the fuel mass reserve for ascent, but also to a decrease in initial thrust-to-weight ratio $n_i = \frac{T_i}{m_i g}$, as well as to an growth in specific aerodynamic load on the reference area. If n_i^∞ is small, then the role of the last two factors noticeably increases, which leads to a sharp decrease of $\frac{\partial m_f}{\partial m_i}$ down to negative values (Fig. 3.49). In area, where $\frac{\partial m_f}{\partial m_i} < 0$, reduction of the initial mass to $m_{i\,\text{opt}}$ contributes to an increase in injected mass. Thus, with a low initial thrust-to-weight ratio ($n_i^\infty \leq 1$) there may be optimal values of initial mass $m_{i\,\text{opt}}$, defined by necessary condition $\left.\frac{\partial m_f}{\partial m_i}\right|_{m_{i\,\text{opt}}} = 0$.

The current mass $m(t)$ can be considered as the initial one for the subsequent segment of the motion, therefore, in the phase space $\{h, v, \theta, m\}$, in the general case, there may be extremal surface

$$\mathfrak{R}_{\text{opt}} = \left\{\mathbf{x} : m = m_{i\text{opt}}(h, v), \theta_{\text{opt}} = \theta_{i\text{opt}}(h, v)\right\},$$

where the optimal mass and flight path angle are determined by the current values of h, v. Surface $\mathfrak{R}_{\text{opt}}$ depends on ASV parameters and the target orbit. It should be emphasized that the existence of $\mathfrak{R}_{\text{opt}}$ in this case, it is in no way related to dependences of ASV structure mass on the fuel mass or other layout restrictions that were not considered in this work, but is due solely to the dynamics of the optimal flight in the atmosphere.

3.5 The Influence of LV Aerodynamic Shape on Optimal Ascent Trajectories and Control Laws

It was shown above that one of the main characteristics that determines the structure of the optimal LV ascent trajectory is the maximum lift-to-drag ratio $K_{\max}$. However, in most cases, changing the layout to increase of $K_{\max}$ also leads to a change in other aerodynamic characteristics. Therefore, research [3–5] were continued in order to study the influence on the structure of optimal solutions and the maximum injected mass of not isolated aerodynamic characteristics, but aerodynamic shapes of the vehicle as a whole.

It should be noted that the main goal of this work is to identify the main regularities of optimal motion in the atmosphere, and not to optimize ASV layouts. Therefore, taking the injected mass as the functional, rather than the payload mass, is reasonable here. Otherwise, the optimal solutions would also depend on specific design decisions in the choice of the overall layout, materials, etc. Further development of the approach proposed in this monograph for solving the problem of complex ASV optimization according to the criterion of maximum efficiency of the target use, taking into account the requirements of aerodynamics, dynamics, and strength, is contained, for example, in [21].

This section describes the results of numerical studies of the influence of the geometric parameters of LV aerodynamic layout on the optimal ascent trajectories [10, 11]. The main classes of LV aerodynamic shapes are considered:

- conical (blunted cone with an elliptical cross section),
- cylindrical (blunted cylinder with an elliptical cross section), and
- winged (blunted circular cylinder with a low elongation wing).

Inside of these types, the aerodynamic shape parameters were changed in order to increase the lifting properties. For that, the "compression factor" (eccentricity) of the elliptical cross section and the wing area were varied. Approximations of the aerodynamic characteristics as functions of the angle of attack and flight modes were constructed using the results of calculations and experimental data [10, 11]. Calculations of optimal ascent trajectories were carried out using *ASTER* software package.

3.5.1 LV Aerodynamic Shapes

Optimal launch trajectories were studied for three types of aerodynamic shapes, covering in general the most common cases:

- conical bodies with elliptical cross section;
- cylindrical bodies with elliptical cross section;
- cylindrical bodies with a delta wing.

For all types of the layouts, the influence of modifications of their geometric shape was studied on the injected mass and optimal ascent trajectories. A detailed description of the geometric parameters of these layouts is given below.

Aerodynamic configurations of the first type are blunt cones with an elliptical cross section. The basis of this family is a circular cone with an apex angle of 11.42° and a ratio of the bluntness radius to the radius of the cone base of 0.05 (Fig. 3.57a). In order to increase the lifting properties, the cone were modified by compressing along the entire length of the body from circular to elliptical cross sections with a given ratio of width a to height b, while maintaining the cross-sectional area. Parameter a/b ("compression factor") varied from 1 to 3 with a step of 0.25.

Aerodynamic layouts of the second type are blunted cylinders with an elliptical cross section. The basis of this family is a circular cylinder with a blunted conical nose (Fig. 3.57b). The geometric dimensions of the cylinder were determined from the condition that its volume is equal to the volume of the basic conical body described above. The ratio of the length of the circular cylinder to the diameter is equal to 12, the apex angle of the conical nose was equal to 20°, and the ratio of the blunt radius to the radius of the cylinder cross section was equal to 0.1. The modification of this layout consisted, as in the first case, in the transition from circular to elliptical cross sections along the entire length of the body. Compression factor varied in the range from 1 (circular cylinder) to 3 with the step of 0.25 while maintaining a constant cross-sectional area.

For the third type of layouts, an increase in the lifting properties was achieved by installing a delta wing on the base model of the second type according to the

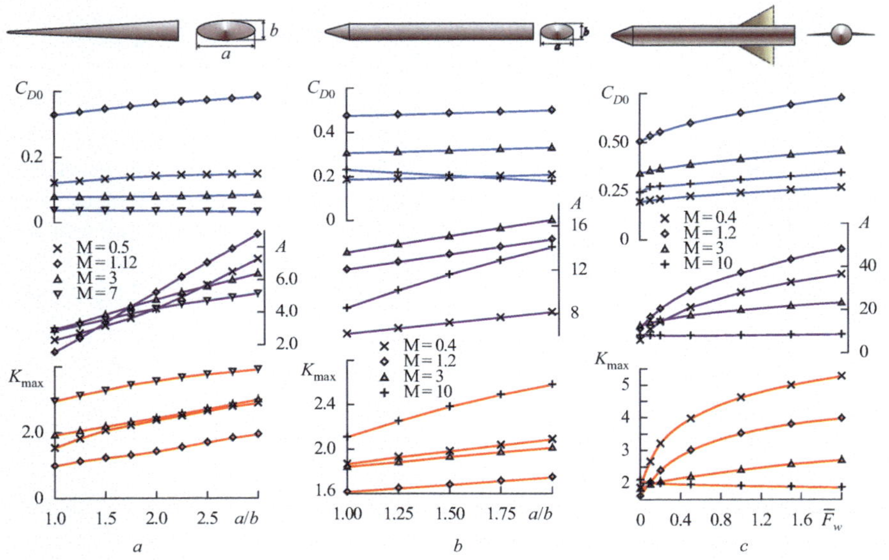

Fig. 3.57 Aerodynamic characteristics of conical (*a*) and cylindrical (*b*) bodies depending on cross section compression factor *a/b* and of cylindrical bodies with a delta wing (*c*) depending on the relative wing panel area $\overline{F}_w$

mid-wing scheme (Fig. 3.57c). The sweep angle of the leading edges of the wing is 60°. The relative area of the wing panels $\overline{F}_w = F_w/F_0$ was specified as the following set of values: 0; 0.1; 0.2 and 0.5.

Characteristics at intermediate points were determined by a linear approximation with respect to parameters *a*/*b* and F_w. In accordance with the model of aerodynamic forces (2.21)–(2.23), the aerodynamic coefficients D_0, D_α, $C_L{}^\alpha$ and their partial derivatives were defined with respect to the Mach number range. To simplify the procedure for the numerical solution of the original problem, these derivatives were continuous. Coefficients D_0 and D_α were calculated using Eqs. (2.23). The dependences of $C_L{}^\alpha$, C_{D0} and A on the Mach number were specified as third-order splines. The splines were created based on the processing of the initial dependences of aerodynamic coefficients C_D and C_L on α and M. For each Mach number corresponding to the nodal point of the spline, the coefficients of aerodynamic polar (2.22) approximating dependences $C_D\ (\alpha)$,$C_L\ (\alpha)$ were calculated. For the approximating polar and initial characteristics, it was necessary to match the values of the maximum lift-to-drag ratio $K_{\max}$, which is determined by formula

$$K_{\max} = \frac{C_L^\alpha}{\sqrt{C_{D0}(C_{D0} + 4A)}} = \frac{C_L^\alpha}{\sqrt{D_0^2 - D_\alpha^2}},$$

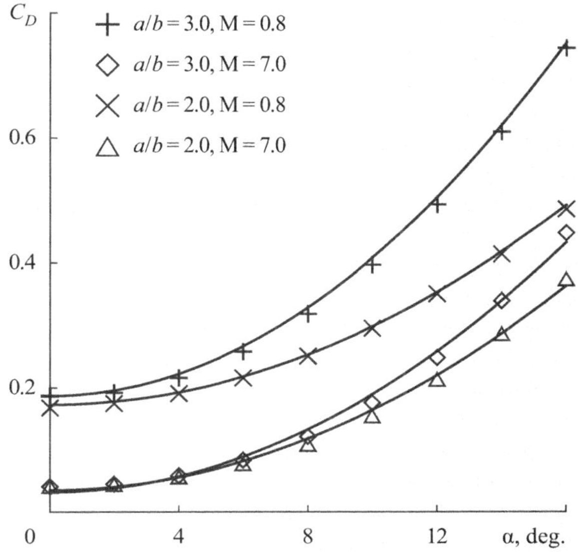

Fig. 3.58 Approximations of the aerodynamic drag coefficient dependence on the angle-of-attack and the experimental data (markers) for various Mach numbers and compression factors *a*/*b*

and of drag coefficient at zero lift C_{D0}. Induced drag factor A was determined from the condition of minimizing the root-mean-square deviation of C_D from the original data. An example of comparing the obtained approximations $C_D(\alpha)$ (solid lines) with the original data (markers) for two Mach numbers and *a*/*b* is shown on Fig. 3.58. After calculating coefficients $C_L{}^{\alpha}$, C_{D0} and A for the nodal points, the third-order spline coefficients were determined by the Mach number. These coefficients were calculated from the condition of minimizing the root-mean-square deviation of $C_L{}^{\alpha}(M)$, $C_{D0}(M)$ and $A(M)$ from piecewise linear dependences passing through the reference points. The specified procedure for constructing the dependences of the coefficients of aerodynamic forces on the angle of attack and Mach number is performed automatically by *ASTER* software package. The obtained dependences of the aerodynamic characteristics are shown in Fig. 3.57.

3.5.2 The Influence of Geometrical Parameters of LV Layout on the Injected Mass

The influence of the geometric parameters of the vehicle layout on the optimal control laws and launch trajectories was studied under the following nominal conditions: initial velocity $v_i = 50$ m/s, initial altitude $h_i = 200$ m, the final orbit is circular with $h_{\text{orb}} = 275$ km, initial thrust-to-weight ratio $n_i = (T/mg)_i = 1.1$, maximum relative mass consumption rate $\mu = 3.83 \cdot 10^{-3}$ s^{-1}, initial specific load on the mid-section $m_i/F_0 = 6 \cdot 10^3$ kg m^{-2} for cylindrical forms and $1.6 \cdot 10^3$ kg m^{-2} for conical forms.

Current LVs with vertical take-off use trajectory control programs that are characterized by two segments: in the initial atmospheric ascent segment, the flight occurs with almost zero angle of attack (g-turn), then, e.g., beginning from the staging, a linear time program for the pitch angle is used, which structure corresponds to the OEL solution obtained without taking into account the atmosphere. In the sequel, the optimal solutions obtained on the basis of PMP are compared with this type of "traditional" solutions with one small edition: here, the "traditional" control after the g-turn of the first stage uses in the following section not approximate (2.16), but the strictly optimal control law based on PMP for the motion model from Sect. 2.2. Thus, the obtained estimates for the functional under the "traditional" control turn out to be better than with the OEL control law (2.16).

As can be seen from Fig. 3.57, an increase in a/b and F_w leads, as a rule, an increase both $K_{\max}$ and C_{D0}. Thus, the change in the injected mass is determined by a compromise between two opposite trends: an increase in aerodynamic lift and drag.

On Fig. 3.59 the change in the relative injected mass with a variation in compression factor a/b of the cross section of the conical LV is shown. Dependences of $\overline{m}_f = m_f(a/b)/m_f(1)$ are qualitatively different for the optimal and "traditional" (with g-turn) control programs. Under the optimal control (upper curve), there is a singular break point, which is due to a bifurcation change in the control structure on the global extremal. To the left of the bifurcation point, an increase in the compression factor leads to a decrease in the injected mass, since the optimal control for such layouts corresponds to traditional programs (type ***B***). The traditional control programs practically do not use LV lifting properties, and an increase in the compression factor causes an increase in aerodynamic drag and, as a result, in aerodynamic losses of the characteristic velocity. For these reasons, the dependences of the injected mass on the compression factor under the optimal and traditional control behave in the same way in the range of parameter *a*/*b* up to the critical (bifurcation) point. However, with a further increase in *a*/*b*, above the critical value $a/b^* \approx 1.285$, the nature of the dependence changes qualitatively: an increase in *a*/*b* leads monotonically to an increase in m_f at the optimal control (corresponds to type ***A***) and to a decrease in m_f at the control with traditional g-turn for the 1st stage. So, with the traditional approach, the best shape of the LV in accordance with Fig. 3.59 would be defined as a circular cone ($a/b = 1$). When comparing different LV shapes using strictly optimal control laws, there is a basis for searching for the best shape in the range of larger values of parameter *a*/*b*.

Note that the bifurcation point corresponds to a relatively small deformation of the circular cone ($a/b^* \approx 1.285$). In a vicinity of this point, several local extremals (***A***, ***B***, ***M***) exist simultaneously, satisfying the necessary conditions of PMP (the corresponding sections are shown in Fig. 3.59 by thin lines).

Derivatives of the injected mass with respect to the parameters of the problem (sensitivity coefficients) are of great importance in a designing process. As can be seen from Fig. 3.60, the sensitivity functions of the injected mass to the compression factor for the conical LV are fundamentally different for the optimal and traditional control programs. The hypothesis of a weak dependence of the sensitivity

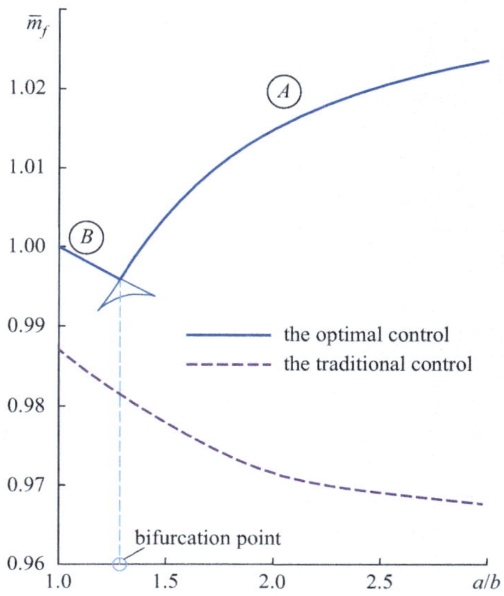

Fig. 3.59 The dependence of the relative injected mass on the compression factor for LV of the conical type

coefficients on a parameter, which is often used at LV designing, is not justified at rigorous objective consideration based on PMP. The results on Fig. 3.60 demonstrate that actual sensitivity functions can contain discontinuities due to the change of the optimal control structure. At the bifurcation point, the derivative $\partial m_f/\partial(a/b)$ can change not only several times in absolute value, but also in sign. The similar jump at the parameter bifurcation value is observed for the sensitivity coefficients of the maximum injected mass to other parameters, e.g., to the initial thrust-to-weight ratio and specific aerodynamic load on the midsection.

On Fig. 3.61 dependencies of the relative maximum injected mass on compression factor a/b of a cylindrical LV has a similar character to that of a conical type. At the same time, the range of changes in the maximum injected mass is noticeably smaller, since the aerodynamic characteristics of a cylindrical vehicle change less with variation of *a*/*b* (Fig. 3.57).

The change in the relative injected mass with a variation in relative wing area $\overline{F}_w = F_w/F_0$ (Fig. 3.62) has the same features as with a variation in the compression factor. There is a critical value $\overline{F}_w^*$ of parameter $\overline{F}_w$, which separates the optimality areas of extremals of types ***A*** and ***B***. The ***B***-extremals are globally optimal for $\overline{F}_w < \overline{F}_w^*$. For $\overline{F}_w > \overline{F}_w^*$, the ***A***-extremals, which use of the lifting properties, become globally optimal. It is important that a qualitative restructuring of the optimal control program and dependences on the parameter occurs at small values of $\overline{F}_w^* \sim 0.025$, which corresponds to the installation of lifting panels, the area of which is only a few percent of the body midsection area, i.e. forming rather a small plumage than a wing.

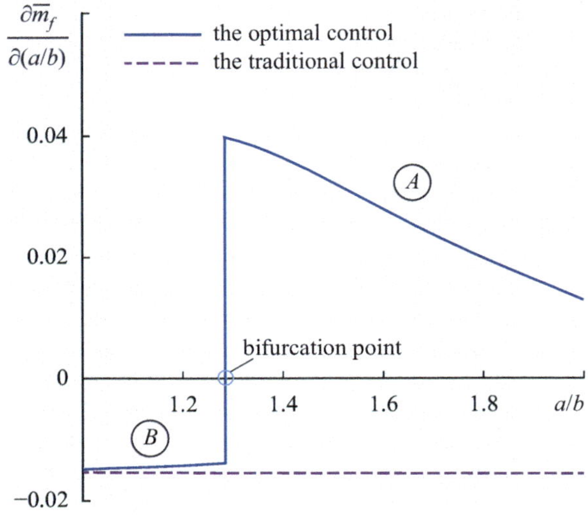

Fig. 3.60 The jump in the sensitivity coefficient of the maximum relative injected mass to variations in the compression factor of the conical LV when changing the type of the global extremal

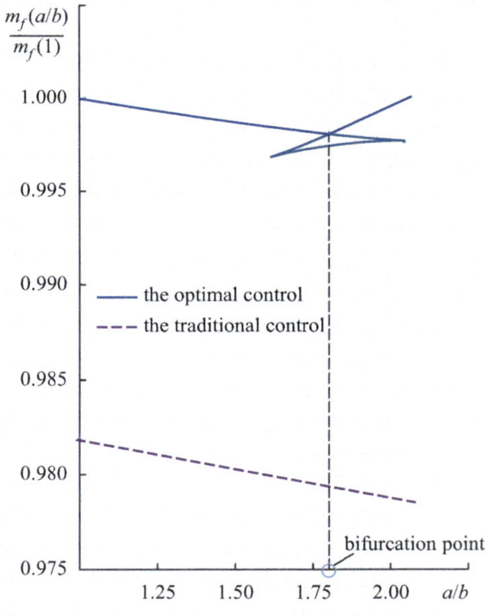

Fig. 3.61 The dependencie of the maximum relative maximum injected mass on the compression factor of the cylindrical-type LV

The above results clearly show that qualitative new features of the optimal control can show up themselves not only for promising "rocket gliders", but already on traditional (ballistic) LV layouts.

The non-monotonic nature of the dependence of the maximum injected mass on parameter $\overline{F}_w$ remains in the practical range of the initial thrust-to-weight ratio $(T/mg)_i$. On Fig. 3.63 the dependences of m_f on $\overline{F}_w$ for various values of $(T/mg)_i$ in the

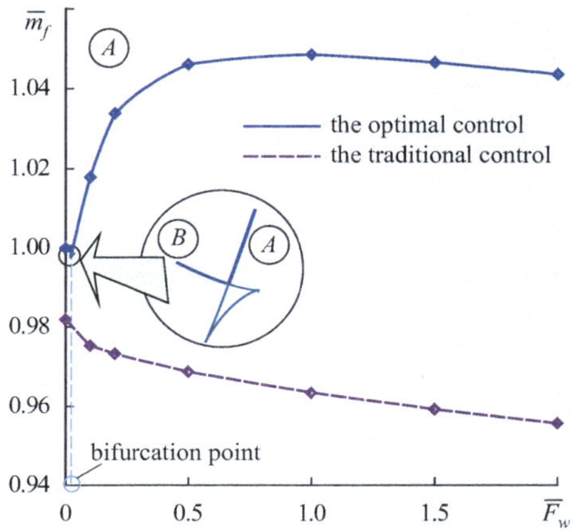

Fig. 3.62 The dependencie of the relative maximum injected mass on the relative LV wing panel area at $(T/mg)_i = 1.1$

range from 1.1 to 1.4 are shown. At $(T/mg)_i \leq 1.15$ these dependencies contain a singular point corresponding to the bifurcation change of the optimal control law. For $(T/mg)_i > 1.15$, extremals of type ***B*** turn into type ***A*** continuously, "evolutionary", with increasing in $\overline{F}_w$. The point of the extremal type change, conditionally coinciding with the minimum of the dependence of m_f on $\overline{F}_w$, shifts to larger $\overline{F}_w$ with an increase in the initial thrust-to-weight ratio.

It is important that the noted qualitative effects are fully preserved when taking into account practical constraints on the admissible loads, dynamic pressure q, and product $q{\cdot}\alpha$. That is due to the following factors. First, the values of $q_{\max}$ and $(q{\cdot}\alpha)_{\max}$ on the optimal trajectory belong to the practical range for current "rocket" layouts.

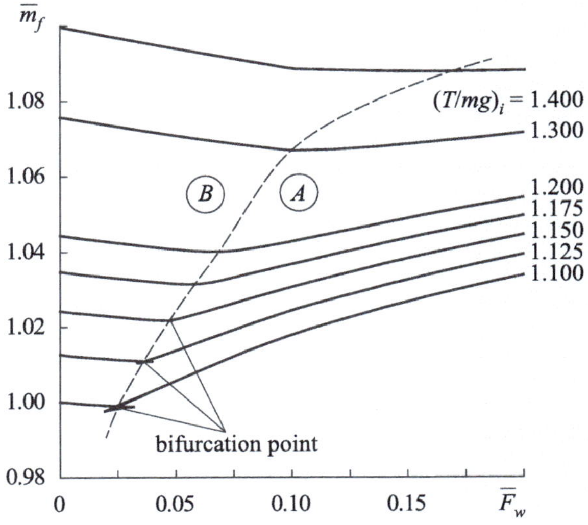

Fig. 3.63 Dependencies of the relative maximum injected mass on the LV relative wing panel area at various initial thrust-to-weight ratios

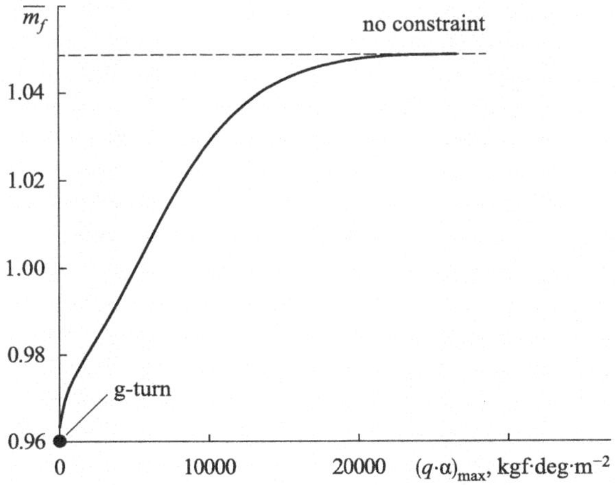

Fig. 3.64 Dependence of the maximum injected mass on the admissible product $(q \bullet \alpha)_{max}$ at relative wing panel area $\overline{F}_w = 1$

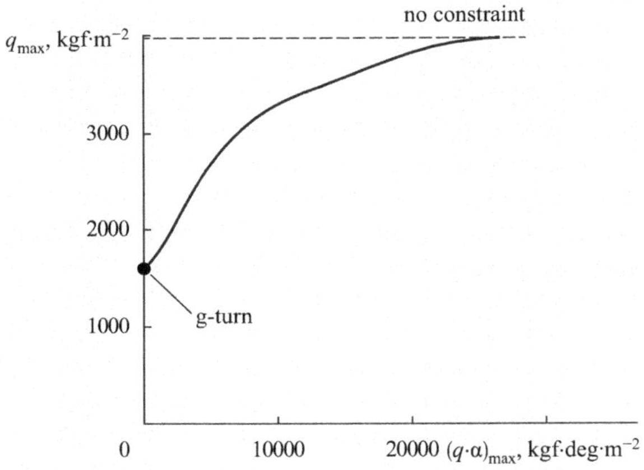

Fig. 3.65 The dependence of the maximum dynamic pressure on the admissible product $(q \bullet \alpha)_{max}$ at the optimal control

Secondly, since the reference solution is strictly optimal, then the losses of the injected mass are values of a higher order of smallness compared to the variations of δq_{max} and $\delta(q{\cdot}\alpha)_{max}$. For example, 30% decrease in the admissible $(q{\cdot}\alpha)_{max}$ as compared to the optimal value (from 26,500 to 18,550 kgf deg m^{-2}) leads to an increase in the injected mass by only 0.17%. At the same time, the maximum dynamic pressure also decreases on the optimal trajectory (Figs. 3.64 and 3.65).

References

1. Okhotsimsky, D.E., Eneev, T.M.: Some variation problems associated with the launch of an artificial Earth satellite. UFN **63**(1a) (1957) (in Russian)

2. Lawden, D.F.: Optimal Trajectories for Space Navigation. Butterworths, London (1963)
3. Filatyev, A.S.: Optimal launch of an artificial Earth satellite using aerodynamic forces. Cosm. Res. **29**(2), 255–271 (1991)
4. Filatyev, A.S.: Optimization of spacecraft ascent using aerodynamic forces. In: IAF-92-0022, 43rd Congress of the International Aeronautical Federation, Washington, DC, Aug 28–Sep 5, 1992
5. Filatyev, A.S.: "Paradoxes" of optimal solutions in problems of space vehicle injection and reentry. Acta Astronaut. **47**(1), 11–18 (2000)
6. Okhotsimsky, D.E.: On the theory of rocket motion. J. Appl. Math. Mech. **10**(2) (1946) (in Russian)
7. Tsien, H.S., Evans, R.C.: Optimum thrust programming for a sounding rocket. J. Am. Rocket Soc. **201**(5), 99–107 (1951)
8. Garfinkel, B.A: Solution of the goddard problem. SIAM J. Soc. Ind. Appl. Math. Ser. Control **1**(3), 349–368 (1963)
9. Filatyev, A.S., Mileev, M.N.: Optimal control of an aircraft in the atmosphere when launching a satellite into orbit. In: V All-Union Conference on Control in Mechanical Systems, 12–14 June 1985 Kazan. Tez. dokl.: AN SSSR (1985) (in Russian)
10. Filatyev, A.S., Yermak, Y.N., Golikov, A.A., Zadonsky, S.M.: New solutions of the optimal injection problem based on complex investigation of dynamics and aerodynamics. In: Proceedings of the 22nd International Congress of Aeronautical Sciences ICAS-2000, Harrogate, UK, ICAS-2000-4.10.5, 27 Aug–1 Sep 2000
11. Filatyev, A.S., Golikov, A.A., Yermak, Y.N., Zadonsky, S.M.: Qualitative changes of optimal launch trajectories due to aerodynamic shape variations. In: 51st Congress of the International Astronautical Federation, Rio de Janeiro, Brazil, IAF-00-A.4.04, 2–6 Oct 2000
12. Filatyev, A.S.: Approximate analytical synthesis of optimal control of hypersonic aircraft when moving in the atmosphere with sub-circular speed, parts 1, 2. Uchenye zapiski TsAGI, v. **11, Nos.** 1, 2 (1980) (in Russian)
13. Petrov, K.P.: Rocket Aerodynamics. Mashinostroenie, Moscow, p. 136 (1977) (in Russian)
14. Petrov, K.P.: Aerodynamics of Bodies of the Simplest Forms. Factorial, Moscow, p. 432 (1998) (in Russian)
15. Project "Air Launch". Polet Airlines. https://web.archive.org/web/20071014064908/http://www.polet.ru/RU/airlaunch.shtml
16. Filatyev, A.S.: A practical way to increase the efficiency of space transportation systems based on the implementation of rigorous through optimization methods. Moscow, Aerospace Technics and Technology Journal, Russian Academy of Engineering, Section "Aerospace", No. 1 (1999) (in Russian)
17. Filatyev, A.S., Golikov, A.A.: New solutions for the problem of optimal injection of hypersonic vehicles with scramjet. In: AIAA Paper 99-4930, AIAA 9th International Space Planes and Hypersonic Systems and Technologies Conference, Norfolk, Virginia, 1–4 November 1999
18. Lopata, J., Rutan, B.: RASCAL—a demonstration of operationally responsive space launch. In: 2nd Responsive Space Conference, Los Angeles, CA, 19–22 April 2004. https://www.researchgate.net/publication/289066762_RASCAL_-_a_demonstration_of_operationally_responsive_space_launch
19. Mileev, M.N., Filatyev, A.S.: On the choice of optimal rocket launch conditions: modern analysis of the Tsiolkovsky's estimates. In: Proceedings of the 23rd Readings Dedicated to the Development of the Scientific Heritage and Development of the Tsiolkovsky's ideas (Kaluga, September 13–16, 1988). Section "Problems of rocket and space technology". M.: IIET AN USSR, 1989, p. 58. (in Russian)
20. Miele, A.: Optimum climbing technique for a rocket-powered aircraft. Jet Propul. **25**(8), 385–391 (1955)
21 Filatyev, A.S., Golikov, A.A., Shanygin, A.N., Voityshen, V.S.: The local distributed criteria method for multidisciplinary optimization of launcher parameters. Acta Astronaut. **65**(3–4), 584–590 (2009)

Chapter 4
The Through Optimization of Branching Trajectories with a Finite Set of Branches

4.1 Introduction

Branching trajectories include trajectories with a variable dimension of the phase vector. In practice, branching processes occur where there is a recombination of objects, i.e. division into several components or, conversely, grouping several objects. In general, in some time moments $\{t^j, j = 1, \ldots, m\}$, k^j objects break up (group) into k^{j+1} objects, while maintaining their number between branch points. Each of the objects can move not only using its own control program (objects can be uncontrolled), but their motion can be described by different equations for different phase vectors.

For example, the through optimization of ASTS launch trajectory with a subsonic CA and an rocket stage (RS) requires consideration of legs of the CA cruise flight and pre-launch maneuver of the CA with RS, the leg of CA and RS separation and the sections of the autonomous flight of CA (return to the cruise flight mode) and RS (insertion into a target orbit). The CA flight leg before RS separation can be considered in the vertical plane ($\mathbf{x} \in \boldsymbol{R}^5$), and RS autonomous flight can be spatial ($\mathbf{x} \in \boldsymbol{R}^7$). Accordingly, the mathematical models of the motion in each of the sections may also differ. Constraints on control and trajectory can be local for each section or global, linking several branches at the same time. Thus, taking into account the requirements for the safe separation of RS and CA limits their mutual position for some time after the break of ties, when each of the objects moves along its own trajectory (a branch).

In more complex problems the set of objects under consideration can form a continuum at some stages of motion. Such problems are discussed in Chaps. 5 and 6.

Methods for optimizing discontinuous systems are considered, for example, in [1–6]. Paper [4] considers the application of the optimal control theory to optimization of launch of several spacecraft into orbits by one LV. The solution of particular problems

A. S. Filatyev and V. G. Petukhov, *Through Optimization of Aerospace Vehicle Trajectories by the Pontryagin Maximum Principle*, Studies in Systems, Decision and Control 579, https://doi.org/10.1007/978-3-031-80756-5_4

of the optimal STS injection with a constraint on SP drop points is considered in [7, 8] using direct methods.

The distinctive features of the methods described below for the PMP optimization of branching trajectories are as follows:

- applications to actual problems of through optimization of a composite ASV launching;
- bringing these methods to "regular enough" numerical algorithms and programs, which made it possible to carry out

  - studies of the qualitative features of branching extremals, as in the case of constraints on the admissible SP drop fields;
  - extensive parametric studies on the optimal launching schemes and parameters for various STS types (see Chap. 6).

The main results of the through optimization of branching trajectories were published earlier in [9–12].

4.2 The Problem Statement

Let us consider a branching trajectory. Branch points can be both on the main trajectory and on side branches, which in turn can also branch. We number all branch points sequentially. Optionally, you can introduce fictitious points. All branch points with the same ordinal number, reached in the general case at different points in time, are combined into one node (Fig. 4.1).

We use the following notation: b_i^j is the i-th branch between (j-1)-th and j-th branch nodes,

$b^j = \left\{b_i^j, 1 \le i \le k^j\right\}, b = \left\{b^j, 1 \le j \le m\right\}.$

Accordingly, $\mathbf{x}_i^j$ is the n_i^j-dimensional phase vector on branch b_i^j; $\mathbf{u}_i^j$ is the p_i^j-dimensional control vector on branch b_i^j; $t_i^j \in \left[\tau_i^j, T_i^j\right]$ is the independent variable on branch b_i^j. Let the motion of the considered system of bodies obeys the following conditions:

(1) The equality-type conditions at nodal points

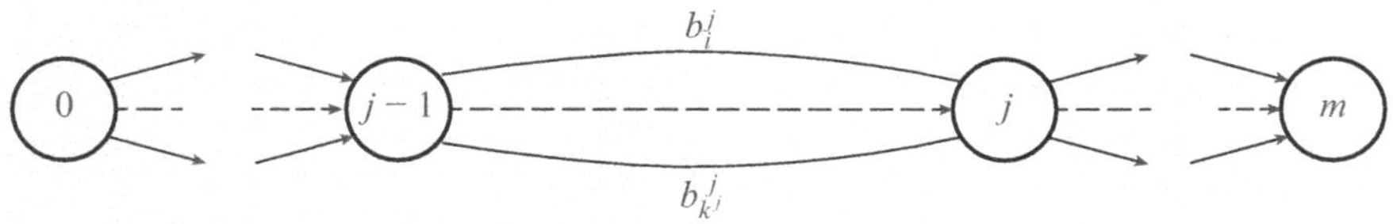

Fig. 4.1 Scheme of a trajectory with a finite set of branches

$$\mathbf{Q}(\mathbf{x}(\tau), \mathbf{x}(T)) = 0, \mathbf{Q} = \{Q_l, 1 \le l \le q\}; \tag{4.1}$$

(2) The inequality-type conditions on the control of individual branches or their groups

$$\mathbf{W}(\mathbf{x}, \mathbf{u}) \le 0, \mathbf{W} = \{W_l, 1 \le l \le r\}; \tag{4.2}$$

(3) Differential or functional links for branches b_i^j:

$$\left\{\varepsilon \frac{d\mathbf{x}}{dt} - \mathbf{f}(\mathbf{x}, \mathbf{u}) = 0, t \in [\tau, T]\right\}_i^j, \tag{4.3}$$

where ε_i^j is the diagonal matrix $n_j^i \times n_j^i$ with elements on set {0,1}.

Let us formulate the optimization problem as follows: define control $\mathbf{u}_{\text{opt}}(t)$, which with relations (4.1)–(4.3) maximizes functional

$$J(\mathbf{x}(\tau), \mathbf{x}(T)) \Rightarrow \max_{\{\mathbf{u}\}}. \tag{4.4}$$

Remark 1 Conditions (4.1) and (4.2) allow covering the cases of phase constraints of the s-th order $[\Omega(\mathbf{x})]_i^j \le 0$. To do this, the point from which the constraint $[\Omega(\mathbf{x})]_i^j = 0$ is valid should be taken as the j-th branch point with imposed constraints of type (4.1):

$$[\Omega(\mathbf{x}(T))]_i^j = 0, \ldots, \left[\frac{d^{s-1}\Omega(\mathbf{x}(T))}{dt^{s-1}}\right]_i^j = 0, \mathbf{x}_i^j(T) - x_i^{j+1}(\tau) = 0, T_i^j - \tau_i^{j+1} = 0,$$

and constraints of type (4.2): $\frac{d^s\Omega}{dt^s} = \Omega^{(s)}(\mathbf{x}, \mathbf{u})_i^{j+1} \le 0$, imposed on control $\mathbf{u}_i^{j+1}$.

Remark 2 Here, m nodes are considered only for the convenience of the geometric representation of the trajectory. To derive optimality conditions and, especially, to construct a universal numerical algorithm, it is sufficient to have two nodes in which all left and right ends of the branches are grouped.

4.3 Optimality Conditions

Let us consider the extended functional

$$\tilde{J} = J + \boldsymbol{\nu}^{\mathrm{T}}\mathbf{Q} + \sum_{i,j}\left\{\int_{\tau}^{T}\left[\mathbf{p}^{\mathrm{T}}(\mathbf{f} - \boldsymbol{\epsilon}\dot{\mathbf{x}}) + \boldsymbol{\lambda}^{\mathrm{T}}\mathbf{W}\right]dt\right\}_i^j \tag{4.5}$$

where $\mathbf{p}_i^j$ is the conjugate vector; $\boldsymbol{\nu}_i^j$, $\boldsymbol{\lambda}_i^j$ are the Lagrange multiplier vectors on the corresponding branch.

Note that it suffices to take into account the term $\lambda_i W_i$ only in one of those integrals in (4.5) where it is significant: $\lambda_i = 0$, if $W_i < 0$.

Analysis of variations of extended functional (4.5) gives equations for conjugate vectors $\mathbf{p}_i^j$

$$\left[\epsilon\frac{d\mathbf{p}}{dt} + \left(\frac{\partial \mathbf{f}}{\partial \mathbf{x}}\right)^{\mathrm{T}}\mathbf{p} + \left(\frac{\partial \mathbf{W}}{\partial \mathbf{x}}\right)^{\mathrm{T}}\boldsymbol{\lambda}\right]_i^j = 0; \tag{4.6}$$

specific transversality conditions at branch points:

$$\left.\begin{array}{c} \left[\epsilon p(\tau) + \left(\dfrac{\partial \mathbf{Q}}{\partial \mathbf{x}(\tau)}\right)^{\mathrm{T}}\nu + \dfrac{\partial J^{\mathrm{T}}}{\partial \mathbf{x}(\tau)}\right]_i^j = 0, \\ \left[\epsilon p(T) - \left(\dfrac{\partial \mathbf{Q}}{\partial \mathbf{x}(T)}\right)^{\mathrm{T}}\nu - \dfrac{\partial J^{\mathrm{T}}}{\partial \mathbf{x}(T)}\right]_i^j = 0, \\ [H(T)]_i^j = 0 \end{array}\right\}; \tag{4.7}$$

where $H_i^j = \mathbf{p}_i^{j\mathrm{T}}\mathbf{f}_i^j$, and the control optimality conditions are following:

$$\left\{\mathbf{u}_{\text{opt}} = \arg\max_{\{\mathbf{u}\}}\left(H + \boldsymbol{\lambda}^{\mathrm{T}}\mathbf{W}\right)\right\}_i^j = 0. \tag{4.8}$$

In the general case, relations (4.6)–(4.8) reduce the original problem to a closed TPBVP for $2n_\Sigma$ Eqs. (4.3) and (4.6), where $n_\Sigma = \sum_{i,j} n_i^j$, with $q + 2n_\Sigma$ boundary conditions (4.1) and (4.7) containing q unknown Lagrange multipliers ν_l, $1 \le l \le q$.

The TPBVP is solved numerically using *ASTER* software package.

4.4 The TPBVP Reduction by Projection of Boundary Conditions from Side Branches

The use of PMP allows us to reduce the original problem of finding the optimal control in the functional space to the solution of TPBVP for ODE systems (4.3), (4.6). The number of varied parameters of such TPBVP does not exceed the dimension of the phase vector. In the case of branching trajectories, each side branch b_j^i increases the dimension of the considered phase space by n_i^j, and the dimension of BVP increases by no more than n_i^j. As will be shown below, in a number of practically important cases, the BVP dimension, increases by only one or two units for a branch. The TPBVP reduction is possible by projecting the boundary conditions (all or part) from the right ends of side branches to the branch point. In this case, the BVP for a branching trajectory is reduced to a BVP with the only (main) branch. The use of this method greatly simplifies and speeds up the procedure for BVP solving, since the fulfillment of the boundary conditions at the right end of a side branch is ensured in this case without expensive computational procedures. The projection of the boundary conditions requires, generally, a twofold integration of the phase and conjugate systems for the side branch. We will assume that the right-hand sides of the phase system for side branches do not depend on conjugate variables (e.g., in the case of fixed control or analytical synthesis of the optimal control). Then this procedure can be replaced by a single integration of only the phase system, and the conjugate vector can be calculated using the transition matrix. When describing the method, consider one of the side branches with one branch point. At the right end of the branch (at time T), condition (4.1) and corresponding transversality condition (4.7) must be satisfied, which determines the connection between the phase and conjugate variables.

Due to the independence of the right-hand sides of the equation of motion on the side branch from the conjugate variables, the values of the conjugate variables at the initial point ($t = \tau$) of the side branch $\mathbf{p}(\tau)$ can be obtained from the solution of the matrix equation

$$\mathbf{p}(\tau) = \tilde{\mathbf{A}}(\mathbf{x}) \cdot \mathbf{p}(T), \tag{4.9}$$

where $\mathbf{p}(T)$ is the conjugate vector at the right end of the side branch, $\tilde{\mathbf{A}}(\mathbf{x})$ is the transition matrix defined as a result of a single integration of the phase system for the side branch. The following is a procedure for obtaining this matrix in the case of using the 4-th order Runge–Kutta method for trajectory integration.

The ODE system for the conjugate vector is a linear with variable coefficients:

$$\dot{\mathbf{p}} = \mathbf{A}(t, \mathbf{x})\mathbf{p},\ \mathbf{A} = -\left[\frac{\partial \mathbf{f}}{\partial \mathbf{x}}\right]^{\mathrm{T}}, \tag{4.10}$$

for which the boundary conditions are given at the right end:

$$\mathbf{p}(T) = \mathbf{p}_T. \tag{4.11}$$

To transfer conditions (4.11) to the beginning of the side branch ($t = \tau$), i.e. to the branch point, it suffices to calculate the transition matrix $\tilde{A}$.

Let us consider the process of integrating system (4.10) by the 4-th order Runge–Kutta method. When integrating by one step h from t_0 to $t_1 = t_0 + h$ we have:

$$\mathbf{p}_1 = \mathbf{p}_0 + \mathbf{k} \cdot h,\ \mathbf{p}_0 = \mathbf{p}(t_0),\ \mathbf{p}_1 = \mathbf{p}(t_1),. \tag{4.12}$$

where

$$\begin{aligned}
\mathbf{k} &= \frac{1}{6}(\mathbf{k}_0 + 2\mathbf{k}_1 + 2\mathbf{k}_2 + \mathbf{k}_3),\\
\mathbf{k}_0 &= \mathbf{A}(t_0, \mathbf{x}_0) \cdot \mathbf{p}_0,\\
\mathbf{k}_1 &= \mathbf{A}(t_0 + h/2, \mathbf{x}_0 + \mathbf{k}_0 \cdot h/2) \cdot (\mathbf{p}_0 + \mathbf{k}_0 \cdot h/2),\\
\mathbf{k}_2 &= \mathbf{A}(t_0 + h/2, \mathbf{x}_0 + \mathbf{k}_1 \cdot h/2) \cdot (\mathbf{p}_0 + \mathbf{k}_1 \cdot h/2),\\
\mathbf{k}_3 &= \mathbf{A}(t_0 + h, \mathbf{x}_0 + \mathbf{k}_2 \cdot h) \cdot (\mathbf{p}_0 + \mathbf{k}_2 \cdot h).
\end{aligned}$$

Denote

$$\begin{aligned}
\mathbf{A}_0 &= \mathbf{A}(t_0, \mathbf{x}_0),\\
\mathbf{A}_{-1/2} &= \mathbf{A}(t_0 + h/2, \mathbf{x}_0 + \mathbf{k}_0 \cdot h/2),\\
\mathbf{A}_{+1/2} &= \mathbf{A}(t_0 + h/2, \mathbf{x}_0 + \mathbf{k}_1 \cdot h/2),\\
\mathbf{A}_1 &= \mathbf{A}(t_0 + h, \mathbf{x}_0 + \mathbf{k}_2 \cdot h).
\end{aligned} \tag{4.13}$$

Then (4.12) takes the form

$$\begin{aligned}
\mathbf{p}_1 = \mathbf{p}_0 + \tfrac{h}{6}\big\{&\mathbf{A}_0\mathbf{p}_0 + 2\mathbf{A}_{-1/2}\big(\mathbf{p}_0 + \tfrac{h}{2}\mathbf{A}_0\mathbf{p}_0\big) + 2\mathbf{A}_{+1/2}\big(\mathbf{p}_0 + \tfrac{h}{2}\mathbf{A}_{-1/2}\big(\mathbf{p}_0 + \tfrac{h}{2}\mathbf{A}_0\mathbf{p}_0\big)\big)\\
&+\mathbf{A}_1\big[\mathbf{p}_0 + h\mathbf{A}_{+1/2}\big(\mathbf{p}_0 + \tfrac{h}{2}\mathbf{A}_{-1/2}\big(\mathbf{p}_0 + \tfrac{h}{2}\mathbf{A}_0\mathbf{p}_0\big)\big)\big]\big\}.
\end{aligned}$$

Whence for vector $\mathbf{p}_1 = \mathbf{p}(t_1)$ at me point $t_1 = t_0 + h$ we obtain matrix equation.

$$\mathbf{p}_1 = \overline{\mathbf{A}}_0\mathbf{p}_0,$$

$$\begin{aligned}
\overline{\mathbf{A}}_0 = \Big\{&E + \tfrac{h}{6}\big(+2\mathbf{A}_{-1/2} + 2\mathbf{A}_{+1/2} + \mathbf{A}_1\big) + \tfrac{h^2}{6}\big(\mathbf{A}_{-1/2}\mathbf{A}_0 + \mathbf{A}_{+1/2}\mathbf{A}_{-1/2} + \mathbf{A}_1\mathbf{A}_{+1/2}\big)\,.\\
&+\tfrac{h^3}{12}\big(\mathbf{A}_{+1/2}\mathbf{A}_{-1/2}\mathbf{A}_0 + \mathbf{A}_1\mathbf{A}_{+1/2}\mathbf{A}_{-1/2}\big) + \tfrac{h^4}{24}\mathbf{A}_1\mathbf{A}_{+1/2}\mathbf{A}_{-1/2}\mathbf{A}_0\Big\},
\end{aligned}$$

where matrices $\mathbf{A}_0$, $\mathbf{A}_{-1/2}$, $\mathbf{A}_{+1/2}$, $\mathbf{A}_1$, and hence matrix $\overline{\mathbf{A}}_0$, are determined as a result of integrating the phase system at step $[t_0, t_1]$.

For vector $\mathbf{p}_2 = \mathbf{p}(t_2)$ at $t_2 = t_1 + h$, we obtain similarly $\mathbf{p}_2 = \overline{A}_1\mathbf{p}_1$, where matrix $\overline{\mathbf{A}}_1$ is determined as a result of integrating the phase system at step $[t_1, t_2]$. And so on up to $t_k = T$. That is

$$\mathbf{p}_T = \prod_{l=0}^{k-1}\overline{\mathbf{A}}_l \cdot \mathbf{p}_0. \tag{4.14}$$

Comparing (4.14) with (4.10) and (4.9), we obtain:

$$\tilde{\mathbf{A}}(\mathbf{x}) = \left(\prod_{l=0}^{k}\overline{\mathbf{A}}_l\right)^{-1}. \tag{4.15}$$

Thus, by integrating the phase system once, boundary conditions (4.7) are transferred from the right end of the side branch to the branch point (to the main branch).

4.5 The Through Optimization of STS Branching Trajectories with Given Landing Areas for Spent Parts

4.5.1 The Problem Statement

Let us consider a branching trajectory as in Fig. 4.2.

The following are indicated here: s_1^0 is the start point, s_1^l is the point of insertion into a target orbit, s_1^j, $1 \le j \le l-1$, are points of SP release, s_2^j are the SP landing/drop points, b_i^j are continuous sections of the trajectory ending at points s_i^j.

We single mass m out from the phase vector $\mathbf{x}_1^j = \{\mathbf{r}, \mathbf{v}\}$, then on the sections b_1^j, $1 \le j \le l - 1$, the phase system will take the form:

$$\dot{\mathbf{x}}_1^j = \mathbf{f}_1^j\left(\mathbf{x}_1^j, \mathbf{u}_1^j\right), t \in \left[\tau_1^j, T_1^j\right], \dot{m}_1^j = -\mu_1^j. \tag{4.16}$$

We will assume that SP motion on branch b_2^j occurs under nominal flight conditions (without random disturbances) under a deterministic control, and the change in phase vector $\mathbf{x}_2^j$ on branch b_2^j is determined by ODE system:

$$\begin{aligned} &\dot{\mathbf{x}}_2^j = \mathbf{f}_2^j\left(\mathbf{x}_2^j\right), 1 \le j \le l-1, \\ &\dot{m}_2^j = 0, t \in \left[\tau_2^j, T_2^j\right]. \end{aligned} \tag{4.17}$$

At points s_1^j, $1 \le j \le l-1$, (at time moments $t_1^j = T_1^j = \tau_1^{j+1}$) the phase vectors are considered continuous:

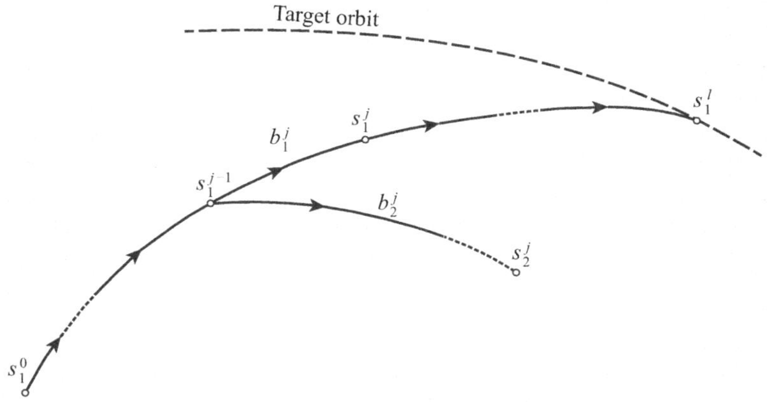

Fig. 4.2 Scheme of STS branching injection trajectory

$$\left[\mathbf{x}_1^j - \mathbf{x}_i^{j+1}\right]_{t_1^j} = 0,\, i = 1, 2, \tag{4.18}$$

while the mass is conserved:

$$\left[m_1^{j-1} - m_1^j - m_2^j\right]_{t=T_1^{j-1}=\tau_1^j} = 0. \tag{4.19}$$

SP masses are given:

$$m_2^j = fix, j \in [1, l-1]. \tag{4.20}$$

Branching moments $\left\{T_1^{j-1} = \tau_1^j = \tau_2^j,\ j \in [1, l-1]\right\}$ are given by scalar conditions:

$$Q^j\left(\mathbf{x}_1^{j-1}, m_1^{j-1}\right) = 0, j \in [1, l-1]. \tag{4.21}$$

At points s_2^j (at the moment of reaching zero altitude) the radius vector $\mathbf{r}_2^j$ is given:

$$T_2^j : h_2^j = 0,\ \mathbf{r}_2^j(t) = fix, j \in [1, l-1]. \tag{4.22}$$

At points $s_1^0$and s_1^l traditional conditions (2.25), (2.49) for the take-off and insertion into given orbit are set.

The functional to be maximized is the injected mass (2.55):

$$J \equiv m\left(s_1^l\right). \tag{4.23}$$

4.5.2 Transversality Conditions

Taking into account conditions (4.14), (4.15), we write the transversality conditions at branch points $\left\{s_1^j,\ 1 \le j \le l-1\right\}$ and at the ends of side branches $\left\{s_2^j,\ 2 \le j \le l-1\right\}$. Following (2.57), we denote $\mathbf{p}_i^j = \left(\mathbf{p}_r, \mathbf{p}_v, p_m\right)_i^j$ as the conjugate vector, H_i^j as the Hamiltonian on branch b_i^j.

According to [10] for constraints (4.18)–(4.22) and functional (4.23) we obtain the following transversality conditions:

$$t = T_2^j,\ \mathbf{p}_{v2}^j = 0,\ p_{m2}^j = 0,\ H_2^j = (\mathbf{p}_r, \mathbf{v})_2^j = 0. \tag{4.24}$$

Since system (4.17) is autonomous and the mass does not explicitly appear on the right-hand sides, then the first integrals take place on side branches b_2^j:

$$H_2^j \equiv 0, \; p_{m2}^j \equiv 0. \tag{4.25}$$

For constraints (4.18)–(4.21), relations (4.7) allow us to obtain transversality conditions at branch points $\left\{s_1^j, \; 1 \le j \le l-1\right\}$:

$$\begin{cases} \mathbf{p}_{r1}^j = \mathbf{p}_{r1}^{j-1} - \mathbf{p}_{r2}^j - \nu_j \left(\dfrac{\partial Q^j}{\partial \mathbf{r}_1^{j-1}} \right)^T, \\ \mathbf{p}_{v1}^j = \mathbf{p}_{v1}^{j-1} - \mathbf{p}_{v2}^j - \nu_j \left(\dfrac{\partial Q^j}{\partial \mathbf{v}_1^{j-1}} \right)^T, \\ p_{m1}^j = p_{m1}^{j-1} - \nu_j \dfrac{\partial Q^j}{\partial m_1^{j-1}}, \end{cases} \tag{4.26}$$

$$H_1^j = H_1^{j-1}. \tag{4.27}$$

The Lagrange multipliers ν_j in (4.26) are found from (4.27):

$$\begin{aligned} &\nu_j \left[\frac{\partial Q^j}{\partial \mathbf{r}_1^{j-1}} \mathbf{v}_1^j + \frac{\partial Q^j}{\partial \mathbf{v}_1^{j-1}} \dot{\mathbf{v}}_1^j + \frac{\partial Q^j}{\partial m_1^{j-1}} \dot{m}_1^j \right] = \\ &= \mathbf{p}_{v2}^{j\mathrm{T}} \left(\dot{\mathbf{v}}_2^j - \dot{\mathbf{v}}_1^j \right) + \mathbf{p}_{v1}^{j-1\mathrm{T}} \left(\dot{\mathbf{v}}_1^j - \dot{\mathbf{v}}_1^{j-1} \right) + p_{m1}^{j-1} \left(\dot{m}_1^j - \dot{m}_1^{j-1} \right). \end{aligned} \tag{4.28}$$

Note that Eq. (4.28) is not linear with respect to ν_j, because this parameter is included in the optimal control and, consequently, in $\dot{\mathbf{v}}_1^j$ and $\dot{m}_1^j$.

In a particular case, when branching moments are determined by reaching the given mass:

$$Q_1^j \equiv Q_1^j \left(m_1^{j-1} \right), \tag{4.29}$$

$$\begin{cases} \mathbf{p}_{r1}^j = \mathbf{p}_{r1}^{j-1} - \mathbf{p}_{r2}^j, \\ \mathbf{p}_{v1}^j = \mathbf{p}_{v1}^{j-1} - \mathbf{p}_{v2}^j, \\ p_{m1}^j = p_{m1}^{j-1} - \nu_j, \end{cases} \tag{4.30}$$

where

$$\nu_j = \frac{\mathbf{p}_{v2}^{j\mathrm{T}} \left(\dot{\mathbf{v}}_2^j - \dot{\mathbf{v}}_1^j \right) + \mathbf{p}_{v1}^{j-1\mathrm{T}} \left(\dot{\mathbf{v}}_1^j - \dot{\mathbf{v}}_1^{j-1} \right) + p_{m1}^{j-1} \left(\dot{m}_1^j - \dot{m}_1^{j-1} \right)}{\dot{m}_1^j} \tag{4.31}$$

(see the notice to (4.28)).

4.5.3 Features of the Boundary Value Problem

Due to the independence of the right-hand sides of Eq. (4.17) from the control, the fulfillment of boundary conditions (4.24) is ensured either by double integration of the phase system in direct time and the conjugate system in reverse time, or by one iteration according to the Newton method, or by solving matrix Eq. (4.9) as a result of a single integration of the phase system on the side branch.

As can be seen from (4.24), additional free parameters of the TPBVP are some components of vector $\mathbf{p}_{r2}^{j}$. Their number depends on the dimension of the phase vector. For the spatial case, the number of varied parameters, and, consequently, the dimension of the TPBVP, increases by two units, and by one unit for a 2D problem. Accordingly, the number of residuals also increases: in the spatial case SP drop point is given by two coordinates on the Earth's surface and by one coordinate in the 2D case. Thus, the boundary value problem remains closed.

To continue the optimal solution with respect to a parameter, one should use, for example, the optimal solution without constraint (4.22) as the reference solution. In this case, the reference conjugate vector on the side branch must be trivial:

$$\mathbf{p}_{r2}^{j}(\tau) = \mathbf{p}_{v2}^{j}(\tau) = 0; \, p_{m2}^{j}(\tau) = 0. \tag{4.32}$$

It is convenient to introduce the continuation parameter as the angular range in the orthodromic plane between the given SP drop point and the one obtained in the reference optimal solution.

4.5.4 The Optimal Solution Features with Constraints on Spent Part Drop Points

The constraints on SP drop points can lead to a qualitative change in the structure of the optimal control of the pitch angle and thrust. Thus, according to [13, 14], in a uniform gravitational field, with the introduction of any essential constraints on the flight range (phasing), the optimal program for the pitch angle tangent is described not by a linear, but by a bi-linear function of time, and the optimal thrust cycloram may contain additional flight sections with a minimum thrust.

The typical optimal pitch angle programs for the exoatmospheric section (after the 1st stage section) with a variation in allowable distance L_{II} of the 2nd stage booster drop point calculated by *ASTER* software for LV with a vertical take-off are shown in Fig. 4.3. Here, $\overline{L}_{II} = \frac{L_{II}}{L_{IIopt}}$, where $L_{II\,\mathrm{opt}}$ is the drop range in absence of constraints on L_{II}. The control program of the 1st stage (in the atmospheric section) corresponded to the g-turn (with zero angle of attack) and was fixed with variations in L_{II}. Permissible pitch rate after the 1st stage separation is $|\omega| \leq 5$ deg/s. It can be seen that the optimal pitch angle program noticeably differs from the linear one as soon as the drop point of the 2nd stage booster ceases to be free ($\overline{L}_{II} \neq 1$). In the presence of essential

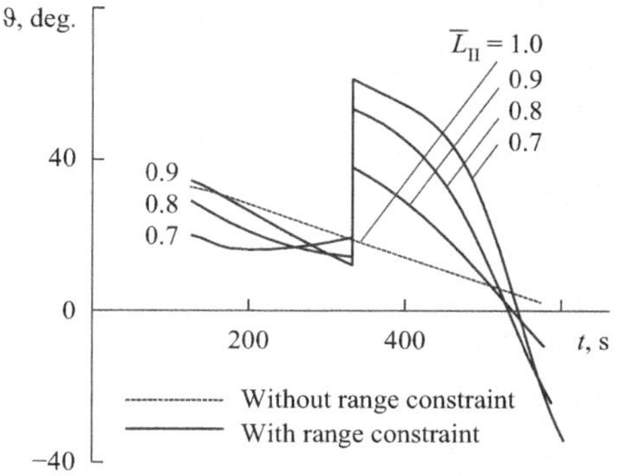

Fig. 4.3 STS optimal pitch angle programs for various relative ranges of drop point of the 2nd stage booster

constraints on SP drop areas, the preservation of the OEL traditional structure of the control program instead of the optimal one leads to losses in the injected mass. Estimates show that these losses can amount to 7–10%.

Constraints of SP drop points can not only qualitatively change the structure of the optimal control program, but also generate non-uniqueness of the solution. To explain the objective reasons for the existence of several local extremals, let us first consider only a side branch, the trajectory of the coast return of the spent 1st stage booster (Fig. 4.4).

It is known that for a fixed initial (at the point of booster release) altitude and velocity, there is such initial flight path angle $\theta_{iL_{\max}}$ at which maximum flight range $L_{\max}$ is reached. If the booster drop range is given $L < L_{\max}$, then there are two different initial angles θ_{iL}^{A} and θ_{iL}^{B}: $\theta_{iL}^{A} < \theta_{iL_{\max}} < \theta_{iL}^{B}$, providing the same drop range.

In view of Chap. 3, it should be expected that larger flight path angles at the separation point will correspond to the global extremal when using STS with low lifting properties (**B**-extremals). On the contrary, smaller angles will correspond to the global optimum for STS with sufficiently high lift-to-drag ratio (**A**-extremals).

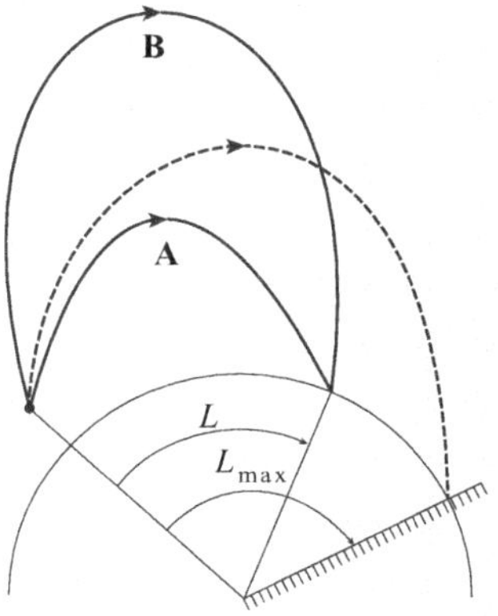

Fig. 4.4 Two qualitatively different types of spent booster return trajectories

Thus, in range $L < L_{\max}$ dependence $\theta_i(L)$ is ambiguous. The upper curve segment corresponds to initial flight path angles θ_{iL}^B, and the lower one corresponds to θ_{iL}^A (Fig. 4.5).

Therefore, the dependence of the maximum injected mass on the drop range of the spent 1st stage booster is also ambiguous and has two segments: $m_{fL_I}^A = m_f\left(\theta_i^A(L_I)\right)$ and $m_{L_If}^B = m_f\left(\theta_i^B(L_I)\right)$, which correspond to two types of local extremals with different altitudes. Depending on the ratio between θ_{iopt} and $\theta_{iL_{max}}$ three types of curves are possible (Fig. 4.6), differing in the extremal type that delivers the global extremum. If $L_I < L_{I\max}$, segments $m_f^B(L)$ and $m_f^A(L)$ can have an intersection point, which is a bifurcation point for the optimal trajectories when the constraint on the drop point of the spent 1st stage booster is varied.

Note that the above arguments about the existence of two types of local extremals are based on the logic of weak variations. Therefore, when using PMP, which is valid

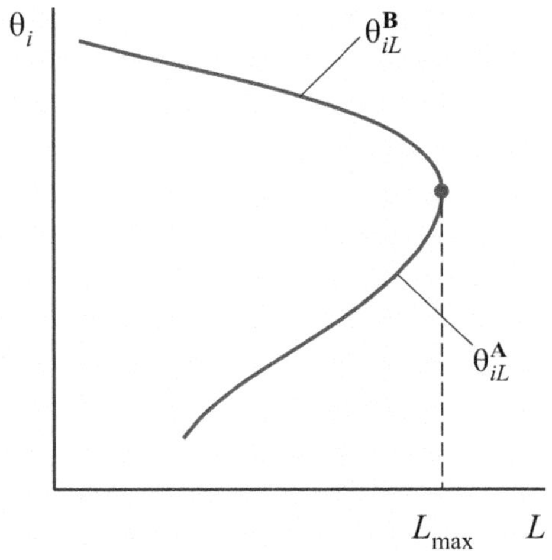

Fig. 4.5 The qualitative relationship between initial flight path angle θ_i and range L of the drop point of the 1st stage booster

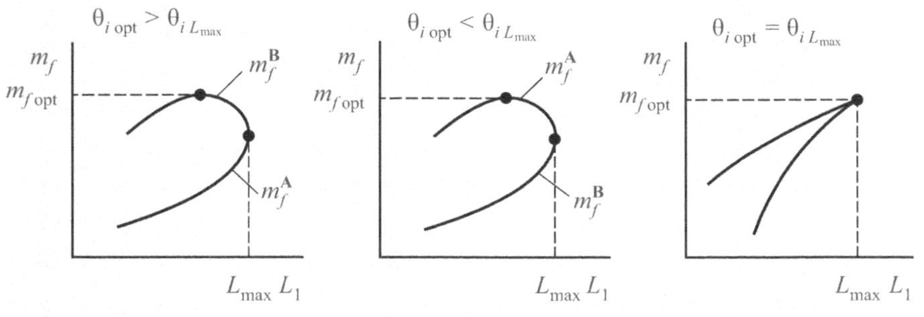

Fig. 4.6 Different types of the dependence of the maximum injected mass on range L_1 of the 1st stage booster drop point

for strong (needle) variations, the second (worst) extremal may be absent (not satisfy PMP conditions).

4.5.5 Local Extremals Caused by Constraints on Spent Part Drop Range

The above qualitative considerations about the existence, as a rule, of more than one local extremal in the problem were confirmed by numerical calculations with *ASTER* program. As an example, a vertical launch of the Proton type LV was considered.

The dependence of the relative injected mass $\overline{m}_f(\overline{L}) = \frac{m_f(\overline{L})}{\max\limits_{\overline{L}} m_f}$ on the relative drop range of the 1st stage booster $\overline{L} = L/L_{\max}$ is shown in Fig. 4.7. The considered LV traditionally belongs to the ballistic type. Nevertheless, the global optimum in this case can be provided by extremals of type ***A***, if the given drop range of the 1st stage booster is greater than bifurcation value L_{bif}. Extremals of type ***B*** are better if the given drop point lies closer to the launch site. That is, the dependence $\overline{m}_f(\overline{L})$ is ambiguous and has two segments $\overline{m}_f^B = \overline{m}_f(\theta_i^B(\overline{L}))$ and $\overline{m}_f^A = \overline{m}_f(\theta_i^A(\overline{L}))$, which correspond to different types of local extremals. The difference in the maximum injected mass between the identified two types of extremals can reach 10%. The change in the type of the global extremal with a variation in the position of the drop point occurs bifurcationally and is accompanied by a jump in the influence function of the booster drop range on the maximum injected mass (Fig. 4.8). The extremals of both types and the optimal pitch angle programs for the bifurcation admissible range of the drop point of the 1st stage booster, where both extremals simultaneously deliver the global optimum, are shown in Fig. 4.9.

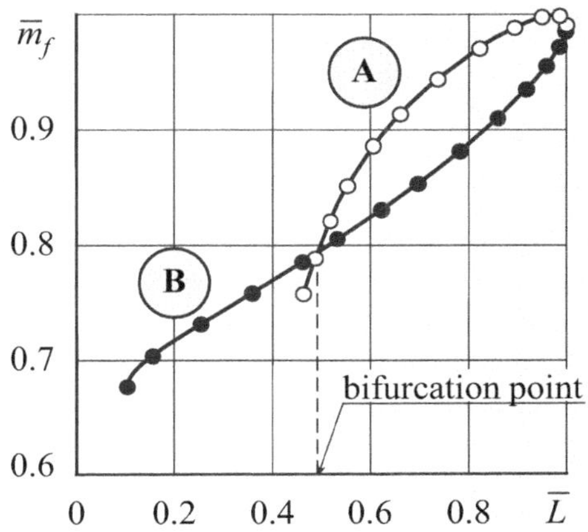

Fig. 4.7 The dependence of relative maximum injected mass $\overline{m}_f$ on relative drop range $\overline{L}$ of the 1st stage booster

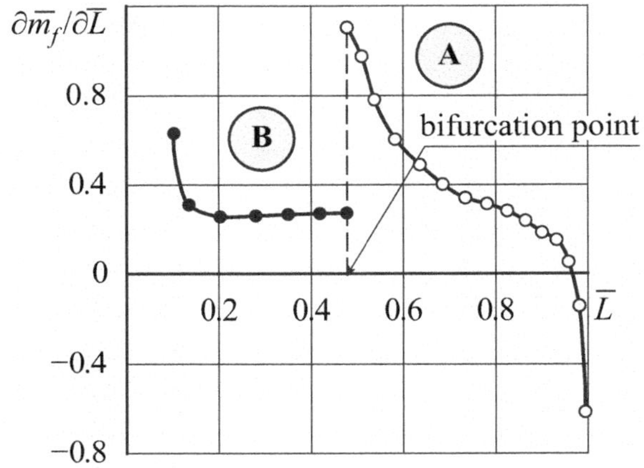

Fig. 4.8 The sensitivity function of relative injected mass $\overline{m}_f$ to relative drop range $\overline{L}$ of the 1st stage booster depending on the relative drop range

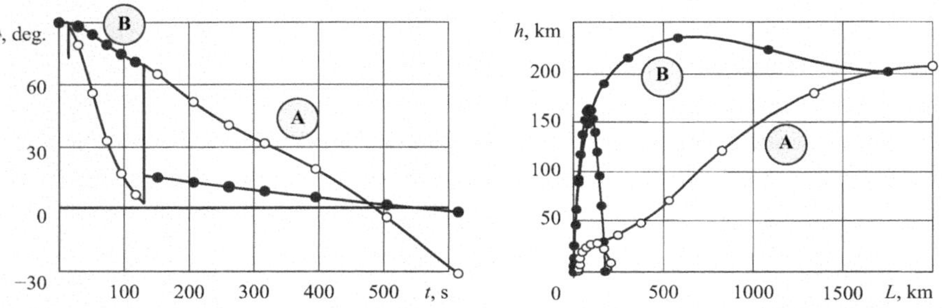

Fig. 4.9 Time-programs of the pitch angle (left) and trajectories (right) for two different global extremals at the bifurcation value $L_{\text{bif}} \approx 0.49 L_{\max}$ of the drop range of the 1st stage booster

4.6 The Through Optimization of ASTS Branching Trajectories with Safe Staging Constraints

Let us consider the problem of branching trajectory optimization on the example of two-stage ASTS taking into account requirements for the safety staging. The subsonic CA with a turbojet is used as the 1st stage for launch the rocket stage (RS). The ASTS trajectory after CA cruise flight with RS to the launch region consists of following main sections: the pre-launch maneuver of CA with RS and the autonomous flight of the RS and CA. Each of the segments corresponds to motion of vehicles with significantly different aerodynamic characteristics, types of propulsion system and constraints. At the same time, a small neighborhood of the "gluing" point of these segments can be distinguished, which corresponds to separation of RS and CA, in which there are specific severe constraints on the control of RS and CA due to separation safety requirements [15]. These constraints can significantly change the optimal control program. Therefore, even with a short duration of these constraints, they have a noticeable effect both on the control programs and trajectories of CA and RS, and on the maximum mass m_{orb} injected into an orbit [16].

The following two approaches for the separation process simulation were considered. In the first one, the motion of CA and RS after breaking the ties is not modeled at the through optimization, but the integrated change of the RS phase vector in this process is taken into account in the form of an abrupt change in the phase vector at the gluing (staging) point. This approach made it possible to actually separate the tasks of the through optimization of the RS ascent trajectory and the study of the dynamics of the processes of safe separation of CA and RS. When studying the separation process, its safety and quality were taken as the main criteria [15, 17], and the connection with the ASTS trajectory optimization was carried out in the form of setting boundary conditions for the separation process.

Another approach is based on the through optimization of the launch trajectory as a branching process with inclusion of all sections of motion into the optimization. On all branches of ASTS trajectory, the constraints on the control and phase vector are considered as that described in Chaps. 1 and 2, and specific constraints on the separation section are added to take into account the safety requirements: the absence of collisions and getting CA into the high temperature zone of the RS main engine plume. The specificity of these constraints is that they connect the controls on the main (RS injection trajectory) and side (CA return trajectory) branches simultaneously.

The expressions for the constraints on the motion of ASTS stages during the separation process in this chapter are given for the two-dimensional case (in the vertical plane). The spatiality of the motion is taking into account in other legs of the branching launch trajectory before and after the staging.

4.6.1 Integrated Consideration of the Staging

Integrated consideration of the change in the RS phase vector during separation makes it possible, without resorting to detailed simulation of the separation process, to carry out a fairly complete analysis of the influence of the staging on parameters of the optimal launch trajectory, to identify new effects and acceptable separation conditions.

With this approach, the lack of prompt consideration of the mutual influence between the optimal launch trajectory and the separation section, as a rule, is compensated by the expansion of the allowable scatter in values of the separation parameters. This can lead to an underestimation of functional m_{orb}.

The separation of CA and RS is a complex dynamic process, in which a number of essential constraints are imposed on the control of both vehicles, including the mutual attitude of RS and CA, on the maximum allowable pitch angular rate of RS when reaching the optimal angles of attack at the beginning of the RS autonomous flight, on the CA normal overload, etc. To a large extent, this circumstance explains the losses of the RS velocity Δv, altitude Δh and flight path angle $\Delta\theta$ in the process of separation of RS and CA.

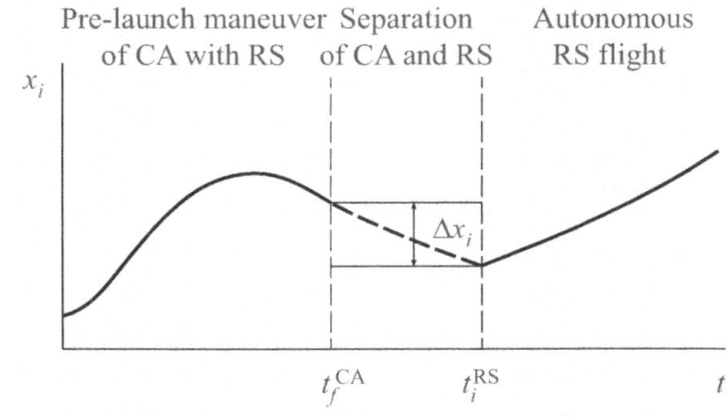

Fig. 4.10 Scheme of ASTS launch trajectory with the change of the phase vector at staging

Integrated consideration of the separation section consists in setting the increments of the phase vector $\Delta\mathbf{x}$ and, first of all, the horizontal Δv_x and vertical Δv_y components of the RS velocity (Fig. 4.10). Due to separation section shortness and flight path angle smallness, the altitude practically does not change during separation.

Let $t = t_f^c$ to be the moment of the end of the CA pre-launch maneuver, $t = t_i^r$ is the start time of the RS autonomous flight. Neglecting the separation segment duration, we assume that.

$$t_f^{CA} = t_i^{RS} = t_*,$$

and at time-point t_* of gluing two flight segments, the phase coordinate breaks are following:

$$\begin{aligned} &\mathbf{v}^+ - \mathbf{v}^- - \Delta\mathbf{v} = 0, \\ &\mathbf{r}^+ - \mathbf{r}^- - \Delta\mathbf{r} = 0, \\ &m^+ - m^- - \Delta m = 0, \end{aligned} \tag{4.33}$$

where superscripts "+" and "–" denote the phase variables at the moments $t_* + 0$ and $t_* - 0$ consequently. Increments $\Delta\mathbf{v}$, $\Delta\mathbf{r}$, Δm are considered to be specified based on the results of the previous study of the safe separation processes in detail.

According to [1, 3], conditions (4.33) do not lead to a change in conjugate vector $\mathbf{p}$ at moment t_*:

$$\mathbf{p}^+ - \mathbf{p}^- = 0. \tag{4.34}$$

We will assume moment t_* (the duration of the pre-launch maneuver of CA with RS) to be given, then the Hamiltonian at the gluing point suffers a break:

$$H^+ - H^- - \Delta H = 0. \tag{4.35}$$

Detailed results of the study of through optimal injection trajectories with integrated consideration of the safe separation process are given in Chap. 6.

4.6.2 *The Through Optimization of ASTS Branching Trajectories with Safe Staging Requirements*

We accept the following assumptions:

- The ASTS motion is described by equations that take into account the spatiality, centrality of the gravitational field, presence of aerodynamic forces, and dependence of the thrust of CA and RS on flight modes;
- Due to a short duration of the CA pre-launch and separation maneuvers, we will consider them in the vertical plane, neglecting the change in RS mass.

We consider the ASTS motion in the coordinate system associated with the start point and introduce notation for the shorted phase vector $\overline{\mathbf{x}} = \{\mathbf{r}, \mathbf{v}\}$, so that $\mathbf{x} = \{\overline{\mathbf{x}}, m\}$. By virtue of the accepted assumptions, the phase vectors $\mathbf{x}$ have the dimension of five in the CA flight segments until the end of the separation, and seven in the subsequent segments. The control is the unit vector $\mathbf{e}_\tau$ directed along the longitudinal axis of the vehicle. The RS thrust vector has a fixed point of application and is directed to the center of mass.

The ASTS launch trajectory is schematically shown in Fig. 4.11, where.

0 is the point of departure from the cruising flight regime of CA with RS;

0–1 is the pre-launch maneuver of the CA with RS, as a result of which it is brought to the area of possible separation;

1 is the beginning of the pre-release maneuver of CA with RS;

1–2 is the pre-release maneuver of CA with RS, characterized by an intensive decrease in the angle of attack to reach negative normal overloads of CA for the separation and an angle-of-attack stabilization before breaking ties;

2–2′is the point of the breaking ties between RS and CA;

2–3 is the beginning leg of RS autonomous flight, where the RS pitch angle is constrained due to a dangerous nearness to CA;

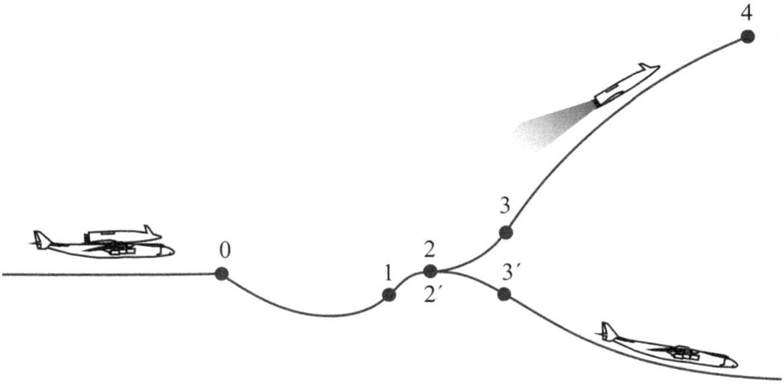

Fig. 4.11 Scheme of ASTS branching injection trajectory

$2'$–$3'$is the beginning leg of CA autonomous flight, corresponding to the section *2–3* with essential constraints on the safe relative position of CA and RS;

3–$3'$is the end of the separation process, when the CA leaves the area of aerodynamic and thermal impact of the RS engine plume, and the RS attitude is no longer limited by the CA position;

3-4 is the RS autonomous flight after the separation;

$3'$-$4'$ is the CA autonomous flight after the separation.

The CA + RS motion in section *0–2* is described by the equation

$$t \in [t_o, t_2) : \frac{d\mathbf{x}_\Sigma}{dt} = \mathbf{f}_\Sigma(\mathbf{x}_\Sigma, \mathbf{u}_\Sigma, t), \tag{4.36}$$

where $\mathbf{x}_\Sigma = \{x_{\Sigma j}, j = 1, \ldots, n_\Sigma\}$ is the phase vector, and $\mathbf{u}_\Sigma$ is the control vector of the CA + RS system.

At point 0, phase vector $\mathbf{x}_\Sigma$ is considered to be fixed:

$$Q_{oo} = t - t_o = 0, \ \mathbf{Q}_o = \mathbf{x}_\Sigma - \mathbf{x}_{\Sigma o} = 0. \tag{4.37}$$

In section *0–2*, there are essential constraints on the normal n_z and longitudinal n_x load factors, Mach number, dynamic pressure q and angle of attack α, which in general can be written in the form (4.2):

$$\mathbf{G}_\Sigma(\mathbf{x}_\Sigma, \mathbf{u}_\Sigma, t) \le 0. \tag{4.38}$$

In section *1–2*, an additional constraint is introduced on the pitch angular rate:

$$G_\Sigma = \vartheta - \vartheta_1^o - \omega\left(t - t_1^o\right) \le 0, \tag{4.39}$$

where ϑ is the pitch angle, ω is the maximum admissible pitch angular rate (in absolute value), ϑ_1^0 and t_1^0 are fixed values ($\vartheta_1^0 \ge \vartheta(t_1), t_1^0 \le \mathrm{t}_1$, where t_1 is the time of reaching the constraint (4.39)). Moment t_1 is determined from condition $G_\Sigma(t_1) = 0$. Time t_2 of the ties release is determined by the time of reaching a given normal load factor n'_{z2} of CA without RS:

$$Q_{20} = n'_z\left(\mathbf{x}'_2\right) - n'_{z2} = 0, \tag{4.40}$$

where $\mathbf{x}'_2 = \mathbf{x}'(t_2), \boldsymbol{x}'$ is the CA phase vector on section *$2'$–$4'$*.

The RS autonomous motion in section *2–4* is described by vector equation

$$\frac{d\mathbf{x}}{dt} = \mathbf{f}(\mathbf{x}, \mathbf{u}), \tag{4.41}$$

where $\mathbf{x} = \{x_j, j = 1, \ldots, n\}$ is the phase vector, $\mathbf{u}$ is the RS control vector.

The autonomous motion of CA in section $2'$–$4'$ is described by vector equation

$$\frac{d\mathbf{x}'}{dt} = \mathbf{f}'(\mathbf{x}', \mathbf{u}'), \tag{4.42}$$

where $\mathbf{x}' = \{x'_j, j = 1, \ldots, n'\}$ is the phase vector, $\mathbf{u}'$ is the CA control vector.

In sections *3–4* and *3′–4′*, constraints of form (4.2) apply to the normal n_z and longitudinal n_x load factors, Mach number M, dynamic pressure q, angle-of-attack α:

$$\mathbf{G}(\mathbf{x}, \mathbf{u}) \le 0 \tag{4.43}$$

and

$$\mathbf{G}'(\mathbf{x}', \mathbf{u}') \le 0. \tag{4.44}$$

In areas 2–3 and 2′-3′, in order to ensure the safe separation, an additional constraint on the relative attitude and position of the separated vehicles must be performed:

$$W(\mathbf{x}, \mathbf{x}', \mathbf{u}, \mathbf{u}', t) \le 0. \tag{4.45}$$

The end of sections 2–3 and 2′–3′ is determined by the time of departure from constraint (4.45):

$$W(t_3) = 0.$$

At the right end of the RS trajectory ($t = t_4$), the conditions for insertion into a given orbit must be met:

$$\mathbf{Q}_4(\mathbf{x}(t_4)) = 0, \; \mathbf{Q}_4 = \{Q_{4j}, j = 1, \ldots, k_4\}. \tag{4.46}$$

One of conditions (4.46) is used to define moment t_4 of the end of the injection trajectory. Without loss of generality, we can put $t_{4'} = t_4$, since section *3–4* of the RS autonomous flight is independent of the CA, so the duration of section *3′–4′* of the CA's branch cannot affect the functional.

Gluing sections *0–1* and *1–2*, *2–3* and *3–4*, *2′–3′* and *3′-4′* should occur without breaking the phase vectors $\mathbf{x}_\Sigma$, $\mathbf{x}$ and $\mathbf{x}'$:

$$\begin{aligned} \mathbf{Q}_1 &= \mathbf{x}_{\Sigma 1}^- - \mathbf{x}_{\Sigma 1}^+ = 0, \\ \mathbf{Q}_3 &= \mathbf{x}_3^- - \mathbf{x}_3^+ = 0, \\ \mathbf{Q}'_3 &= \mathbf{x}_3'^- - \mathbf{x}_3'^+ = 0, \end{aligned} \tag{4.47}$$

where $()_j^- = ()_{t=t_j-0}$, $()_j^+ = ()|_{t=t_j+0}$

The velocities v_Σ, v, v', the radius vector of the center of mass $\mathbf{r}_\Sigma$ and mass m_Σ of the system at branching point 2 are continuous, that is, in the general case, the following conditions must be met:

$$\mathbf{Q}_2\left(\mathbf{x}_{\Sigma 2}, \mathbf{x}_2, \mathbf{x}_2', t_2\right) = 0. \tag{4.48}$$

Thus, we consider the control problem of the vehicles maneuvers, the motion of which is described by Eqs. (4.36), (4.41), (4.42) taking into account constraints (4.38), (4.39), (4.43)–(4.45) and discontinuity of the phase vector and right-hand parts of motion equations at the staging point, from initial point (4.30) to final manifold (4.41) with the minimum mass consumption:

$$J \equiv m(t_4) - m(t_0) \Rightarrow \max_{\{\mathbf{u}_\Sigma, \mathbf{u}, \mathbf{u}'\}}. \tag{4.49}$$

4.6.3 Optimality Conditions

According to [1], the extended functional is written as

$$\begin{aligned}\tilde{J} &= J(\mathbf{x}_4) + \nu_{00} Q_{00} + \mathbf{n}_0^{\mathrm{T}} \mathbf{Q}_0(\mathbf{x}_{\Sigma 0}, t) + \mathbf{n}_2^{\mathrm{T}} \mathbf{Q}_2\left(\mathbf{x}_{\Sigma 2}, \mathbf{x}_2, \mathbf{x}_2', t_2\right) + \\ &+ \mathbf{n}_4^{\mathrm{T}} \mathbf{Q}_4(\mathbf{x}_4) + \nu_{20} Q_{20}\left(\mathbf{x}_2'\right) + \int\limits_{t_o}^{t\bar{2}} \left[\mathbf{p}_\Sigma^{\mathrm{T}}(\dot{\mathbf{x}}_\Sigma - \mathbf{f}_\Sigma) - \boldsymbol{\lambda}_\Sigma^{\mathrm{T}} \mathbf{G}_\Sigma - \pi_\Sigma W_\Sigma\right] dt + \\ &+ \int\limits_{t_2^+}^{t_4} \left[\mathbf{p}^{\mathrm{T}}(\dot{\mathbf{x}} - \mathbf{f}) - \boldsymbol{\lambda}^{\mathrm{T}} \mathbf{G} + \mathbf{p}'^{\mathrm{T}}\left(\dot{\mathbf{x}}' - \mathbf{f}'\right) - \boldsymbol{\lambda}'^{\mathrm{T}} \mathbf{G}' - \pi W\right] dt,\end{aligned} \tag{4.50}$$

where π, $\boldsymbol{\nu}$, $\mathbf{n}$, $\boldsymbol{\lambda}$ are the Lagrange multipliers, $\mathbf{p}$ is the conjugate vector, the indices point at belonging to the corresponding sections of the branching trajectory.

Below, the optimality conditions for the controls $\mathbf{u}_\Sigma$, $\mathbf{u}$, $\mathbf{u}'$ and the transversality conditions at the boundary points of the trajectory sections are obtained based on the analysis of the increment $\tilde{J}$ for weak variations of the control [1].

The variation of the extended functional (4.50) taking into account (4.36)–(4.49) has the form

$$\begin{aligned}d\tilde{J} &= \frac{\partial J}{\partial \mathbf{x}_4} d\mathbf{x}_4 + \mathbf{n}_0^{\mathrm{T}} d\mathbf{x}_{\Sigma_0} + \nu_{00} dt_0 \\ &+ \mathbf{n}_2^{\mathrm{T}}\left(\frac{\partial \mathbf{Q}_2}{\partial \mathbf{x}_{\Sigma_2}} d\mathbf{x}_{\Sigma_2} + \frac{\partial \mathbf{Q}_2}{\partial \mathbf{x}_2} d\mathbf{x}_2 + \frac{\partial \mathbf{Q}_2}{\partial \mathbf{x}'_2} d\mathbf{x}'_2 + \frac{\partial \mathbf{Q}_2}{\partial t_2} dt_2\right) + \mathbf{n}_4^{\mathrm{T}} \frac{\partial \mathbf{Q}_4}{\partial \mathbf{x}_4} d\mathbf{x}_4 + \nu_{20} \frac{\partial Q_{20}}{\partial \mathbf{x}'_2} d\mathbf{x}' \\ &+ \left(\mathbf{p}_\Sigma^{\mathrm{T}} \dot{\mathbf{x}}_\Sigma - H_\Sigma\right)\Big|_{t_2} dt_2 - \left(\mathbf{p}_\Sigma^{\mathrm{T}} \dot{\mathbf{x}}_\Sigma - H_\Sigma\right)\Big|_{t_0} dt_0 + \left[\left(\mathbf{p}^{\mathrm{T}} \dot{\mathbf{x}} - H\right) + \left(\mathbf{p}'^{T} \dot{\mathbf{x}}' - H'\right)\right]\Big|_{t_4} dt_4 \\ &- \left[\left(\mathbf{p}^{T} \dot{\mathbf{x}} - H\right) + \left(\mathbf{p}'^{T} \dot{\mathbf{x}}' - H'\right)\right]\Big|_{t_2} dt_2 + \mathbf{p}_\Sigma^{T} \delta \mathbf{x}_\Sigma\Big|_{t_0}^{t_2} + \mathbf{p}^{T} \delta \mathbf{x}\Big|_{t_2}^{t_4} + \mathbf{p}'^{T} \delta \mathbf{x}'\Big|_{t_2}^{t_4}\end{aligned}$$

$$
-\int_{t_0}^{t_2}\left[\frac{\partial H_\Sigma}{\partial \mathbf{x}_\Sigma}\delta\mathbf{x}_\Sigma+\frac{\partial H_\Sigma}{\partial \mathbf{u}_\Sigma}\delta\mathbf{u}_\Sigma+\dot{\mathbf{p}}_\Sigma^{\mathrm{T}}\delta\mathbf{x}_\Sigma+\pi_\Sigma\left(\frac{\partial W_\Sigma}{\partial \mathbf{x}_\Sigma}\delta\mathbf{x}_\Sigma+\frac{\partial W_\Sigma}{\partial \mathbf{u}_\Sigma}\delta\mathbf{u}_\Sigma\right)\right]dt
$$

$$
-\int_{t_2}^{t_4}\left[\frac{\partial H}{\partial \mathbf{x}}\delta\mathbf{x}+\frac{\partial H}{\partial \mathbf{u}}\delta\mathbf{u}+\dot{\mathbf{p}}^{\mathrm{T}}\delta\mathbf{x}+\pi\left(\frac{\partial W}{\partial \mathbf{x}}\delta\mathbf{x}+\frac{\partial W}{\partial \mathbf{u}}\delta\mathbf{u}\right)\right]dt
$$

$$
-\int_{t_2}^{t_4}\left[\frac{\partial H'}{\partial \mathbf{x}'}\delta\mathbf{x}'+\frac{\partial H'}{\partial \mathbf{u}'}\delta\mathbf{u}'+\dot{\mathbf{p}}'^{\mathrm{T}}\delta\mathbf{x}'+\pi\left(\frac{\partial W}{\partial \mathbf{x}'}\delta\mathbf{x}'+\frac{\partial W}{\partial \mathbf{u}'}\delta\mathbf{u}'\right)\right]dt, \tag{4.51}
$$

where $H=\mathbf{p}^{\mathrm{T}}\mathbf{f}+\boldsymbol{\lambda}^{\mathrm{T}}\mathbf{G}$. The independent increment $d\mathbf{x}=\delta\mathbf{x}+\dot{\mathbf{x}}dt$. From the condition of vanishing the variation of functional $\tilde{J}$ we obtain

$$
\begin{cases}
(\mathbf{n}_0+\mathbf{p}_\Sigma)|_{t_o}=0,\ H_\Sigma|_{t_o}+\nu_{00}=0,\ \left(\mathbf{n}_2^{\mathrm{T}}\dfrac{\partial \mathbf{Q}_2}{\partial \mathbf{x}_{\Sigma 2}}+\mathbf{p}_\Sigma^{\mathrm{T}}\right)\Bigg|_{t_2}=0,\\
\left(\mathbf{n}_2^{\mathrm{T}}\dfrac{\partial \mathbf{Q}_2}{\partial \mathbf{x}_2'}-\mathbf{p}'^{\mathrm{T}}+\nu_{20}\dfrac{\partial Q_{20}}{\partial \mathbf{x}_2'}\right)\Bigg|_{t_2}=0,\\
\left(\mathbf{n}_2^{\mathrm{T}}\dfrac{\partial \mathbf{Q}_2}{\partial \mathbf{x}_2}-\mathbf{p}^{\mathrm{T}}\right)\Bigg|_{t_2}=0,\ \left(H_\Sigma-H-H'-\mathbf{n}_2^{\mathrm{T}}\dfrac{\partial \mathbf{Q}_2}{\partial t_2}\right)\Bigg|_{t_2}=0,\\
\left(\dfrac{\partial J}{\partial \mathbf{x}_4}+\mathbf{n}_4^{\mathrm{T}}\dfrac{\partial \mathbf{Q}_4}{\partial \mathbf{x}_4}+\mathbf{p}'^{\mathrm{T}}+\mathbf{p}^{\mathrm{T}}\right)\Bigg|_{t_4}=0,\\
\dot{\mathbf{p}}_\Sigma=-\left(\dfrac{\partial H_\Sigma}{\partial \mathbf{x}_\Sigma}\right)^{\mathrm{T}}-\pi_\Sigma\left(\dfrac{\partial W_\Sigma}{\partial \mathbf{x}_\Sigma}\right)^{\mathrm{T}},\ \dot{\mathbf{p}}=-\left(\dfrac{\partial H}{\partial \mathbf{x}}\right)^{\mathrm{T}}-\pi\left(\dfrac{\partial W}{\partial \mathbf{x}}\right)^{\mathrm{T}},\\
\dot{\mathbf{p}}'=-\left(\dfrac{\partial H'}{\partial \mathbf{x}'}\right)^{\mathrm{T}}-\pi\left(\dfrac{\partial W}{\partial \mathbf{x}'}\right)^{\mathrm{T}},\\
\dfrac{\partial H_\Sigma}{\partial \mathbf{u}_\Sigma}+\pi_\Sigma\dfrac{\partial W_\Sigma}{\partial \mathbf{u}_\Sigma}=0,\ \dfrac{\partial H}{\partial \mathbf{u}}+\pi\dfrac{\partial W}{\partial \mathbf{u}}=0,\\
\dfrac{\partial H'}{\partial \mathbf{u}'}+\pi\dfrac{\partial W}{\partial \mathbf{u}'}=0,\ (H+H')|_{t_4}=0.
\end{cases} \tag{4.52}
$$

If the phase vector remains continuous at the branch point:

$$
Q_{2j}=\left(x_\Sigma^j-x^j\right)\Big|_{t_2}=0, j=1,\ldots,n,
$$

$$
Q_{2j+n}=\left(x_\Sigma^j-x'^j\right)\Big|_{t_2}=0, j=1,\ldots,n',
$$

then, taking into account (4.48),

$$
\mathbf{n}_2^{\mathrm{T}}\frac{\partial \mathbf{Q}_2}{\partial x_{\Sigma 2}^j}=\nu_2^j+\nu_2^{j+n},\ \mathbf{n}_2^{\mathrm{T}}\frac{\partial \mathbf{Q}_2}{\partial x_2^j}=-\nu_2^j,\ \mathbf{n}_2^{\mathrm{T}}\frac{\partial \mathbf{Q}_2}{\partial x_2'^j}=-\nu_2^{j+n},
$$

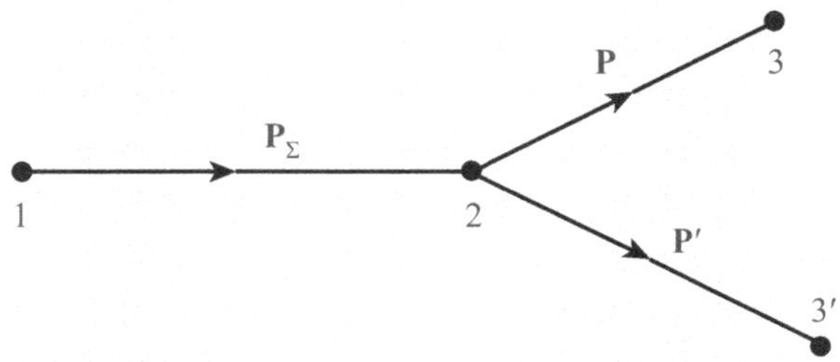

Fig. 4.12 Illustration of preservation of the "total" conjugate vector, when passing through a branching point, similarly to the charge conservation law in an electric circuit

and the transversality condition at this point takes the following form:

$$\mathbf{p}_{\Sigma} = \mathbf{p} + \mathbf{p}' + \nu_{20}\left(\frac{\partial Q_{20}}{\partial \mathbf{x}_2'}\right)^{\mathrm{T}} \tag{4.53}$$

If branching time t_2 is fixed, that is $\frac{\partial Q_{20}}{\partial \mathbf{x}_2'} = 0$, then

$$\mathbf{p}_{\Sigma} = \mathbf{p} + \mathbf{p}',$$

i.e. similar to the law of charge conservation in an electric circuit, when passing through a branch point, the "total" conjugate vector $\mathbf{p}$ is preserved (Fig. 4.12).

From analysis of (4.52), the optimization rules for branching trajectories [1, 3, 10] are following:

- Phase and conjugate systems of equations are integrated separately for each branch in accordance with PMP. The composition of phase variables and equations of motion on each branch can be set independently;
- At each branch point, special conditions for gluing conjugate variables must be met, depending on the conditions for gluing phase trajectories;
- When a new branch is added, the dimension of the BVP increases by no more than the dimension of the phase vector on the branch.

4.6.4 Control Constraints at the ASTS Separation Process

When the RS with operating engine is separating from CA, such conditions must be provided under which there are no collisions of the RS and CA, and the thermal and aerodynamic effects on the CA from the RS engine plume do not exceed the permissible limits [10, 15–17].

The development of control algorithms of safety separation is an independent task. However, with through optimization of the launch trajectory, we can consider the separation process in a simplified statement, taking into account those constraints on the motion of the CA and RS, which can significantly affect the trajectories and functional. Based on these considerations, in order to simplify the mathematical notation of the constraints, the following assumption was made: the motion of ASTS until the end of the separation process is considered in the vertical plane, and further motion is described by a spatial model.

Next, three types of control constraints on the RS are formulated, which ensure the safe motion of the center of mass and attitude. First, the most general form of such constraints is considered, and then some simplifying assumptions regarding the shape of the RS engine plume and the conditions of their action on the CA are introduced.

4.6.5 A General View of Separation Constraints

To formulate constraints on the RS control that ensure the fulfillment of these conditions, consider the coordinate systems: Oxy is associated with the Earth, $O'\,x'\,y'$ is associated with the RS, $O''\,x''\,y''$ is associated with CA (Fig. 4.13). Denote $\mathbf{r}$, $\mathbf{r}'$ and $\mathbf{r}''$ as radius vectors in the corresponding coordinate systems.

We will assume that distribution of heat fluxes (temperatures) $q_T(\mathbf{r}')$ around the RS with operating engines is known, the parametric equations $\mathbf{r}' = \mathbf{r}'(\xi')$ and $\mathbf{r}'' = \mathbf{r}''(\xi'')$ of the outer contours L' and L'' of RS and CA surfaces are given, the distribution of admissible heat fluxes $q_{\mathrm{Tadm}}(\xi'')$ along the outer contour of CA surface is determined.

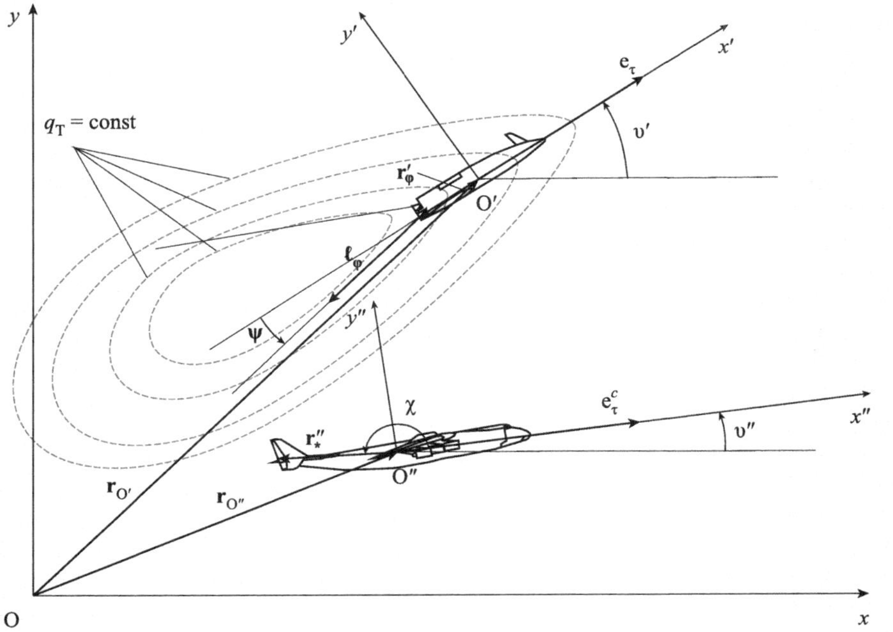

Fig. 4.13 Scheme of ASTS separation

If the current positions of the centers of mass of RS $\mathbf{r}_{O'}$ and CA $\mathbf{r}_{O''}$ and CA pitch angle ϑ'' are known, then the maximum admissible RS pitch angle ϑ' at which the level of heat fluxes from the RS on the CA surface does not exceed admissible values can be determined [10].

Let us designate $\mathbf{r}''_* = \mathbf{r}''(\xi''_*)$ as the critical point on the CA surface, in which at the maximum admissible RS pitch angle $q_{\mathrm{T}}(\mathbf{r}') = q_{\mathrm{Tadm}}(\xi''_*)$. This means that [19]

$$
\begin{aligned}
&a)\ W_1(\xi''_*, \vartheta') = q_T(-\mathbf{r}''_{O'} + \mathbf{A}\mathbf{r}''(\xi''_*)) - q_{\mathrm{Tadm}}(\xi''_*) = 0,\\
&b)\ W_2(\xi'_*, \vartheta') = \frac{\partial W_1}{\partial \xi''_*} = \left.\frac{\partial q_T}{\partial \mathbf{r}'}\right|_{W_1=0} \cdot \mathbf{A} \left.\frac{\partial \mathbf{r}''}{\partial \xi''_*}\right|_{\mathbf{r}''\in L''} - \frac{\partial q_{\mathrm{Tadm}}}{\partial \xi''_*} = 0,\\
&\left(\frac{\partial W_1}{\partial \vartheta'} = \frac{\partial q_T}{\partial \mathbf{r}'}\frac{\partial \mathbf{A}}{\partial \vartheta'}\mathbf{r}''(\xi'_*) \geq 0\right),
\end{aligned}
\tag{4.54}
$$

where $\mathbf{A} = \mathbf{A}(\Delta\vartheta) = \begin{pmatrix} \cos\Delta\vartheta & \sin\Delta\vartheta \\ -\sin\Delta\vartheta & \cos\Delta\vartheta \end{pmatrix}$, $\Delta\vartheta = \vartheta' - \vartheta''$, $\mathbf{r}''_{O'}$ is the radius vector from point O″ to point O′.

Functions W_1 and W_2 determine the RS control on the boundary of the admissible field (4.45). Condition a) in (4.54) requires that at the critical point on the CA surface, the heat flux of the RS engines should be equal to the admissible value at this point. Condition b) means touching the CA surface at the critical point of the contour line of the RS heat flux $q_{\mathrm{T}}(r') = q_{\mathrm{Tadm}}(\xi''_*)$.

4.6.6 The Frozen Jet Model

Let us consider the case when the admissible heat fluxes q_{Tadm} are of the same level on the entire CA surface.

$$
q_{\mathrm{Tadm}}(\xi'') \equiv const = \min_{\xi''_*} q_{\mathrm{Tadm}}(\xi''_*)
$$

Let the parametric equation of the L'_T contour line be given $q_{\mathrm{T}}(r') \equiv q_{\mathrm{Tadm}}$:

$$
L'_{\mathrm{T}} = \left\{\mathbf{r}'(\xi'), \xi' \in [\xi'_O, \xi'_1] : q_{\mathrm{T}}(\mathbf{r}'(\xi')) = q_{\mathrm{Tadm}}\right\}. \tag{4.55}
$$

Then the constraint a) in (4.54) takes the following form [19]:

$$
\left.
\begin{aligned}
&\begin{pmatrix} W_1(\vartheta', \xi'_*, \xi''_*) \\ W_2(\vartheta', \xi'_*, \xi''_*) \end{pmatrix} = -\mathbf{r}''_{O'} + \mathbf{A}\mathbf{r}''(\xi''_*) - r'(\xi'_*) = 0,\\
&W_3(\vartheta', \xi'_*, \xi''_*) = \det A \left.\frac{\partial \mathbf{r}''}{\partial \xi''_*}\right|_{\mathbf{r}''\in L''} \left.\frac{\partial \mathbf{r}'}{\partial \xi'_*}\right|_{\mathbf{r}'\in L'_{\mathrm{T}}} = 0.
\end{aligned}
\right\}.
\tag{4.56}
$$

The last equation in (4.56) corresponds to the condition that contour line L'_T touches the CA surface at the critical point $\mathbf{r}''(\xi''_*)$.

Remark 1 Relations (4.56) determine conditions for separation from the CA of the object, the shape $L'_{\rm T}$ of which is given by parametric Eq. (4.55). Therefore, the external contour of RS with "frozen jet", the boundary of which is determined by the condition $q_{\rm T}(\mathbf{r}') \equiv q_{\rm Tadm}$, can be taken as $L'_{\rm T}$.

Remark 2 Curve $L'_{\rm T}$ can be understood in the general case as boundary L'_J of a safe (from the point of view of a criterion J) distance from the RS of some elements of CA. The contours of the CA elements that are critical to this criterion will be denoted by L''_J. In order to take into account several risk factors, one should compose a vector constraint (4.45) on the control of RS and CA from conditions like (4.55), (4.56), where instead of $L'_{\rm T}$ and L'' one should understand L'_J and L''_J.

In the same way, it is possible to take into account the requirements of shockless separation, constraints on the thermal effect of the RS jet on the CA, constraints on the aerodynamic effect of the RS propulsion on the CA, etc.

Remark 3 It is desirable that curves L'_J, L''_J to be smooth. For piecewise-smooth contours L'_J, L''_J conditions like (4.55), (4.56) will have to be written separately for each of the smooth sections of these contours.

4.6.7 The Conical Jet Model

We will assume that the area of hazardous impact of the RS engine jet is limited by a cone (on the plane O′x′y′ by an angle) with the apex at point $\mathbf{r}'_\Phi = \{x'_\Phi, 0\}$, and the line (generatrix of the cone) in the direction $l\ell_\Phi(|l\ell_\Phi| = 1)$, passing through the critical point $\mathbf{r}''_*$ on the CA surface, form angle ψ with the RS longitudinal axis (Fig. 4.13).

In this case, conditions (4.55), (4.56) in coordinate system Oxy take the form (the tangency condition $W_3 = 0$ is absent in this case):

$$\mathbf{W} = \mathbf{r}_{O'} + x_\Phi \mathbf{e}_\tau + \lambda \mathbf{A}^{\rm T}(\psi)\mathbf{e}_\tau - \mathbf{r}_{O''} - r''_* \mathbf{A}^{\rm T}(\chi)\mathbf{e}^c_\tau = 0, \tag{4.57}$$

where $\mathbf{e}_\tau$, $\mathbf{e}^c_\tau$ are unit vectors in the directions of the longitudinal axes RS and CA respectively, λ is the distance of the critical point from the jet apex, χ is the angle between r''_* and axis $O''x''$.

From (4.57) one can explicitly express the constraint on the RS attitude, where λ is found from the norming condition for $\mathbf{e}_\tau$:

$$\mathbf{e}_{\tau adm} = \frac{\left[x'_\Phi \mathbf{E} + \lambda \mathbf{A}^T\right]\boldsymbol{\mu}}{|\mu|^2}, \tag{4.58}$$

where $\lambda = -x'_\Phi \cos\phi - \sqrt{|\mu|^2 - x'^2_\Phi \sin^2\psi}$, $\boldsymbol{\mu} = \mathbf{r}_{O''} - \mathbf{r}_{O'} + r''_* \mathbf{A}^T(\chi)\mathbf{e}^c_\tau$, $\mathbf{E}$ is the identity matrix, $(x'_\Phi, 0)$ are coordinates of the jet apex in system O′x′y′.

Constraint (4.45) in case (4.58) can be written as a constraint on the RS angle of attack:

$$G \equiv f_\tau(\mathbf{x}) - (\mathbf{e}_\tau, \mathbf{e}_v), f_\tau = (\mathbf{e}_{\tau\text{adm}}, \mathbf{e}_v). \quad (4.59)$$

4.6.8 Features of the Numerical Problem Solving

The through optimization of ASTS launch trajectories, taking into account the constraints on RS and CA motion during separation, was carried out using the *ASTER* software package [19, 20]. Constraints on the RS in the separation process were given in form (4.58), (4.59).

In accordance with the method described above, when optimizing a branching trajectory, the order of the corresponding TPBVP increases. Since it was assumed that during the separation process, the flight of CA and RS occurs in the vertical plane, then the dimension of the phase vector on branch 2′–3′ is equal four. The TPBVP order increases by the same number. The components of conjugate vector $\mathbf{p}'$ at the beginning of section 2′ - 3′ were taken as additional varied parameters. The residuals were the components of vector $\mathbf{p}'$ at time t_3 of the end of the separation process, since at the moment of the end of the effect of CA on the RS, condition $\mathbf{p}'(t_3) = 0$ must be met.

Since section 2′–3′ is short, it is advisable to organize the solution of the BVP in this section inside the main BVP. Calculations have shown that a decrease in residual norm $\|\mathbf{p}'(t_3)\|$ by 6–7 orders of magnitude when using Newton's method for solving BVP is achieved in 2–3 iterations. To speed up the solution procedure, the method of reducing the BVP described in Sect. 4.3 based on the projection of the boundary conditions from the right ends of the side branches can also be used.

4.6.9 Examples of Calculations for the Conical Jet Model

The BVP was solved numerically using *ASTER* software package. Due to the short duration of the separation section, it is convenient to divide the general BVP into outer and inner ones. The outer BVP is connected with the boundary conditions for the branches b_1^1 and b_1^3, and the inner BVP for $b_{1,2}^2$. The inner TPBVP is solved in the calculation of each trajectory in the iterative process of the outer loop. In this case, the taking into account the branching nature of the trajectory increases slightly (by 1–3%) the time for solving the problem.

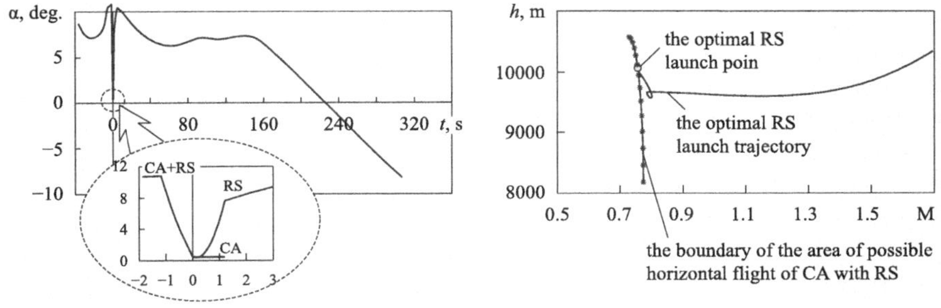

Fig. 4.14 The optimal angle-of-attack program (left) and launch trajectory (right) of ASTS An-225/Interim HOTOL

Using the described method and *ASTER* software package, extensive parametric studies were carried out on projects MAKS [21, 22] and An-225/Interim HOTOL [23].

As an example, on Fig. 4.14 the optimal launch trajectories and control programs for one of the An-225/Interim HOTOL layout variant are shown.

ASTER program made it possible with minimal computational costs to investigate the influence of the main layout parameters and RS propulsion cycloram on the maximum functional. The average number of iterations (in the outer BVP) did not exceed $2 \div 3$ in such calculations with a relative error in fulfillment of the boundary conditions that not exceed 10^{-6}.

4.7 Optimization of Carrier Aircraft Steady-State Flight Mode Before Rocket Stage Launch

Studies of the optimal RS injection trajectories show that for the optimal RS launch, CA must reach altitudes and velocities close to the limit. The RS launch is preceded by a fairly long section of acceleration, climb, and, perhaps, cruising flight of the CA.

Many theoretical and computational works have been devoted to the optimization of these CA flight modes. In particular, the classical results of Miele are well known in this field. According to [24], the optimal program for aircraft climb contains three characteristic sections (Fig. 4.15):

(a) degenerate section (1–2), independent of the boundary conditions;
(b) Section (0–1) from the start point to the touching the degenerate section;
(c) Section (2–3) of leaving the degenerate section and moving to the final point.

If the RS launch is preceded by a CA cruise flight at a given range, then the flight profile of the CA can also be schematically divided into three sections:

(a) Section (1–2) of cruising flight, independent of the boundary conditions;

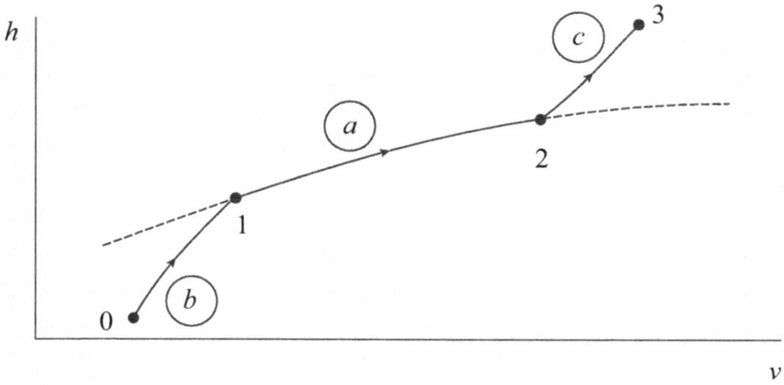

Fig. 4.15 Scheme of optimal aircraft climb trajectory with degenerate segment according to Miele [24]

(b) Section (0–1) from the start point to the touching the cruising section;
(c) Section (2–3) of leaving the cruising section and moving to the final point.

Consequently, the section of the CA dynamic maneuver before the RS launch is always preceded by some degenerate, i.e. independent of the boundary conditions, flight segment. In addition to the two types considered above, a degenerate segment can also correspond to a flight along the boundary of an admissible phase area.

Thus, the optimization of the ASTS trajectory is divided into two independent problems: the problem of optimal (according to some criterion) transfer of the CA from the initial point to a degenerate segment (for example, the cruise one) and the problem of insertion into the target orbit from the degenerate flight segment. In this section, when speaking about the through optimization of the RS injection trajectory, we mean the second task, in which the cruise flight optimization at point 2 is an important component.

Let us note some principal difficulties of the through solution of the optimization problem.

The functional (the mass injected into the orbit) weakly depends (or does not depend at all) on the duration of the maneuver of CA with RS and is determined mainly by the phase vector at the end of this maneuver. In particular, by the beginning of any "optimal" maneuver of CA with RS, it is possible to adjust the stationary flight segment, and the newly obtained trajectory of the CA with RS will provide the same value of the functional as the "optimal" one.

Therefore, if the duration of the CA maneuver with RS is not limited before the RS launch, then for the chosen functional, the through problem turns out to be degenerate (or ill-conditioned). To regularize the problem, one can, for example, limit the duration of the CA maneuver with RS or bind the RS mass injected into orbit with the CA fuel consumption.

A technique has been developed to resolve these difficulties in the through optimization of launch trajectories, namely: to determine the best steady-state flight mode of the CA with RS, from which the dynamic maneuver of the CA to launch the RS should begin.

4.7.1 The Area of Possible Carrier Aircraft Steady-State Flight Modes

Typical area D of possible CA steady-state flight modes in the plane of velocity v_i and altitude h_i at fixed path angle θ is shown in Fig. 4.16.

The boundary ∂D of the area D has the following characteristic smooth segments:

(0-1) is the constraint on the admissible dynamic pressure,
(1-2) is the constraint on the Mach number,
(2-3) is the maximum reachable altitudes, at which stationary flight is possible,
(3-4) is the maximum altitudes of a possible steady-state flight with the maximum admissible angle of attack,
(4-0) is the minimum admissible altitude ($h = h_{\min}$).

On Fig. 4.16 outer normals **n** to each segment of boundary $\partial D|_{\theta=\text{fix}}$ are shown. Due to the physical nature of these segments, the normals to them for almost any

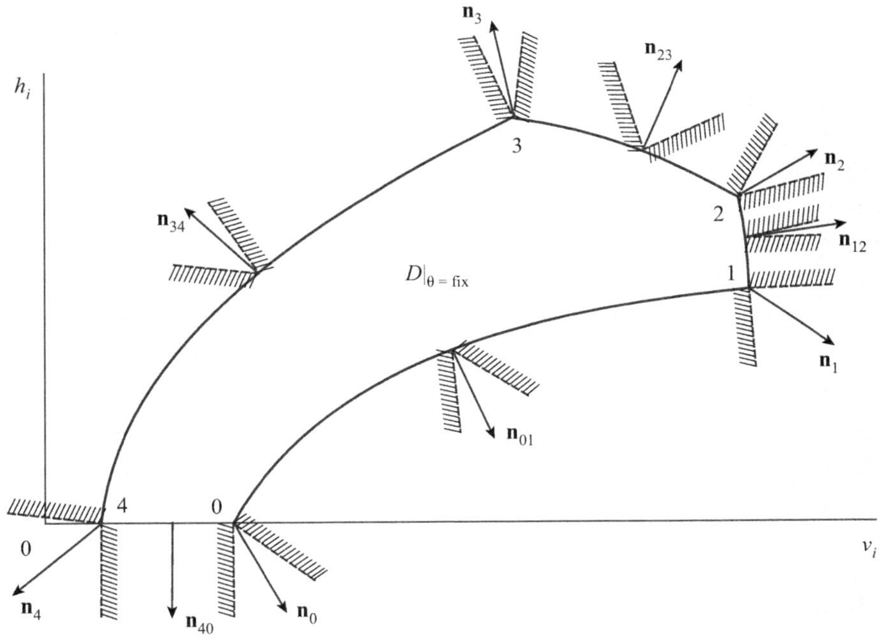

Fig. 4.16 Typical area D of possible steady-state flight of carrier aircraft

aicraft are characterized by a quite definite sector of possible directions: $\mathbf{n}_{01}$ lies in the fourth quadrant, $\mathbf{n}_{12}$ and $\mathbf{n}_{23}$ are in the first quadrant, $\mathbf{n}_{34}$ belongs to the second quadrant, $\mathbf{n}_{40}$ is antiparallel to the h-axis. At corner points the normals are characterized by following directions: $\mathbf{n}_0$ lies in the fourth quadrant, $\mathbf{n}_1$ is in the first or fourth quadrant, $\mathbf{n}_2$ is in the first, $\mathbf{n}_3$ belongs to the first or second quadrant, $\mathbf{n}_4$ is in the second or third quadrant.

4.7.2 Optimization of Initial Conditions for Carrier Aircraft Maneuver with Rocket Stage

As shown in the previous sections, the initial conditions for the flight of CA with RS before the maneuver for the RS launch are limited on the plane of variables $\{v_0, h_0\}_{\theta=0}$ or $\{M_0, h_0\}_{\theta=0}$ by the area D of a possible steady-state flight. The optimal control theory implies the existence of an optimal regime $\xi_{\text{opt}} = \{M_0, h_0\}_{\text{opt}}$ of a horizontal flight of CA with RS such that the gradient of the functional $\mathbf{G} = \left\{\frac{\partial m_{orb}}{\partial M_0}, \frac{\partial m_{orb}}{\partial h_0}\right\}$ at point ξ_{opt} is directed along the outer normal to smooth boundary ∂D of area D (Fig. 4.17).

If ξ_{opt} is the corner point of boundary ∂D, then at ξ_{opt} the gradient $\mathbf{G}$ must lie inside the angle formed by the outward normals to the boundary segments to the right and left of ξ_{opt}. Taking into account that the conjugate variables are the influence functions:

$$\mathbf{p}_r = \left(\frac{\partial m_{\text{orb}}}{\partial \mathbf{r}}\right)^{\mathrm{T}}, \mathbf{p}_v = \left(\frac{\partial m_{\text{orb}}}{\partial \mathbf{v}}\right)^{\mathrm{T}}, P_m = \frac{\partial m_{\text{orb}}}{\partial m},$$

and

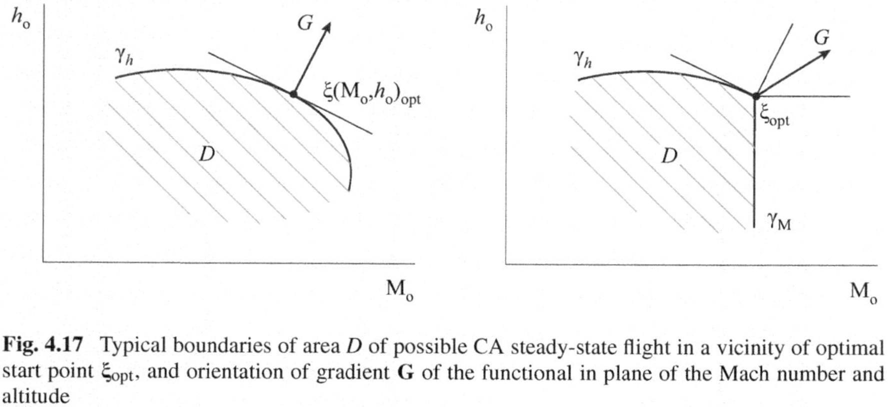

Fig. 4.17 Typical boundaries of area D of possible CA steady-state flight in a vicinity of optimal start point ξ_{opt}, and orientation of gradient $\mathbf{G}$ of the functional in plane of the Mach number and altitude

$$\delta v_0 = a_s \delta M_0 + M_0 \frac{\partial a_s}{\partial h} \delta h_0, \ \mathbf{v}_0 = (v_0, 0), \ \mathbf{r}_0 = (0, h_0)$$

(a_s is the sound speed), we obtain in coordinates (M_0, h_0):

$$\mathbf{G} = \left\{ a_s p_{v_x 0}, \ p_{y0} + M_0 \frac{\partial a_s}{\partial h} p_{v_x 0} \right\}, \tag{4.60}$$

or in coordinates (v_0, h_0):

$$\mathbf{G} = \left(p_{v_x 0}, p_{y0} \right). \tag{4.61}$$

Due to the physical meaning of conjugate variables as the influence functions, the following inequalities take place for typical aircraft: $p_{v_x 0} > 0, p_{y0} > 0$, (see Sects. 3.3, 3.4). Then the optimal start point can correspond either to section γ_h of boundary ∂D obtained from the condition of equality of the required and maximum available thrust, or to section $\gamma_M \in \partial D$, corresponding to the constraint on the admissible Mach number. Since normal $\mathbf{n}_M$ to boundary γ_M in the plane $\{M_0, h_0\}$ has coordinates $\mathbf{n}_M = (1, 0)$, then inclusion $\xi_{opt} \in \gamma_M$ is only in the case of strict equality $p_{yo} + M_0 \frac{\partial a_s}{\partial h} p_{v_x 0} = 0$, which is practically impossible. Therefore, for the considered class of inference problems $\xi_{opt} \in \gamma_h$.

Boundary γ_h is given by a system of parametric equations [25]:

$$\gamma_h : \begin{cases} \frac{T(h_0, M_0)}{G} - K^{-1}(\alpha, M_0) = 0 \\ C_L(\alpha, M_0) q_0 F_{ref}/G - 1 = 0 \end{cases}, \tag{4.62}$$

where K is the lift-to-drag ratio, q is the dynamic pressure, G/F_{ref} is the specific load on the wing, T is the thrust (all for the pre-launch configuration of CA with RS).

Normal $\mathbf{n}_h$ to boundary γ_h has coordinates

$$\mathbf{n}_h = \left\{ \frac{1}{G} \frac{\partial T}{\partial M} - \frac{\partial K^{-1}}{\partial M} + \frac{2c_y}{M} \frac{\partial K^{-1}}{\partial c_y}; \frac{1}{G} \frac{\partial T}{\partial h} + \frac{\partial K^{-1}}{\partial c_y} c_y \left(\frac{1}{\rho} \frac{\partial \rho}{\partial h} + \frac{2}{a_s} \cdot \frac{\partial a_s}{\partial h} \right) \right\}. \tag{4.63}$$

If $\xi_{opt} \in \gamma_h \backslash \partial \gamma_h$, then

$$\boldsymbol{\xi}_{opt} : \mathbf{n}_h (\boldsymbol{\xi}_{opt}) \uparrow\uparrow \mathbf{G}(\boldsymbol{\xi}_{opt}). \tag{4.64}$$

Substituting (4.60), (4.61), (4.63) into (4.64), we get:

$$\begin{gathered} \boldsymbol{\xi}_{opt} : \left(\frac{1}{G} \frac{\partial T}{\partial M} - \frac{\partial K^{-1}}{\partial M} + \frac{2c_y}{M} \frac{\partial K^{-1}}{\partial c_y} \right) \left(p_y + M p_{v_x} \frac{\partial a_s}{\partial h} \right) \\ - \left[\frac{1}{G} \frac{\partial T}{\partial h} + \frac{\partial K^{-1}}{\partial c_y} c_y \left(\frac{1}{\rho} \frac{\partial \rho}{\partial h} + \frac{2}{a_s} \cdot \frac{\partial a_s}{\partial h} \right) \right] a_s p_{v_x} = 0 \end{gathered} \tag{4.65}$$

or

$$\xi_{\text{opt}}: \; p_y\left(\frac{1}{G}\frac{\partial T}{\partial M} - \frac{\partial K^{-1}}{\partial M} + \frac{2c_y}{M}\frac{\partial K^{-1}}{\partial c_y}\right) -$$
$$-p_{v_x}a_s\left[\frac{M}{a_s}\frac{\partial a_s}{\partial h}\left(\frac{\partial K^{-1}}{\partial M} - \frac{1}{G}\frac{\partial T}{\partial M}\right) + \frac{1}{G}\frac{\partial T}{\partial h} + c_y\frac{\partial K^{-1}}{\partial c_y}\frac{1}{\rho}\frac{\partial \rho}{\partial h}\right] = 0.$$

If the boundary γ_h is known, then by approximating its section in the vicinity of ξ_{opt} by an ellipse (Fig. 4.18)

$$\gamma_h: \frac{(M_0 - M_*)^2}{a^2} + \frac{(h_0 - h_*)^2}{b^2} = 1, \tag{4.66}$$

one can obtain fairly simple expressions for the transversality condition (4.65):

$$\begin{aligned} \varphi_{\text{opt}} &= \arctan\frac{b^2\left(p_y + Mp_{v_x}\frac{\partial a_s}{\partial h}\right)}{a^2 p_{v_x} a_s}, \\ M_{0\text{opt}} &= M_* + r(\varphi_{\text{opt}}) \cdot \cos\varphi_{\text{opt}}, \\ h_{0\text{opt}} &= h_* + r(\varphi_{\text{opt}}) \cdot \sin\varphi_{\text{opt}}, \end{aligned} \tag{4.67}$$

where $r(\varphi) = \frac{a \cdot b}{\sqrt{a^2 \sin^2\varphi + b^2\cos^2\varphi}}$.

Due to the known stability of the influence functions with variation of the phase vector, optimal point $\{M_0, h_0\}_{\text{opt}}$ according to nonlinear relations (4.67) are calculated rather quickly even when using the method of simple iterations:

$$\varphi_{\text{opt}}^k = \varphi_{\text{opt}}\left(M_0\left(\varphi_{\text{opt}}^{(k-1)}\right), h_0\left(\varphi_{\text{opt}}^{(k-1)}\right)\right), \tag{4.68}$$

where k is the iteration number.

Note that the inclusion of both the control and initial conditions in the optimization process does not increase the dimension of the corresponding BVP and, as a result, does not lead to any noticeable increase in calculations when using a rigorous approach based on PMP.

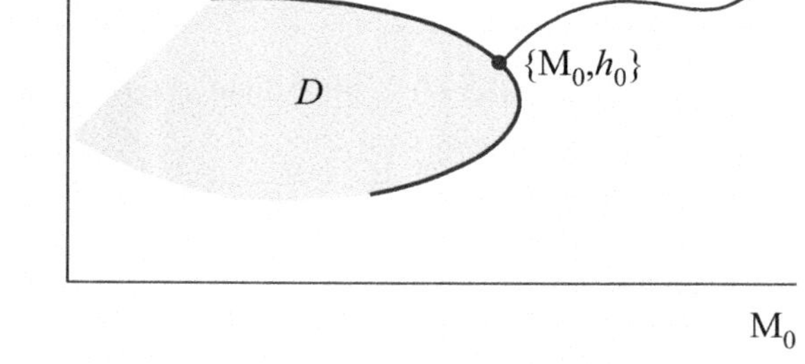

Fig. 4.18 Elliptic approximation of the boundary of area D of possible CA steady-state flight in a vicinity of the optimal RS launch point

References

1. Bryson, A.E., Ho, Y.C.: Applied Optimal Control. Hemisphere Publishing Corporation, Washington, DC (1975)
2. Velichenko, V.V.: Optimal control of composed systems. Doklady AN USSR **176**(4), 754–756 (1967). (in Russian)
3. Ashchepkov, L.T.: Optimal Control of Discontinuous Systems. Nauka Publisher, Novosibirsk (1987). (in Russian)
4. Vincent, T.L., Mason, J.D.: Disconnected optimal trajectories. JOTA **3**(4), 263–281 (1969)
5. Morozov, S. F., Sumin, M. I.: On a class of control problems for dynamical systems with a discontinuous right-hand side. Cybernetics, **3** (1985)
6. Gera, G.: Branched trajectory optimization by the projected gradient technique. AIAA P No. 69–917 (1969)
7. Sokolov, N.L., Tsybulsky, G.A.: Research on optimizing the design and ballistic parameters of an ambulance complex based on intercontinental ballistic missiles. Cosmic Research, **34**(3), 325–331 (1996)
8. Nagorski, R.P.: Application of the direct injection method for the launch from the Vandenberg launch site, AIAA P. 84-2040
9. Ilyin, V.A., Filatyev, A.S.: Synthesis of optimal trajectories for launching into orbit, from any point of which a descent in the atmosphere is possible with the fulfillment of specified constraints. Cosmic Research, **XXIII**(1), 37 (1985)
10. Filatyev, A.S.: Optimization of branched trajectories for aerospace transport systems, ICAS-94-5.2.3, 19th ICAS Congress, 18–23 September 1994, Anaheim CA, USA
11. Filatyev, A.S., Yanova, O.V.: Optimization of space launching with limitations on fall zones for spent components, IAF-96-A.3.01, The 47th IAF Congress, 1996, Beijing, China
12. Filatyev, A.S., Yanova, O.V.: The optimization of ascent trajectories with continuum of branches, IAF-94-A.2.019, 45th Congress of the International Astronautical Federation, October 9–14, 1994, Jerusalem, Israel
13. Okhotsimskii, D.E., Eneev, T.M.: Some variational problems connected with the launching of artificial Earth satellites. UFN, **63**(1a) (1957) (in Russian)
14. Lawden, D.F.: Optimal trajectories for space navigation. Butterworths, London (1963)
15. Leutin, A.P., Demeshkina, V.V.: Separation of stages of aerospace systems. J Aerospace Eng Technol **2** (1998) (in Russian)
16. Filatyev, A.S., Yanova, O.V.: Optimization of branching trajectories in the presence of global constraints. In: Proceedings of the 7th conference "Control in Mechanical Systems", Sverdlovsk 1990 (in Russian)
17. Kolesnikov, K.S., Kozlov, V.I., Kokushkin, V.V.: Dynamics of stage separation of aircraft. Mashinostroenie (1977) (in Russian)
18. Filatyev, A.S.: "Paradoxes" of optimal solutions in problems of space vehicle injection and reentry. In: The 48th Congress of the International Astronautical Federation, October 6–10, 1997/Turin, Italy, IAF-97-A.6.09
19. Golikov, A.A., Filatyev, A.S., Yanova, O.V.: Optimization of spacecraft launch trajectories taking into account three levels of flight safety requirements. The International Scientific and Technical Symposium "Aviation Technologies of the XXI Century", Zhukovsky, Moscow region, Russia, August 17-22, 1999, Proceedings of TsAGI, No. 2640 (1999) (in Russian)
20. Filatyev, A.S.: Optimal launch of an artificial Earth satellite using aerodynamic forces. Cosmic Research **29**(2), 225–271 (1991)
21. Dudar, E.N.: Flight Dynamic Analysis of Aerospace System with Subsonic Carrier-plane, Russian-Ukrainian-German Symposium, Munich (1993)
22. Lozino-Lozinsky, G.E., Dudar, E.N.: The development of aviation and the prospects for the development of aerospace transport systems. Polyot, **1** (1998) (in Russian)
23. Parkinson, R.: The An-225/Interim Hotol launch vehicle. In: The AIAA 3rd International Aerospace Planes Conference (1991). https://doi.org/10.2514/6.1991-5006

24. Miele, A.: Optimum climbing technique for a rocket-powered aircraft. Jet Propulsion **25**(8), 385–391 (1955)
25. Ostoslavsky, I.V., Strazheva, I.V.: Flight dynamics. Aircraft trajectories. Moscow, Oborongiz (1963) (in Russian)

Chapter 5
The Through Optimization of Branching Trajectories with a Continuum of Branches

5.1 Introduction

When solving a number of problems, it becomes necessary to study a class of branching trajectories containing a continuum of branches adjacent to the main branch. An example is the launch trajectory, from any point of which the return of a reentry vehicle (RV) is possible without violating given restrictions. In this case, imaginary emergency RV trajectories form a continuous veil of side branches descending from the main launch trajectory. This problem was apparently considered for the first time in [1]. Further generalizations are made in [2–4].

Another class of problems, leading to the consideration of a continuum set of branches, is associated with taking into account the action of random factors. In this case, each implementation of a random process is considered as a separate branch. Optimization of such a class of problems with the help of PMP is considered in [5, 6].

This chapter is devoted to the optimization method for such branching trajectories.

5.2 Problem Statement with Fail-Safety Conditions

Let us consider trajectory b^1 of the STS center of mass, determined by phase vector $\mathbf{x}^1(t), t \in \left[t_i^1, t_f^1\right]$, with the functional depending on the phase vector at the trajectory right end (at time t_f^1):

$$J\left(\mathbf{x}^1\left(t_f^1\right)\right) \Rightarrow \max_{\{\mathbf{u}\}}, \tag{5.1}$$

A. S. Filatyev and V. G. Petukhov, *Through Optimization of Aerospace Vehicle Trajectories by the Pontryagin Maximum Principle*, Studies in Systems, Decision and Control 579, https://doi.org/10.1007/978-3-031-80756-5_5

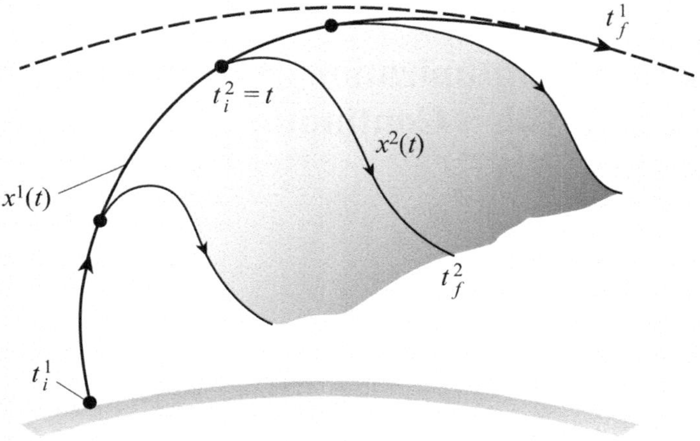

Fig. 5.1 LV injection trajectory and continuum of imaginary emergency RV return trajectories

taking into account possible constraints on control and phase vector at internal trajectory points, which together allows consideration of the constraints of the s-th order on the phase vector.

In addition to the "standard" statement, we require that the launch trajectory satisfy the following fail-safety requirements: at any point of the trajectory, it is possible to abort the STS ascent program, separate the RV and return it to the Earth with the fulfillment of specified constraints. The RV role can play a special vehicle (orbital vehicle, payload, crew escape vehicle, etc.) or current STS stage as a whole.

Let the RV autonomous flight can begin from a certain set of points of the launch trajectory $\{\mathbf{x}^1(t)\}$. The RV trajectory is determined by phase vector $\mathbf{x}^2$ (Fig. 5.1):

$$\mathbf{x}^2 = \mathbf{x}^2(t),\ \mathbf{x}^2\left(t_i^2\right) = \mathbf{x}^1(t) + \Delta\mathbf{x}^1(t),\ t \in \left[t_i^2 = t, t_f^2\right]. \tag{5.2}$$

Phase vector jump $\Delta\mathbf{x}^1$ is due to a possible change in mass, as well as taking into account changes in the velocity vector and position in the separation process (see Chap. 4).

The RV trajectory is almost always subject to constraints of the form (4.1), (4.2), typical examples of which are followings (Fig. 5.2):

- Constraints on the maximum dynamic pressure and heat fluxes:

$$\Omega \equiv \max_{\tau} q - q_{\text{adm}} \leq 0. \tag{5.3}$$

In practical problems, we can restrict ourselves to checking the fulfillment of inequality (5.3) at a certain moment t_f^2, determined from the local extremum condition of $q(t)$:

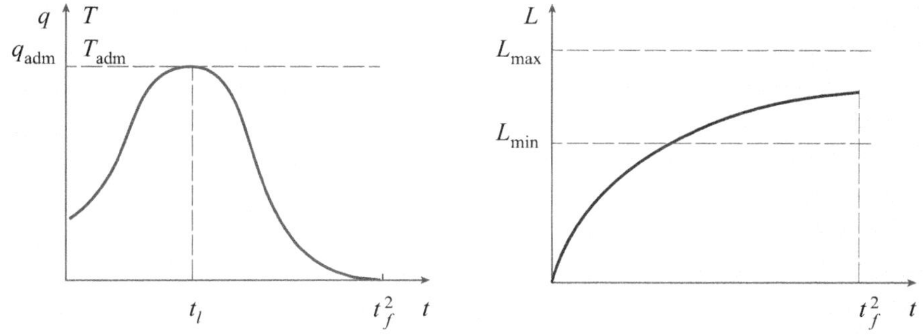

Fig. 5.2 Typical time-dependences of the dynamic pressure q, temperature T and range L on RV return trajectories

$$\omega \equiv \frac{dq}{dt}\left(t_f^2\right) = 0. \tag{5.4}$$

- Constraints on the RV final flight range

$$L_{\min} \leq L_f \leq L_{\max}, \tag{5.5}$$

which can be rewritten as inequality

$$\Omega = \left(L_f - L_{\min}\right)\left(L_f - L_{\max}\right) \leq 0. \tag{5.6}$$

Note that the problem of rescue after an accident at the insertion into an orbit, as a rule (although not always), requires deceleration of RV, so that the landing velocity is much less than at the abortion. In addition, the most common is the vertical landing, at which the horizontal speed is close to zero. Since the horizontal velocity is a derivative of the longitudinal range, then in the indicated cases, a condition similar to the extremum necessary condition (5.4) is satisfied at moment t_f^2:

$$\omega \equiv \frac{dL}{dt}\left(t_f^2\right) \cong 0. \tag{5.7}$$

According to (5.2), the RV trajectory depends on the phase coordinates on the launch trajectory at their common point. Therefore, the fulfillment of conditions (5.3)–(5.6) with a fixed or optimal reentry control program leads to a restriction of the admissible phase vector on the ascent trajectory $\mathbf{x}^1(t)$, and, in the general case, requires a change in the ascent trajectory compared to the optimal trajectory constructed without these limitations.

The above consideration allows us to give the following general statement of the problem of the STS injection trajectory optimization with constraints on RV return trajectory.

The motion of the LV center of mass is described by ODE system (2.18):

$$\frac{d\mathbf{x}^1}{dt} - \mathbf{f}^1\left(\mathbf{x}^1, \mathbf{u}^1\right) = 0, \ t \in \left[t_i^1, t_f^1\right]. \tag{5.8}$$

For definiteness, we will assume that the initial phase vector $\mathbf{x}^1\left(t_i^1\right)$ is fixed, and conditions (2.49) of entering the target orbit must be satisfied at the right end t_f^1.

Let the motion of the RV center of mass be described by ODE of the same form (5.8):

$$\frac{d\mathbf{x}^2}{dt} - \mathbf{f}^2\left(\mathbf{x}^2, \mathbf{u}^2\right) = 0, \ t \in \left[t_i^2, t_f^2\right] \tag{5.9}$$

under the initial condition

$$\mathbf{x}^2\left(t_i^2\right) - \mathbf{x}^1(t) = \Delta\mathbf{x}(t). \tag{5.10}$$

The jump in the phase vector is due to the possible change in the mass, as well as the velocity and position vectors in the separation process.

The following conditions must be met at the endpoint:

$$\mathbf{g}\left(\mathbf{x}_f^2\right) = 0, \mathbf{g} = \left(g^1, \ldots, g^{k_f}\right)^{\mathrm{T}}. \tag{5.11}$$

The following constraints are imposed on the RV trajectory:

$$\Omega^j\left(\mathbf{x}^2(t)\right) \le 0, \ j = 1, \ldots, K, \tag{5.12}$$

each of which is checked at moments t_j of reaching the extremum of function $\Omega\left(\mathbf{x}^2(t)\right)$:

$$\omega\left(t_j\right) = 0. \tag{5.13}$$

It is required to determine control functions $\mathbf{u}^1(t)$, $\mathbf{u}^2(t)$ that maximize functional (5.1) taking into account (5.8)–(5.13).

In the general case, this problem can be solved using optimization methods for systems with distributed parameters [7]. In this paper, we propose the method based on the projection of the constraints specified on side branches onto the main trajectory [2, 4]. As a result, the problem is reduced to an ordinary optimization problem with phase constraints.

5.3 The TPBVP Reduction by Projection of Boundary Conditions from Side Branches

Let us consider the following auxiliary problem. Let us find set $\mathbf{X}_{\text{adm}}^{2i}$ of such admissible initial phase vectors of the RV return trajectory $\mathbf{x}_{\text{adm}}^{2i}$, for which the further RV reentry occurs without violating constraints (5.3), (5.6).

To form this set, the RV control can be fixed, as, for example, in the case of deterministic control or uncontrolled reentry, or it can be optimal according to the criterion of the minimum of the maximum dynamic pressure or heat flow or the landing point range [8–10].

Let us consider some k-th constraint from (5.12) at moment t_k determined by condition (5.13):

$$\Omega^j\big(\mathbf{x}^2(t)\big) \leq 0,\ j = 1, \ldots, K. \tag{5.14}$$

Phase vector $\mathbf{x}^2(t_k)$ for some given RV control $\mathbf{u}^2(t)$ is a function of initial vector $\mathbf{x}^1(t)$:

$$\mathbf{x}^2(t_k) = \mathbf{x}^2\big(\mathbf{x}^1(t), \mathbf{u}^2(t)\big),\ t \in \big[t_i^2, t_k\big]. \tag{5.15}$$

For the k-th constraint of type (5.14), we call the optimal control $\mathbf{u}_{\text{opt}}^{2,k}(t)$ such control that delivers a minimum to functional Ω^k under the condition that all other constraints (5.12) are satisfied:

$$\mathbf{u}_{\text{opt}}^{2,k}(t) : \min_{v(t)\in V} \Omega^k\big(\mathbf{x}^2(t_k), t_k\big) \equiv \Omega_{\min}^k\big(\mathbf{x}^1(t)\big),$$

$$\Omega^j\Big(\mathbf{x}^1(t), \mathbf{u}_{\text{opt}}^{2,k}(t)\Big) \leq 0,\ j \neq k. \tag{5.16}$$

If there is a set of controls satisfying conditions (5.16), then in the further analysis we will take any fixed control from this set as $\mathbf{u}_{\text{opt}}^{2,k}(t)$. Denote by $\tilde{\mathbf{X}}$ the set of all admissible phase vectors $\mathbf{x}(t)$ on the launch trajectory, so that when reentering from any point of this set, there is control $\mathbf{u}^2(t) \in \mathbf{U}^2$, under which constraints (5.12) are satisfied:

$$\tilde{\mathbf{X}} = \big\{\mathbf{x}(t) : \Omega^j\big(\mathbf{x}^1(t), \mathbf{u}^2(t)\big) \leq 0,\ \mathbf{u}^2(t) \in \mathbf{U}^2,\ j = 1, 2, \ldots, K\big\}. \tag{5.17}$$

Denote by $\hat{\mathbf{X}}_k$ a similar set for constraints (5.16):

$$\hat{\mathbf{X}}_k = \Big\{\mathbf{x}(t) : \Omega_{\min}^k\big(\mathbf{x}^1(t)\big) \leq 0, \Omega^j\Big(\mathbf{x}^1(t),\ \mathbf{u}_{\text{opt}}^{2,k}(t)\Big) \leq 0,\ j \neq k\Big\}. \tag{5.18}$$

Since $\tilde{\mathbf{X}}$ is considered on all admissible controls $\mathbf{u}^2(t) \in \mathbf{U}^2$, and set $\hat{\mathbf{X}}_k$ is considered on some fixed control $\mathbf{u}_{\text{opt}}^{2,k}(t) \in \mathbf{U}^2$, the following inclusion is obvious:

$$\hat{\mathbf{X}}_k \subset \tilde{\mathbf{X}}. \tag{5.19}$$

On the other hand, from inequality

$$\Omega^k_{\min}\left(\mathbf{x}^1(t)\right) \leq \Omega^k\left(\mathbf{x}^1(t), \mathbf{u}^2(t)\right) \tag{5.20}$$

it follows that if point $\mathbf{x}(t) \in \tilde{\mathbf{X}}$, i.e., for it there is some control $\mathbf{u}^2(t) \in \mathrm{U}^2$, then for the same point there is also optimal control $\mathbf{u}^{2,k}_{\text{opt}}(t) \in \mathrm{U}^2$. Therefore,$\mathbf{x}(t) \in \hat{\mathbf{X}}_k$, and we have inclusion

$$\tilde{\mathbf{X}} \subset \hat{\mathbf{X}}_k. \tag{5.21}$$

From (5.19), (5.21) it follows that

$$\tilde{\mathbf{X}} = \hat{\mathbf{X}}_k. \tag{5.22}$$

If this choice of $\mathbf{u}^{2,k}_{\text{opt}}(t)$ is made for each k ($k = 1, \ldots, K$), then from (5.22) we obtain

$$\cap^J_{k=1}\hat{\mathbf{X}}_k \equiv \hat{\mathbf{X}} = \tilde{\mathbf{X}}. \tag{5.23}$$

where, by virtue of (5.18), for points $\mathbf{x}(t) \in \hat{\mathbf{X}}$ under the choice of appropriate controls $\mathbf{u}^{2,j}_{\text{opt}}(t)$ the following inequalities are valid: $\Omega^j_{\min}\left(\mathbf{x}^1(t)\right) \leq 0 (j = 1, 2, \ldots, K)$. But then obviously $\max\limits_{\mathbf{x}\in\hat{\mathbf{X}}} J\left(\mathbf{x}^1_f, t^1_f\right) = \max\limits_{\mathbf{x}\in\tilde{\mathbf{X}}} J\left(\mathbf{x}^1_f, t^1_f\right)$.

The meaning of optimal controls $\mathbf{u}^{2,j}_{\text{opt}}(t)$ (5.16) is that with such a choice of control, the set of admissible phase vectors $\mathbf{x}(t)$, according to (5.22), (5.23), does not narrow, and control $\mathbf{u}^{2,j}_{\text{opt}}(t)$ allows you to get the same maximum value of functional $J\left(\mathbf{x}^1_f, t^1_f\right)$ as in the original problem.

Let us formulate a rule for choosing the RV optimal control.

(1) If $\mathbf{x}(t) \in \hat{\mathbf{X}}\backslash\partial\hat{\mathbf{X}}$ ($\partial\hat{\mathbf{X}}$ is the boundary of set $\hat{\mathbf{X}}$), then there is a set of controls $\mathrm{U}^2(t) = \left\{\mathbf{u}^2(t)\right\}, t \in \left[t^2_i, t^2_f\right]$ such that (5.12), (5.13) are valid for any control $\mathbf{u}^2(t) \in \mathrm{U}^2(t)$. As a RV control, you can take any $\mathbf{u}^2(t) = fix, \mathbf{u}^2(t) \in \mathrm{U}^2(t)$, including one of optimal $\mathbf{u}^{2,j}_{\text{opt}}(t), j = 1, 2,\ldots, K$.

(2) If $\mathbf{x}(t) \in \partial\hat{\mathbf{X}}$, then as a RV control, one should take $\mathbf{u}^2(t) = \mathbf{u}^{2,k}_{\text{opt}}(t),\ t \in \left[t^2_i, t^2_f\right]$. In case $\mathbf{x}(t) \in \partial\hat{\mathbf{X}}_{k_1} \cap \partial\hat{\mathbf{X}}_{k_2} \ldots \cap \partial\hat{\mathbf{X}}_{k_s}$, as a RV control, one can take any of $\mathbf{u}^{2,k_i}_{\text{opt}}(t)$ ($i = 1, \ldots, s$).

Having defined controls $\mathbf{u}^{2,j}_{\text{opt}}(t)$, we can find functions (5.16) $\Omega^j_{\min}\left(\mathbf{x}^1(t)\right)$ ($j = 1, 2, \ldots, K$). Then the original problem (5.1), (5.8)–(5.13) reduces to the following. For dynamic system (5.8), (5.9) it is required to define control $\mathbf{u}(t) \in \mathbf{U}$ so that the following inequalities to be valid:

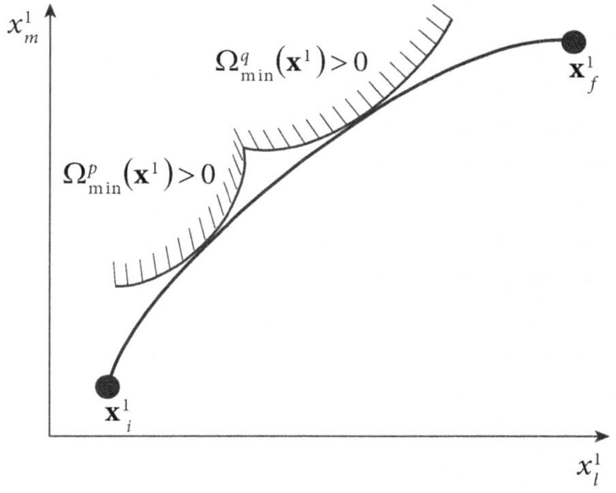

Fig. 5.3 The fail-safety constraints and ascent trajectory

$$\Omega_{\min}^j\left(\mathbf{x}^1(t)\right) \le 0 (j = 1, 2, \ldots, K), \tag{5.24}$$

and functional (5.1) reaches its maximum (Fig. 5.3).

Thus, the original problem is reduced to the standard optimal control problem with first-order phase constraints (see Sect. 1.5).

Let the safety-driven admissible set $\mathbf{X}_{\mathrm{adm}}^1$ for phase vector $\mathbf{x}^1$ be described by inequality

$$Q\left(\mathbf{x}^1\right) \le 0, \tag{5.25}$$

and its boundary $\partial \mathbf{X}_{\mathrm{adm}}^1$ is given by surface equation

$$Q\left(\mathbf{x}^1\right) = 0. \tag{5.26}$$

Differentiating $Q\left(\mathbf{x}^1\right)$ with respect to t, we obtain the control condition for moving along boundary $\partial \mathbf{X}$:

$$\frac{dQ}{dt} = \frac{\partial Q\left(\mathbf{x}^1\right)}{\partial \mathbf{x}} \mathbf{f}^1\left(\mathbf{x}^1, \mathbf{u}^1\right) = G\left(\mathbf{x}^1, \mathbf{u}^1\right) = 0. \tag{5.27}$$

We will assume that, in the general case, the control enters explicitly into stationarity condition (5.27), i.e., the fail-safety constraint is a first-order phase constraint.

We introduce, as before, Hamiltonian

$$H = \mathbf{p}^{1\,\mathrm{T}} \mathbf{f}^1\left(\mathbf{x}^1, \mathbf{u}^1\right) + \lambda G\left(\mathbf{x}^1, \mathbf{u}^1\right), \tag{5.28}$$

and conjugate system

$$\left(\frac{d\mathbf{p}^1}{dt}\right)^{\mathrm{T}} = -\frac{\partial H}{\partial \mathbf{x}^1} = -\mathbf{p}^{1\mathrm{T}}\frac{\partial \mathbf{f}}{\partial \mathbf{x}^1} - \lambda\frac{\partial G}{\partial \mathbf{x}^1},$$

where $\mathbf{p}^1,\ \lambda$ are the conjugate vector and Lagrange multiplier ($\lambda(t) = 0$, if $Q < 0$).
At time t_Q of entrance to the constrain segment:

$$Q\big(\mathbf{x}^1(t_Q)\big) = 0, \tag{5.29}$$

the following jump conditions for the conjugate vector are satisfied:

$$\mathbf{p}^1(t_Q + 0) - \mathbf{p}^1(t_Q - 0) = \nu\frac{\partial Q}{\partial \mathbf{x}}^{\mathrm{T}}\big(\mathbf{x}^1(t_Q)\big),\ H(t_Q + 0) = H(t_Q - 0),$$

where ν is the Lagrange multiplier according to condition (5.29). Moment t_Q is the initial time for the RV return trajectory, where the following criterion can be considered as a functional:

$\Omega\big(\mathbf{x}^2(t)\big) \Rightarrow \min$. Therefore gradient $\frac{\partial Q}{\partial \mathbf{x}^1}\big(\mathbf{x}^1(t_Q)\big) = \frac{\partial Q}{\partial \mathbf{x}^2}\big(\mathbf{x}^2(t_Q)\big)$ can be interpreted as a sensitivity function corresponding to the conjugate vector on the side branch at the branching point:

$$\mathbf{p}^2(t_Q) = \frac{\partial Q}{\partial \mathbf{x}^1}^{\mathrm{T}}\big(\mathbf{x}^1(t_Q)\big). \tag{5.30}$$

Optimal control $\mathbf{u}^1_{\mathrm{opt}}$ is determined from condition

$$H\left(\mathbf{u}^1_{\mathrm{opt}}\right) = \max_{\mathbf{u}\in\mathbf{U}} H\left(\mathbf{x}^1, \mathbf{u}^1, \mathbf{p}^1, \lambda\right). \tag{5.31}$$

Moreover, when moving along boundary (5.29), condition

$$\mathbf{p}^{2\,\mathrm{T}}\mathbf{f}^1\left(\mathbf{x}^1, \mathbf{u}^1\right) = 0 \tag{5.32}$$

is added to (5.31) with (5.30).

The left side of equality (5.32) looks formally like the Hamiltonian of system (5.28). However, here conjugate vector $\mathbf{p}^2$ and right-hand side function $\mathbf{f}^1(\mathbf{x}^1, \mathbf{u}^1)$, although they are calculated at the same phase point, are defined for different dynamic systems: $\mathbf{x}^1$ and $\mathbf{u}^1$ are the phase vector and control at the current point of the STS injection trajectory (the main branch), and $\mathbf{p}^2(t)$ is the conjugate vector at the initial point of the RV return trajectory (the side branch).

Taking into account the equations of LV motion (2.18)–(2.27), expression (5.32) determines the constraint on the LV control of type (2.40):

$\mathbf{b}^{\mathrm{T}}\mathbf{e}_\tau - f_q = 0,$

$$\mathbf{b} = \left(T + \tilde{c}_y^{\alpha}\right)\mathbf{p}_v^2 - \left(\tilde{c}_y^{\alpha} + \tilde{D}_{\alpha}\right)(\mathbf{p}_v^2, \mathbf{e}_v)\mathbf{e}_v,\ f_q = \mathbf{p}_v^{2\mathrm{T}}\left(\tilde{D}_o\mathbf{e}_v - m(\mathbf{g} + \Omega)\right) - m\left(\mathbf{p}_r^2, \mathbf{v}\right), \tag{5.33}$$

where conjugate vector $\mathbf{p}^2$ corresponds to the initial point $\mathbf{x}^2(t_i^2) = \mathbf{x}^1(t)$ of the RV return trajectory, the remaining variables and parameters correspond to the LV injection trajectory. (Here, in accordance with notation (2.18), $\boldsymbol{\Omega}$ is the acceleration vector due to the non-inertiality of the coordinate system related to the rotated Earth).

5.4 Through Optimization of STS Trajectories with Fail-Safety Requirements

Two methods for the synthesis of the fail-safe control are considered below. The first one is based on the approximate analytical solution for the conjugate system for the RV nonequilibrium reentry with subcircular velocities [9, 10]. The second one, conventionally called "rigorous", determines the boundary of admissible set $\mathbf{X}_{\mathrm{adm}}^{2i}$ based on numerical simulation of RV return trajectories using sufficiently complete motion models (2.18). The optimization problem is then solved using PMP with appropriate phase constraints. In this case, the RV control is quasi-optimal in accordance with the analytical synthesis obtained in [9, 10], which minimizes the maximum dynamic and thermal loads during nonequilibrium reentry.

5.4.1 Approximate Analytical Synthesis of the Optimal Fail-Safe Trajectory Control

RV Return Trajectory

Firstly, consider the side branch of the RV return at the most important case, when the admissible dynamic pressure and specific heat flux are constrained. Let us wright the constrained in following universal form:

$$q \equiv \rho^{k_h} v^{k_v} \le q_{\mathrm{adm}}, \tag{5.34}$$

where v is the velocity; ρ is the atmosphere density depending on altitude h; k_h and k_v are constants. Form (5.34) makes it possible to take into account various types of aerodynamic load constraints by assigning different values to these constants. For example, the values of $k_h = 1,\ k_v = 2$ correspond to the dynamic pressure; $k_h = 0.5,\ k_v = 3.25$ correspond to one of possible specific heat fluxes [11, 12].

Let us accept

$$\Omega\left(\mathbf{x}^2\right) \equiv \max_t (q - q_{\mathrm{adm}}) = q\left(t_f\right) - q_{\mathrm{adm}}, \tag{5.35}$$

where time t_f is defined from the extremum necessary condition:

$$\left.\frac{1}{q}\frac{dq}{dt}\right|_{t=t_f} \equiv \left.\left(\frac{k_h}{\rho}\frac{d\rho}{dt} + \frac{k_v}{v}\frac{dv}{dt}\right)\right|_{t=t_f} = 0. \tag{5.36}$$

RV optimal control $\mathbf{u}^2_{\text{opt}}(t)$ is determined from condition

$$\mathbf{u}^2_{\text{opt}}(t) : \min_{\mathbf{u}^2} \max_{t} q - q_{\text{adm}} = Q\left(\mathbf{x}^1\right). \tag{5.37}$$

To define $\mathbf{u}^2_{\text{opt}}(t)$ and analytical solutions for the phase and conjugate systems of equations, we use the approximate analytical results [8–10, 13, 14, 16, 17].

The RV trajectories of reentry into the atmosphere with subcircular velocities are characterized by the presence of pronounced maxima of the dynamic pressure and specific heat flux corresponding to the first immersion into the atmosphere. Therefore, the RV trajectory (Fig. 5.4) until reaching the maximum q (5.34) can be divided into two sections corresponding to the initial motion in rarefied atmosphere layers, where the influence of aerodynamic forces is small (the "upper" section), and motion in dense atmosphere layers, where aerodynamic forces are predominant compared to the gravitational (the "lower" section). Under this assumption, the upper section is practically a Keplerian orbit, and the motion in the lower section is fairly well described by the Allen–Eggers equations [18].

We will assume that the phase vector $\mathbf{x}^2 = \{v, \theta, R\}$ at the initial point is given

$$\mathbf{x}^2\left(t_i^2\right) = \mathbf{x}_i^2, \tag{5.38}$$

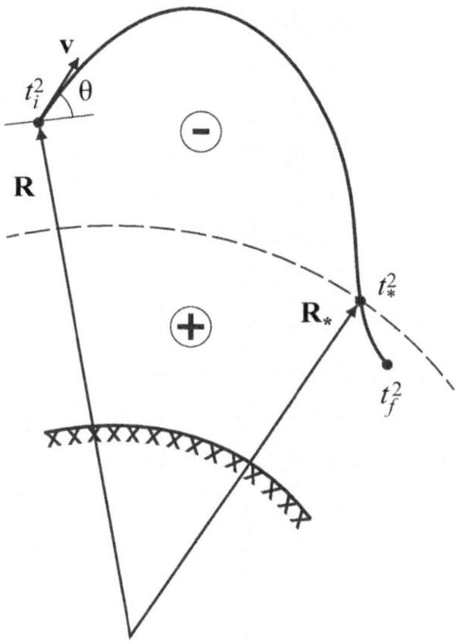

Fig. 5.4 The RV return trajectory

and the gluing of the upper and lower sections is performed at time moment t_*^2 of the RV entry into the dense atmosphere layers, determined, for example, by the condition of equality of the gravitational (mg) and aerodynamic $|\mathbf{A}|$ forces:

$$t_*^2 : \frac{|\mathbf{A}|}{mg} - 1 = 0, \tag{5.39}$$

with continuous phase vector:

$$\mathbf{x}^{2-}(t_*^2) = \mathbf{x}^{2+}(t_*^2) \tag{5.40}$$

The right end of the trajectory will be defined in accordance with condition (5.36) of reaching the maximum point of function $q(t)$.

We will also assume that RV aerodynamic coefficients are constant in the lower section, the nominal flight occurs without thrust, and the atmosphere is exponential.

Then, for the model of motion in the vertical plane, the conjugate system has the following approximate solution [9, 10, 17]:

$$p_v^{2-} = A\frac{Bv^2 - 1}{v},\ p_\theta^{2-} = A \cdot \tan\theta,\ p_R^{2-} = A\frac{B - R}{R^2}, \tag{5.41}$$

$$p_v^{2+} = q_{\max}\frac{k_v}{v},\ p_\theta^{2+} = \frac{q_{\max}k_v}{K}\left(1 - \frac{\sin\theta}{\sin\theta_f}\right),\ p_R^{2+} = -q_{\max}k_h\beta\frac{\rho}{\rho_f}, \tag{5.42}$$

where conjugate vector $\mathbf{p}^2 = \left(p_v^2, p_\theta^2, p_R^2\right)^{\mathrm{T}}$ corresponds to phase vector $\mathbf{x}^2 = (v, \theta, R)^{\mathrm{T}}$; $K = C_L/C_D$ is the RV lift-to-drag ratio; $\beta = -\frac{1}{\rho}\frac{\partial\rho}{\partial h}$; superscripts "-" and " + " mark variables on the upper and lower sections respectively; subscript "f" corresponds to the point of reaching maximum $q_{\max}$ of $q(t)$. At $K \to 0$ singularity in (5.42) is removable:

$$K = 0 \Rightarrow p_v^{2+} = q_{\max}\frac{k_v}{v},\ p_\theta^{2+} = q_{\max}k_h\left(1 - \frac{\rho^+}{\rho_f}\right)\cot\theta^+,\ p_R^{2+} = -q_{\max}k_h\beta\frac{\rho}{\rho_f}. \tag{5.43}$$

Constants $A > 0$ and $B > 1$ are uniquely determined [9, 10, 17] by the initial conditions. The gluing conditions for the phase and conjugate variables at altitude h_* (5.39) are as follows:

$$h = h_* : p_v^- = p_v^+ - \frac{2\xi}{\beta v_*}, p_\theta^- = p_\theta^+, p_h^- = p_h^+ + \xi, p_m^- = p_m^+ + \frac{\xi}{\beta}; (\mathbf{p}, \mathbf{f})^+ = (\mathbf{p}, \mathbf{f})^-,$$
$$\xi = q_{\max} \cdot k_v\frac{1 + æ\rho_*/(\omega\rho_f) + (1 - v_*^2R_*)D}{v_*^2R_*^2(1 + \omega)}, \tag{5.44}$$

$$
\begin{aligned}
&A = q_{\max}k_vD > 0,\ B = \frac{1 + v_*^2R_*\omega + \left(1 - æ\rho_*/\rho_f\right)/D}{v_*^2(1+\omega)} > 1,\\
&C = q_{\max}k_h\left[1 + \omega\frac{1 - æ\rho_*/\rho_f + \left(1 - v_*^2R_*\right)D}{æ(1+\omega)}\right] > q_{\max}k_h,\\
&D = \begin{cases}\cot\theta_*\left(1 - \sin\theta_*/\sin\theta_f\right)/K, & \text{if } K \neq 0,\\ \cot^2\theta_*\left(1 - \rho_*/\rho_f\right)æ/2, & \text{if } K = 0,\end{cases}\\
&\omega = \frac{2}{\beta v_*^2R_*^2},\ R_* = R_E + h_*,\ æ = \frac{2k_h}{k_v}.
\end{aligned}
\tag{5.45}
$$

Subscript "*" marks values of variables at the point matching the upper and lower sections.

Substituting conjugate vector $\mathbf{p}^{2-}$ (5.41) for the initial section of the RV return trajectory into equality (5.32), we obtain the equation for the optimal STS control on the fail-safety boundary. By solving it, we obtain the synthesis of the STS quasi-optimal fail-safe control.

Approximate Synthesis of the Optimal Fail-Safe Trajectory Control

Using the formulas for the influence functions on RV trajectories, we can obtain relations for the STS fail-safe control.

As shown in [14, 15], the fail-safety constraints become essential when initial part of the RV trajectory begins in rarefied atmosphere layers. Therefore, as a first approximation, we can consider the STS motion in the hazard area ignoring aerodynamic forces. In this case, the model of STS flight in the vertical plane under the action of thrust vector $\mathbf{T}$ in the Newtonian gravitational field is described by the following equations:

$$
\begin{cases}
\dfrac{dv}{dt} = \dfrac{T}{m}\cos\alpha - \dfrac{\sin\theta}{R^2} \equiv f_v,\\
\dfrac{d\theta}{dt} = \dfrac{T}{mv}\sin\alpha + \dfrac{v\cos\theta}{R} - \dfrac{\cos\theta}{vR^2} \equiv f_\theta,\\
\dfrac{dR}{dt} = v\sin\theta \equiv f_R,
\end{cases}
\tag{5.46}
$$

where α is the angle between the thrust and velocity vectors, $\mathbf{f}^1 = (f_v, f_\theta, f_R)^{\mathrm{T}}$. We will assume that thrust magnitude $T \equiv \text{const}$, and control is performed only by changing the unit thrust vector.

Let us use an approximate analytical representation of conjugate vector (5.41). Then the optimal angle of attack is related to the current values of the phase variables as follows [1, 2, 15]:

$$
\tan\alpha = -\frac{vp_v^{2-}}{p_\theta^{2-}} = \frac{1 - Bv^2}{\tan\theta}. \tag{5.47}
$$

Let us illustrate it on a simplified example, when the motion of a two-stage STS occurs in a vertical plane, and the dynamic pressure is the critical factor for the RV emergency return trajectory:

$$q \equiv \rho v^2 \leq q_{\mathrm{adm}}. \tag{5.48}$$

The STS control program consists of a g-turn (with zero angle of attack) at the 1st stage flight segment and the OEL classical approximate optimal control with a linear pitch angle tangent program (2.16b) at the 2nd stage flight segment:

$$1^{\mathrm{st}}\text{stage: } \alpha \equiv 0; \;\; 2^{\mathrm{nd}}\text{stage: } \tan\vartheta = at + b. \tag{5.49}$$

The constants a and b are selected from the condition of the 2nd stage insertion into a target circular orbit. Thus, the 1st stage trajectory segment is fixed. We assume that the RV can return from this section of the ascent trajectory without violation of fail-safety constraint (5.48), but the fail-safety constraint becomes essential at the beginning of the 2nd stage trajectory section with quasi-optimal control law (5.49). To ensure fail-safety, the 2nd stage control law was modified using an approximate analytical synthesis (5.47). Fig. 5.5 shows the quasi-optimal fail-safe LV trajectory and the quasi-optimal 2nd stage trajectory, which does not take into account fail-safety requirement (5.48).

The maximum dynamic pressure $\max\limits_t q$, which is realized during the RV return from various points of the LV trajectories, is shown in Fig. 5.6, where v is the current relative LV velocity, which equals to the initial RV velocity in the case of its emergency return to the Earth. Values of $\max\limits_t q$ were obtained not from an approximate analytical solution, but by numerical integration of the system of equations of RV motion in the central gravitational field with the exponential atmosphere for non-rotating Earth and with quasi-optimal control law [9, 10, 17]. It can be seen that

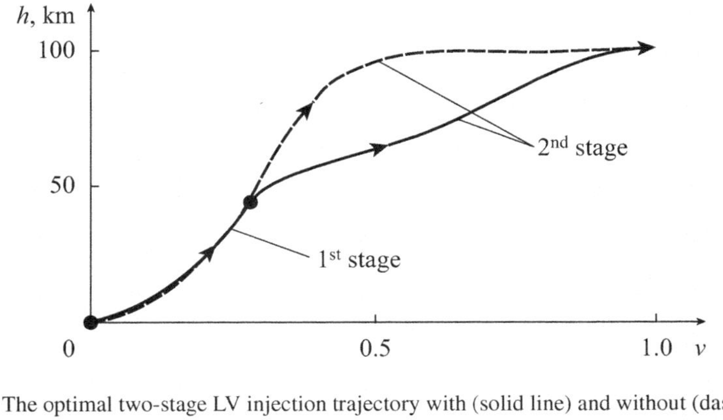

Fig. 5.5 The optimal two-stage LV injection trajectory with (solid line) and without (dashed line) the fail-safety constraint

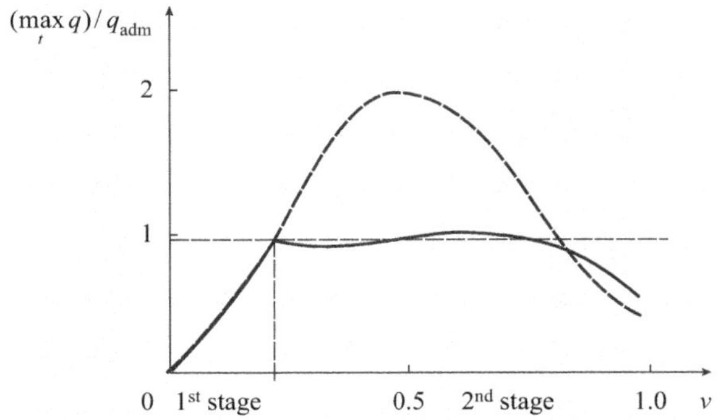

Fig. 5.6 The relative maximum dynamic pressure on the quasi-optimal RV emergency return trajectories vs the relative LV velocity at an imaginary failure point

the approximate analytical synthesis of the optimal fail-safe LV control law (5.47) makes it possible to sufficiently accurately maintain the LV motion along the trajectory segment, from any point of which the RV can return with the admissible dynamic pressure.

As can be seen from the example in Fig. 5.5, the fail-safety constraint causes a significant change in the optimal LV trajectory. However, as calculations show in this case, as well as in many other practical problems for various types of LV, modification of the LV ascent trajectory to take into account the requirements of fail-safety leads to a relatively small decrease in the injected mass if the trajectory is optimized as a whole. As practice shows, the reduction in the injected mass is a few percent for LV with a vertical take-off, and fractions of a percent for air-launched ones (see Sects. 6. 2 and 6.3).

5.4.2 *Rigorous Optimization of LV Fail-Safe Trajectory*

In the previous section, the optimization of the LV thrust vector control with the fail-safety constraint was solved under some simplifying assumptions regarding the RV motion model and the structure of the optimal LV control law. This section presents a technique for numerical solving this problem based on PMP without serious simplifications in equations of RV motion. The reduction of the original optimization problem with distributed parameters to the standard statement with phase constraints is used as that described in Sect. 5.3.

Problem Statement

Let us denote by $\boldsymbol{D}$ the safe area of the initial RV phase vector, at which an autonomous return to the Earth is possible without exceeding the constraints on dynamic and thermal loads and hinge moments on the attitude control surfaces. As a rule, an upper boundary of the safe area passes in the region of a low dynamic pressure, where the motion is described well by the Kepler laws. Therefore, the set of initial phase variables $\{v_i, h_i, \theta_i\}$ can be replaced [9, 16] by the set of osculating parameters

$(v_0, h_0)_{\theta_i=0}$ corresponding to the velocity and altitude at apogee of the osculating orbit:

$$h_0 = R_0 - 1 = \frac{1 + \sqrt{1 - WK^2}}{W} - 1, \ v_0 = \frac{WK}{1 + \sqrt{1 - WK^2}},$$

where $W = \frac{2}{R} - v^2, \ K = Rv\cos\theta$.

The osculating parameters are uniquely determined from the initial conditions $\{v_i, h_i, \theta_i\}$ according to the Kepler laws. Thus, the dimension of the space, in which the upper boundary of the safe area is calculated, is reduced from three to two.

Area $\boldsymbol{D}$ has two continuous, generally non-smooth, boundaries (Fig. 5.7). Here, the upper boundary Γ^- corresponds to such values (v_0, h_0) at which the maximum dynamic pressure or overload or hinge torques is reached at an inner point of the RV descent trajectory and is equal to admissible values. Lower boundary Γ^+ corresponds to initial velocity and altitude, at which the RV admissible dynamic or thermal loads are reached just at the initial point, i.e., in fact, on the LV injection trajectory. Thus, the necessary condition for the RV rescue is to reduce the RV initial flight altitude, at least, to boundary Γ^-.

In order to meet the fail-safety constraints during the LV ascent, the LV trajectory must lie entirely in the admissible area $\boldsymbol{D}$, which we will call the "ascent corridor" by analogy with the problem of reentry with escape velocity.

Let us consider the problem of optimizing the LV injection trajectory through the admissible ascent corridor, i.e. with an additional constraint:

$\mathbf{x} \in \boldsymbol{D}$,

which can be interpreted as following altitude constraints:

$$h - h_{\max}(v, \theta)|_{\Gamma^-} \leq 0, \tag{5.50}$$

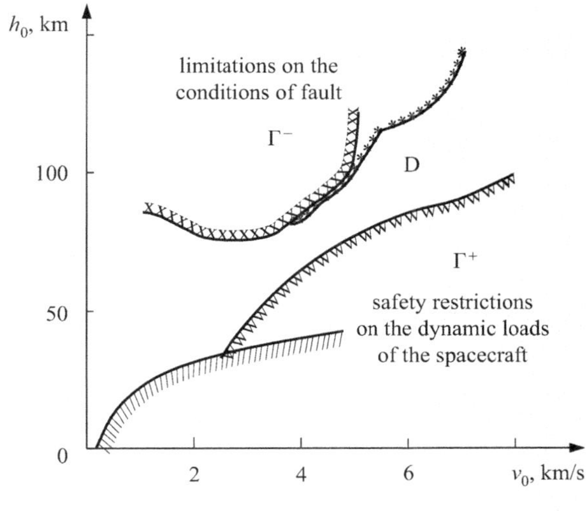

Fig. 5.7 Fail-safety area $\boldsymbol{D}$ of the osculating initial velocity and altitude, at which the RV return occurs without exceeding given constraints

$$h_{\min}(v,\theta)|_{\Gamma^+} - h \leq 0. \tag{5.51}$$

Since constraint (5.51) is imposed on the dynamic pressure and specific heat fluxes at the current point, it is taken into account according to the technique described in Sect. 2.7.2. New, in comparison with the previously considered phase constraints, is the inclusion of the upper boundary Γ^- given by expression (5.50). This inequality in the osculating variables takes the form [3]:

$$Q = h_0 - h_0(v_0)_{\max}|_{\Gamma^-} \leq 0,\ \ h_{0\max} = h_q - a_q(v_0 - v_q)^2, \tag{5.52}$$

where the constants h_q, a_q, v_q are selected from the condition of the best approximation of boundary Γ^-, according to the tabular dependences of the maximum dynamic pressure, overloads, hinge torques and so on. (see Sect. 6.2).

Optimality Conditions

Phase constraint (5.52) is of the first order, that is, the first time derivative contains explicitly the control. In accordance with PMP, at time t^* of reaching the upper boundary of area ***D*** (Fig. 5.8), following transversality conditions must be satisfied (see Sect. 2.5):

$$\mathbf{p}^+ - \mathbf{p}^- - \nu\left(\frac{\partial Q}{\partial\,\mathbf{x}}\right)^{\mathrm{T}} = 0, \tag{5.53}$$

$$H^+ - H^- = \left(\mathbf{p}^+, \mathbf{f}^+\right) - \left(\mathbf{p}^-, \mathbf{f}^-\right) = 0, \tag{5.54}$$

where ν is the Lagrange multiplier, defined so that the condition of motion along the boundary of safe area (5.52) is satisfied:

$$Q \equiv 0. \tag{5.55}$$

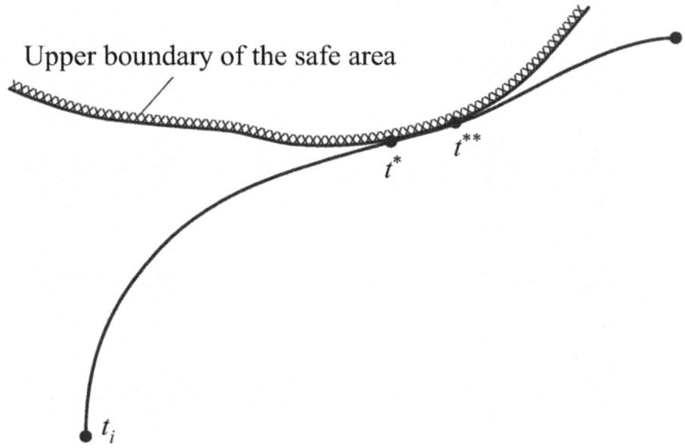

Fig. 5.8 Scheme of optimal injection trajectory with fail-safety constraint

The optimal control when moving inside area $\boldsymbol{D}$ according to PMP is as follows:

$$\mathbf{u}_{\text{opt}} = \arg\max_{\mathbf{u}\in\mathbf{U}} H, \tag{5.56}$$

where $\boldsymbol{U}$ is the admissible control set.

Boundary control $\tilde{\mathbf{u}}$ (5.55) is determined by:

$$\tilde{\mathbf{u}} : G \equiv \dot{Q}(\mathbf{x}, \tilde{\mathbf{u}}) = 0. \tag{5.57}$$

At the time of reaching the boundary:

$$\mathbf{u}_{\text{opt}}(t^*) = \tilde{\mathbf{u}}(t^*)$$

and according to (5.57)

$$G(\mathbf{x}, \mathbf{u}_{\text{opt}}) = 0. \tag{5.58}$$

From (5.58) it follows that function $Q(\mathbf{x})$ under optimal control touches smoothly the boundary at point t^*.

Moment t^{**} of leaving boundary (5.55) is determined by condition:

$t^{**} : G(\mathbf{x}, \mathbf{u}_{\text{opt}})\big|_{t^{**}} = 0.$

When moving along boundary (5.55), Hamiltonian

$$H = \mathbf{p}^{\mathrm{T}}\mathbf{f}(\mathbf{x}, \mathbf{u}) + \lambda G(\mathbf{x}, \mathbf{u}),$$

where λ is determined from condition:

$\max\limits_{\mathbf{u}} H(\mathbf{x}, \mathbf{p}, \mathbf{u}, \lambda) = H(\mathbf{x}, \mathbf{p}, \tilde{\mathbf{u}}, \lambda).$

The conjugate system when moving along boundary (5.55) transforms as follows:

$$\dot{\mathbf{p}}|_{Q=0} = \dot{\mathbf{p}} - \lambda\left(\frac{\partial G}{\partial \mathbf{x}}\right)^{\mathrm{T}}. \tag{5.59}$$

The constraint on LV orientation $\mathbf{e}_\tau$ at boundary (5.55) can be represented as follows [3]:

$$G \equiv (\mathbf{b}, \mathbf{e}_\tau) - f_q = 0,$$

$$\mathbf{b} = (T + \tilde{C}_L^\alpha)\left(\frac{\partial Q}{\partial \mathbf{v}}\right)^{\mathrm{T}} - (\tilde{C}_L^\alpha + \tilde{D}_\alpha)\mathbf{e}_v\frac{\partial Q}{\partial \mathbf{v}}\mathbf{e}_v,$$

$$f_q = \left(\tilde{D}_0\frac{\partial Q}{\partial \mathbf{v}} - mv\frac{\partial Q}{\partial \mathbf{r}}\right)\mathbf{e}_v + m\frac{\partial Q}{\partial \mathbf{v}}\left(\frac{\mathbf{e}_R}{R^2} - \Omega\right),$$

$$\frac{\partial \mathbf{b}}{\partial \mathbf{r}} = -\beta(h)\left[\tilde{C}_L^\alpha\left(\frac{\partial Q}{\partial \mathbf{v}}\right)^{\mathrm{T}} - (\tilde{C}_L^\alpha + \tilde{D}_\alpha)\mathbf{e}_v\frac{\partial Q}{\partial \mathbf{v}}\mathbf{e}_v\right]\mathbf{e}_R^{\mathrm{T}}$$

$$
-\frac{\tilde{C}_L^\alpha}{C_L^\alpha}\frac{dC_L^\alpha}{dh}\left(\frac{\partial Q}{\partial \mathbf{v}}\right)^{\mathrm{T}}\mathbf{e}_R^{\mathrm{T}}+\frac{\partial T}{\partial h}\left(\frac{\partial Q}{\partial \mathbf{v}}\right)^{\mathrm{T}}\mathbf{e}_R^{\mathrm{T}}
$$

$$
+\frac{\tilde{C}_L^\alpha+\tilde{D}_\alpha}{C_L^\alpha+D_\alpha}\frac{d(C_L^\alpha+D_\alpha)}{dh}\left(\mathbf{e}_v\frac{\partial Q}{\partial \mathbf{v}}\mathbf{e}_v\right)\mathbf{e}_R^{\mathrm{T}}
$$

$$
+\left[(T+\tilde{C}_L^\alpha)\mathbf{E}-(\tilde{C}_L^\alpha+\tilde{D}_\alpha)\mathbf{e}_v\mathbf{e}_v^{\mathrm{T}}\right]\frac{\partial}{\partial \mathbf{r}}\left(\frac{\partial Q}{\partial \mathbf{v}}\right)^{\mathrm{T}},
$$

$$
\frac{\partial \mathbf{b}}{\partial \mathbf{v}}=\frac{C_L^\alpha}{v}\left(2+\frac{M}{C_L^\alpha}\frac{\partial C_L^\alpha}{\partial M}\right)\left(\frac{\partial Q}{\partial \mathbf{v}}\right)^{\mathrm{T}}\mathbf{e}_v^{\mathrm{T}}+(T+\tilde{C}_L^\alpha)\frac{\partial}{\partial \mathbf{v}}\left(\frac{\partial Q}{\partial \mathbf{v}}\right)^{\mathrm{T}}
$$

$$
-\left(\frac{\tilde{C}_L^\alpha}{C_L^\alpha}\frac{\partial C_L^\alpha}{\partial M}+\frac{D_\alpha}{D_\alpha}\frac{\partial D_\alpha}{\partial M}\right)\left(\frac{\partial Q}{\partial \mathbf{v}},\mathbf{e}_v\right)\frac{\mathbf{e}_v\mathbf{e}_v^{\mathrm{T}}}{a_s}
$$

$$
-(\tilde{C}_L^\alpha+D_\alpha)\mathbf{e}_v\left(\frac{1}{v}\frac{\partial Q}{\partial \mathbf{v}}+\mathbf{e}_v^{\mathrm{T}}\frac{\partial}{\partial \mathbf{v}}\left(\frac{\partial Q}{\partial \mathbf{v}}\right)^{\mathrm{T}}\right),\quad \frac{\partial \mathbf{b}}{\partial m}=0,
$$

$$
\frac{\partial f_q}{\partial \mathbf{r}}=\frac{\partial Q}{\partial \mathbf{v}}\left(-\tilde{D}_0\left(\beta(h)+\frac{1}{D_0}\frac{dD_0}{dh}\right)\mathbf{e}_v\mathbf{e}_R^{\mathrm{T}}+\frac{m\mathbf{E}}{R^2}-\frac{3m}{R^3}\mathbf{e}_R\mathbf{e}_R^{\mathrm{T}}\right)
$$

$$
+\left(\tilde{D}_0\mathbf{e}_v^{\mathrm{T}}+\frac{m\mathbf{e}_R^{\mathrm{T}}}{R^2}-m\boldsymbol{\Omega}^{\mathrm{T}}\right)\frac{\partial}{\partial \mathbf{r}}\left(\frac{\partial Q}{\partial \mathbf{v}}\right)^{\mathrm{T}}-m\mathbf{v}^{\mathrm{T}}\frac{\partial}{\partial \mathbf{r}}\left(\frac{\partial Q}{\partial \mathbf{r}}\right)^{\mathrm{T}},
$$

$$
\frac{\partial f_q}{\partial \mathbf{v}}=\frac{\tilde{D}_0}{v}\frac{\partial Q}{\partial \mathbf{v}}\left(\left(1+\frac{M}{D_0}\frac{\partial D_0}{\partial M}\right)\mathbf{e}_v\mathbf{e}_v^{\mathrm{T}}+\mathbf{E}\right)
$$

$$
+\left(\tilde{D}_0\mathbf{e}_v^{\mathrm{T}}+\frac{m\mathbf{e}_R^{\mathrm{T}}}{R^2}-m\boldsymbol{\Omega}^{\mathrm{T}}\right)\frac{\partial}{\partial \mathbf{v}}\left(\frac{\partial Q}{\partial \mathbf{v}}\right)^{\mathrm{T}}-m\mathbf{v}^{\mathrm{T}}\frac{\partial}{\partial \mathbf{v}}\left(\frac{\partial Q}{\partial \mathbf{r}}\right)^{\mathrm{T}},
$$

$$
\frac{\partial f_q}{\partial m}=\frac{\partial Q}{\partial \mathbf{v}}\left(\frac{\mathbf{e}_R}{R^2}-\boldsymbol{\Omega}\right)-v\frac{\partial Q}{\partial \mathbf{r}}\mathbf{e}_v
$$

$$
\mathbf{e}_{\tau\,\mathrm{opt}}=\frac{\boldsymbol{\tau}_*}{\tau_*},\ \boldsymbol{\tau}_*=p_v\boldsymbol{\tau}+\lambda m\mathbf{b},\ \boldsymbol{\tau}=(T+\tilde{C}_L^\alpha)\mathbf{e}_S-(\tilde{C}_L^\alpha+\tilde{D}_\alpha)\mathbf{e}_v\mathbf{e}_S^{\mathrm{T}}\mathbf{e}_v,\ \lambda=\frac{\tau_* f_q-p_v(\boldsymbol{\tau},\mathbf{b})}{mb^2},
$$

$$
\tau_*=p_v\sqrt{\frac{\tau^2-(\boldsymbol{\tau},\mathbf{e}_b)}{1-(f_q/b)^2}},\ \mathbf{e}_b=\frac{\mathbf{b}}{b},\ b^2=(\mathbf{b},\mathbf{b}),\ \mathbf{e}_S=\frac{\mathbf{p}_v}{p_v}.
$$

Features of the Numerical Procedure for Solving the Boundary Value Problem

Phase constraint (5.52) leads to an increase in the dimension of the BVP by one. The Lagrange multiplier ν in (5.53) is used as an additional varied parameter. Let us note some essential features of solving the BVP, which must be taken into account in order to ensure the convergence of the iterative procedure in a fairly wide range of varied parameters.

To determine time t^* of reaching the constraint boundary, conditions (5.52) cannot be used directly. The point is that in the iterative shooting procedure of solving the BVP this condition may not be satisfied at all on some of the "trial" trajectories, especially in a vicinity of the solution, when function $Q(\mathbf{x})$ touches the boundary. In this case, one of the residuals would be undefined. If the local maximum of function $Q(\mathbf{x})$ is clearly pronounced, then to determine t^* it is convenient to use condition (5.58). In this case, the transversality condition (5.54) will be satisfied automatically (see (5.53)):

$$\mathbf{p}^{-\mathrm{T}}\mathbf{f}^{+} - \mathbf{p}^{-\mathrm{T}}\mathbf{f}^{-} + \nu\left(\frac{\partial Q}{\partial \mathbf{x}}\right)^{\mathrm{T}}\mathbf{f}^{+} = 0,$$

since according to (5.57):

$\left(\frac{\partial Q}{\partial \mathbf{x}}\right)\mathbf{f}^{+} \equiv \left(\frac{\partial Q}{\partial \mathbf{x}}\right)\frac{d\mathbf{x}}{dt} \equiv \left.\frac{dQ}{dt}\right|_{t^{*}+0} = 0.$

Therefore (5.54) cannot be used as an additional residual. In this case, as the residual, it is convenient to take the maximum of $Q(\mathbf{x})$ on the ascent trajectory and achieve its zeroing:

$$\max Q(\mathbf{x}) = 0 \tag{5.60}$$

5.5 Through Optimization of STS Injection Trajectory with Dropping Spent Parts to Specified Areas Under Random Atmosphere Disturbances

Let us consider the problem of through optimization of STS branching spatial trajectories, taking into account atmosphere perturbations (wind and variations in thermodynamic parameters) and constraints on the admissible regions of dropping spent parts (SP) [5, 6]. The solution is based on PMP. The systematic component of disturbances is taken into account when forming the equations of motion and optimal control. The impact of random perturbations is determined on the basis of the minimax approach [19, 20], in which perturbations are considered as an additional control vector function of nature with antagonistic goals in relation to the STS ascent. An analytical synthesis of the optimal control of nature is obtained. Its physical meaning consists in setting the profile of critical (worst) perturbations for the vehicle at a given event probability. Verifying the modified software package for this problem solution and regulating the optimal solution accuracy is described on base of the known first integrals and physical meaning of conjugate variables as influence functions. The effectiveness of the developed methodology and software is demonstrated on specific examples of the through optimization of STS branching trajectories. In particular, it is shown that the solution of the BVP does not require an intervention of the researcher to the solving process even in the absence of a good initial estimation for varied parameters.

5.5.1 Minimax Problem Statement

The problem of through optimization of STS branching trajectories is considered with simulation of external disturbances. The constraints on the control and flight modes are taken into account in all the branches, including the requirements for

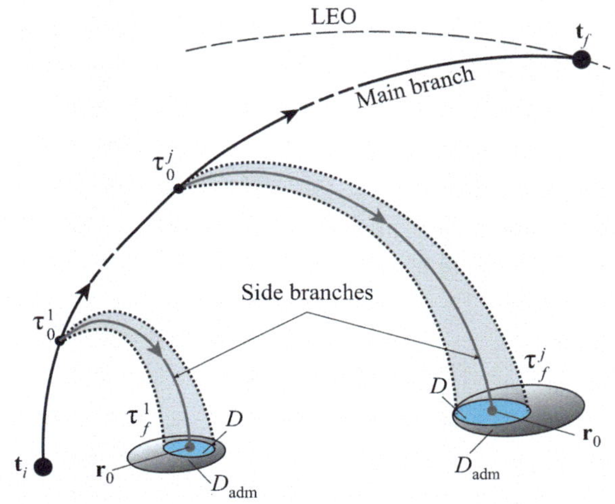

Fig. 5.9 Scheme of STS branching trajectory with continual sets of random SP return trajectories

SP landing (dropping) in the specified regions. The main branch corresponds to the payload insertion into a target orbit, and the side branches correspond to SP return (drop).

The branches can correspond to both real physical processes and be virtual (imaginary), reflecting the possible realizations of SP motion under the influence of random disturbances or emergency situations. The set of branches is continuous, and we come to the problem of optimization of branching trajectories with continuums (cones) of side branches (Fig. 5.9).

To solve the problem, PMP is used. The choice of an optimization method is determined by the advantages of PMP (see Chap. 1), which are especially useful for solving the problem under consideration:

- The structure of the optimal control program is not given a priori, but obtained objectively from the solution of the problem.
- The conjugate system solution makes it possible to define the effect on the functional of small variations of parameters (aerodynamic layout, propulsion system, boundary conditions, constraints, etc.) and systematic and random external disturbances (variations of atmosphere thermodynamic parameters, wind, etc.) acting along the trajectory.

The main idea of the proposed approach is to reduce the problem to a more traditional one with a finite set of branches, when one or several trajectories are selected from a continuum of random realizations, which are the most critical (at a given event probability) for meeting the given boundary conditions. Such a possibility is the use of the minimax approach, in which nature is considered as a player with antagonistic interests in relation to the purpose of the flight. To guarantee the presence of a saddle point in the game problem, a number of assumptions are made that do not go beyond the limits accepted in the practice of ballistic analysis of actual ASVs.

Due to the complexity of the considered mathematical model of motion, a special attention is paid to the development of a multi-level verification system for the software package and to ensuring objective control of the accuracy of the optimal solution. Verification methods are universal, since they rely on the use of known integrals of optimal motion and the physical meaning of conjugate variables as influence functions.

The application of the developed method is demonstrated by the example of trajectory optimization of the Proton-type LV [21], taking into account the constraints on the admissible regions of SP drop and atmospheric disturbances (wind and variations in thermodynamic parameters). The types of local extremals caused by these constraints and the stability of the operation of the algorithm for TPBVP solving by the continuation method are shown.

On the actual injection trajectory, random disturbances are compensated by a control system. Therefore the effect of random disturbances is considered only in the uncontrolled side branches. The systematic component of disturbances is taken into account both on the side and main branches.

The motion is described by a normal ODE system in the start coordinate system. On the main branch:

$$\frac{d\mathbf{x}}{dt} = \mathbf{f}\big(\mathbf{x}, \mathbf{u}, \boldsymbol{\epsilon}_{\text{sys}}\big), \tag{5.61}$$

and on side branches:

$$\frac{d\mathbf{x}}{dt} = \mathbf{f}(\mathbf{x}, \boldsymbol{\epsilon}), \tag{5.62}$$

where $\boldsymbol{\epsilon} = \boldsymbol{\epsilon}_{\text{sys}} + \Delta\boldsymbol{\epsilon}$ is the disturbance vector function, $\boldsymbol{\epsilon}_{\text{sys}}$ and $\Delta\boldsymbol{\epsilon}$ are the systematic and random components consequently.

It is assumed that the thrust is directed along the longitudinal axis of the vehicle, characterized by unit vector $\mathbf{e}_\tau$. In this case, control is carried out by changing the orientation of the vehicle and thrust magnitude T:

$$\mathbf{u} = (\mathbf{e}_\tau, T). \tag{5.63}$$

At the initial, final and, perhaps, intermediate points of the branches, boundary conditions are set, including the matching conditions. Constraints are imposed on the control and phase vectors on all branches.

In the deterministic statement ($\Delta\boldsymbol{\epsilon} = 0$), it is required that the SP hit the given aiming point $\mathbf{r}_0 \in D_{\text{adm}}$ on the Earth's surface:

$$\mathbf{Q}\big(\mathbf{x}(\tau_f)\big) = \mathbf{r}(\tau_f) - \mathbf{r}_0 = 0. \tag{5.64}$$

In the general case, the conditions for the SP drop into admissible region D_{adm} must be met on the entire set of random trajectories corresponding to a given level of event probability (Fig. 5.9):

$$D \subset D_{\text{adm}}, \tag{5.65}$$

where D is the area of SP drop points dispersion (e.g., "dispersion ellipse").

Functional J is the mass injected into the target orbit:

$$J \equiv m(t_f) \Rightarrow \max_{\mathbf{u}}. \tag{5.66}$$

The unconventionality of the problem consists in the presence of random perturbations and inclusion conditions of form (5.65). To regularize it in frames of PMP, we replace the continuum of random trajectories with one or several side branches that are critical from the point of view of the fulfillment of conditions (5.65).

In the general case, if a deterministic statement is used when choosing the aiming point (see (5.64)), condition (5.65) may not be satisfied. On Fig. 5.10 an example of intersecting sets D and D_{adm} is shown. Let us call ***critical*** the direction $\mathbf{e}_L^{crit}$ (or several directions) of the largest displacement dL of the dispersion ellipse D with respect to the admissible region D_{adm}. If we assume that with a small change in the aiming point $\mathbf{r}_0$ the shape of the dispersion ellipse does not change, then to fulfill condition (5.65), it suffices to shift the position of the aiming point $\mathbf{r}_0$ by $d\mathbf{L} = dL\ \mathbf{e}_L^{crit}$. Thus, the solution algorithm is reduced to the following steps:

1. Calculation of the nominal ($\Delta\boldsymbol{\epsilon} = 0$) optimal trajectory.
2. Calculation of the boundary of the dispersion region D.
3. If condition (5.65) is not satisfied, the critical direction $\mathbf{e}_L^{crit}$ and the necessary displacement $d\mathbf{L}$ of the aiming point are determined.
4. Step 1 is repeated for the shifted aiming point.

Note that the entire boundary of region D is not necessary to solve the problem. Usually, checking from two to four points corresponding to the maximum longitudinal and lateral displacements is enough. Nevertheless, region D may be of independent interest. It will be shown below that the developed technique makes it possible to construct the complete boundary of D practically without spending computational resources (they are orders of magnitude less than integrating the trajectory).

To solve the problem, the approach is used that is ideologically close to the minimax approach, in which random perturbations $\Delta\boldsymbol{\epsilon}(\boldsymbol{\xi})$ are considered as the effect of a player "nature" with control $\boldsymbol{\xi}$, which has antagonistic interests in the flight target. This assumes that:

- the probability P_k of a set of random factors is given;
- the optimal control vector of nature $\boldsymbol{\xi}_{\text{opt}}$ (the worst in terms of the functional J) corresponds to the maximum shift (miss) $\Delta L_{\max}$ of the SP drop point from $\mathbf{r}_0$ in the direction $\mathbf{e}_L$;
- vector $\boldsymbol{\xi}_{\text{opt}}$ does not depend on $\mathbf{u}_{\text{opt}}$;
- random perturbations are small, so that their influence on $\Delta L_{\max}$ can be estimated by the Bliss formula [22], which is linear in $\boldsymbol{\xi}$.

Then we come to the game problem, where vectors $\mathbf{u}_{\text{opt}}(t)$ and $\boldsymbol{\xi}_{\text{opt}}$ are the optimal control of STS and nature and are determined from the condition:

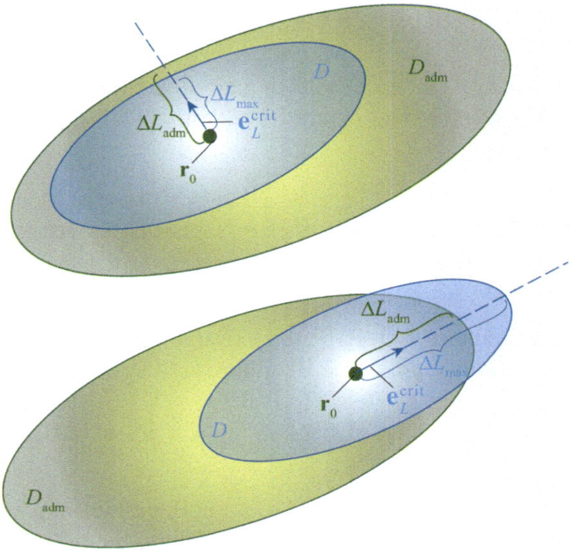

Fig. 5.10 Illustration of choosing the critical direction $\mathbf{e}_L^{\text{crit}}$ to the SP drop point in the dispersion field

$$\left\{\mathbf{u}_{\text{opt}}(t), \boldsymbol{\xi}_{\text{opt}}\right\} = \arg \max_{\mathbf{u}} \min_{\boldsymbol{\xi}} J. \tag{5.67}$$

Under the assumptions made, the problems of finding $\boldsymbol{\xi}_{\text{opt}}$ and $\mathbf{u}_{\text{opt}}(t)$ can be separated. The iterative procedure is used for the solution (see above), at each iteration of which $\mathbf{u}_{\text{opt}}(t)\big|_{\boldsymbol{\xi}=fix}$ and $\boldsymbol{\xi}_{\text{opt}}\big|_{\mathbf{u}=fix}$ are defined as

$$\boldsymbol{\xi}_{\text{opt}} = \arg \min_{\boldsymbol{\xi}\in\mathbf{I}_\sigma} J(\boldsymbol{\xi}) = \arg \max_{\boldsymbol{\xi}\in\mathbf{I}_\sigma} \Delta L(\boldsymbol{\xi}), \tag{5.68}$$

where $\mathbf{I}_\sigma$ is the set determined by the given probability level.

5.5.2 *Perturbation Model*

Deviations of atmosphere thermodynamic parameters (density ρ and sound speed a_s) and horizontal wind with velocity vector $\mathbf{W}$ are considered as perturbations:

$$\begin{gathered}
\boldsymbol{\epsilon}^{\mathrm{T}} = \left\{\rho, a_s, \mathbf{W}^{\mathrm{T}}\right\}, \boldsymbol{\epsilon}_{\text{sys}}^{\mathrm{T}} = \left\{\rho_{\text{sys}}, a_{s\text{sys}}, \mathbf{W}_{\text{sys}}^{\mathrm{T}}\right\}, \\
\Delta\boldsymbol{\epsilon}^{\mathrm{T}} = \left\{\Delta\rho, \Delta a_s, \Delta\mathbf{W}^{\mathrm{T}}\right\}, \Delta\mathbf{W} = \Delta W_\lambda \mathbf{e}_\lambda + \Delta W_\phi \mathbf{e}_\phi,
\end{gathered} \tag{5.69}$$

where ΔW_λ and ΔW_ϕ are the latitudinal (to the East) and meridional (to the North) projections of the wind velocity; $\mathbf{e}_\lambda$, $\mathbf{e}_\phi$ are the corresponding unit vectors.

Random components are given in the form of canonical expansions:

$$\Delta\boldsymbol{\epsilon} = \mathbf{A}_{\varepsilon}(\mathbf{x}, n_t) \cdot \boldsymbol{\xi}, \tag{5.70}$$

where $\mathbf{A}_{\varepsilon}(\mathbf{x}, n_t)$ is the known matrix function, n_t is the flight date, $\boldsymbol{\xi}$ is the vector of independent random numbers with normal distribution and unit dispersion.

5.5.3 Calculation of the Nominal Optimal Branching Trajectory

In accordance with PMP, the STS optimal control is determined from the condition for the maximum of the Hamiltonian of systems (5.61), (5.62):

$$\mathbf{u}_{\text{opt}} = \arg \max_{\mathbf{u}(t) \in \mathrm{U}} H, \tag{5.71}$$

where $H(\mathbf{p}, \mathbf{x}, \mathbf{u}) = \mathbf{p}^{\mathrm{T}}\mathbf{f}(\mathbf{x}, \mathbf{u})$ is the Hamiltonian, $\mathbf{p} = (\mathbf{p}_r, \mathbf{p}_v, p_m)^{\mathrm{T}}$ is the conjugate vector with coordinates corresponding to the radius vector $\mathbf{r}$, velocity vector $\mathbf{v}$ and mass m, respectively, and satisfies the system of equations:

$$\frac{d\mathbf{p}}{dt} = -\left(\frac{\partial H}{\partial \mathbf{x}}\right)^{\mathrm{T}} = -\left(\frac{\partial \mathbf{f}}{\partial \mathbf{x}}\right)^{\mathrm{T}} \mathbf{p}, \tag{5.72}$$

with boundary conditions [23] (see Chap. 2). In particular, at the branch points, the conditions for gluing the phase vectors of the branches lead to the following jump conditions for conjugate variables (transversality conditions):

$$\begin{aligned} \mathbf{p}^{+} &= \mathbf{p}^{-} - \mathbf{p}_s - \lambda\left(\frac{\partial Q_0}{\partial \mathbf{x}}\right)^{\mathrm{T}}, \\ H^{+} &= H^{-} - H_s, \end{aligned} \tag{5.73}$$

where $Q_0(\mathbf{x}(\tau_0)) = 0$ is the condition for implicitly determining branching time τ_0, the subscript "s" here corresponds to the side branch.

Constraint (5.64) on the position of the SP drop point reveals itself under the conditions of transversality at the right end of the side branch:

$$t = \tau_f : \begin{cases} \mathbf{p}_{rs} = \text{var}, \\ \mathbf{p}_{vs} = 0, \\ p_{ms} = 0, \\ H_s = \mathbf{p}_{rs}^{\mathrm{T}}\mathbf{v} = 0, \end{cases} \tag{5.74}$$

where τ_f is implicitly determined by condition (5.64), since it corresponds to reaching the Earth's surface at altitude h_0:

$$\tau_f : Q_h = h(\tau_f) - h_0 = 0. \tag{5.75}$$

The use of PMP makes it possible to reduce the original problem of finding the optimal control in a functional space to the solution of BVP for ODE systems (5.61), (5.62), (5.72). The number of variable parameters of such BVP does not exceed the dimension of the phase vector taking into account the branching (each j-th -branch increases the dimension of the phase space by dimension n_j of the phase vector on the branch). However, as shown above, in the case of SP non-controlled motion on the side branches, it is possible to reduce the boundary value problem so that the side branch with condition (5.64) increases the dimension of the boundary value problem not by six, but for two units only [5].

5.5.4 Calculation of the Spent Part Drop Dispersion Region

To calculate the boundary point of the dispersion region in direction $\mathbf{e}_L$, we optimize the vector function of random factors $\boldsymbol{\xi}$ with displacement $\Delta L(\Delta\boldsymbol{\epsilon}(\boldsymbol{\xi}))$ of the drop point as the functional. Thus, it is necessary to find such "optimal" combination of random factors that leads to the maximum displacement of the drop point in direction $\mathbf{e}_L$:

$$\Delta L_{\max} = \max_{\boldsymbol{\xi}} \Delta L. \tag{5.76}$$

Solving the specified problem for various vectors $\mathbf{e}_L$, whose hodograph describes a circle on the local horizontal plane with the center at the reference drop point $\mathbf{r}_0$, we obtain the corresponding boundary points of the dispersion region.

To determine the transversality conditions at time τ_f, let us consider the changes in ΔL and Q_h (5.75) under variations of the right end of the branch:

$$\begin{cases} d(\Delta L) = \nabla(\Delta L) \cdot \delta\mathbf{x}_f + \nabla(\Delta L) \cdot \dot{\mathbf{x}}(\tau_f)d\tau_f \\ dQ_h = \nabla Q_h \cdot \delta\mathbf{x}_f + \nabla Q_h \cdot \dot{\mathbf{x}}(\tau_f)d\tau_f = 0 \end{cases}. \tag{5.77}$$

In view of (5.77), the transversality condition at the SP drop point is as follows:

$$\mathbf{p}_{sf}^L = \mathbf{p}_s^L(\tau_f) = \nabla(\Delta L) - \frac{\nabla(\Delta L) \cdot \dot{\mathbf{x}}(\tau_f)}{\nabla Q_h \cdot \dot{\mathbf{x}}(\tau_f)} \cdot \nabla Q_h. \tag{5.78}$$

The superscript L is used to emphasize the dependence of the conjugate vector on $\mathbf{e}_L$.

From (5.78), taking into account (5.75), we obtain:

$$\mathbf{p}_{rs}^L(\tau_f) = \mathbf{e}_L - \frac{\mathbf{e}_L^{\mathrm{T}}\mathbf{e}_v}{\mathbf{e}_{r_0}^{\mathrm{T}}\mathbf{e}_v} \cdot \mathbf{e}_{r_0}, \mathbf{p}_{vs}^L(\tau_f) = 0, \mathbf{p}_{ms}^L = 0, \tag{5.79}$$

where $\mathbf{e}_{r_0}$ is the unit vector from the center of the Earth to the nominal SP drop point.

Note that due to the linearity of conjugate system (5.72) with respect to the conjugate vector, numerical integration of (5.72) for the boundary conditions (5.79) can be replaced by a linear transformation with a transition matrix calculated by integrating the corresponding side branch of the nominal trajectory (see Sect. 4.4).

It is assumed that random perturbations of the trajectory are small, and their influence on the SP drop point at the right end of branch $\mathbf{x}_f = \mathbf{x}(\tau_f)$ can be estimated by the Bliss formula:

$$\Delta L = \left(\mathbf{p}_s^L, \delta\mathbf{x}\right)_i + \int_{\tau_0}^{\tau_f} \left(\mathbf{p}_s^L, \delta_\varepsilon \mathbf{f}\right) dt,$$

$$\delta_\varepsilon \mathbf{f} = \mathbf{f}(\mathbf{x}, \Delta\varepsilon) - \mathbf{f}(\mathbf{x}, 0), \tag{5.80}$$

where $\mathbf{p}_s^L$ is the solution of the Cauchy problem (5.72), (5.79), and the phase vector $\mathbf{x}$ corresponds to the nominal solution ($\Delta\boldsymbol{\epsilon} = 0$).

From (5.80), taking into account (5.70), we obtain the expression for ΔL in the form of a linear dependence on random vector $\boldsymbol{\xi}$:

$$\Delta L = \mathbf{L}_\xi^{\mathrm{T}} \boldsymbol{\xi},$$

$$\mathbf{L}_\xi = \int_{\tau}^{\tau_f} \mathbf{p}_{vs}^{\mathrm{T}} \frac{\partial \mathbf{f}}{\partial \boldsymbol{\epsilon}} \mathbf{A}_\varepsilon dt. \tag{5.81}$$

Let the set of random perturbations $\mathbf{I}_\sigma = \{\boldsymbol{\xi} : \boldsymbol{\xi} \leq \kappa_\sigma\}$ be a hypersphere with radius κ_σ determined by the given event probability P_k. Then expressions (5.68) - (5.81) allow one to obtain the analytical synthesis of the nature's optimal control, which maximizes deviation ΔL of the drop point in direction $\mathbf{e}_L$:

$$\boldsymbol{\xi}_{\mathrm{opt}} = \frac{\kappa_\sigma}{||\mathbf{L}_\xi||} \mathbf{L}_\xi \tag{5.82}$$

The physical meaning of vector $\boldsymbol{\xi}_{\mathrm{opt}}$ is that it sets the critical spatial profile of random perturbations $\Delta\boldsymbol{\epsilon}(\mathbf{x}, \boldsymbol{\xi}_{\mathrm{opt}})$, which is the worst for achieving the stated flight goal.

5.5.5 *Regularity of the Numerical Procedure for Solving the Boundary Value Problem*

One of the main criteria for the effectiveness of the developed numerical procedures is their regularity in the studied range of parameter values. Here, regularity is understood as the ability of the program to find the solution of the problem with the required accuracy without the researcher intervening in the solving process, in the absence of any serious requirements for the closeness of the given initial estimation to the solution.

The demonstration of the numerical procedure regularity is carried out on the example of calculation of the dependence of the maximum injected mass m_f on the given drop range L_1 of the 1st stage booster. As can be seen from Fig. 5.11, dependency $m_f(L_1)$ is ambiguous. The dependence segments correspond to local extremals of ***A***- and ***B***-type (see Sect. 4.5). They qualitatively differ in the structure of trajectories. The ***A***-trajectories are "flat" and ***B***-trajectories are "steep". Qualitatively, these extremals correspond to two different throwing angles known from Keplerian mechanics to achieve the same range (provided that it is less than the maximum one).

Branching trajectories corresponding to extremals of each type for the same drop range of the 1st stage ($\overline{L}_1 \cong 0.42$) are shown in Fig. 5.12. In this case, the global optimum for the functional is provided by the upper segment corresponding to the "flat" extremals. In other cases, these may interchange or have one or more intersection (bifurcation) points. There, the same value of the functional is reached for qualitatively different trajectories and optimal control structures.

On Fig. 5.13 the required number of iterations of the procedure for the BVP solving for different L_1 are shown. In accordance with the procedure of solution continuation on a parameter, the solution of the optimization problem for each value of L_1 is based on the previously obtained optimal solutions. The maximum range $L_{1\,max}$ is the singular point for this procedure, but the convergence at this point is not violated.

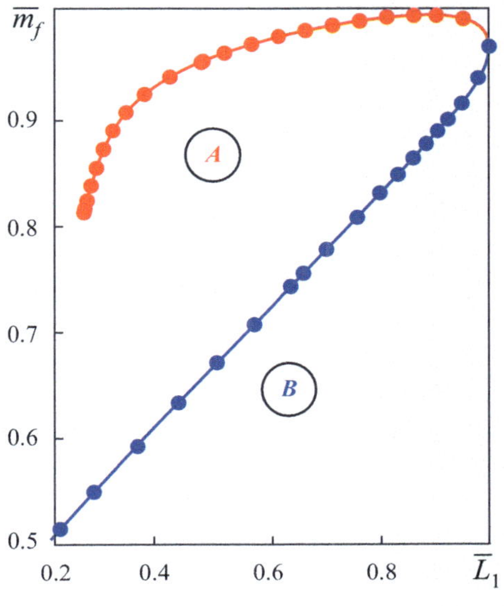

Fig. 5.11 Relative maximum injected mass $\overline{m}_f$ depending on the relative SP drop range of the 1ts stage booster $\overline{L}_1 = L_1/L_{1\,max}$

Fig. 5.12 Optimal STS branching trajectories of two types with given drop range $L_1 \approx 0.42 L_{1\,max}$ for the 1st stage booster

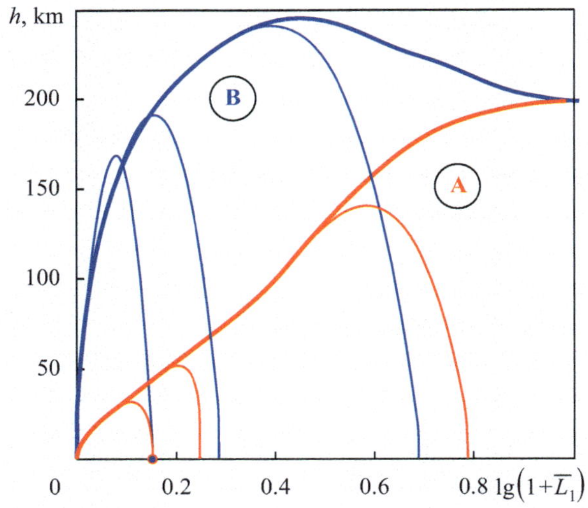

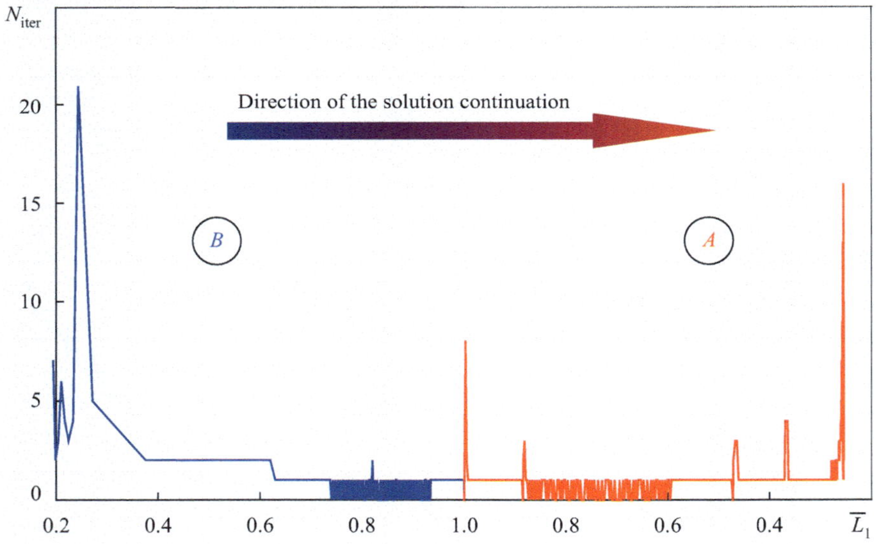

Fig. 5.13 Number of iterations for solving the BVP depending on the given relative SP drop range with continuous passing through the singular point ($\overline{L}_1 = 1$) dividing two qualitatively different types of extremals

5.5.6 Numerical Results of the STS Branching Trajectory Optimization with Constraints on Spent Part Drop Points in Disturbed Atmosphere

The application of the developed method is shown by the example of calculating through optimal injection trajectories of the Proton-type LV, taking into account constraints on SP drop regions and atmosphere disturbances (thermodynamic parameters and wind). The random vector $\boldsymbol{\xi}$ has dimension of 24.

The critical profiles of random wind $\Delta\mathbf{W}(\mathbf{x}, \boldsymbol{\xi}_{\text{opt}})$ with event probability $P_k = 0.9973$ for the atmosphere in January [24] are shown in Fig. 5.14. The wind vector distribution in physical space is assumed to be frozen during the SP return. Each profile is represented as the worst random horizontal wind speed vectors for SP flight altitude h, changing with longitude λ and latitude φ on the SP trajectory. The worst wind means its distribution along the SP trajectory that maximizes deviations from the nominal drop point in the selected direction. The four wind profiles in Fig. 5.14 correspond to four directions: two of them maximize and minimize the longitudinal range of the SP drop points, and two more maximize lateral deviations to the right and left of the launch plane.

If unit vector $\mathbf{e}_L$ is rotated around the reference point $\mathbf{r}_0$ in the local horizontal plane then vector $\Delta\mathbf{L} = \Delta L_{\max}\mathbf{e}_L$, calculated in accordance with procedure (5.58)–(5.82), describes the boundary of dispersion area D of SP drop points (Fig. 5.14), (Fig. 5.15). We note that in order to obtain this boundary in accordance with the proposed method, the integration of perturbed trajectories and the conjugate system (5.72), (5.79) for a fan of directions $\mathbf{e}_L$ is not required. Since, firstly, the perturbation of phase variables (deviation of the SP drop point from $\mathbf{r}_0$) is estimated by the Bliss formula (5.80), which includes phase variables only for the unperturbed ($\Delta\boldsymbol{\epsilon} = 0$) nominal trajectory. Secondly, due to the linearity of (5.72) with respect to the conjugate vector, the integration of the conjugate system (5.72) for the boundary conditions (5.79) changing after the change of $\mathbf{e}_L$, can be replaced by a linear transformation with the same transition matrix. This matrix is also determined once when integrating the nominal trajectory (see Sect. 4.5).

On the whole, the proposed approach makes it possible to reduce the amount of computations by ~$10^5 \div 10^7$ times in comparison, e.g., with the widespread engineering Monte Carlo method. Additionally, the proposed method provides a higher accuracy of determination of the dispersion area, since it actually has a zero confidence interval.

The boundary of the dispersion area of drop points of the 1st stage booster is shown in Fig. 5.15. It was calculated in accordance with the proposed minimax method for $P_k = 0.9973$ at the launch in January. The same figure shows drop points (marked with crosses) obtained as a result of the statistical Monte Carlo simulation of 10,000 SP trajectories. This number of trajectories is dictated by specified probability P_k. It is evident that the results agree very well with each other. However, the proposed minimax method is incomparably more efficient in terms of the calculation time and volume. In addition, it makes it possible to study not only the total effect of

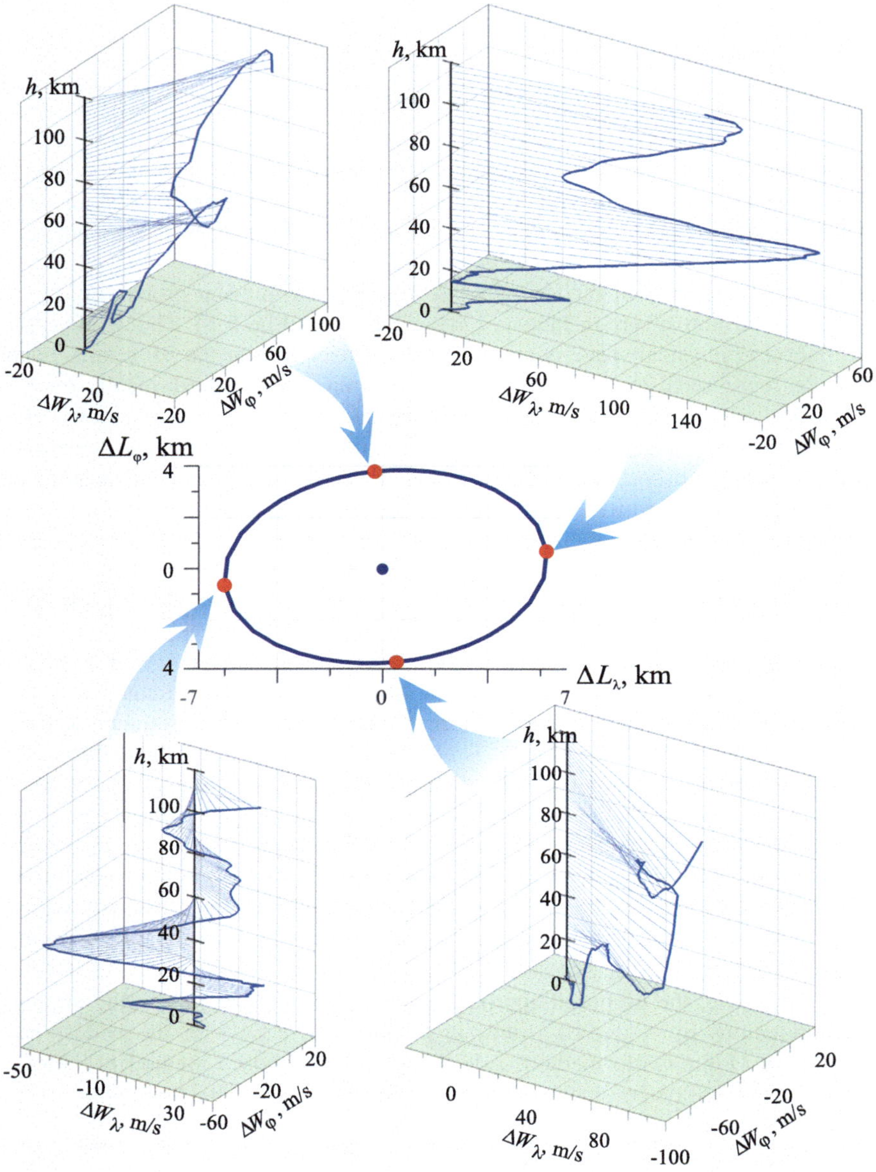

Fig. 5.14 In the center: The boundary of the dispersion area of probable (with a probability of 0.9973) SP drop points, calculated by the proposed minimax method. Around: Spatial profiles of the worst wind distribution during the SP return, maximizing deviations of the drop point from the nominal one in the longitudinal and lateral directions

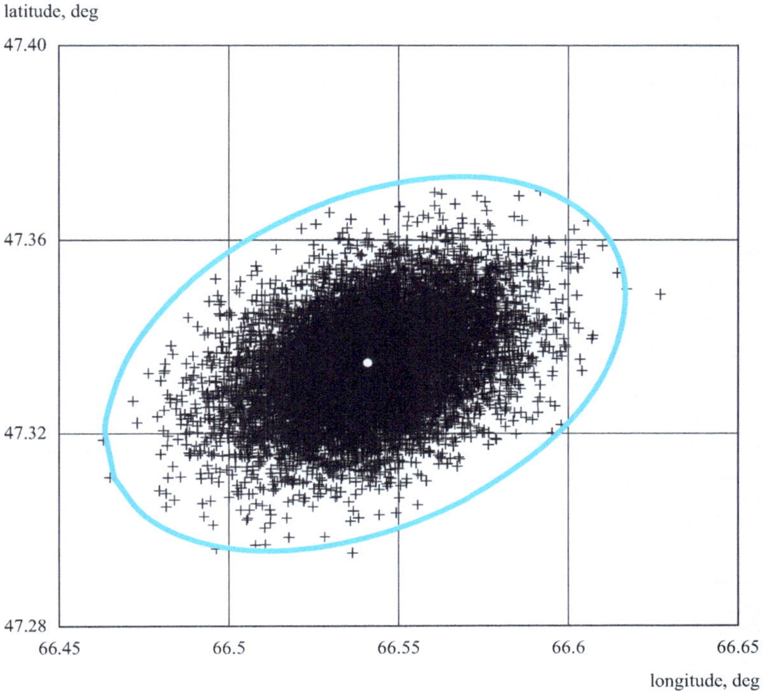

Fig. 5.15 Comparison of the dispersion area boundary (blue line) of SP drop points, calculated by the proposed minimax method on the base of the only nominal SP trajectory, and random drop points (crosses), simulated by the Monte Carlo method with integration of 10,000 SP trajectories

disturbances, but also the contribution of each of them. In real problems, hundreds of random factors have to be taken into account.

5.5.7 Some Conclusions to Sect. 5.5

Based on PMP and the minimax approach, the method for solving optimization problems has been developed, which allows:

- to calculate the nominal (taking into account only the systematic component of the disturbances) optimal trajectory with dropping the SP to the admissible area;
- to form critical profiles of disturbances, the influence of which leads to the maximum (for a given event probability) deviation from the given constraints and terminal conditions;
- to calculate dispersion areas of SP drop points, due to the influence of an arbitrary number of random factors.

The analytical synthesis of "the optimal control" of random disturbances is obtained, leading to the worst deviations of trajectories from the required conditions. This made it possible to reduce the time and volume of calculations of trajectory dispersion areas by 5–7 orders of magnitude and to increase the accuracy compared to the Monte Carlo method.

References

1. Filatyev, A.S.: Features of atmospheric reentry trajectories with a subcircular velocity. Proceedings of The 6th Readings dedicated to the development of the scientific heritage and ideas of F.A. Tsander (Dnepropetrovsk, June 12–15, 1979). Section "Astrodynamics". M., 1980, p. 92; In: Ideas of F.A. Zander and the development of rocket and space science and technology. Moscow, Nauka (1983)
2. Ilyin, V.A., Filatyev, A.S.: Synthesis of optimal trajectories for launching into orbit, from any point of which a descent in the atmosphere is possible with the fulfillment of specified restrictions. Cosm. Res. **23**(1), 37 (1985)
3. Filatyev, A.S., Yanova, O.V.: The optimization of ascent trajectories with continuum of branches.In: The 45th Congress of the International Astronautical Federation, October 9–14, 1994, Jerusalem, Israel, IAF-94-A.2.019.
4. Ilyin, V.A., Filatyev, A.S.: Synthesis of optimal trajectories for launching into orbit, from any point of which a safe descent in the atmosphere is possible. In: Proceedings of the 5th All-Union Congress on Theoretical and Applied Mechanics, Alma-Ata, 1981
5. Filatyev, A.S., & Yanova, O.V.: Optimization of space system launching with limitations on fall zones for spent components. In: The 47th Congress of the International Astronautical Federation, October 7–11, Beijing, China, IAF-96-A.3.01 (1996)
6. Filatyev, A.S., Yanova, O.V.: Through optimization of branching injection trajectories by the Pontryagin maximum principle using stochastic models. Acta Astronaut. **68**(7–8), 1042–1050 (2011)
7. Butkovsky, A.G.: The theory of optimal control for systems with distributed parameters. Nauka Publishing House, Moscow (1965). (in Russian)
8. Filatyev, A.S.: Reduction of maximum dynamic and thermal loads during passive motion of hypersonic vehicles with the help of small control actions. Uchenye Zapiski TsAGI, **9**(2) (1978) (in Russian)
9. Filatyev, A.S.: Approximate analytical synthesis of the optimal control of hypersonic aircraft when moving in the atmosphere with a sub-circular speed. Parts 1 and 2. Uchenye Zapiski TsAGI **11**(1, 2) 1980 (in Russian)
10. Filatyev, A.S.: Paradoxes of optimal solutions in problems of space vehicle injection and reentry. Acta Astronaut **47**(1), 11–18 (2000). (See also Filatyev, A.S.: "Paradoxes" of optimal solutions in problems of space vehicle injection and reentry. The 48th Congress of the International Astronautical Federation, October 6–10, 1997, Turin, Italy, IAF-97-A.6.09).
11. Detra, R., Kemp, N., Riddell, F.: Addendum to "Heat Transfer To Satellite Vehicles Re-entering the Atmosphere", Jet Propulsion, pp 1256 (1957)
12. Kemp, N., Riddell, F.: Heat transfer to satellite vehicles reentry atmosphere. Jet Propulsion, 21(2) (1957)
13. Filatyev, A.S.: Synthesis of optimal control of aircraft during descent in the atmosphere on essentially non-stationary trajectories. In: Abstracts of the 8th All-Union Conference on Control in Mechanical Systems of the USSR Academy of Sciences, Lvov, 1988 (in Russian)
14. Filatyev, A.S.: Space vehicle safety problem: reentry with subcircular speeds. IAA. 6.1-93-731. Space Safety and Rescue 1993 (ed. by GW Heath). Science and Technology Series, vol. 87 American Astronautical Society Publication 1993 31 40

15. Filatyev, A.S.: Fail-Safe Ascent of Space Transport Systems. IAA 94–676. Space Safety and Rescue 1994 (ed. by GW Heath). Science and Technology Series, Vol. 88 American Astronautical Society Publication 1994 49 56
16. Filatyev, A.S.: Critical reentry conditions. ISTS 94-c-32. In: The 19th International Symposium on Space Technology and Science, Yokohama, Japan, May 15–24 (1994)
17. Filatyev, A.S., Yanova, O.V.: Nonequilibrium reentry. ICAS-2.2-1. In: The 21st ICAS Congress, September 7–14, 1996, Sorrento, Italy.
18. Allen, H.J., Eggers, A.J.: A Study of the motion and aerodynamic heating of missile entering the Earth atmosphere at high supersonic speeds, NASA Rep. No.1381 (1958)
19. Malyshev, V.V., Kibzun, F.I., Karp, K.A.: Minimax approach for complex technical system stochastic modeling. AMSE Press. **1**(3) (1988)
20. Kuzmin, V.P., Yaroshevsky, V.A.: Estimation of phase coordinates large deviation of dynamic system caused by random perturbations. Moscow, Nauka, Publishing Company Fizmatlit (1995) (in Russian)
21. Kobelev, V.N., Milovanov, A.G.: Spacecraft launch vehicles. Moscow: RESTART, p 528. ISBN 978-5-904348-01-4 (in Russian) (2009)
22. Bliss, G.A.: Mathematics for Exterior Ballistics. NY (1944)
23. Filatyev, A.S.: Optimization of branched trajectories for aerospace transport systems. ICAS-94–5.2.3. In: The 19th ICAS Congress, 18–23 September 1994, Anaheim CA, USA
24. GOST 4401-81. The standard atmosphere. Parameters. Moscow, Publishing House of Standards (1981)

Chapter 6
Investigation of Optimal Branching Trajectories of Various Types of Space Transportation Systems

6.1 Introduction

The developed methods and programs for the through optimization of branching trajectories for spacecraft injection into orbits based on PMP, described in previous chapters, were used at various stages of studying analogues of the following space transport systems: reusable STS Energia-Buran, LV Proton-M and Angara, ASTS MAKS and An-225/Interim HOTOL, ASTS AirLaunch, reusable STS Soyuz-7 and MRKS-1, etc. Thanks to work on specific projects, new problem statements arose that required permanent improvement of optimization methods and their software implementation. Some of the results of such studies are demonstrated below in this chapter. The author did not strive for accurate reproduction of the values of design parameters, especially since this is practically impossible. The main attention is paid to parametric studies in wide ranges, identification and physical explanation of qualitatively new solutions, which became possible to obtain only after the development of the methods and corresponding software described in the monograph. It also seemed important for the author to show the relative quantitative effect that the integral approach gives with the combination of all possible flight stages into a single complex and their rigorous optimization based on PMP. Examples of using PMP for strict optimization of branching trajectories in deterministic and stochastic statements with finite and continual sets of branches are given. Restrictions on the phase vector and control, requirements for safe separation and fault tolerance are considered. The high efficiency of the minimax approach for assessing the influence of random disturbances, based on the physical meaning of conjugate variables as functions of influence, is shown, which makes it possible to reduce computational costs and errors by several orders of magnitude compared to the Monte Carlo method widely used in an engineering practice.

The purpose of this chapter is to demonstrate the effectiveness of the use of PMP and specifically *ASTER* software for a qualitative analysis of the behavior of optimal solutions not only on simplified models as in Chap. 3, but also for wide parametric

A. S. Filatyev and V. G. Petukhov, *Through Optimization of Aerospace Vehicle Trajectories by the Pontryagin Maximum Principle*, Studies in Systems, Decision and Control 579, https://doi.org/10.1007/978-3-031-80756-5_6

studies of optimal trajectories of various STS classes, described by detailed mathematical models with a high calculation accuracy as is customary for ballistic analysis when designing aerospace vehicles. In particular, in most of the problems considered below, it was required that total calculation errors, including errors of integrating the ODE systems and BVP solutions, should not lead to an error in the injected mass (the functional) of more than 0.01–0.10 kg. Given that the STS launch mass is measured in hundreds of tons, the relative total error did not exceed ~ 10^{-8}–10^{-7}. The high accuracy of the calculations is especially evident in the study of changes in the optimal trajectory, control program and functional as a result of variations in the problem parameters. Therefore, parametric analysis is given special attention from a methodological point of view. It is important to note that thousands of optimal branching trajectories were calculated without any problems with convergence of corresponding BVPs, the dimensions of which could reach several tens. In all cases, the solutions of BVPs were obtained automatically, without a researcher intervention.

Taking into account this goal, the author did not seek to accurately reproduce the source data of the objects under study, which may differ from characteristics of current and developed STS. In practice, hardly anyone in research institutes manages to track in real time the changes in the source data of aerospace systems, except, perhaps, the developers themselves.

6.2 The Optimal Fail-Safe Injection of STS with a Vertical Take-Off

In an emergency situation at STS launching, the rescue of the reentry vehicle (RV) can be carried out in two ways. The first way is an autonomous rescue, when RV is separated from the STS immediately after detecting an emergency situation that does not allow the completion of the flight task. The second way is STS return maneuver, when the STS energy is used to bring RV into such phase region, from which the safe return of the RV and/or crew is possible. In any case, it is required to study the admissible area of the initial conditions for the RV emergency return, under which it can pass through the dense atmosphere layers with the fulfillment of specified load constraints. The widest admissible area is realized when RV uses the optimal control law [1–7].

Below, there are some results of research carried out by the author on the problem of autonomous rescue of the Buran-type RV. Since, as will be shown in Sect. 6.2.2, the problem of RV return in emergency situations, apparently, cannot be solved by means of only RV, then the problem arises of a comprehensive optimization of all segments of LV ascent and RV emergency return with constraints on the fail-safety conditions (see Chap. 5). The solution of this problem was obtained both with the use of approximate optimal STS control laws in the injection segment (Sect. 5.4.1) and on the basis of the rigorous application of PMP (Sect. 5.4.2).

6.2.1 Source Data for the Reentry Vehicle

Balancing values of RV aerodynamic force coefficients C_D, C_L, the deflection angles of ailerons δ_a, flaps δ_f and corresponding hinge torques $M_{h\,a}$, $M_{h\,f}$ (referenced to the dynamic pressure $q = 1000$ kgf/m^2) were set based on the interpolation of tabular data for Mach numbers from 0.2 to 10 with step 0.2 and for angles of attack α from 0° to 39° with step 2°. The aerodynamic characteristics were taken constant for $M \geq 10$. Interpolation here and below was carried out with splines using the original technique that provides the minimum integral discrepancy with a piecewise linear approximation. Such interpolation criterion makes it possible to automatically avoid unwanted high-frequency oscillations, which are characteristic of high-order interpolation polynomials, and to close to approximations that are typical for an engineering practice. The last property makes it more correct to compare the rigorous solutions based on PMP and other solutions based on simplified engineering approaches.

Constraints on the angle of attack correspond to the upper boundary $\alpha_{\max}(M)$ of the investigated range of angles of attack (Figs. 6.1 and 6.2). The maximum admissible angles of attack α_n and α_h, due to the requirements of maintaining a given level of the load factor and actuator hinge torque, respectively, were determined in advance and given in tabular form as a function of the Mach number M and dynamic pressure q. The following constraints were accepted in the base case:

$$n_{\text{adm}} = 3.5,\; q_{\text{adm}} = 3500\ \text{kgf} \cdot \text{m}^{-2}, \tag{6.1}$$

$$|M_h|_{\text{adm}} = 20\text{tm}. \tag{6.2}$$

Nominal RV engine characteristics were taken as follows:

$$\left.\begin{array}{l}\text{relative fuel mass } \overline{m}_T = 0.092,\\ \text{initial thrust-to-weight ratio } n_i = 0.2,\\ \text{exhaust velocity } c = 3500\ \text{m/s}.\end{array}\right\} \tag{6.3}$$

It was assumed that the RV engines can operate only at a low dynamic pressure not exceeding $q_{\text{power}} = 100$ kgf $\cdot$ m^{-2}. The integration step was set depending on the initial flight speed, which generally determines the duration of the RV return trajectory.

The initial phase coordinates of the RV trajectory corresponding to ones at the moment of release from LV were determined by the nominal STS ascent trajectory in accordance with Table 6.1.

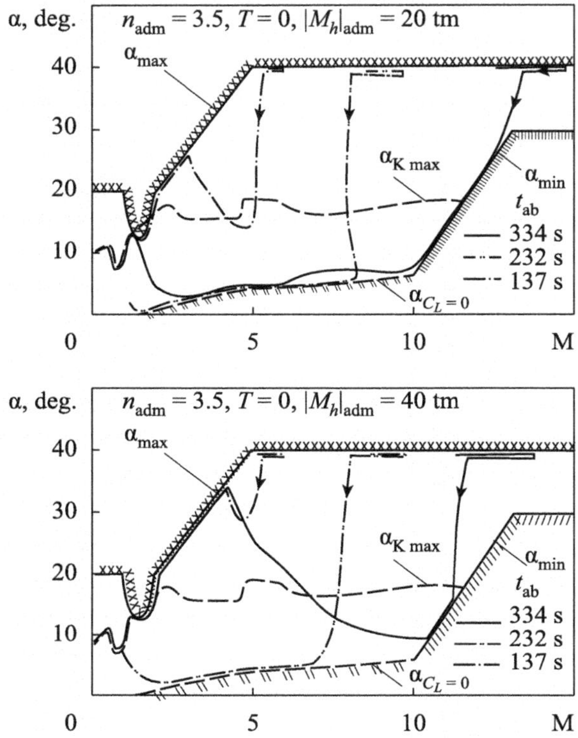

Fig. 6.1 Constraints and quasi-optimal control programs for RV angle of attack after STS ascent abortion at various times for two admissible hinge torques

6.2.2 Aerodynamic Loads on RV Emergency Return Trajectories

The RV return after STS ascent abortion occurs by essentially non-equilibrium trajectories. Therefore, the fundamental problem is to reduce loads upon reentry into dense atmosphere layers [1–3].

RV motion in the return phase is described by ODE system (2.18–2.26). In this case, the following assumptions were made: the gravitational field is central, the Earth is motionless and spherical, the atmosphere is standard [8], and the slip angle is zero.

The control is carried out by changing angle-of-attack α and bank angle γ, as well as using the thrust. It has to ensure minimization of the maximum dynamic pressure $q_{\max}$ for given constraints in accordance with [1–6].

Dependencies of maximum load factor $n_{\max}$, maximum dynamic pressure $q_{\max}$, actuator hinge torques $|M_h|_{\max}$ and temperature in a critical point $T_{\max}$ from ascent abortion time t_{ab} are shown for nominal data (6.2), (6.3) in Fig. 6.3. Let us note the characteristic features of these dependencies that determine the time intervals on the ascent trajectory in which RV autonomous return is possible. Dependencies

Fig. 6.2 Constraints and quasi-optimal control programs for the RV angle of attack with optimal using the thrust after STS ascent abortion at various times for two admissible hinge torques

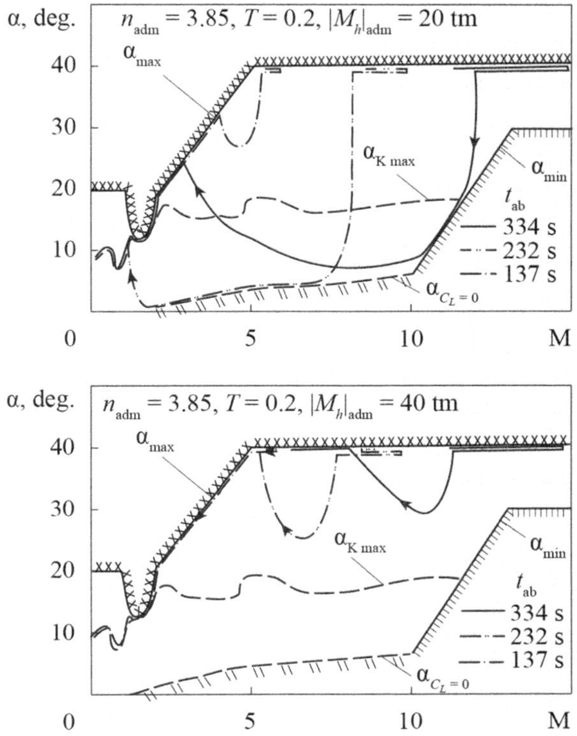

$n_{max}(t_{ab})$ and $q_{max}(t_{ab})$ have two local maxima. The first of them (for $t_{ab} \leq 105$ s) corresponds to the case when the RV separation occurs in dense atmosphere layers, and the highest load factor and dynamic pressure are reached at the initial point of the RV autonomous flight. The second maximum ($t_{ab} \approx 250$–300 s) corresponds to the loads when RV returning to the dense atmosphere layers with critical initial velocities of about 3 km/s [4]. In this case, the maximum load factor and dynamic pressure are realized at the inner point of the return trajectory during the first dive into dense atmosphere layers.

It can be seen that given admissible load factor $n_{adm} = 3.85$ cannot be maintained by controlling the angle of attack at 190 s < t_{ab} < 330 s because the load factor exceeds the admissible one even at angles of attack corresponding to zero lift (Figs. 6.1 and 6.2). (In these calculations it was assumed that the admissible load factor can be increased by 10% compared to the nominal limit (6.1) in emergency situations).

The maximum hinge torque occurs during the RV return flight from the ascent segment according to $t_{ab} \approx 300$–400 s (Fig. 6.3). This is explained by the fact that angles of attack necessary to maintain the admissible hinge torque at the large dynamic pressure become less than the minimum admissible one (Figs. 6.1 and 6.2).

Table 6.1 The initial RV trajectory parameters depending the ascent abortion time

t_{ab}, s	80.59	89.2	100.7	115.3	123.5	137.7	191.6	232.1	272.7	333.6	384.0	414.7	445.8
v, km/s	0.625	0.803	1.046	1.40	1.60	1.83	2.20	2.60	3.07	3.96	4.90	5.59	6.41
θ, deg.	45.2	41.3	34.8	28.7	24.3	21.2	11.5	6.6	4.8	0.1	− 0.6	− 0.6	− 0.15
h, km	17	21.3	27.6	37.0	42.8	53.1	84.9	101	111	117	116	114	113

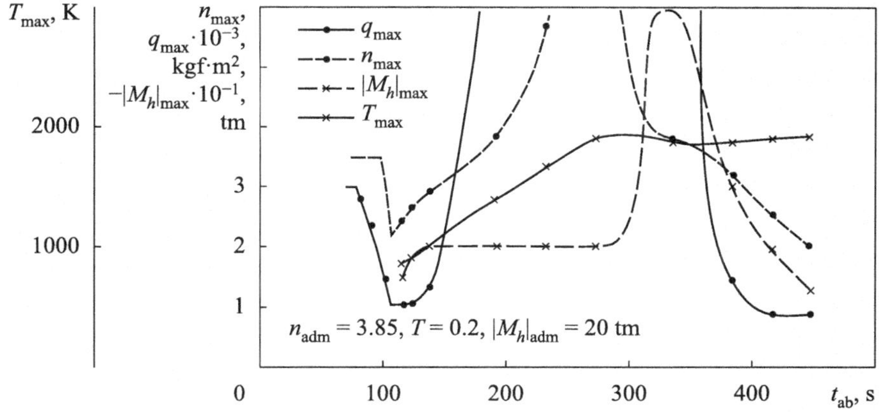

Fig. 6.3 The maximum dynamic pressure, load factor, hinge torque and temperature at RV return trajectories with optimal using the thrust as a function of STS ascent abortion time

Thus, there is range $t \in [t^*, t^{**}]$, in which RV autonomous safe return is impossible. Let us name the STS ascent segment, from which the RV rescue is impossible, as the unsafe segment. As can be seen from Figs. 6.3 and 6.4 left border t^* of the unsafe segment is determined by reaching the admissible dynamic pressure, and right border t^{**} is determined by reaching the admissible hinge torque.

For nominal constraints (6.2), when using the RV propulsion system, the unsafe segment duration is $\Delta t^* \cong 250$ s ($160 \text{ s} \leq t \leq 410$ s), i.e. about 50% of the total duration of the powered phase of the STS ascent.

The most critical limitation is the constraint on the hinge torque. Figure 6.5 shows the dependence of the duration of the unsafe segment on the admissible hinge torque.

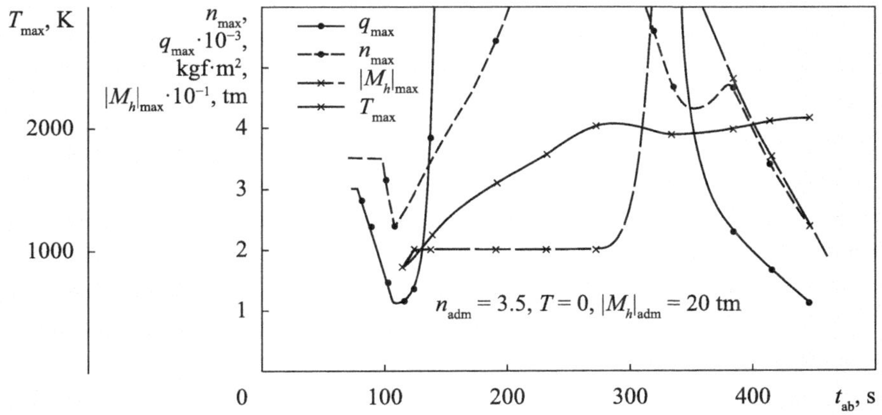

Fig. 6.4 The maximum dynamic pressure, load factor, hinge torque and temperature at RV unpowered return trajectories as a function of STS ascent abortion time

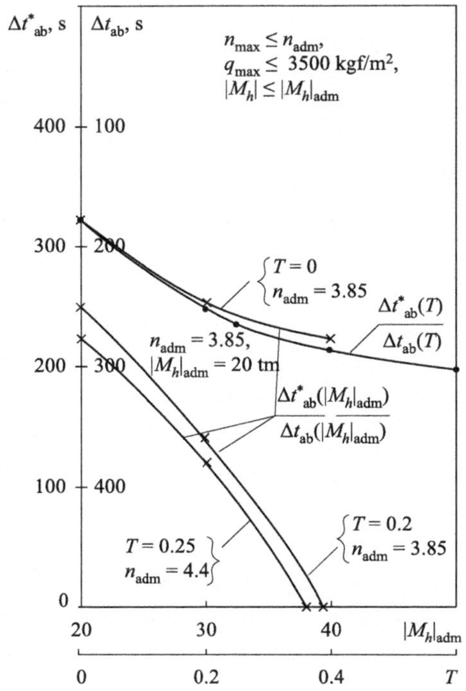

Fig. 6.5 The duration of the STS ascent unsafe segment vs the RV admissible hinge torque and thrust

It can be seen that with doubling of the admissible hinge torque, the RV autonomous return is possible from any point of the STS ascent trajectory.

6.2.3 Influence of Fail-Safety Requirements on the Maximum Injected Mass, Optimal Trajectory and Control Program

As noted in Chap. 5, in order to zero the duration of the unsafe segment, the STS ascent trajectory must lie entirely in ascent corridor *D*. Below are the results of calculations of the optimal STS fail-safe trajectories with the simplified control laws using g-turn and OEL solutions, and in the rigorous statement based on PMP.

The admissible range of speed v_0 and altitude h_0 at the apogee of the osculating return trajectory was determined by the spline interpolation of given data $q_{max}(v_0, h_0)$, $n_{max}(v_0, h_0)$, $|M_h|_{max}(v_0, h_0)$, $L_{max}(v_0, h_0)$. In this case, for the basic (nominal) variant of constraints (6.1), (6.2), two cases of powered and unpowered RV descent were considered.

On the calculations performed, admissible areas *D* were constructed in osculating variables, from which further RV autonomous descent occurs without exceeding the

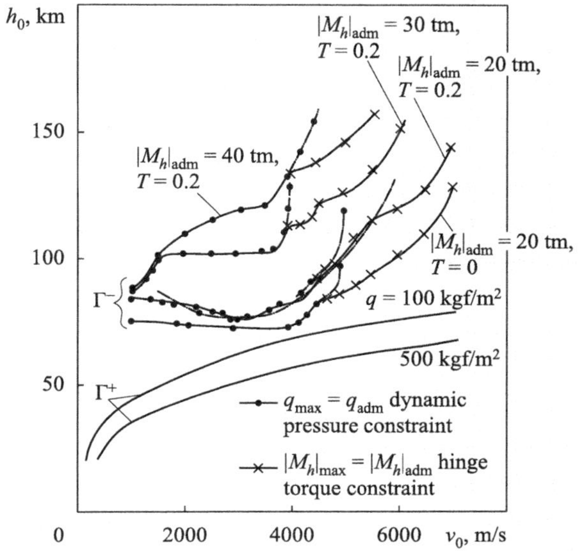

Fig. 6.6 Upper and lower boundaries of the admissible area for fail-safe STS ascent (the "ascent corridor")

specified dynamic pressure, load factor and actuator hinge torque (Fig. 6.6). Upper boundaries Γ^- of areas D consist of two segments. At lower speeds, the constraint on the dynamic pressure is essential. At higher speeds, the constraint on the hinge torque becomes significant.

The lower boundary of D is determined by constraints on dynamic and thermal loads at the initial point of the RV return trajectory, which belongs to the STS ascent trajectory.

Approximate synthesis of the optimal fail-safe trajectory control

The STS ascent trajectories were calculated to ensure the possibility of RV autonomous return from any of their points (dashed lines in Figs. 6.7 and 6.8). In accordance with the control algorithm formulated in Chap. 5, the optimal pitch angle program consists of several sections with a quasi-linear law when moving inside fail-safety ascent corridor D:

$$\vartheta = \arctan(at + b), \tag{6.4}$$

and sections along constraints on the dynamic pressure and hinge torque.

Coefficients a, b in (6.4) are generally different in each of the sections and must be chosen from the boundary conditions and maximize the functional (injected mass m_f or payload mass m_{pl} that is equivalent since the structure mass is given).

In numerical calculations, the 2nd stage ascent trajectory was divided into two successive parts. Values of a and b for the second part were determined from the boundary conditions at the right end of the ascent trajectory, and, for the first part, they were determined from maximization of the injected mass. As can be seen from

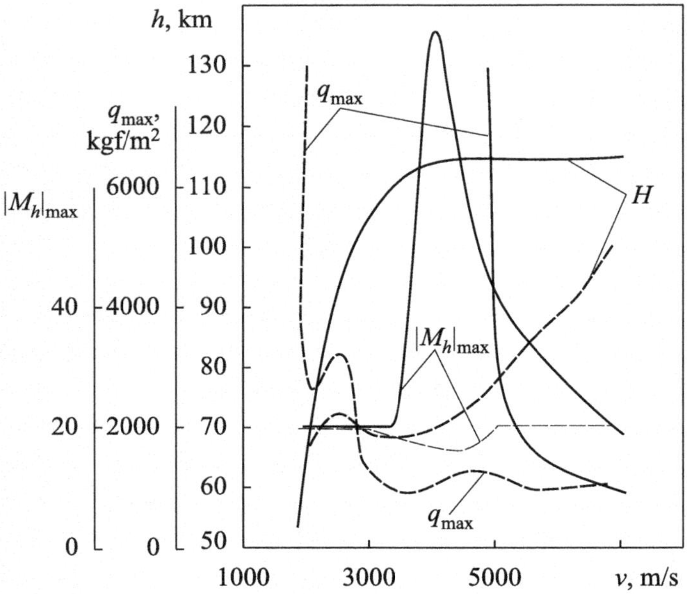

Fig. 6.7 Nominal (solid line) and fail-safe (dashed line) optimal ascent trajectories and corresponding maximum dynamic pressure and hinge torque at unpowered RV emergency return; RV constraints: $n_{adm} = 3.5$, $q_{adm} = 3500$ kgf m^{-2}, $|M_h|_{adm} = 20$ tm, $T = 0$

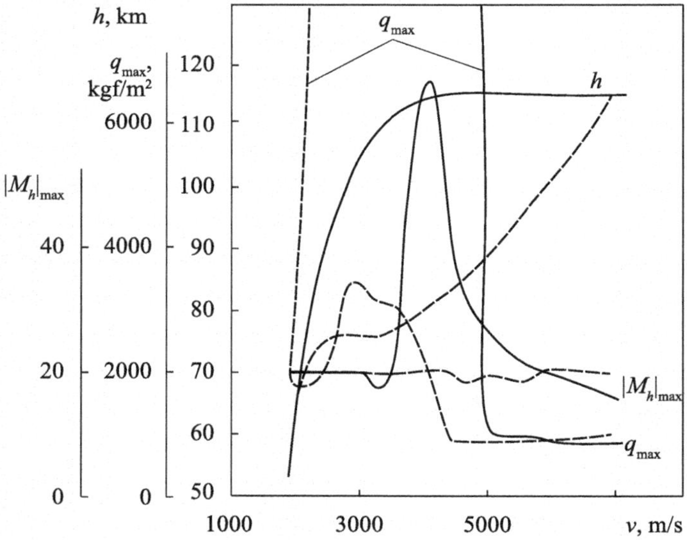

Fig. 6.8 Nominal (solid line) and fail-safe (dashed line) optimal ascent trajectories and corresponding maximum dynamic pressure and hinge torque at powered RV emergency return; RV constraints: $n_{adm} = 3.5$, $q_{adm} = 3500$ kgf m^{-2}, $|M_h|_{adm} = 20$ tm, $T = 0.2$

Figs. 6.6 and 6.9, upper boundary Γ^- at speeds $v_0 = 1.5$–3 km/s, which are reached at the first segment of the 2nd stage ascent trajectory, passes at fairly moderate altitudes h_0, and therefore the ascent trajectory at the beginning reaches quickly the boundary Γ^-. To move along the boundary Γ^- it is necessary to adjust variables v, h, θ. But at the initial phase, due to still low thrust-to-weight ratio of the 2nd stage, the values of v_0, h_0 at boundary Γ^- can be changed mainly by reducing the flight path angle θ, therefore, at the beginning of the second stage motion, the trajectory practically reaches its apogee ($\theta = 0$). However, in a vicinity of $v_0 = 2$–3.5 km/s, the maximum admissible values of h_0 continue to decrease to the minimum value at v_0^*. If $\theta \cong 0$ and the motion occurs along the boundary Γ^-, then the decrease in altitude is impossible. Therefore, to keep the given constraints, this trajectory must reach $\theta \cong 0$ in a vicinity of $v_0 = v_0^*$, i.e. the trajectory presented in osculating coordinates (v_0, h_0) must either touch or be located below the point $h_0^* = h_0(v_0^*)$ on boundary Γ^- (Fig. 6.9). As a result, the ascent trajectory is naturally divided into two parts: before and after the point of reaching v_0^*. As calculations have shown, the optimal altitude at $v_0 \approx v_0^*$ (from the condition of maximum m_f) practically coincides with h_0^* (in fact, it is somewhat less, since the ascent trajectory with an approximate quasi-linear control law (6.4) in this problem reaches its maximum altitude at an intermediate point (Figs. 6.7, 6.8 and 6.9). Thus, in the initial phase of the ascent trajectory (up to $v_0 \approx v_0^*$), the optimal values of a and b in control law (6.4) can be determined from the boundary condition: $\theta = 0$, $v \cong v_0^*$, $h = h_0^*$.

Approximate optimal control programs, taking into account the requirement of fail-safety in this way, are shown in Fig. 6.10. It can be seen that the additional condition that the trajectory belongs to the ascent corridor D leads to a significant change in the control program compared to the nominal one. However, the payload mass does not change so much. Figure 6.11 shows a diagram of the change in the

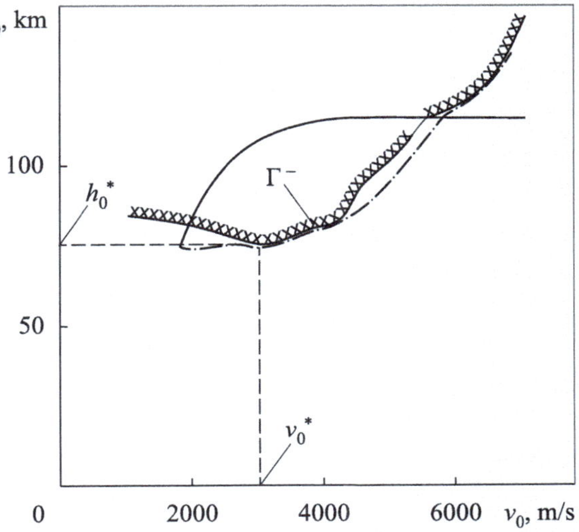

Fig. 6.9 Nominal (solid line) and fail-safe (dashed line) ascent trajectories and the upper fail-safety boundary Γ^- in the osculating phase coordinates; RV constraints: $|M_h|_{adm} = 20$ tm, $T = 0.2$.

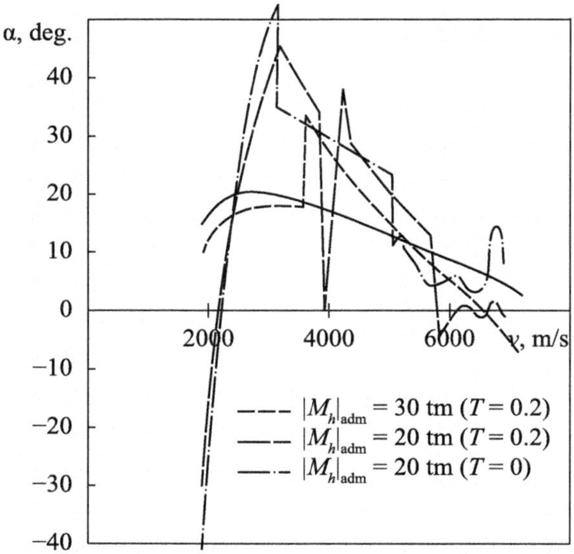

Fig. 6.10 Quasi-optimal LV angle-of-attack programs, taking into account the fail-safety requirements under three variants of the hinge torque constraint and thrust-to-weight ratio at RV emergency return

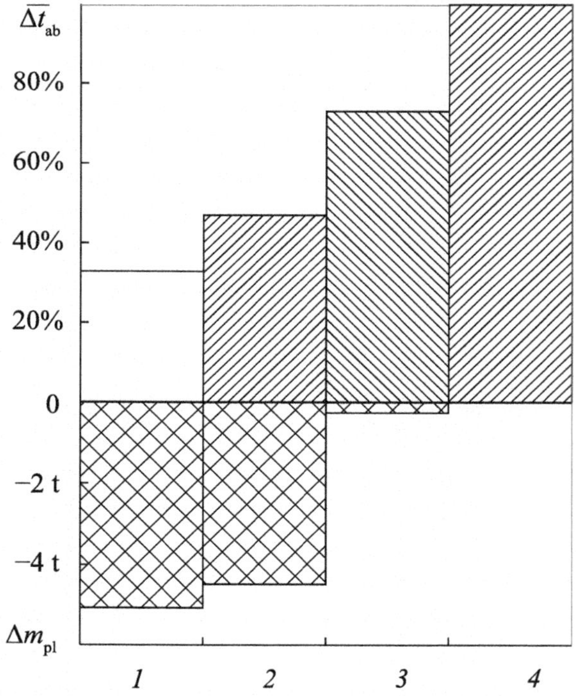

Fig. 6.11 Diagram of the relative duration of fail-safe ascent segment and corresponding payload mass loss for four variants of the RV hinge torque constraint and thrust-to-weight ratio;
1. $|M_h|_{adm} = 20$ tm, $T = 0$;
2. $|M_h|_{adm} = 20$ tm, $T = 0.2$;
3. $|M_h|_{adm} = 30$ tm, $T = 0.2$;
4. $|M_h|_{adm} = 40$ tm, $T = 0.2$

payload mass inserted into the circular orbit with $h_{orb} = 200$ km, $i_{orb} = 52.25°$, depending on constraints on the hinge torque with fixed constraints on dynamic loads (6.1). For comparison, the relative duration of the nominal (without fail-safety requirements) STS ascent trajectory segments, from which RV return is possible, are also given.

As follows from Fig. 6.11, it is possible to form such ascent trajectory, from any point of which the RV descent occurs without exceeding specified constraints (6.1), (6.2). In this case, the payload mass loss is ~ 16%. If the admissible hinge torque increases for emergency situations by 1.5 times compared with the regular case, the fulfillment of the fail-safe requirements practically does not lead to a decrease in the functional: the payload mass loss is less than 0.1%.

The optimal fail-safe solution for the rigorous statement

The optimal fail-safe ascent trajectory in the rigorous statement based on PMP (Sect. 5.4.2) was obtained for STS with a vertical take-off. It was assumed that the hypothetical RV has the characteristics of the Buran-type. Upper boundary Γ^- of the fail-safe area D in the osculating variables was approximated by a function of the following form (dashed lines in Fig. 6.12):

$$h_{max} = h_* + a_*(v - v_*)^2;\ h_*, a_*, v_* = \text{const}, \tag{6.5}$$

where the parameters have following nominal values (correspond to the lowest dashed line in Fig. 6.12):

$$a_* = 0.006\ \text{m/s}^2,\ h_* = 75000\ \text{m},\ v_* = 3000\ \text{m/s}. \tag{6.6}$$

The problem was solved for a spatial model of the STS motion (the dimension of the extended phase vector is seven). The corresponding TPBVP had the same dimension. The variable parameters consist of six components of the initial conjugate

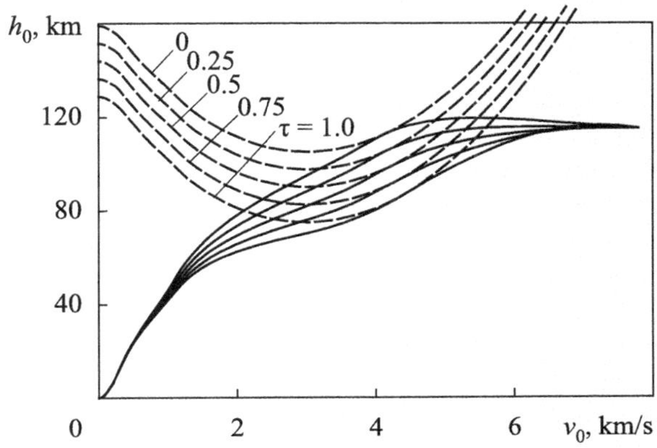

Fig. 6.12 Optimal fail-safe trajectories (solid lines) and upper boundaries of the fail-safe area (dashed lines) in osculating phase coordinates for various values of continuation parameter τ

vector (the seventh component, p_m, was determined from the equality of the Hamiltonian to zero) and the Lagrange multiplier in the condition of the jump of conjugate variables at point of reaching the fail-safety constraint (Sect. 5.3). The TPBVP solution was based on the continuation method with the nominal optimal solution without fail-safety constraints as an initial approximation. Continuation parameter τ (Sect. 1.10.2) was artificially introduced into the equation for the upper boundary of area D:

$$h_{\max}(v, \tau) = h_{\max}(v, 0)(1 - \tau) + h_{\max}(v, 1)\tau \qquad (6.7)$$

where τ varies from 0 to 1. At $\tau = 0$ the upper boundary of area D touches the reference ascent trajectory in osculating variables $h(v, 0)_{\max}$, obtained without taking into account the fail-safety conditions (Fig. 6.12). At $\tau = 1$ boundary $h_{\max}(v, 1)$ coincides with (6.5), (6.6). A value of parameter τ specifies the level of admissible dynamic and thermal loads for the corresponding RV trajectories.

Figure 6.12 shows the optimal STS ascent trajectories with fail-safety condition (6.7) for successive values of continuation parameter τ. It should be noted, as in the case of the implementation of the constraint on the dynamic pressure, the phase constraint for fail-safety are carried out in accordance with PMP by changing the attitude control program along the entire ascent trajectory without throttling the thrust. In this case, a smooth meeting the boundary is realized on the optimal trajectory (see Sects. 2.7 and 5.4). Optimal angle-of-attack program $\alpha(t)$ for various values of continuation parameter τ are shown in Fig. 6.13.

The change in the maximum injected mass, taking into account the fail-safety constraint, is shown in Fig. 6.14. It can be seen that the loss of the injected mass due to the fail-safety requirements is about 1.8%. The loss can be significantly reduced if the RV control limits or the admissible loads for emergency situations are extended.

Thus, the optimal consideration of the fail-safety conditions when forming the ascent trajectory makes it possible to significantly increase the STS reliability, while practically maintaining the payload mass.

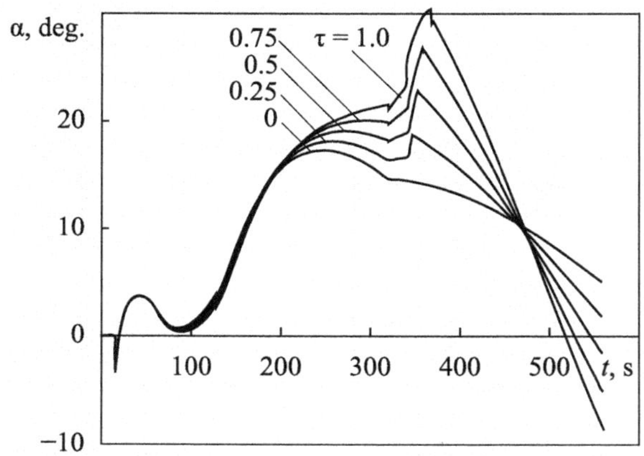

Fig. 6.13 Optimal time-programs of the STS angle-of-attack taking into account fail-safety constraints with various values of continuation parameter τ

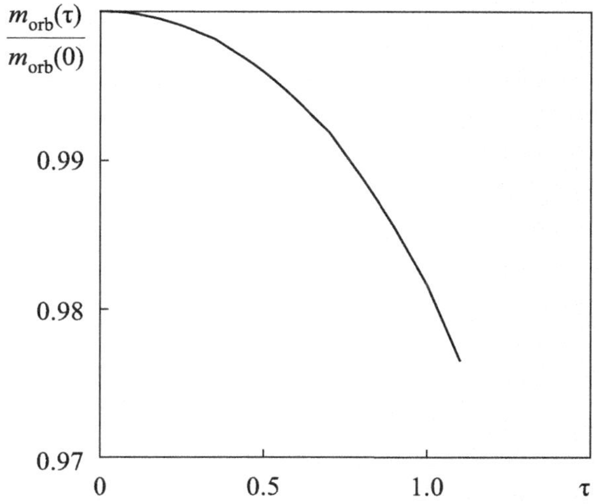

Fig. 6.14 Relative decrease in the maximum injected mass with tightening the fail-safety constraint by increasing continuation parameter τ

6.3 Through Trajectory Optimization of ASTS with An-225 Carrier Aircraft

In accordance with the approach described in Chaps. 2 and 4, the through optimization of branching trajectories of two ASTSs that consist of the subsonic An-225 carrier aircraft (CA), as the first stage, and a rocket stage (RS) was carried out. ASTS-1 corresponds to the Interim HOTOL project [9], and ASTS-2 close to the MAKS project [10].

The optimization was carried out in the rigorous statement based on PMP using *ASTER* software. The following results were defined: the optimal program for the angle of attack and bank angle, optimal engine cyclograms and trajectory at all flight segments (ASTS, CA and RS); optimal initial conditions (altitude, velocity) of the pre-launch maneuver of CA with RS from the set of possible modes of ASTS horizontal flight and the optimal RS launch azimuth.

6.3.1 Source Data

The launch trajectories of the ASTS are considered starting from the steady level flight mode of the CA with RS, and consist of the following successive sections:

- The dynamic pre-launch maneuver of CA with RS of a given duration;
- The short-term pre-release maneuver of CA with RS to reach a negative acceleration before breaking ties between RS and CA;

- Flight sections of RS and CA during the separation (from breaking the ties until RS reaches the angles of attack that are optimal for the successive autonomous flight);
- RS autonomous flight until the end of the main engine operation;
- The passive flight in an transfer orbit until touching a target circular orbit at the apogee;
- The RS velocity impulse (with the appropriate mass consumption) at apogee to reach the local circular velocity and, thus, stabilization in the target circular orbit.

The control of CA and RS was optimized with taking into account all the main constraints on phase and control variables:

- on the maximum and minimum admissible angles of attack of CA and RS;
- on the maximum admissible CA Mach number;
- on the maximum admissible RS longitudinal and lateral accelerations;
- on the maximum admissible RS dynamic pressure;
- on the pitch angular rate of the CA with RS

$$|\dot{\vartheta}| \leq \omega_{\text{adm}} \cong 10\,\text{deg}/\text{s}; \tag{6.8}$$

- on the increase rate of RS thrust when the engine is turned on:

$$\frac{1}{T_{\max}}\frac{dT}{dt} \leq 0.25\ \text{s}^{-1}; \tag{6.9}$$

- on the angle-of-attack rate before breaking ties between RS and CA;
- on the thrust moment balance.

The thermodynamic parameters used in the calculations correspond to the unperturbed standard atmosphere [11]. At the separation process, the motion of RS and CA was optimized taking into account constraints on their mutual position, excluding their collision and dangerous impact of high-temperature RS engine flame on CA. The hazardous influence area of RS engine jets on CA was limited by a cone, the lower (closer to CA) boundary of which makes an angle ψ with the RS longitudinal axis (see Sect. 4.6.2). It was assumed that RS is placed on CA so that at the start of the RS engines, the plume cone does not cross the CA surface. The RS relative location on the CA is schematically shown for the case of ASTS-1 in Fig. 6.15. The intermediate level of RS thrust T_* at separation, varied from $T_* = 0$ to $T_* = T_{\max}$, where $T_{\max}$ is the RS maximum thrust. It was assumed that the RS fuel is consumed during the RS main engine operation.

Calculations of optimal ASTS branching trajectories were carried out using *ASTER* software. The extended phase vectors had a variable dimension. On the segments of the joint flight of CA with RS, it was equal to four (for 2D motion model) or six (for 3D model). When simulating the separation of CA and RS, the

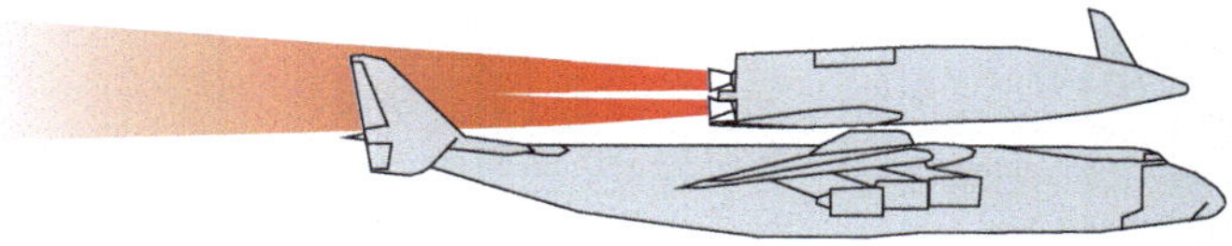

Fig. 6.15 Scheme of RS placement on CA in ASTS-1

dimension of the phase vector was nine, including five for RS and four for CA. In the area of autonomous RS injection into the orbit, the phase vector had a dimension of seven. Thus, the total dimension of the phase space was twenty or twenty-two. At the same time, the total dimension of the TPBVP was eleven, of which seven varied parameters and residuals formed an outer TPBVP (at the left and right ends of the injection trajectory) and four variable parameters and residuals formed an inner TPBVP (at the ends of the separation section). In the case when the constraint on the Mach number is essential for CA, the dimension of the BVP increases to twelve. The initial values of the conjugate variables were taken as the variable parameters of the outer TPBVP (due to time constraints, the Hamiltonian at the initial point is not defined). In the inner TPBVP, the variable parameters were the conjugate variables at the beginning of the only side branch corresponding to the separation segment. In the case of optimizing the azimuth of the RS launch direction in the outer TPBVP, initial azimuth was taken as varied parameter instead of initial conjugate variable p_{v_z}. The residuals were the conditions at the right ends of the main (for RS) and side (for CA) branches.

The following source data were used in the calculations.

ASTS-1 data

Carrier aircraft

Nominal initial altitude $h_i = 10{,}000$ m, velocity $v_i = 227$ m/s and flight path angle $\theta_i = 0$ correspond to the conditions of the horizontal flight of CA with RS (Fig. 6.16). Initial CA mass $m_i^{\mathrm{CA}} = 303{,}000$ kg, the propulsion system consists of eight engines D18T. Dependences of aerodynamic coefficients C_D and C_L on the Mach number are shown in Figs. 6.17 and 6.18 for reference area $F_{\mathrm{ref}} = 905$ m^2. Admissible Mach number is $M_{\mathrm{adm}} = 0.8$; admissible normal aerodynamic load factor range is $-1 \leq n_z \leq 1.7$; admissible pitch rate is $\omega_{\mathrm{adm}} = 10$ deg/s. The constraint on the angle of attack is given in Table 6.2:

Rocket stage

The initial RS mass is $m_i = 250{,}000$ kg. The RS propulsion system consists of four liquid-propellant rocket engines with the thrust in vacuum $T_\infty = 91{,}138$ kgf, specific impulses in vacuum/at sea level $J_\infty / J_0 = 463.8/298.6$ s. The maximum admissible longitudinal load factor $n_{x\,\max} = 3.5$, the admissible angle-of-attack range is $-12.5° \leq \alpha \leq 25°$. Dependences of the aerodynamic coefficients C_L, C_D on the Mach number are shown in Figs. 6.19 and 6.20 for reference area $F_{\mathrm{ref}} = 308$ m^2.

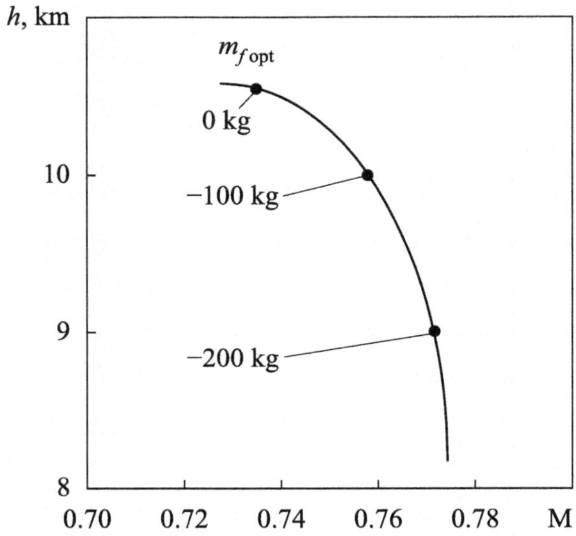

Fig. 6.16 The upper boundary of the area of possible ASTS-1 horizontal flight just before RS launch; and the maximum injected mass losses due to deviation of the launch point relative the optimal one

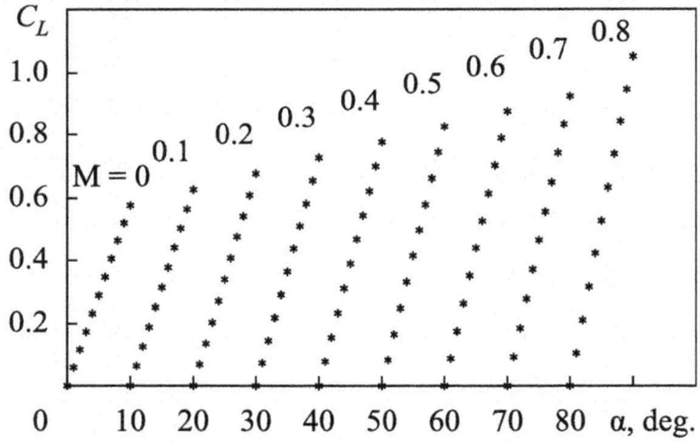

Fig. 6.17 Dependences of the CA aerodynamic lift coefficient on the angle of attack for various Mach numbers (each curve is shifted by 10° along the angle-of-attack axis)

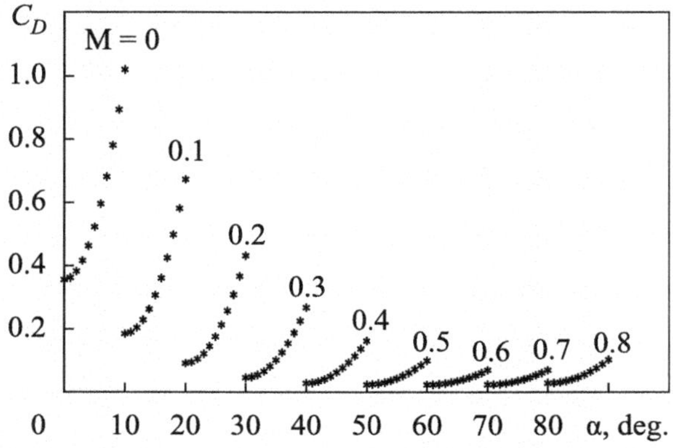

Fig. 6.18 Dependences of the CA aerodynamic drag coefficient on the angle of attack for various Mach numbers (each curve is shifted by 10° along the angle of attack)

Table 6.2 The CA maximum admissible angle of attack depending on the Mach number

M	0.6	0.7	0.75	0.83
$\alpha_{\max}$, deg.	13.7	13.6	11.6	8.5

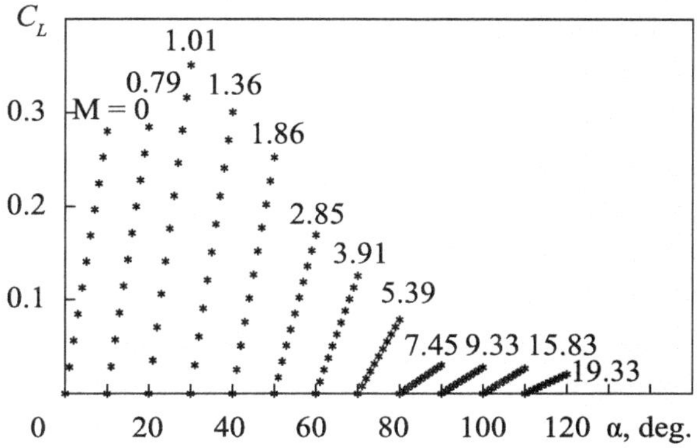

Fig. 6.19 Dependences of the RS aerodynamic lift coefficient on the angle of attack for various Mach numbers (each curve is shifted by 10° along the angle of attack)

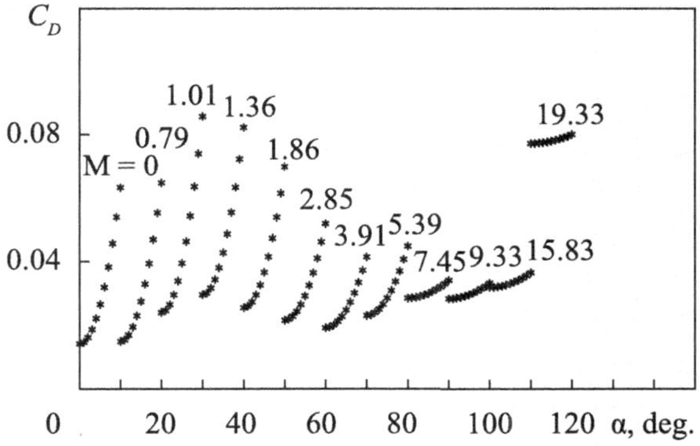

Fig. 6.20 Dependences of the RS aerodynamic drag coefficient on the angle of attack for various Mach numbers (each curve is shifted by 10° along the angle of attack)

ASTS-2 Data

Rocket stage

The initial mass is $m_i = 250{,}000$ kg. Combined RS engine parameters have the following values: the thrust in mode 1 is $T_{\infty}^{1} = 2 \times 204$ t, the specific impulse in mode 1 is $J_{\infty}^{I}/J_{0}^{I} = 415.4/329.7$s; the thrust in mode 2 is $T_{\infty}^{2} = 2 \times 81$ t, the specific impulse in mode 2 is $J_{\infty}^{2}/J_{0}^{2} = 461.17/219.04$ s. The maximum admissible longitudinal load factor, aerodynamic lift, angle of attack and dynamic pressure are $n_{x\max} = 3.5$, $L_{\max} = 145$ t, $\alpha_{\max} = 25°$ and $q_{\text{adm}} = 3200\ kgf \cdot m^{-2}$ consequently. Aerodynamic coefficients in the body axes are given in Tables 6.3 and 6.4 with nodes by Mach numbers and angles of attack for reference area $F_{\text{ref}} = 31.97$ m^2.

Table 6.3 RS aerodynamic drag coefficient in the body axis system depending on the Mach number

M	0	0.6	0.8	0.9	0.95	1	1.1	1.2	1.5	1.7	2	3	4	27
C_D	0.217	0.217	0.232	0.296	0.400	0.481	0.608	0.631	0.580	0.549	0.515	0.436	0.393	0.393

Table 6.4 RS aerodynamic lift coefficient as a function of angle of attack and Mach number

M / α, deg	0	0.6	0.8	0.9	0.95	1	1.1	1.2	1.5	1.7	2	3	4	27
−15	−1.06	−1.06	−1.08	−1.12	−1.14	−1.15	−1.18	−1.23	−1.30	−1.24	−1.16	−0.97	−0.8	−0.8
0	0	0	0	0	−0.02	−0.03	−0.07	−0.06	−0.05	−0.04	−0.04	−0.04	−0.03	−0.03
5	0.41	0.41	0.41	0.41	0.41	0.42	0.42	0.46	0.45	0.44	0.42	0.34	0.26	0.26
10	0.82	0.82	0.84	0.88	0.87	0.86	0.92	0.93	0.92	0.9	0.87	0.7	0.56	0.56
15	1.3	1.3	1.32	1.34	1.31	1.27	1.31	1.28	1.31	1.33	1.30	1.04	0.89	0.89

Carrier aircraft

The initial mass is $m_i = 303{,}000$ kg, the reference area is $F_{\text{ref}} = 905\ \text{m}^2$, the maximum admissible Mach number is $M_{\text{adm}} = 0.75$, the admissible normal load factor range is $-1 \le n_z \le 1.7$. The constraint on the angle of attack is given in Table 6.2.

The dependence of the thrust on velocity v and altitude h was given by the following approximate equation:

$$T = T_0 + T_h(\rho(h) - \rho_*) + (T_{vh}(\rho(h) - \rho_*) + T_v)(v - v_*),\ T_0, T_h, T_{vh}, T_v = \text{const.}$$

The comparison of this approximate equation with the source data is shown in Fig. 6.21.

Carrier aircraft with rocket stage

The initial mass is $m_i = 553{,}000$ kg, the reference area is $F_{\text{ref}} = 905\ \text{m}^2$, the admissible pitch rate corresponds to (6.6). Dependence $C_D(C_L, M)$ was set in the following approximate form:

$$C_D = C_{D0}(M) + A(C_L, M)(C_L - C_{L*}(M))^2,$$

where $C_{D0}(M), C_{L*}(M), A(C_L, M)$ are determined by spline interpolating the source data at the nodal points (Tables 6.5 and 6.6).

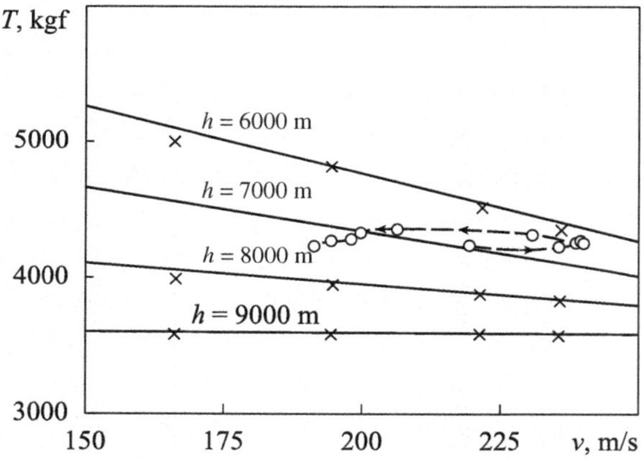

Fig. 6.21 The approximate dependence of the CA thrust on the velocity for four altitudes (solid lines) in comparison with the source data (crosses), and an example of the thrust change on the CA nominal optimal trajectory (dashed line)

Table 6.5 Dependences of aerodynamic coefficients for CA with RS on Mach number

M	0.4	0.5	0.6	0.7	0.75	0.8
C_{D0}	0.0226	0.0250	0.0268	0.03	0.0327	0.0342
C_{L*}	0	0	0.025	0.025	0.1	0.1

Table 6.6 Dependence of aerodynamic drag parameter $A(C_L, M)$ for CA with RS

M C_L	0.4	0.5	0.6	0.7	0.75	0.8
0.4	0.0413	0.0394	0.0468	0.0482	0.0689	0.08
0.6	0.0497	0.0486	0.0535	0.0565	0.0816	0.0924
0.8	0.0550	0.0558	0.0677	0.0712	0.1067	0.1067
0.9	0.0612	0.0711	0.0890	0.1067	0.1067	0.1067

6.3.2 *Influence of CA Steady Flight Modes at Beginning Pre-launch Maneuver*

To study the influence of the initial conditions of the maneuver of the CA with RS on the ASTS efficiency, an essential segment of the boundary of the area of possible steady flight modes was approximated by an ellipse (4.66). With the help of the technique described in the Sect. 4.7.2, the optimal steady flight mode of CA with RS, i.e. the optimal initial conditions for the pre-launch maneuver of CA with RS, was obtained.

Figures 6.22 and 6.23 show the change of injected mass m_{orb} for ASTS-2, as well as the main parameters of the optimal through trajectory at the release moment. It also shows the effect of variations in the initial point of the maneuver of the CA with RS on the injected mass and aerodynamic loads. The optimal point of the beginning of the pre-launch maneuver of the CA with RS corresponds to the conditions of the steady level flight of the CA with RS at altitude $h_{0\ \mathrm{opt}} = 6.89$ km and Mach number $M_{0\ \mathrm{opt}} = 0.7$. The optimal time-programs of the RS thrust and angles of attack of the CA with RS, CA and RS at all branches are shown in Fig. 6.24. The change in time of phase variables $\{v, h, \theta\}$, as well as aerodynamic lift L, load factor n_z, and dynamic pressure q are shown in Figs. 6.25 and 6.26.

The CA with RS, especially at high altitudes in a vicinity of the upper boundary of the area D of possible steady-state flight, has very limited (compared with RS) maneuvering capabilities. Therefore, for a noticeable change in the trajectory before RS launch, a dynamic maneuver of a long duration is required. In particular, in Chap. 2, the existence of extremals of the oscillatory structure was pointed out, realizing large altitudes and flight path angles due to "swinging" in dense atmosphere layers. However, for a number of performance requirements, as well as for reasons of launch accuracy, the maneuver time of the CA before the RS separation should be limited. The influence of duration τ of CA with RS joint maneuver on maximum injected mass m_{orb} is shown in Fig. 6.27. It can be seen that the reduction in the joint flight duration τ less than $\tau^* \approx 12$ s leads to a sharp decrease in the ASTS efficiency. The optimal change of phase variables (the altitude, velocity and flight path angle) at the time of the RS launch with variations of τ is shown in Fig. 6.28. From Fig. 6.28 and taking into account that the sensitivity of the functional to variation of the flight path angle is several orders of magnitude higher than to variations in other trajectory variables at the given time (Figs. 6.29, 6.30, 6.31 and 6.32), we can conclude that

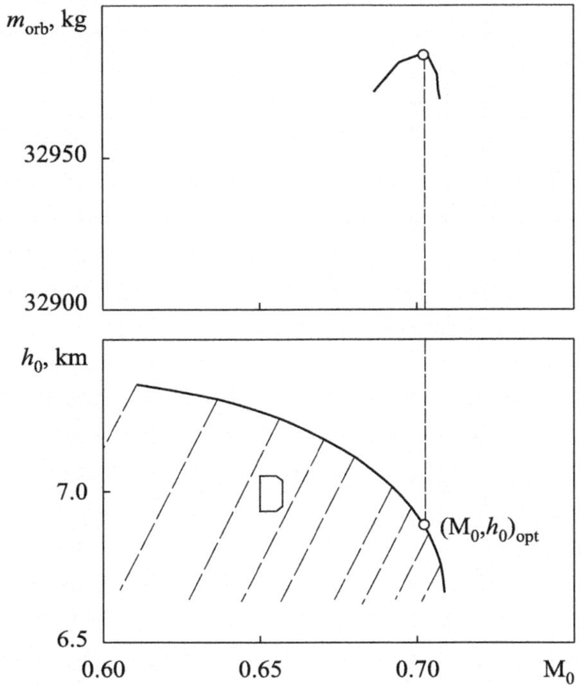

Fig. 6.22 The change in the maximum injected mass (upper graph) depending on the steady-state level flight mode of ASTS-2 before the start of the pre-launch maneuver at the boundary of admissible area *D* (lower graph)

such a drop in efficiency is due to a small flight path angle by the moment of the RS separation.

The manner of RS launch without the pre-launch maneuver of CA has been studied. At the manner, a short-term (~ 1–2 s) maneuver was maintained, which ensured the transfer of the CA with RS from the angles of attack corresponding to horizontal flight to the angles of attack corresponding to load factor $n_z = 0.6$, at which, according to the separation safety conditions, the ties between CA and RS are broken. When RS is started from stationary conditions, the CA control algorithm is significantly simplified: the accuracy of maintaining the nominal conditions for separation of CA and RS is increased; an emergency situation forced by non-release of RS from CA after the pre-launch maneuver ceases to be critical for ASTS safety. However, this manner of RS launch is accompanied by an increase in the RS maximum aerodynamic lift (by 31.5 tons), the dynamic pressure (by 360 kgf/m) and a slight decrease in the injected mass (by 70 kg). The optimal ASTS-1 injection trajectories with and without an optimal pre-launch maneuver are shown in Fig. 6.33.

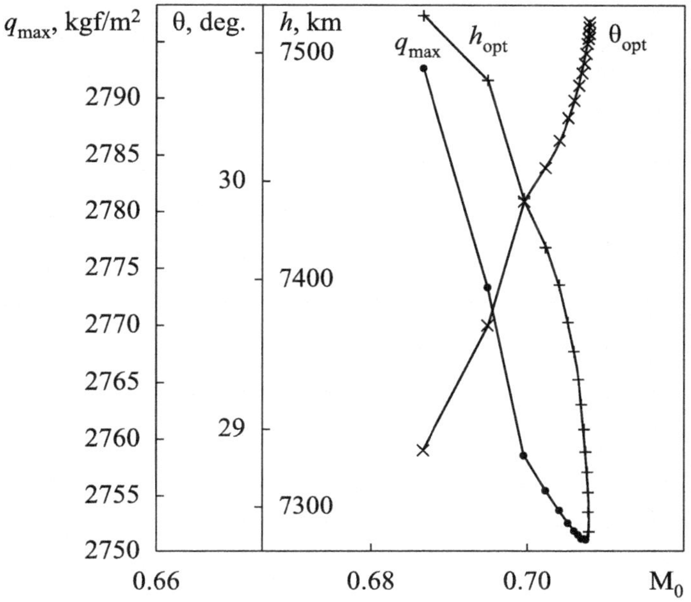

Fig. 6.23 The optimal altitude and flight path angle at the moment of breaking ties of CA with RS, as well as the maximum dynamic pressure on the optimal RS injection trajectory, depending on the initial Mach number that determines the ASTS-2 steady-state level flight mode at the boundary of admissible area *D*

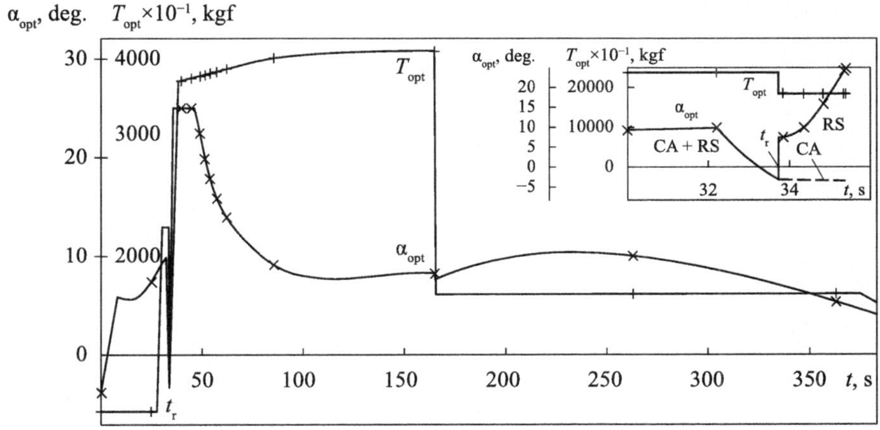

Fig. 6.24 Optimal time-programs of the RS thrust and angles of attack at all ASTS-2 trajectory branches

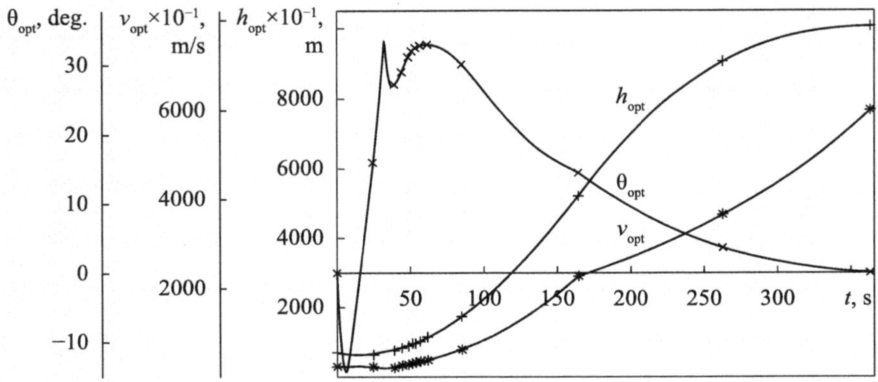

Fig. 6.25 Time-programs of phase variables on the optimal ASTS-2 trajectory

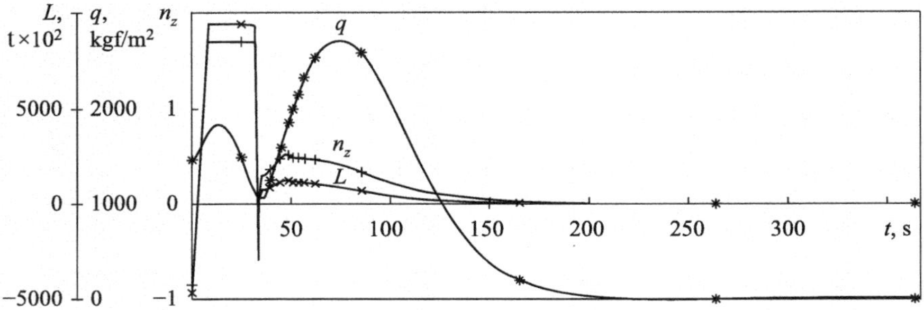

Fig. 6.26 Time-programs of the lift, dynamic pressure and normal load factor on the optimal ASTS-2 trajectory

Fig. 6.27 The influence of the duration of the pre-separation maneuver of CA with RS on the maximum injected mass increment

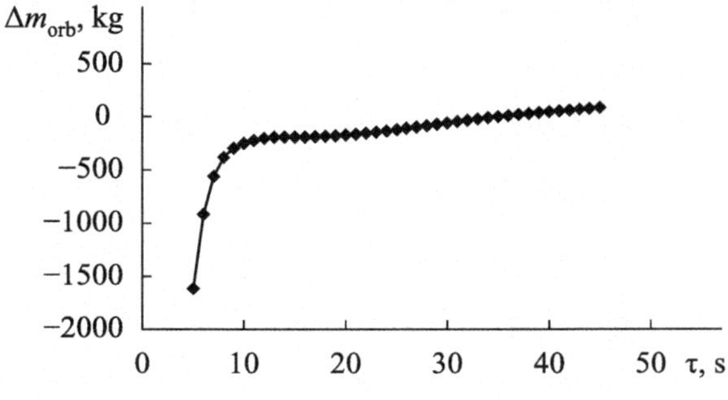

6.3.3 *Integrated Accounting for the Separation Process*

The study of the influence of the separation process on the optimal injection trajectories taking into account only the total change in the phase vector during the separation

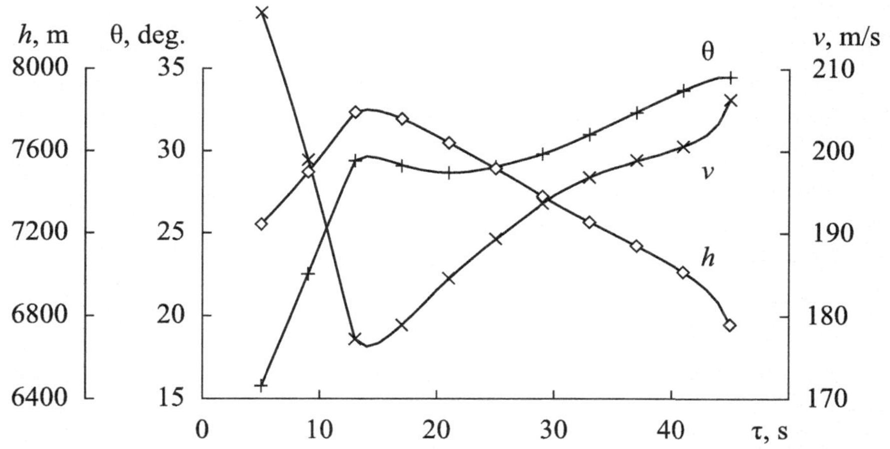

Fig. 6.28 The influence of the duration of the pre-separation maneuver of CA with RS on the optimal phase variables at the moment of breaking ties

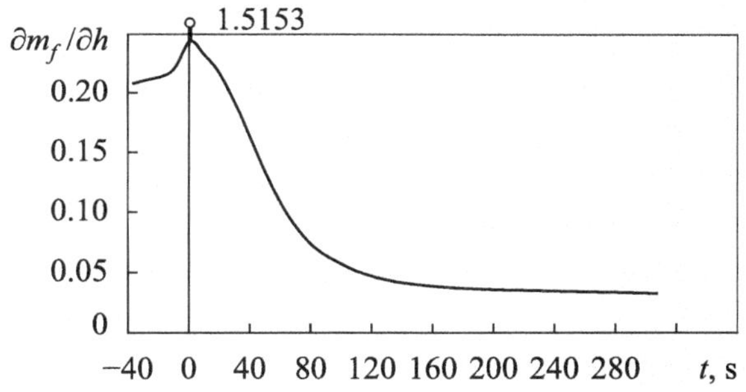

Fig. 6.29 Time-dependence of the influence function of the altitude on the maximum injected mass at the through optimal ASTS-1 injection trajectory

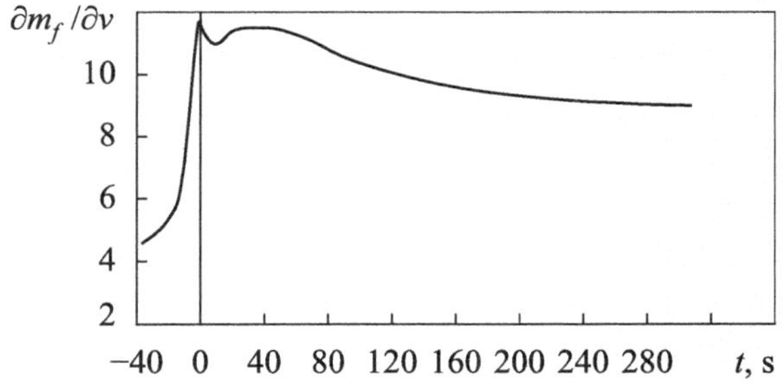

Fig. 6.30 Time-dependence of the influence function of the velocity on the maximum injected mass at the through optimal ASTS-1 injection trajectory

(see Sect. 4.6.1) was carried out for ASTS-1. In this case, the duration of the separation segment was assumed to be negligible, which means a jump in the phase vector at the release moment.

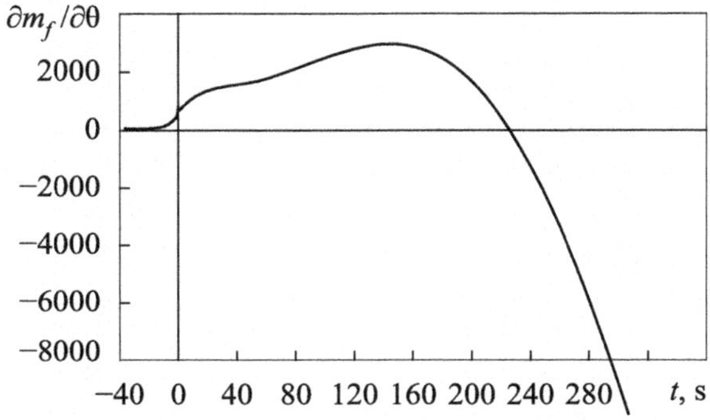

Fig. 6.31 Time-dependence of the influence function of the flight path angle on the maximum injected mass at the through optimal ASTS-1 injection trajectory

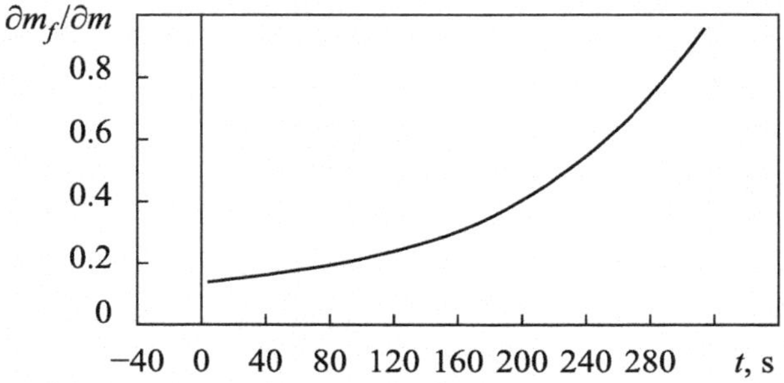

Fig. 6.32 Time-dependence of the influence function of the current mass on the maximum injected mass at the through optimal ASTS-1 injection trajectory

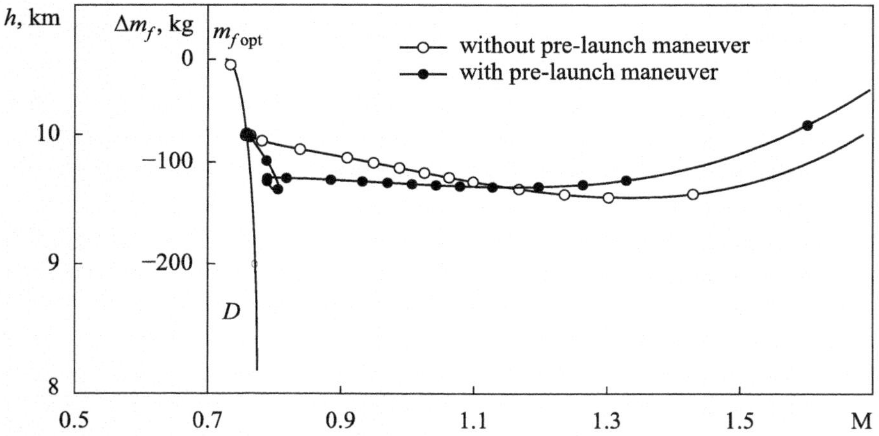

Fig. 6.33 Optimal ASTS-1 injection trajectories with and without optimal pre-launch maneuver

The change in the horizontal Δv_x and vertical Δv_y components of the velocity during the separation of CA and RS varied in the ranges:

$$-30 \text{ m/s} \le \Delta v_x \le 50 \text{ m/s}, \ -110 \text{ m/s} \le \Delta v_y \le 0. \tag{6.10}$$

These ranges were chosen taking into account the estimates [12] of increments of velocity Δv and flight path angle $\Delta\theta$ during the non-impact safe separation of RS and CA without using (manner 1) and using 50% the maximum RS thrust (manner 2). These ranges correspond to the following changes in phase variables during separation:

$$\text{Manner 1}: \begin{array}{l} \Delta v^1 = -26 \text{ to} - 28 \text{ m/s}, \ \Delta\theta^1 = -19^\circ \text{ to} - 31^\circ, \\ \left(\Delta v_x^1 = 15 \text{ to } 0 \text{ m/s}, \qquad \Delta v_y^1 = -105 \text{ to} - 70 \text{ m/s}\right), \end{array} \tag{6.11}$$

$$\text{Manner 2}: \begin{array}{l} \Delta v^2 = 15 \text{ to } 20 \text{ m/s}, \qquad \Delta\theta^2 = -11^\circ \text{ to} - 14^\circ, \\ \left(\Delta v_x^2 = 25 \text{ to } 35 \text{ m/s}, \ \Delta v_y^2 = -30 \text{ to} - 50 \text{ m/s}\right). \end{array} \tag{6.12}$$

The change in the RS injected mass m_f for various combinations of Δv_x, Δv_y and the corresponding Δv, $\Delta\theta$ is shown in Fig. 6.34. The loss of the injected mass Δm_f due to the loss of the velocity and flight path angle during separation of RS and CA without the use of the RS thrust (manner 1) does not exceed

$$\left|\Delta m_f^1\right| \le 500 \text{ kg}. \tag{6.13}$$

In calculations, it was assumed that increments Δv_x, Δv_y at the separation segment are not associated with a change in the RS initial mass. Therefore, as it can be seen in Fig. 6.34, if $\Delta v > 0$ (manner 2), the injected mass m_f increases. As the velocity increase at separation is possible only through the use of a thrust, then inserted mass correction Δm_f^2 has to be made:

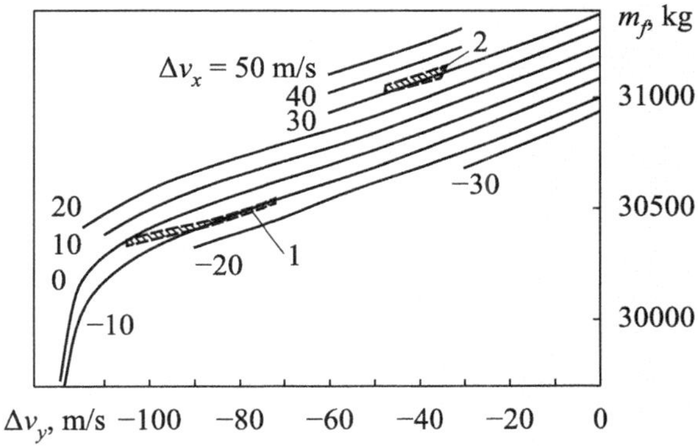

Fig. 6.34 The maximum injected mass for various combinations of losses of horizontal and vertical components of the velocity vector in the process of separation of CA and RS without (manner 1) and with (manner 2) the use of RS thrust

$$\Delta m_f^2 = -\frac{\partial m_f}{\partial m_i}\Delta m_i,$$

where Δm_i is the RS fuel mass consumption in the separation segment. Due to the physical meaning of the conjugate variable p_m:

$$\frac{\partial m_f}{\partial m_i} = p_m(t_i).$$

The separation of RS with thrust *T*, specific impulse *J* and duration $\Delta t \approx 7$–8 s [12] is accompanied by fuel mass consumption $\Delta m_i \cong \frac{T}{Jg}\Delta t$. Taking into account the above expressions, we obtain an estimate of the injected mass loss due to the fuel consumption in the separation process:

$$\Delta m_f^2 \cong -260 \text{ to } -320 \text{ kg}. \tag{6.14}$$

Thus, the use of the RS propulsion during the separation allows getting a functional gain.

Losses of velocity (Δv_x and Δv_y) in the separation segment reduce injected mass m_f. If $|\Delta v_y|<100$ m/s than dependence $m_f(\Delta v_y)$ is close to linear with proportionality coefficient

$$\frac{\Delta m_f}{\Delta v_y} \cong 7.5\frac{\text{kg}}{m/\text{s}}.$$

For greater vertical velocity losses ($|\Delta v_y| > 115$ m/s) the intensity of the decrease in the injected mass raises more than an order of magnitude (Fig. 6.34):

$$\frac{\Delta m_f}{\Delta v_y} \cong 90\frac{\text{kg}}{\text{m/s}}.$$

Thus, the vertical velocity during separation should not decrease by more than $|\Delta v_y| \approx 100$ m/s to avoid significant losses of the injected mass.

The level lines $\Delta v = \text{const}$ and $\Delta\theta = \text{const}$ are plotted in Fig. 6.35 by dashed lines. It can be seen that the changes of $(v, h, \theta)_f$ are determined mainly by the loss of flight path angle $\Delta\theta$ and weakly depend on the loss of velocity Δv. At the same time, the optimal parameters of the RS autonomous flight branch, such as the RS optimal initial angle of attack α_i, and the maximum dynamic pressure $q_{\max}$, significantly depend on the losses Δv_x, Δv_y: at $-100 \text{ m/s} \le \Delta v_y \le 0$ values of α_i and $q_{\max}$ change in ranges

$$13° \le \alpha_i \le \alpha_{\max} = 25°, 2500 \text{ kgf} \cdot \text{m}^{-2} \le q_{\max} \le 4000 \text{ kgf} \cdot \text{m}^{-2}.$$

The difference in the influence of velocity losses during the separation of the RS and CA on the optimal CA pre-launch maneuver and RS autonomous injection is explained by significantly greater maneuverability of RS compared to CA.

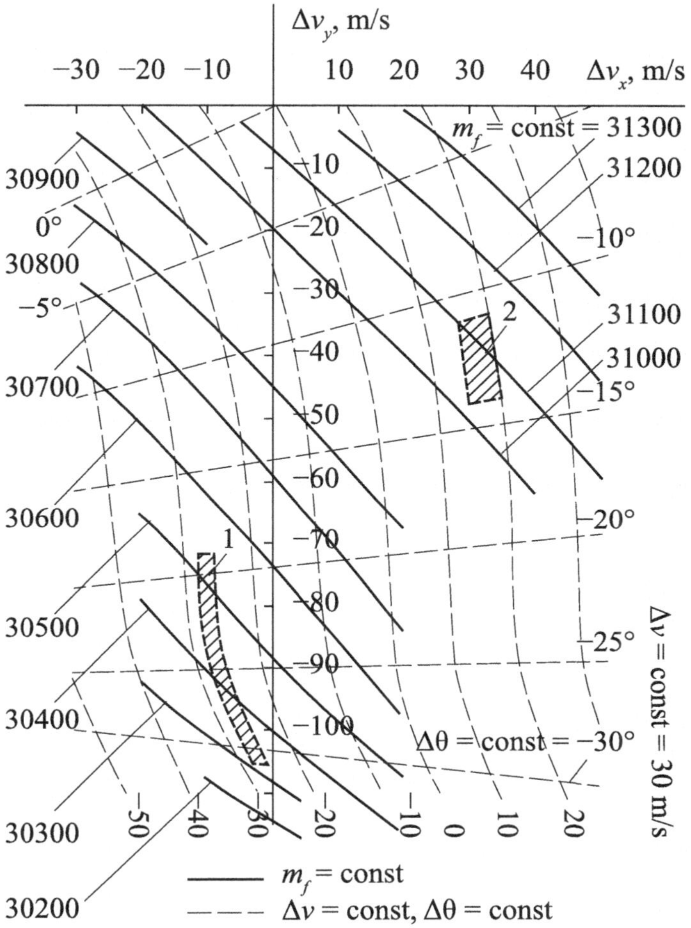

Fig. 6.35 Maximum injected mass contour lines in coordinates of losses of velocity and flight path angle in the separation process of CA and RS without (manner 1) and with (manner 2) the use of RS thrust

6.3.4 Through Optimization of ASTS Trajectories Including Staging Process

The study of the effect of separation of CA and RS on the injected mass, optimal control and injection trajectory was carried out in accordance with the developed methodology for through optimization of branching trajectories, described in Sect. 4.6.2.

The time-dependencies of phase variables on the main branch of the ASTS-1 optimal trajectory are presented in Figs. 6.36, 6.37, 6.38 and 6.39. Figure 6.40 shows the optimal time-programs of angle of attack $\alpha(t)$ on all branches of the trajectory, including the separation of RS and CA. Time $t = 0$ corresponds to the moment of ties breaking. Conditions for the start of the pre-launch maneuver were not set in advance, but were optimized (according to the criterion of the maximum injected mass) from the area of possible steady-level flight. The duration of the joint maneuver of CA and RS was set $\tau = 35$ s. The RS engines were fired on 2 s before the beginning the pre-release maneuver, which is characterized by an intense pitch change, taking into account the constraint on the admissible angular rate.

The use of PMP allows obtaining influence functions of current phase variables on the injected mass m_{orb} without additional calculations of optimal trajectories and loss of accuracy in a numerical differentiation. They are recalculated by the known solution of the conjugate system and are shown in Figs. 6.29, 6.30, 6.31 and 6.32. We note a significant increase in the influence of v, h, θ (especially h) on m_{orb} in the separation section, which is associated with constraints on the mutual position

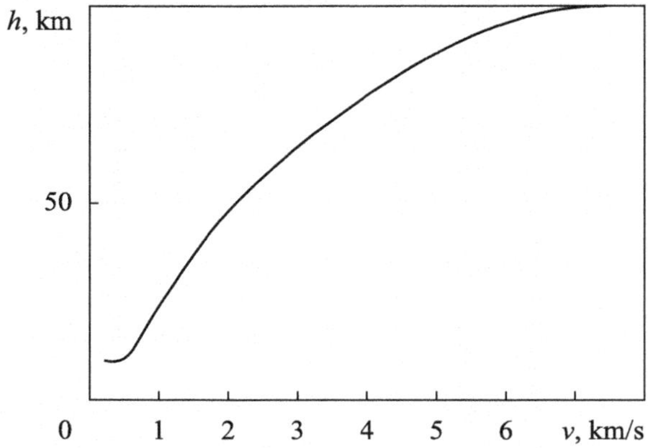

Fig. 6.36 The ASTS-1 optimal ascent trajectory

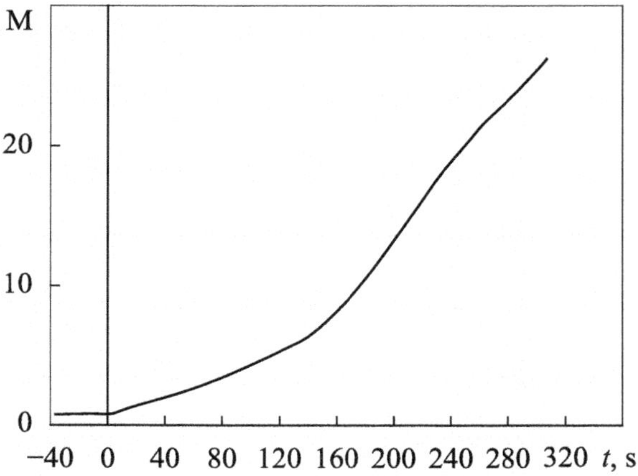

Fig. 6.37 Time-dependence of Mach number on the ASTS-1 optimal ascent trajectory

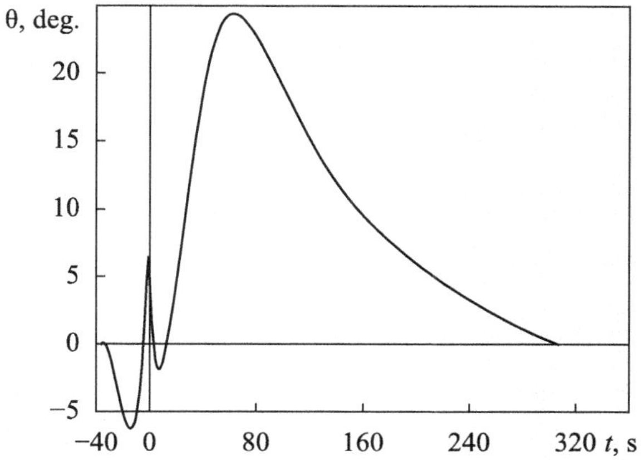

Fig. 6.38 Time-dependence of the flight path angle on the ASTS-1 optimal ascent trajectory

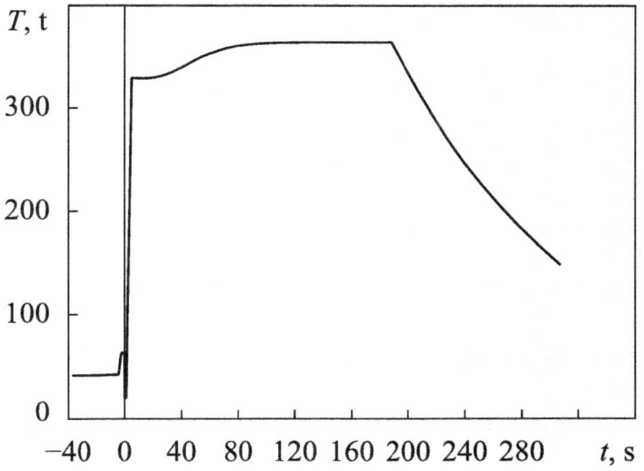

Fig. 6.39 Time-dependence of the thrust on the ASTS-1 optimal ascent trajectory

of CA and RS in the separation process, due to the safety requirements. Despite the relatively short duration of the separation section, its inclusion in the overall process of through optimization turns out to be very important for the formation of the ASTS optimal launch trajectory.

Below are the results of studies of the influence on the control, trajectory and functional of factors associated with the ASTS staging, such as the start time of RS main engines, the angle of deviation of the thrust vector from the longitudinal axis, inertia of actuators of the attitude control system.

Optimization of RS engine ignition moment

One of the important factors affecting the safety and efficiency of ASTS launch is the timely activation of RS engines. Cases of using RS engine before and after separation of RS and CA are considered. It was assumed that only one of the four engines is turned on at 50% of the nominal thrust if it occurs before RS separation.

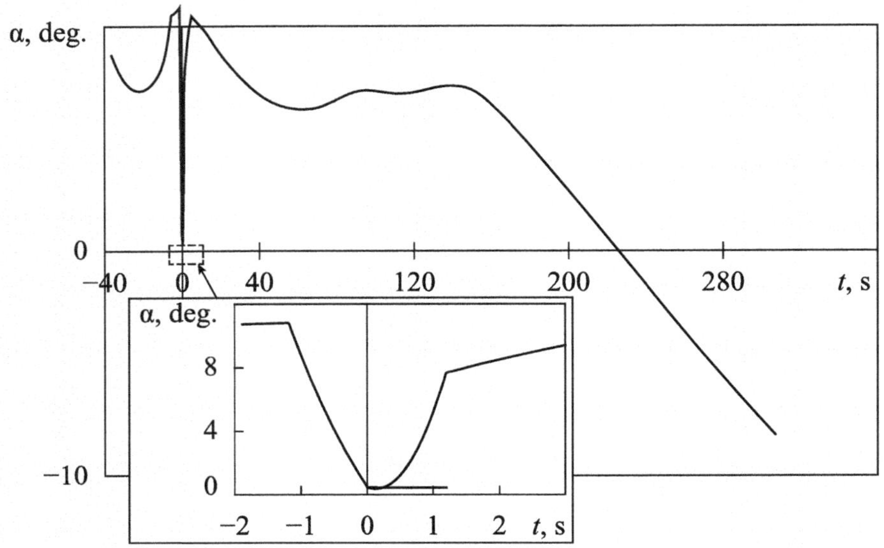

Fig. 6.40 Time-dependence of the angle of attack on all branches of the ASTS-1 optimal ascent trajectory

The remaining engines are fired after the separation. When RS engines are turned on after the separation, all of them are fired simultaneously with constraint on the thrust rise rate. The change in the injected mass with a variation in the engine ignition moment is shown in Fig. 6.41. The use of RS engines before breaking the RS ties with CA leads to a loss of the injected mass. Therefore, an early of RS engine ignition can make sense only in order to check its performance for increasing mission safety. The "cold" separation of RS allows slightly increasing the injected mass and simplifying separation safety. Figure. 6.41 shows also the dependence of the maximum dynamic pressure, which is realized on the ASTS optimal launch trajectory, on the moment of the RS engine ignition. The engine ignition at the end of the separation process is optimal in terms of the injected mass and, at the same time, takes into account some additional safety requirements.

Influence of RS thrust vector deflection angle

Longitudinal balancing of RS (compensation of the thrust longitudinal moment) is achieved by deflecting the thrust vector relative to the longitudinal axis by angle φ_E to ensure the direction of the RS thrust to the current center of mass. In this case, the angle ψ between the bottom (the nearest to CA) boundary of the engine plume also changes (Fig. 4.13): $\psi = \psi|_{\varphi_E=0} + \varphi_E$. In Sect. 2.8.2, relations are given that make it possible to take into account the balancing condition when optimizing launch trajectories.

The dependence of ASTS-2 injected mass on the deflection angle φ_E is shown in Fig. 6.42. It also indicates the time of RS motion with the maximum admissible

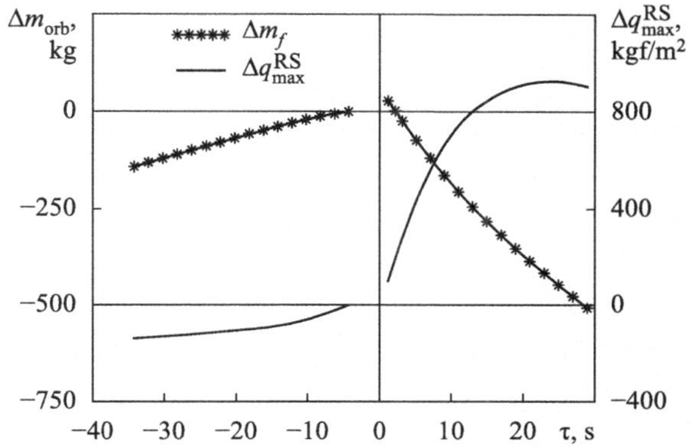

Fig. 6.41 Increments of the maximum injected mass and duration of flight at the maximum admissible dynamic pressure vs the RS engine ignition time relative to the moment of breaking ties between RS and CA

dynamic pressure. It can be seen that the presence of a non-zero deflection angle has a noticeable effect on the injected mass, and an increase in φ_E (in absolute value) leads to increase in the dynamic pressure level and the duration of motion along boundary $q = q_{adm.}$ This is explained by following reasons:

- The RS admissible angle of attack is reduced by a value of φ_E in the sections of the RS autonomous flight and separation, which leads to an increase in the duration of motion under constraints, and, consequently, to a decrease in the injected mass;
- While maintaining the optimal thrust vector control program, the presence of a non-zero deflection angle leads to a decrease in the effective (from the aerodynamics point of view) angle of attack and, consequently, the aerodynamic lift, which also causes a decrease in the injected mass.

Influence of ASTS actuator inertia and a delay in RS turn after release

In the calculations considered in previous sections, the simulation of separation process on the RS launch trajectories was carried out ignoring actuator dynamics. But in the phase of the pre-separation maneuver, which requires an intensive change in pitch angle ϑ with the maximum admissible angular rate, the actuator dynamics

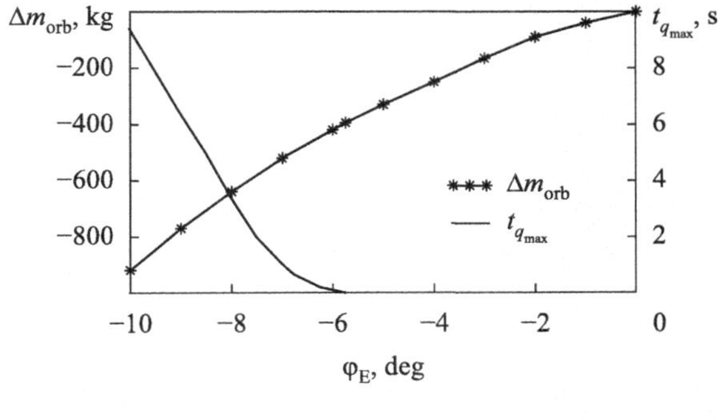

Fig. 6.42 Increments of the ASTS-2 maximum injected mass and duration of flight at the maximum admissible dynamic pressure vs the RS thrust vector deflection angle

can make a correction to the ASTS motion scheme, since the breaking ties between RS and CA is carried out at zero pitch rate for separation safety [12].

When calculating the optimal through launch trajectories, the influence of the actuator inertia was modeled by introducing a motion segment with a constant load factor before breaking ties (Fig. 6.43). In this case, the ASTS motion on the prelaunch maneuver is modeled according to the following scheme:

1. In interval $[0,\ \tau)$, there is an optimal motion of CA with RS; $\tau = fix$, $t_1^0 = \tau$ in (4.39);
2. In interval $[\tau,\ t_{\omega\, i})$, the optimal motion of CA with RS continues; $t_1 = t_{\omega\, i}$ in (4.39);
3. In interval $[t_{\omega i}, t_{\omega f})$, a turn occurs with the maximum pitch angular rate: $\vartheta = \vartheta_1^0 + \omega\left(t - t_1^0\right)$; the end time $t_{\omega f}$ is determined by the moment when the specified normal load factor n'_{z2} is reached: $Q_{20} = n'_z\left(\mathbf{z}'\right) - n'_{z2} = 0$ (for comparison, see (4.40));
4. In interval $[t_{\omega f}, t_2]$, the flight occurs with constant negative load factor $n_{z\Sigma 2}$.

Moments $t_{\omega\, i}$, $t_{\omega\, f}$ are determined from conditions:

$$
\begin{aligned}
t_{\omega i} &: \vartheta\left(t_{\omega i}\right) = \vartheta_1^o + \omega\left(t_{\omega i} - t_1^o\right),\\
t_{\omega f} &: n_z\left(\mathbf{x}_\Sigma, t_{\omega f}\right) = n_{z\Sigma 2}.
\end{aligned}
\tag{6.15}
$$

According to (6.15), the duration of segment 2 (the separation segment) is defined by given pitch angle ϑ_1^0, and this increases actually duration τ of the joint flight of RS with CA. The duration of segment 3 is almost constant. The duration of the pre-release maneuver is mainly determined by the duration of segment 4: $\Delta t_d = t_2 - t_{\omega\, f}$. The dependence of ASTS-1 injected mass increment on Δt_d is shown in Fig. 6.44. It can be seen that when the pre-release maneuver is tightened, the injected mass loss increases with the intensity of $\frac{\Delta m_{\text{orb}}}{\Delta t_d} \approx -300$ to -400 kg $\cdot$ s^{-1}. Thus, the decrease in the actuator inertia at the pre-release maneuver is one of the noticeable ways for increasing ASTS efficiency.

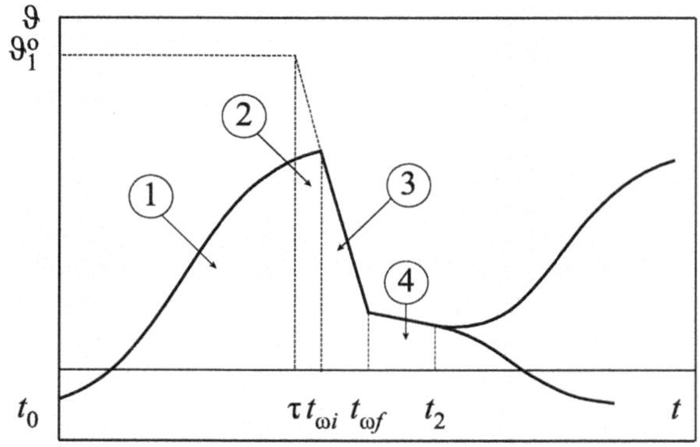

Fig. 6.43 The structure of the optimal pitch angle control program with the approximate consideration of the actuator inertia

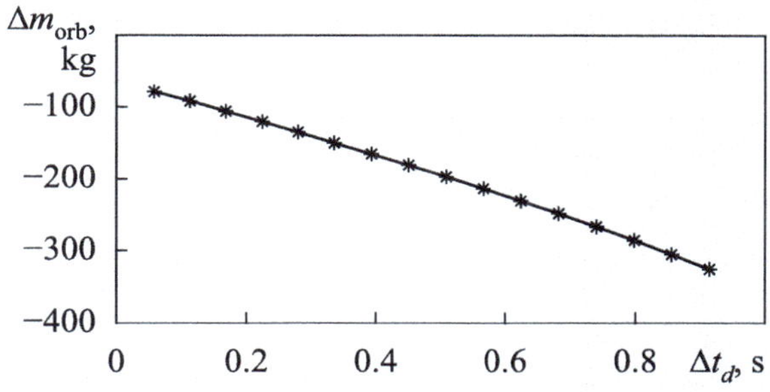

Fig. 6.44 The dependence of the ASTS-1 injected mass change on duration Δt_d of the pre-release maneuver

Within the framework of the proposed (see Sect. 4.6.2) separation control algorithm, the RS turn delay was modeled by artificially delaying the separation process due to the formal moving the critical point $\mathbf{r}_*''$ back from CA by $\Delta\mathbf{r} \parallel \mathbf{r}_*''$ (Fig. 6.45). In this case, the critical point must remain on the lower generatrix of the cone, which limits the area of hazardous impact of the RS plume. The change in the difference $\Delta\vartheta$ between pitch angles of RS and CA in the separation segment as a function of time for various values of $r_*'' = \left|\mathbf{r}_*''\right|$ is shown in Fig. 6.46. If $r_*'' = r_{*0}''$, there is no turn delay. It can be seen in Fig. 6.47 that after the RS release, delay $\Delta\tau$ of RS turn relative to CA leads to a noticeable decrease in the maximum injected mass (Fig. 6.48):

$$\left|\frac{\Delta m_{\mathrm{orb}}}{\Delta\tau}\right| \approx 400\ \mathrm{kg}\cdot\mathrm{s}^{-1}.$$

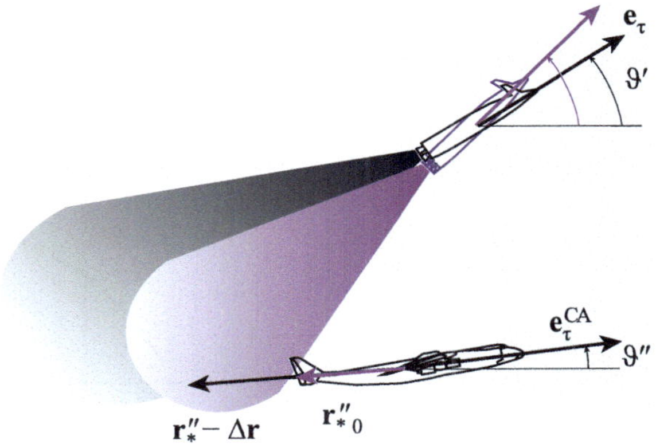

Fig. 6.45 Scheme for separation modeling with a delay in the RS turn

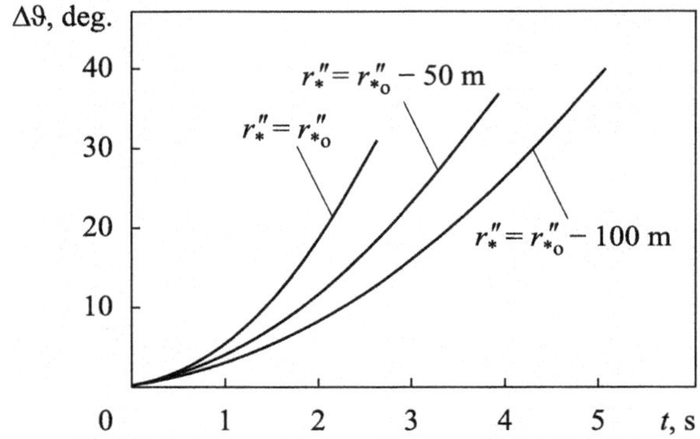

Fig. 6.46 Differences between the pitch angles of RS and CA depending the separation duration for various distances r''_* between the RS engine flame and conventional critical point on CA

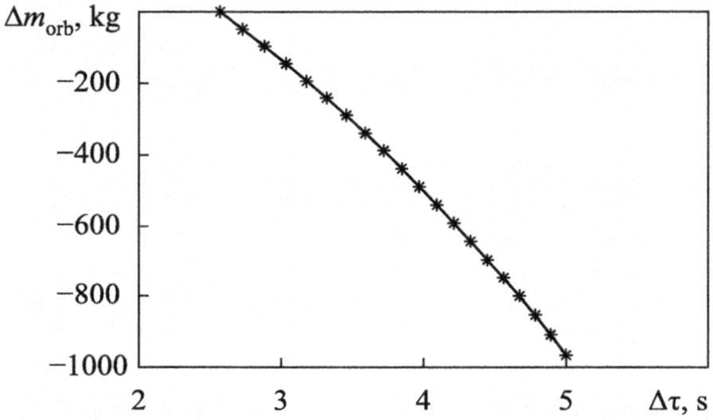

Fig. 6.47 The change in the maximum injected mass due to the delay of the RS turn during separation from CA

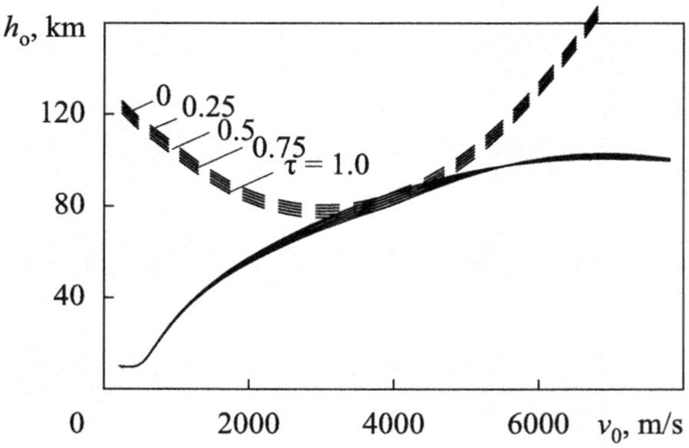

Fig. 6.48 Deformation of the ASTS-1 optimal ascent trajectory in osculating velocity-altitude coordinates in the process of solution continuation from parameter $\tau = 0$ (no fail-safety requirements) to $\tau = 1$ (the optimal fail-safe trajectory)

6.3.5 *Influence of Fail-Safety Requirements*

The features of ASTS optimal fail-safe injection trajectories are studied in comparison with the ballistic STS with a vertical take-off. The optimization was carried out on the basis of PMP using *ASTER* software.

It was assumed that the area of admissible initial conditions for the RV emergency return from the ascent trajectory is the same as in the case considered above. The interpolation curve for the upper boundary of the admissible area D of initial conditions for the safe return is shown in Fig. 6.6 by dashed line.

To obtain a solution of the problem, taking into account the fail-safety conditions, as in Sect. 6.2, optimal solution continuation parameter τ in (6.7) was introduced such that at $\tau = 0$ the boundary of the admissible area touches the optimal ascent trajectory calculated without the fail-safety constraint, and at $\tau = 1$ we have the actual boundary of area D. The changes in flight altitude in terms of the velocity and range and the changes in the angle of attack in time on the ASTS-1 optimal fail-safe ascent trajectories for various values of parameter τ are shown in Figs. 6.48, 6.49 and 6.50.

Calculations have shown that the influence of the fail-safety constraints on ASTS through optimal launch trajectory turns out to be weaker than in the case of a vertically launched STS (Figs. 6.12, 6.13, 6.14, 6.15, 6.16 and 6.17). This is explained by the fact that ASTS-1 has a higher lift-to-drag ratio (in ~ 2.5 times) in flight modes in a vicinity of the start point and, in accordance with the classification given in Sect. 3.2, the ASTS optimal launch trajectory belongs to the qualitatively different type of extremals (type ***A***) than the optimal trajectory of a vertically launched STS (type ***B***). Without consideration of the fail-safety constraints for ascent trajectory optimization, only 16% of the ASTS-1 trajectory duration passes in the area from which the safe RV return is impossible. The change in the ASTS-1 injected mass due to the fail-safety requirements, is shown in Fig. 6.51. It can be seen that the fail-safety requirements for ASTS-1 leads to the maximum injected mass loss

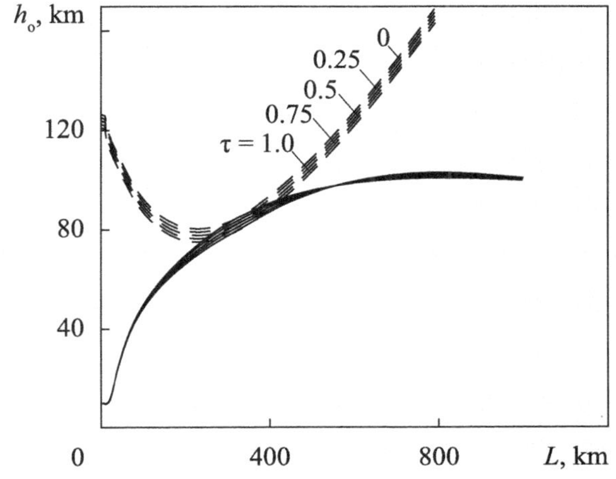

Fig. 6.49 Deformation of the ASTS-1 optimal ascent trajectory in osculating range-altitude coordinates in the process of solution continuation from parameter$\tau = 0$ (no fail-safety requirements) to $\tau = 1$ (the optimal fail-safe trajectory)

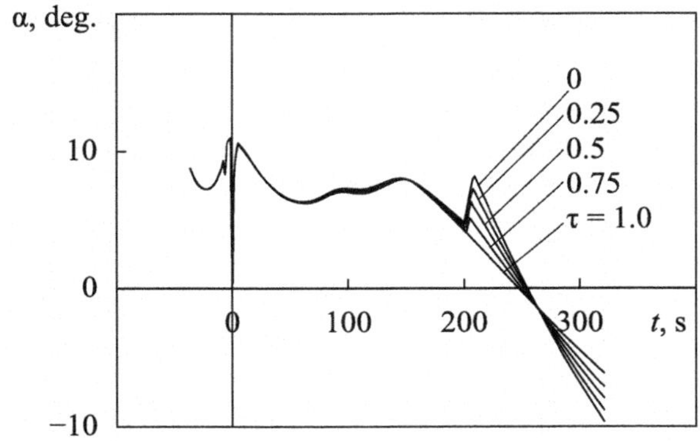

Fig. 6.50 Deformation of the ASTS-1 optimal angle-of-attack program in the process of solution continuation from parameter $\tau = 0$ (no fail-safety requirements) to $\tau = 1$ (the optimal fail-safe trajectory)

of ~ 0.2%, which is an order of magnitude less than in the previously considered case of STS with vertical take-off. The weak influence of the fail-safety requirements on the functional with a relatively short duration of the dangerous segment of ascent trajectory is explained by the general property of optimal solutions: small variations of the optimal trajectory have an impact on the functional of the second (or higher) order of smallness with respect to the norm of these variations.

Thus, a comprehensive consideration of the fail-safety requirements, including the optimal modification of the ascent trajectory, makes it possible to increase the mission safety at practically the same injected mass.

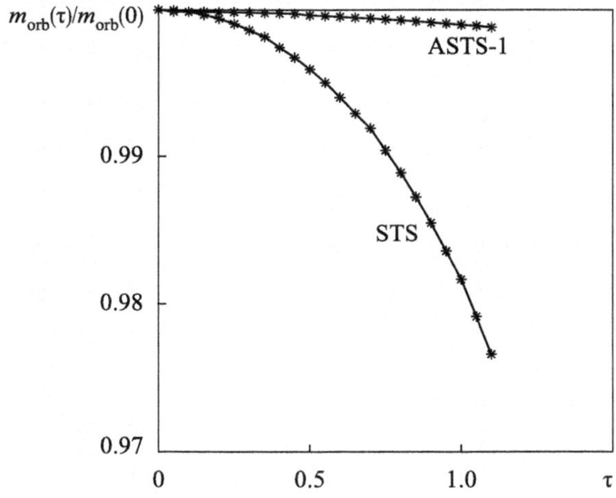

Fig. 6.51 The related maximum injected mass depending on the optimal solution continuation parameter from $\tau = 0$ (no fail-safety requirements) to $\tau = 1$ (the fail-safe trajectory) for ASTS-1 with a horizontal take-off and STS with a vertical take-off

6.3.6 Analysis of Optimal Solution Accuracy

In this section, based on the methodology described in Sect. 1.9, an analysis of admissible perturbations at various segments of the ASTS trajectory is carried out. Particular attention is paid to the maneuver of CA with RS before separation.

The study is based on the analysis of the influence functions (conjugate variables) on the through optimal injection trajectory.

Influence of phase disturbances on all branches of the ASTS optimal injection trajectory

With current phase vector $\mathbf{x}$ known, control $\mathbf{u}_{\text{opt}}$, which ensures the insertion of the maximum mass m_f into the orbit, is determined from the solution of the corresponding optimization problem based on PMP. As shown in Sect. 1.8, the conjugate vector on the optimal trajectory is the influence function of current phase vector variation $\delta\mathbf{x}$ on maximum functional m_f:

$$\mathbf{p} \equiv \left(\frac{\partial m_f}{\partial \mathbf{x}}\right)^{\mathrm{T}}. \tag{6.16}$$

If the optimal trajectories are obtained with constraints on the control and phase vector, then influence functions (6.16) automatically take them into account. In particular, up to the moment t_2 of breaking ties between CA and RS, influence function (6.16) characterizes the change in functional m_f with the variation of the phase vector $\mathbf{x}$, taking into account the fulfillment of the requirements for the safe separation of CA and RS.

Dimensionless influence functions (6.16) for the ASTS-2 optimal injection trajectory are shown in Fig. 6.52. Equation (2.20) are used to pass to dimensionless influence functions. As can be seen, the sensitivity of functional m_f to variations of the phase vector changes significantly during ascent to orbit. Three areas can be highlighted:

(a) $t > 200$ s, where the influence of aerodynamic forces is small (dynamic pressure $q < 100$ kgf/m^2);
(b) $t < 200$ s, where the influence of aerodynamic forces is noticeable ($q > 100$ kgf/m^2);
(c) a short segment of separation (31 s $< t <$ 35 s), where RS and CA controls are severely limited by safety requirements.

In segment a, the influence functions $\frac{\partial m_f}{\partial v_x}, \frac{\partial m_f}{\partial v_y}$ are practically constant, $\left|\frac{\partial m_f}{\partial x}\right| \ll \left|\frac{\partial m_f}{\partial y}\right|, t \cdot \left|\frac{\partial m_f}{\partial x}\right| \ll \left|\frac{\partial m_f}{\partial v_x}\right|$, and $\frac{\partial m_f}{\partial v_y}$ almost linearly dependents on time. This behavior qualitatively coincides with the optimal injection trajectories in the uniform gravitational field, studied in details in the classical OEL works (see Sect. 2.2). In the separation segment, the sensitivity of m_f to variations of radius vector $\mathbf{r}$ increases by several orders of magnitude due to a significant increase in the influence on the

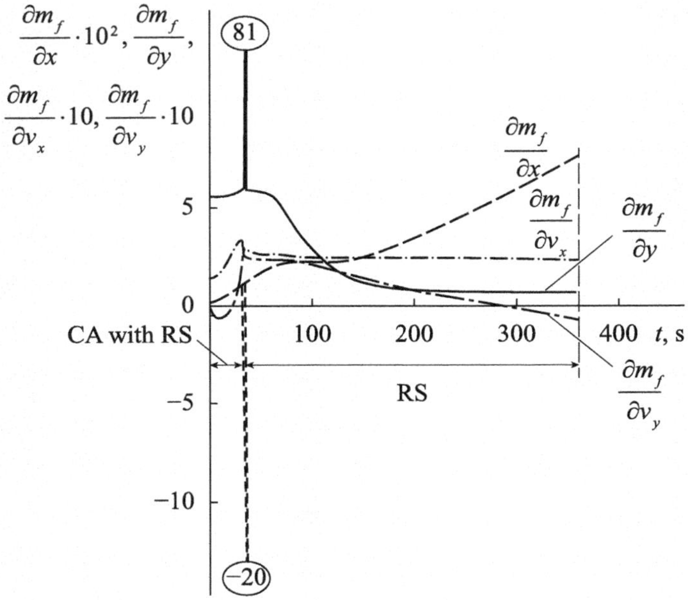

Fig. 6.52 Dimensionless influence functions of phase variables on the maximum injected mass for the ASTS-2 through optimal injection trajectory with RS engine ignition after separation

admissible control of the RS position relatively to the CA immediately after RS release time t_2.

The influence functions can be used to estimate, in the first (linear) approximation, the increment of the maximum functional with variations in the phase vector:

$$\delta m_f \cong \frac{\partial m_f}{\partial \mathbf{x}} \delta \mathbf{x} \equiv \mathbf{p}^T \delta \mathbf{x}, \tag{6.17}$$

in a fairly wide range of $\delta \mathbf{x}$.

If phase disturbances are specified in the air-path axis system, then the following formulas are used to recalculate the influence functions:

$$\frac{\partial m_f}{\partial v} = \frac{\partial m_f}{\partial v_x} \cos\theta + \frac{\partial m_f}{\partial v_y} \sin\theta, \quad \frac{\partial m_f}{\partial \theta} = v\left[-\frac{\partial m_f}{\partial v_x} \sin\theta + \frac{\partial m_f}{\partial v_y} \cos\theta\right]. \tag{6.18}$$

In the Table 6.7 dimensional influence functions and phase variables at reference points of the ASTS-2 optimal injection trajectory are shown.

Here,

$t_0 = 0$ is the end of cruising flight of CA with RS, the beginning of the pull-up maneuver;

Table 6.7 Influence functions and phase variables at reference points of the ASTS-2 optimal injection trajectory

t	$\frac{\partial m_f}{\partial x}$, kg/m	$\frac{\partial m_f}{\partial y}$, kg/m	$\frac{\partial m_f}{\partial v_x}$, kg/(m/s)	$\frac{\partial m_f}{\partial v_y}$, kg/(m/s)	$\frac{\partial m_f}{\partial v}$, kg/(m/s)	$\frac{\partial m_f}{\partial \theta}$, kg/deg	h, m	v, m/s	θ, deg.
t_0	0.00006	0.186	4.5	− 0.1	4.5	− 0.3	9000	200	0
t_1	0.0004	0.196	9.5	5.9	10.8	11.4	7970	232	17.5
t_2-0	0.0004	0.199	8.5	9.4	11.1	24.3	8130	230	19.3
t_2+0	− 2.240	4.370	6.2	10.2	9.2	30.4	8130	230	19.3
t_3	0.0004	0.200	7.8	7.4	9.4	21.0	8280	236	15.1
t_q	0.001	0.029	7.0	2.2	7.3	45.6	57,000	2486	9.2
t_4	0.002	0.025	6.8	− 1.9	6.8	− 261	100,000	7873	− 0.0

$t_1 = 30$ s is the end of the pull-up maneuver, the beginning of the pre-release maneuver of CA with RS;
$t_2 \cong 32$ s is the moment of breaking ties between RS and CA;
$t_3 \cong 35$ s is the end of the separation process, the beginning of the RS autonomous flight;
$t_q \cong 200$ s is the moment of exit from the dense atmosphere layers ($q \cong 100$ kgf/m^2);
$t_4 \cong 357$ s is the end of the powered phase of the RS ascent.

In Table 6.7, the values of the influence functions are rounded. In the calculations, their relative error was the same as that of the phase variables that is no more than 10^{-7}.

Estimates of deviations of the phase variables from the optimal values, leading to the same decrease in the injected mass $\Delta m_f = -100$ kg, are shown in Table 6.8 for the reference points of the ASTS-2 optimal injection trajectory.

Table 6.8 Increments of phase variables at various points of the optimal trajectory, leading to the decrease in the injected mass − 100 kg

Flight segment		$\Delta m_f = -100$ kg		
	t	Δh, m	Δv, m/s	$\Delta\theta$, deg
CA with RS	t_0	180	7.4	unlimited ($\partial m_f/\partial\theta = 0$)
	t_1	170	3.1	2.9
	t_2-0	170	3.0	1.4
Separation of RS and CA	t_2+0	8	3.6	1.1
	t_3	170	3.5	1.6
RS	t_q	1100	4.6	0.7
	t_4	1300	4.9	0.1

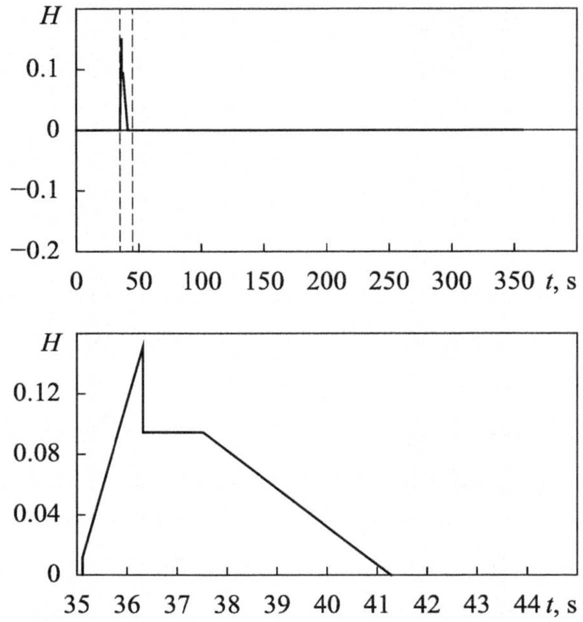

Fig. 6.53 Hamiltonian time-history on the ASTS-2 through optimal ascent trajectory

Verification of the optimal solution accuracy

To control the calculation accuracy, the compliance with the first integrals (1.53), (1.54) was carried out: constancy of the Hamiltonian in the flight segments where the constraints that contain explicitly the time (constraints on the pitch angular rate, on the thrust change rate, etc.) are not essential. An example of the time-dependence of the Hamiltonian on the ASTS-2 optimal injection trajectory is shown in Fig. 6.53. The permissible error in solving the optimization problem (taking into account errors of the numerical integration and BVP solution) corresponded to the injected mass error $|\Delta m_{\text{orb}}| < 0.1$ kg.

Therefore, in accordance with the methodology described in Sect. 1.10, the maximum deviations of the Hamiltonian from a constant function were allowed within:

$$|\Delta H| < \frac{\Delta m_{\text{orb}}}{m_i} \frac{\sqrt{R_0/g_0}}{t_f - t_i} \approx \frac{0.1}{250000} \cdot \frac{806}{360} \sim 10^{-6}.$$

6.4 Through Optimization of STS Trajectories with Constraints on Spent Part Drop Points

Based on the methodology described in Chap. 4 (see also [13]) and using *ASTER* software package, a study of through optimal trajectories and control laws for a heavy-class LV of Proton type (*P*-type) [14] and light-class LV of Rokot type (*R*-type) [14] was carried out. Spent parts (SP) of the 1st, 2nd and 3rd stages, including the half-fairings (HF), must be dropped into specified drop fields (DF).

The influence of a position of SP specified drop points on the maximum injected mass has been studied. The possibility of a significant reduction in DF of the 1st stage booster for a wide range of orbit inclinations by optimizing spatial LV maneuvers during injection has been studied. In particular, there is considered the possibility of dropping the 1st stage booster to one point for a wide range of orbit inclinations and dropping various SPs, separated at different times, to one point.

6.4.1 Source Data

LV of R-type

LV of *R*-type consists of two stages and an upper stage (US). At time $t = 7.453$ s after the vertical take-off ($t = 0$), the launching lugs are dropping. The HFs are released on the 2nd stage flight segment at the moment when the dynamic pressure decreases to $\approx$10 kgf/m^2. US engines are firing in 10 s after the 2nd staging.

The aerodynamic coefficients are given in the body axis system in a tabular form as a function of angle of attack α, altitude h and Mach number.

On the atmospheric phase of the 1st stage, after the vertical flight segment and the short-term pitch turn, the g-turn program is carried out with $\alpha = 0$.

Two options for the launch point are considered:

$$\text{a) } \varphi = 63^\circ, \lambda = 41^\circ; \tag{6.19}$$

$$\text{b) } \varphi = 50^\circ, \lambda = 128^\circ. \tag{6.20}$$

The parameters of the target circular orbit are varied: altitude $h_{\text{orb}} \in$ [200 km, 35,896.8 km], inclination $i_{\text{orb}} \in [51^\circ, 101^\circ]$.

LV of P-type

LV of *P*-type consists of three stages. The initial masses of the stages are assumed to be fixed, while SP masses are varied. The fuel consumption is given in the form: $\dot{m} = -\frac{T}{c} + \Delta\dot{m}$, where $\Delta\dot{m} = \text{const}$ for each stage.

The aerodynamic coefficients are given in the body axis system in a tabular form as a function of angle of attack α, altitude h and Mach number M.

The following constraints were taken into account: on the 1st stage $|\alpha| \le 5^\circ$, $|q\alpha| \le 9000$ kgf/m^2; on the 2nd stage $|q\,\alpha| \le 9000$ kgf/m^2.

The launch point has coordinates: $\varphi = 46^\circ$, $\lambda = 63^\circ$. The target circular orbit has altitude $h_{orb} = 200$ km. Three variants of the orbit inclination are considered: $i_{orb} = 51.64^\circ$, $64.6^\circ$and 72.6°.

Modified LV of P-type

Modified LV of *P*-type consists of three stages and US.

The aerodynamic coefficients are given in the body axis system in a tabular form as a function of angle of attack *α*, altitude *h* and Mach number *M*.

In the atmospheric ascent segment of the 1st stage, after the end of the vertical ascent section and the short pitch turn, the angle of attack is assumed to be zero (g-turn program). The pitch rate in all ascent sections is limited to $\omega \le 3$ deg/s.

When turning the stage engines on and off, the thrust rate limitation is taken into account.

The launch point has coordinates: $\varphi = 46^\circ$, $\lambda = 63^\circ$. The target circular orbit has altitude $h_{orb} = 160$ km and inclination $i_{orb} = 51.6^\circ$.

The admissible DFs are set by the geographic coordinates of their centers. The designations of these DFs with the highlighting of the main ones for each stage are represented in Table 6.9.

Calculations of optimal spatial launch trajectories were carried out using *ASTER* program. The extended phase vectors had a variable dimension (the total dimension was 46), including seven in each the injection segments of the 1st, 2nd, 3rd and 4th (US) stages, and six in DFs for spent boosters and HFs. The total dimension of the BVP is equal to 25. This BVP includes eleven varied parameters of outer TPBVP, two parameters of inner TPBVP for segments of the ascent and return of the 1st stage boosters, six parameters of inner TPBVP for the segment of the 2nd stage booster return and six parameters of inner TPBVP for the segments of the 3rd stage booster and HFs return. The varied parameters of the outer BVP were as follows: seven conjugate variables at the end of the vertical take-off section (due to time constraints, the Hamiltonian at the initial point is not defined) plus four conjugate variables p_x and p_z at the drop points of boosters of the 2nd and 3rd stages (or HFs).

Table 6.9 Designations of drop fields for various separated parts

Stage number	Drop point designation
1st	S_1^1
	S_2^1 (main)
	S_3^1
2nd	S_1^2 (main)
	S_2^2
	S_3^2
3rd	S_1^3 (main)
	S_2^3

The varied parameters of the inner BVPs were as follows: the flight path angle and azimuth at the beginning of the 1st stage g-turn, six conjugate variables at the initial point of the 2nd stage booster return and six conjugate variables at the initial point of the 3rd stage booster (or HFs) return. The residuals were the conditions at the right ends of the main and three side branches.

6.4.2 *Influence of the Drop Point Constraints on Optimal Trajectories of* R-*type LV*

Figures 6.54, 6.55 and 6.56 show the optimal control programs and the change in phase variables on the optimal branching trajectories of *R*-type LV without constraints on drop points, that is, when SPs drop to the optimal points.

Let us consider the features of optimal injection into orbits with various inclinations at the optimal launch azimuth. To do this, the initial azimuth is assumed to be free in the statement of the optimization problem.

The dependence of the maximum mass inserted into a circular orbit with altitude $h_{orb} = 295$ km on the orbit inclination at the optimal launch azimuth is shown in Fig. 6.57. The inclusion of the initial azimuth (see Sect. 2.3.4) into the trajectory optimization procedure does not lead to a change in the BVP order and, accordingly, in the volume of calculations when using PMP. The same figure shows the optimal launch azimuth depending on the orbit inclination. In this case, both local extremals are considered, corresponding to the launch into the ascending and descending legs of the target orbit. Insertion into the descending orbital leg provides more payload mass

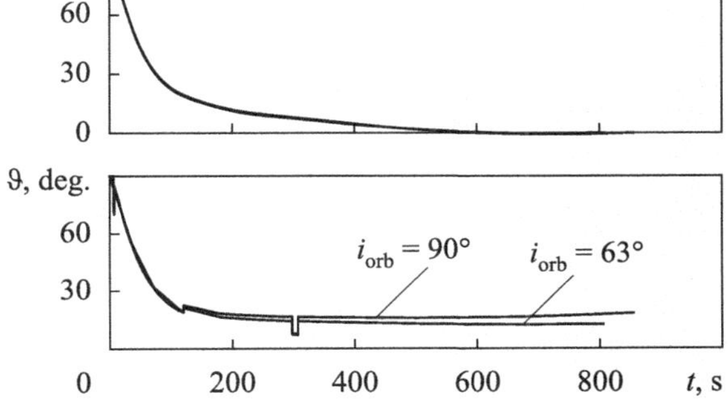

Fig. 6.54 Time-histories of the angle of attack, flight path angle and pitch angle on the optimal branching injection trajectories of the R-type LV without constraints on SP drop points for two orbit inclinations

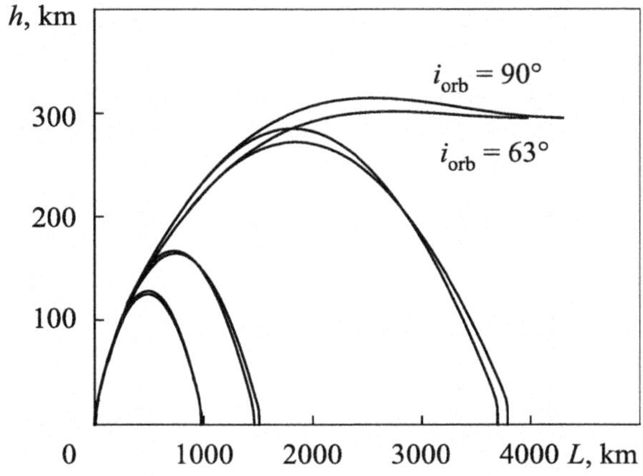

Fig. 6.55 Optimal branching trajectories of the R-type LV in plane of altitude vs. range without constraints on SP drop points for two orbit inclinations

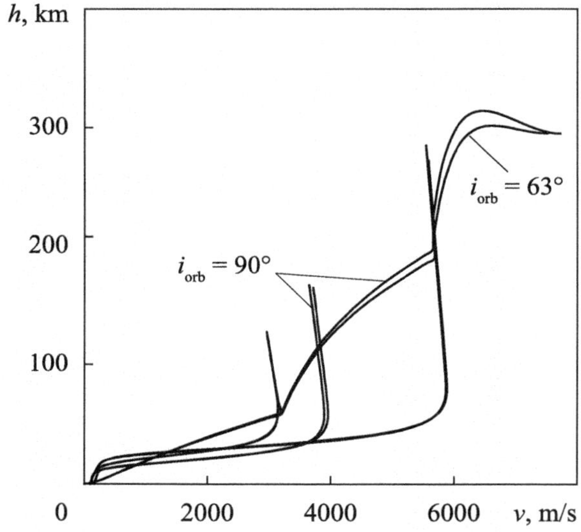

Fig. 6.56 Optimal branching trajectories of the R-type LV in plane of altitude vs. speed without constraints on SP drop points for two orbit inclinations

in the considered inclination range. (However, these trajectories may be unacceptable from the point of view of location of SP drop points.)

To assess the impact of SP drop point constraints on the maximum injected mass, optimal control programs and launch trajectories, parametric studies of through optimal trajectories were carried out using *ASTER* complex. At the same time, the problem statement presented in Chap. 3 was somewhat modified as follows.

Let us represent the geographical coordinates (φ, λ) of the SP drop point S in a parametric form:

$$(\varphi, \lambda) = (\varphi, \lambda)_{S_{\text{opt}}} + r(\cos\psi, \sin\psi),$$

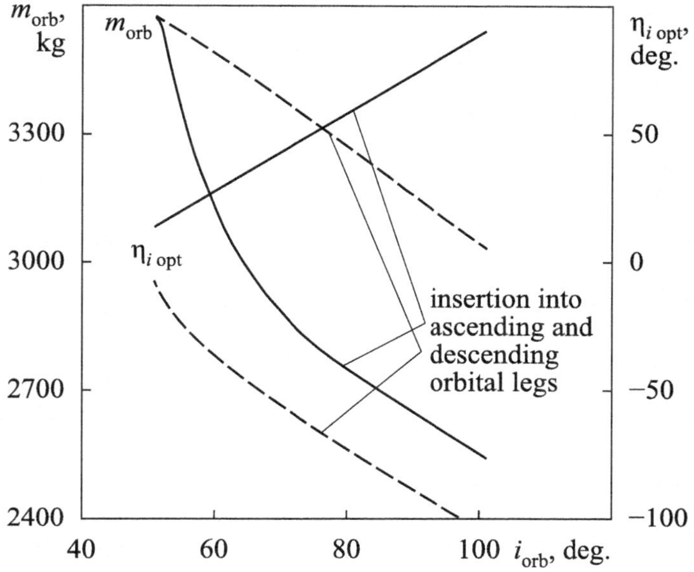

Fig. 6.57 Dependences of the maximum injected mass and optimal launch azimut of the *R*-type LV on the orbit inclination without constraints on SP drop points for two local extremals

where S_{opt} is the optimal SP drop point, that is, obtained by optimizing the branching trajectory without the constraint on the drop point. Let the drop point not be fixed, but only the angle ψ that determines the direction of deviation of the specified drop point from the optimal (nominal) one. We also fix the loss of the maximum injected mass Δm_{orb}, induced by this deviation, and represent the distance r of the given fall point from the optimal one as the functional to be maximized.

As a result, we come to a mutual optimization problem, the solution of which determines the extremal reachable position of the SP drop point, as far as possible from S_{opt} in a direction given by angle ψ, for a given injected mass loss. By changing parameter ψ in interval [0, 2π], we obtain the level line of injected mass losses in the plane of geographical coordinates:

$$\begin{cases} \varphi|_{\Delta m_{\text{orb}}} = \varphi(\psi, \Delta m_{\text{orb}} = fix) \\ \lambda|_{\Delta m_{\text{orb}}} = \lambda(\psi, \Delta m_{\text{orb}} = fix) \end{cases} \quad 0 \le \psi \le 2\pi,$$

which is the boundary of the area of the Earth's surface into which SP can be dropped, providing the loss of the injected mass not exceed Δm_{orb} compared to the maximum one.

Below are the results of calculating such level lines for two launch points: (6.19) and (6.20). To draw each level line, an average of about 360 through optimal injection trajectories was calculated that corresponds to the change of continuation parameter ψ with step of ~ 1°. The small step was chosen for indirect check of the high accuracy of obtained solutions due to absence of visual chaotic oscillation of the dependence on the parameter (even under the highest possible plot resolution).

Figure 6.58 shows the boundaries of feasible areas (level lines) for nominal (without influence of random factors [15–19]) drop points of the 1st stage booster, inserting the payload into the target orbit with altitude of $h_{\text{orb}} = 295$ km. It can be seen that, in general, the shape of the level lines changes slightly with variation in the target orbit inclination. However, when the orbit inclination close to the latitude of the launch point, the level lines are deformed significantly. This is explained by the existence of two local extremals that are close in the functional and correspond to the insertion into the ascending and descending orbital traces. The dependence of the maximum injected mass on the initial azimuth for orbit inclination $i_{\text{orb}} = 51°$ is shown in Fig. 6.59, where two local maxima, which almost coincide in value, correspond to the optimal initial azimuths for two local extremals. Similar level lines for drop areas of the 2nd stage booster and HF for given losses of the injected mass are shown in Fig. 6.60.

Figure 6.61 shows the dependence of relative maximum injected mass $\overline{m}_{\text{orb}}(\overline{L}) = \frac{m_{\text{orb}}(\overline{L})}{\max\limits_{\overline{L}} m_{\text{orb}}}$ on relative admissible drop point range $\overline{L} = L/L_{\max}$ for the 1st and 2nd stage boosters and HF.

The possibility of reducing number of drop fields by dropping SPs, separated at different points of LV trajectory, to the same point has been studied. Figure 6.62 shows the relative loss of the injected mass under the condition that the 1st stage

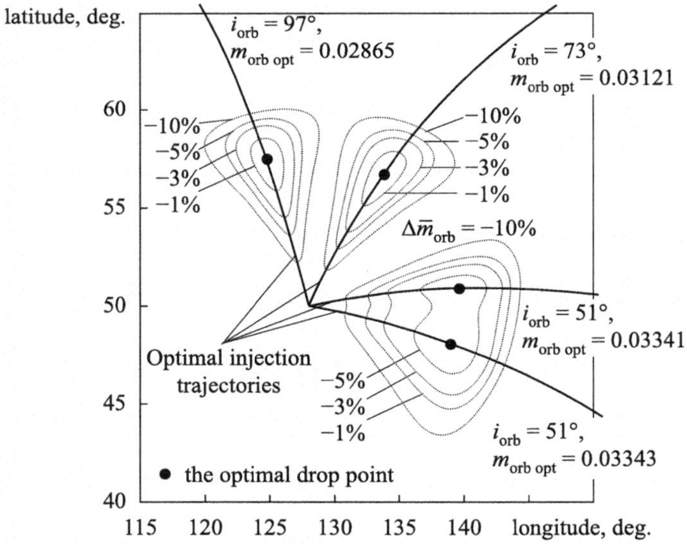

Fig. 6.58 Feasible drop areas at an uncontrolled return of the spent 1st stage booster using optimal spatial maneuvers of the *R*-type LV at the cost of a given relative loss in percentage of the maximum mass inserted into target orbits of four inclinations

Fig. 6.59 Dependence of the maximum injected mass on the initial azimuth for orbit inclination of 51°

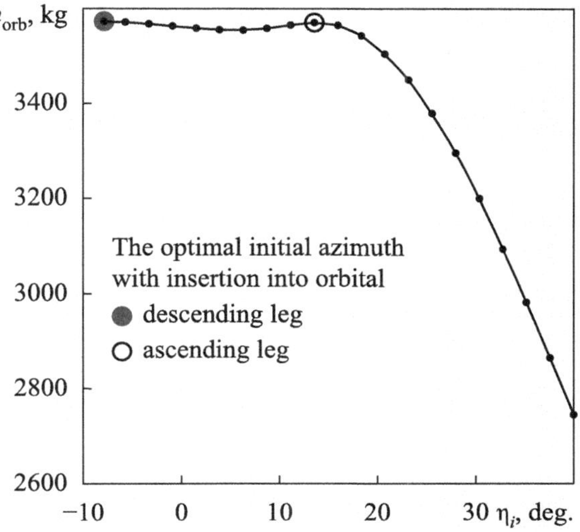

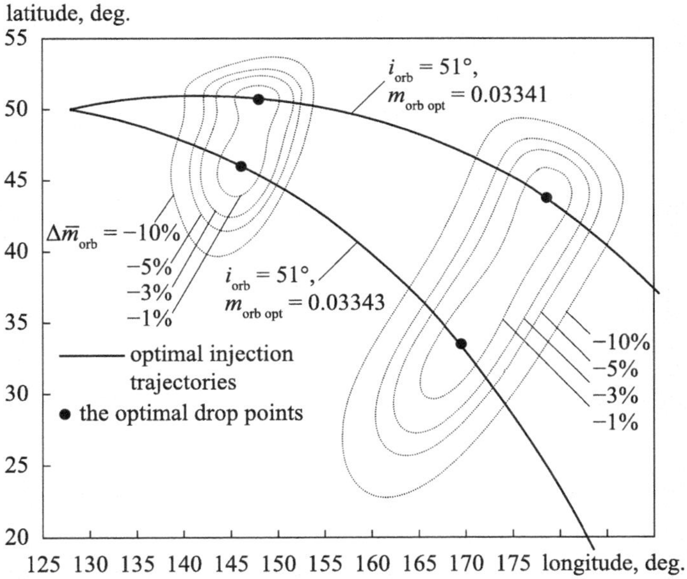

Fig. 6.60 Feasible drop areas at an uncontrolled return of the spent 2rd stage booster and HF using optimal spatial maneuvers of the *R*-type LV at the cost of a given relative loss in percentage of the maximum injected mass for two local optimal initial azimuths

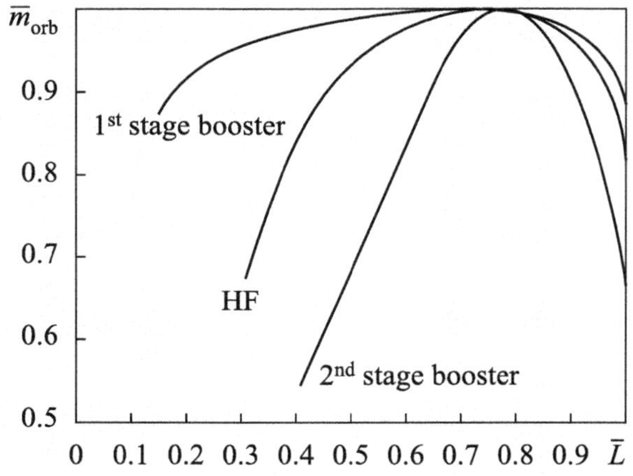

Fig. 6.61 Dependences of the relative maximum injected mass on the relative admissible drop point range for the 1st and 2nd stage boosters and HF

booster and HF drop to the same point, depending on the distance of this common drop point from the launch site. There are also shown dependences of the maximum injected mass on the limitation of the drop point range for only the 1st stage booster or only HF (dashed lines). Thus, the release of one drop field costs approximately 8% of the maximum injected mass loss.

If the launch site is located at higher latitude (6.19) than in the previous case, the change in the maximum injected mass with constraints on the 1st stage booster drop point is shown in Fig. 6.63. It also shows a relatively small shift in the optimal drop point when the target orbit altitude increases from 295 to 35,896.8 km. It is

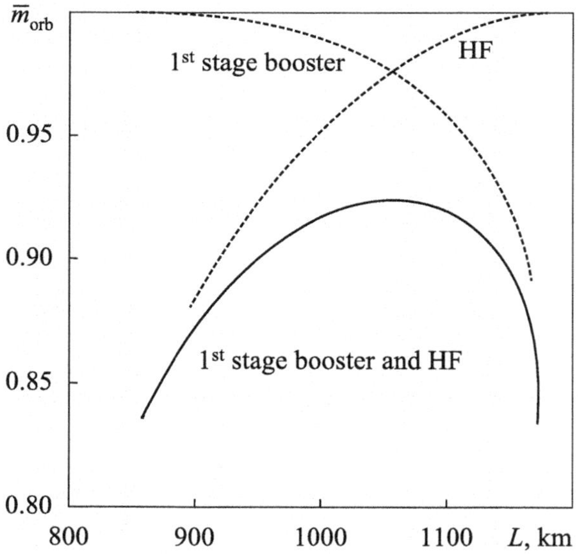

Fig. 6.62 Dependences of the relative maximum injected mass on the distance of SP drop points from the launch site under three conditions: drop points of the 1st stage booster and HF are given the same (solid line), and these points are constrained separately (dashed lines)

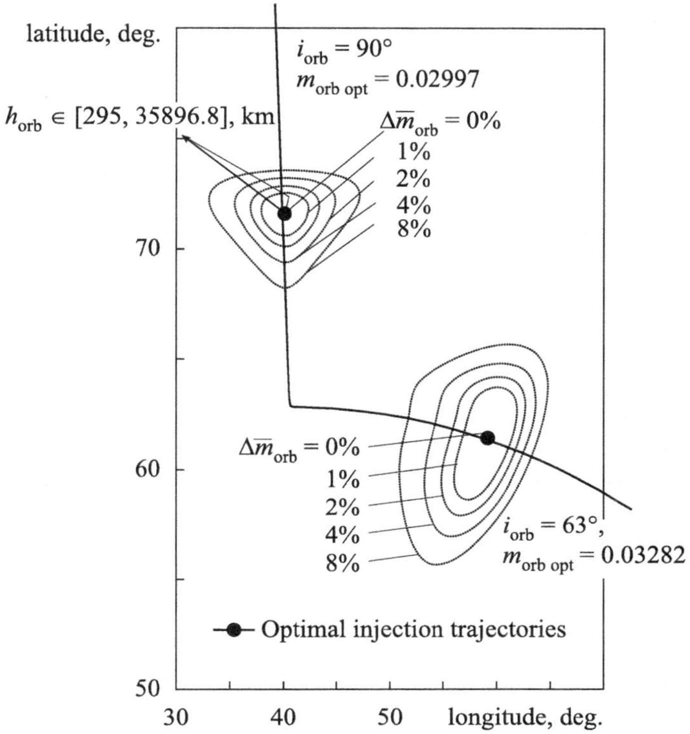

Fig. 6.63 Feasible drop areas at an uncontrolled return of the spent 1st stage booster using optimal spatial maneuvers of the *R*-type LV at the cost of a given relative loss in percentage of the maximum mass inserted into orbits of two inclinations for the high-latitude launch site

obvious that the feasible drop areas will undergo approximately the same equidistant displacement.

6.4.3 Influence of Drop Point Constraints on Optimal Trajectories of P*-type LV*

The optimal trajectories and control laws of *P*-type LV for the target circular orbit of altitude $h_{orb} = 200$ km and inclination $i_{orb} = 51.64°$ without constraints on SP drop points are shown in Figs. 6.64 and 6.65. In this case, a scheme of "direct" insertion into the target orbit is implemented, when LV powered flight ends in the target orbit ($h_f = h_{orb}$).

Let us analyze qualitatively the optimal trajectories and control laws when adding constraints on drop point of the spent 1st stage booster. If the specified drop point lies away from the nominal launch path, then the optimal initial azimuth changes in such

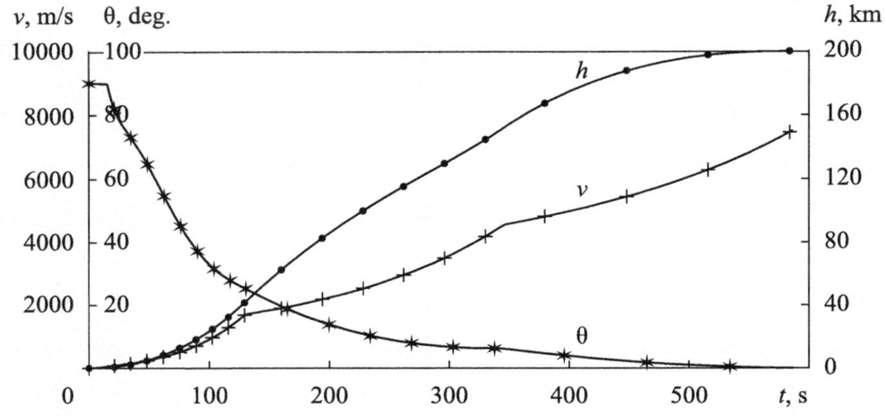

Fig. 6.64 Time-histories of the altitude, velocity, and flight path angle on the *P*-type LV optimal trajectory without constraints on SP drop points

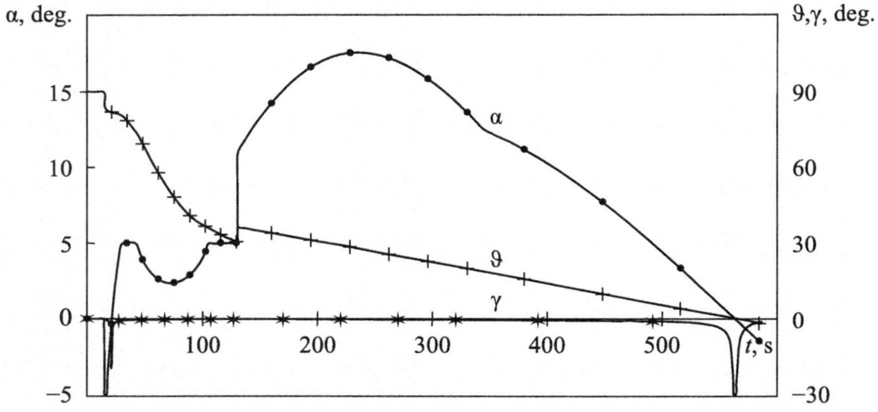

Fig. 6.65 Time-histories of the angle of attack, pitch and bank angles on the *P*-type LV optimal injection trajectory without constraints on SP drop points

a way as to orient the velocity vector at the 1st staging to the specified booster drop point (Fig. 6.66). After that, the 2nd stage performs an intensive optimal spatial turn so that the inclination of the launch plane almost coincides with the inclination of the target orbit plane. At the same time, the optimal pitch angle program practically does not change (Fig. 6.67).

Dropping the spent 1st stage booster not at the optimal point leads to some decrease in the maximum injected mass Δm_{f}. However, Δm_{f} has the second (or higher) order of smallness with respect to the distance of the specified drop point from the optimal one. Therefore, the small deviation of the drop point has almost no effect on the

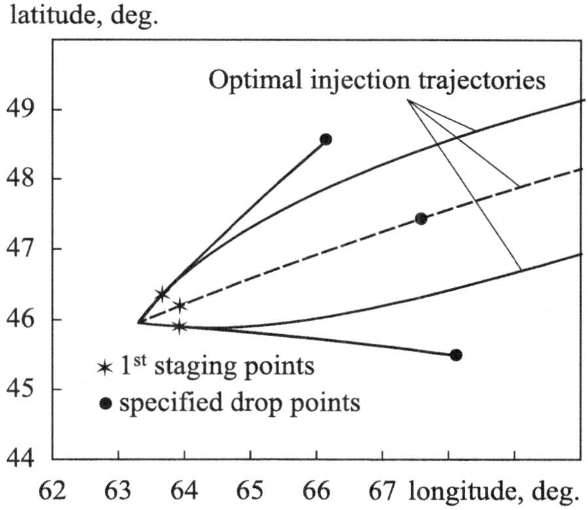

Fig. 6.66 Optimal injection trajectories of the P-type LV and side branches of the 1st stage booster uncontrolled return to specified drop points

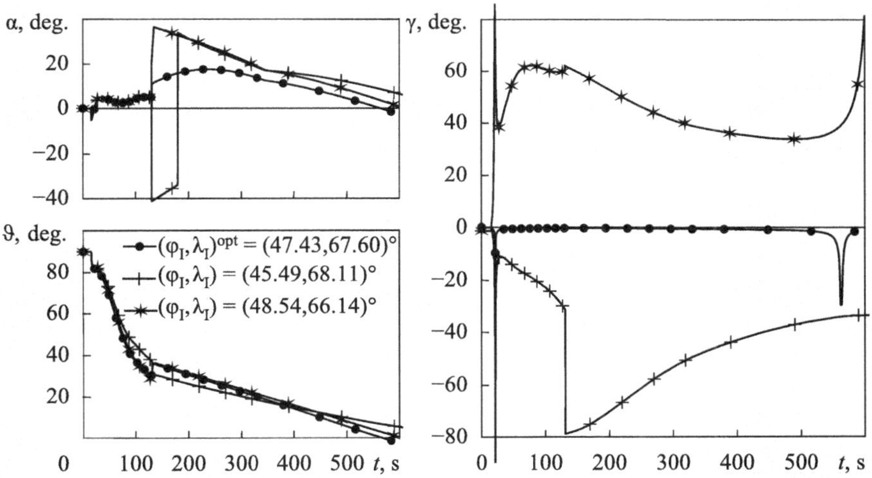

Fig. 6.67 Optimal time-histories of the angle of attack, pitch and bank angles of the P-type LV for various 1st stage booster drop points

injected mass. The question is only in the quantitative assessment of the term of "small".

In geographic coordinates the level lines of the maximum injected mass are shown in Fig. 6.68 for varying the given drop points of the 1st stage booster for orbit inclinations $i_{orb} = 51.64°$ and 72.6°. It can be seen that the optimal drop point lies near the feasibility bound (the shaded line). This means that the optimal 1st stage branch, delivering the maximum payload, is close to the trajectory with the maximum booster drop range. Thus, the drop range of the 1st stage booster L_1 is fundamentally limited from above: $L_1 - L_{1\ opt} < 11$ km, where $L_{1\ opt}$ is the distance of the drop point at the optimal injection trajectory without constraints on SP drop points. The deviation of the booster drop point in azimuth (at the same range) leads to a noticeably smaller (almost twice) loss of the injected mass m_{orb} than drop point deviation along the launch path. The level lines are practically invariant to orbit inclination i_{orb}. The deformation of the level line shape in geographical coordinates at the orbit inclination variation can be mainly explained by the fact that the longitude and latitude axes at the considered latitudes have linear scales, which differ by cosφ times at the same angular scale.

Based on the results shown in Fig. 6.68, it is possible to define the maximum injected mass depending on the specified drop point of the 1st stage booster, and the optimal location of such the drop point, which is the same for servicing all orbit inclinations in the range of $51.64° \leq i_{orb} \leq 72.6°$. For example, when choosing a common drop point in a vicinity of the optimal one for insertion into orbit with $i_{orb} = 64.6°$ ($\varphi_1 = 48.57$, $\lambda_1 = 66.22$), relative maximum mass $m_{orb} > 0.03045$ (referred to the LV initial mass) is provided for any inclination in the considered range.

The dependences of maximum injected mass m_{orb} on the orbit inclination is shown in Fig. 6.69 at the condition that the only drop point uses for the entire range of

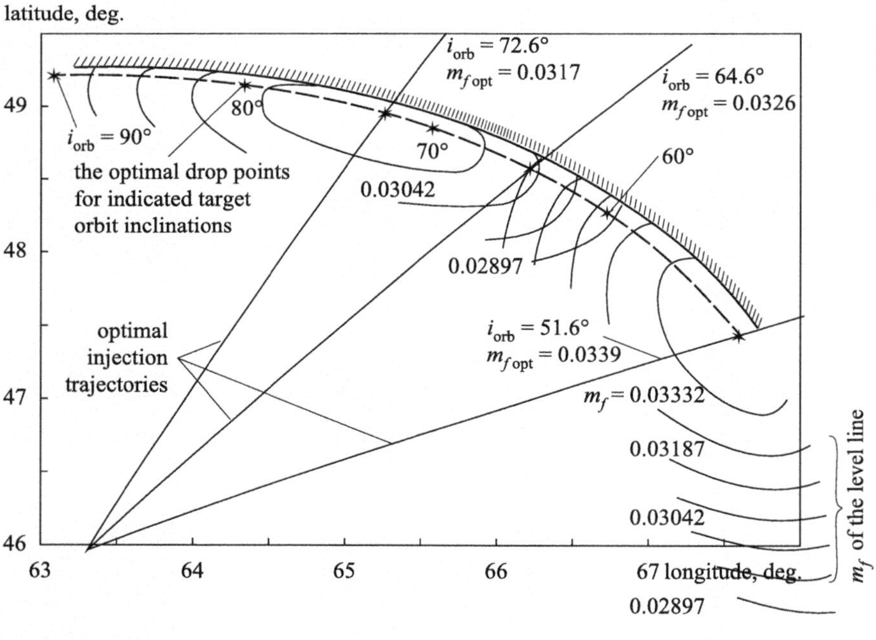

Fig. 6.68 Feasible drop areas at an uncontrolled return of the spent 1st stage booster using optimal spatial maneuvers of the *P*-type LV at the cost of a decrease of the maximum mass inserted into orbits of three inclinations

inclinations: $(\varphi_1, \lambda_1) = \underset{51.64^\circ \le i_{orb} \le 72.6^\circ}{\text{fix}}$. The position of this common point was varied on the set of optimal drop points (dashed line). Using Fig. 6.69, it is convenient to determine the best common booster drop point if a distribution of the nomenclature of payloads (their masses) for orbit inclinations is known. Similarly, it is possible to construct dependences $m_{orb}(i_{orb})$ for various combinations of available admissible drop points for rationale choice of DFs. Inclusion of the requirement of dropping the booster in one or several specified points makes it possible to increase the flexibility of launch service, to ensure the all-azimuth launch (here, in the range of $51.64^\circ \le i_{orb} \le 72.6^\circ$).

As for the decrease of the injected mass with a reduction in the number of permissible DFs, it is necessary to keep in mind the following.

(1) With the current launch scheme, the actual DF does not always include the optimal drop point. Therefore, while maintaining traditional guidance algorithms, the maximum injected mass is still not achieved.
(2) The flexible choice of the launch azimuth makes it possible to decrease the characteristic velocity losses of orbital transitions and, in addition, the time to reach a given orbit.
(3) The efficiency of LV is estimated, as a rule, by the mass inserted into some one reference orbit. If we take the neighborhood of the optimal drop point upon injection into this orbit as a single area serving all possible orbit inclinations, then the "passport" efficiency of such LV does not decrease at all.

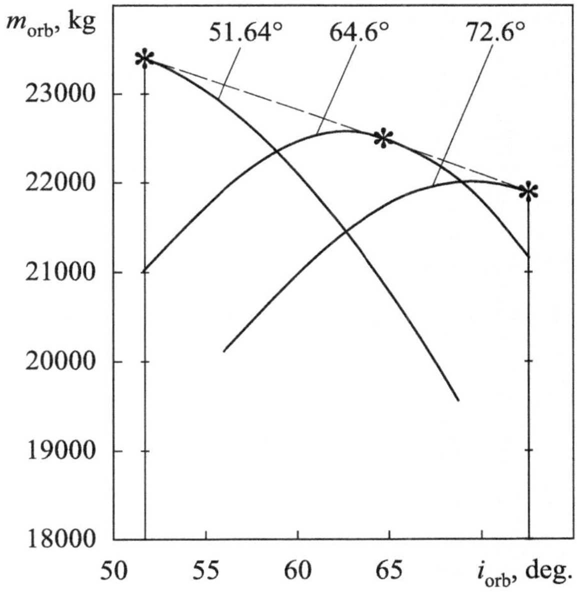

Fig. 6.69 Dependences of the maximum injected mass on the orbit inclination without limitation of SP drop points (dashed line) and under the condition that the 1st stage booster drops into one common point corresponding to the optimal one for orbit inclinations of 51.64°, 64.6° and 72.6°

6.4.4 Influence of Drop Point Constraints on Optimal Trajectories of Modified P-type LV

Optimal branching injection trajectories of a modified *P*-type LV are studied. Thirty two variants of DF combinations for 1st, 2nd and 3rd stage boosters from the given list (Table 6.9) are considered. The optimal points S^1_{opt}, S^2_{opt} and S^3_{opt} (without constraints on admissible drop points) are additionally considered as the ideal for comparison. The control programs was determined from a rigorous solution of the through optimization problem based on PMP using *ASTER* program, taking into account the above constraints, including those limiting the 1st stage control of the g-turn program. The influence on the injected mass, optimal trajectories and control programs of such factors as the DF location, target orbit parameters, pitch angle rate, thrust changing rate, distribution of fuel mass over stages, etc., has been studied. Table 6.10 shows the relative masses injected into the given circular orbit ($h_{orb} = 160$ km, $i_{orb} = 51.6°$), depending on the choice of drop point of the 2nd and 3rd stage boosters when the 1st stage booster drops into S^1_2.

Optimal trajectories and control programs are presented for all combinations of drop points from Table 6.9. Due to the coincidence of the main DF center with the optimal 1st stage booster drop point with sufficient accuracy, we will assume in further calculations that 1st stage booster has always to drop to the center of the main admissible DF S^1_2.

Figures 6.70, 6.71 and 6.72 show the optimal branching trajectories with a free (optimal) drop point for the 3rd stage booster S^3_{opt} (Fig. 6.70), with the drop of the 3rd stage booster to the center of S^3_1 (Fig. 6.71) and S^3_2 (Fig. 6.72) with variation of drop points for the 2nd stage booster: S^2_{opt}, S^2_1, S^2_2, S^2_3.

Figures 6.73, 6.74, 6.75 and 6.76 show the optimal branching LV trajectories with a free (optimal) drop point of the 2nd stage booster S^2_{opt} (Fig. 6.73), with the drop of the 2nd stage booster to the center of S^2_1 (Fig. 6.74), S^2_2 (Fig. 6.75), and S^2_3 (Fig. 6.76) with variation of drop points of the 3rd stage booster: S^3_{opt}, S^3_1, S^3_2.

The optimal branching LV trajectories are presented in Figs. 6.77 and 6.78

(a) with free (optimal) drop points of the 2nd and 3rd stage boosters,
(b) when SPs drop into the main fields S^1_2, S^2_1 and S^3_1,

Table 6.10 The relative maximum injected mass for specified SP drop points. The numbers in *Italic* correspond to imaginary best drop points. The numbers in **bold** correspond to the best combinations of the available drop fields

		The 2nd stage booster drop point			
		S^2_1	S^2_2	S^2_3	S^2_{opt}
The 3rd stage booster drop point	S^3_1	0.94881	0.95629	0.95629	*0.95633*
	S^3_2	0.95989	**0.96741**	**0.96741**	*0.96741*
	S^3_{opt}	*0.96912*	*0.99937*	*0.99042*	*1*

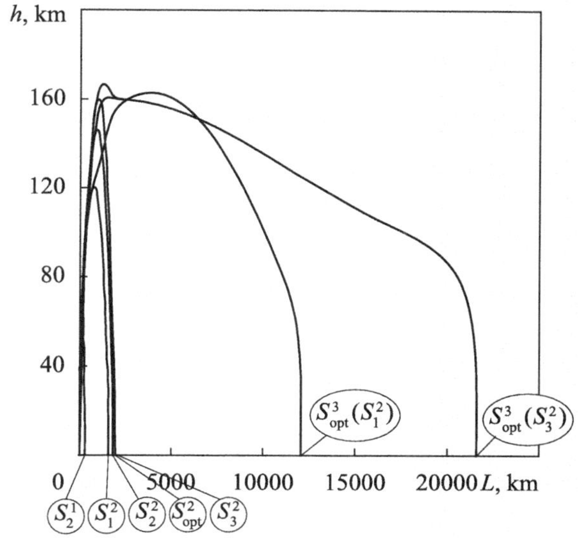

Fig. 6.70 Optimal branching injection trajectories of the *P*-type LV when the 1st stage booster drops in point S_2^1 and the 3rd stage booster drops in the optimal (free) points for various drop points of the 2nd stage booster

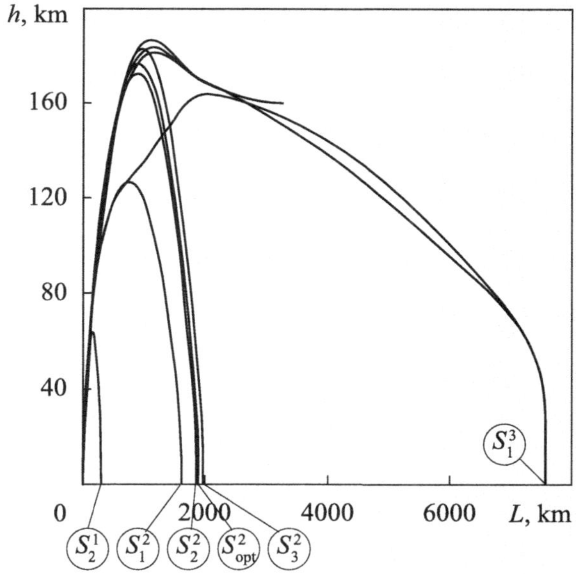

Fig. 6.71 Optimal branching injection trajectories of the *P*-type LV when the 1st stage booster drops in point S_2^1 and the 3rd stage booster drops in point S_1^3 for various drop points of the 2nd stage booster

(c) when SPs drop into the fields S_2^1, S_2^2 and S_2^3 that provides the maximum injected mass.

The maximum dynamic pressure on the LV trajectories for the specified combinations of DFs (a, b, c) does not change. This is explained by the fact that the maximum

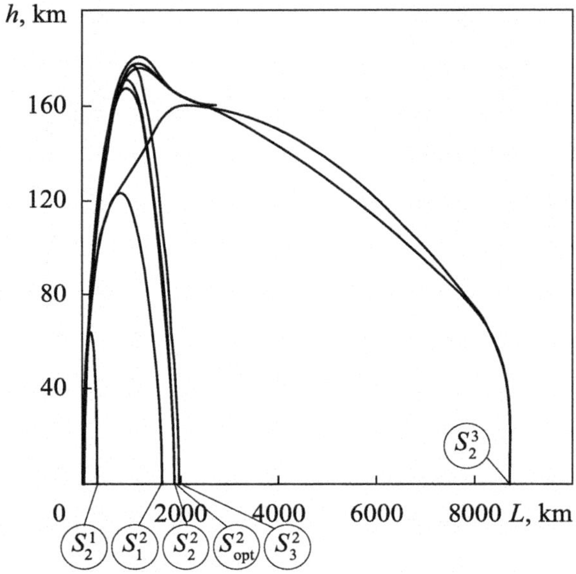

Fig. 6.72 Optimal branching injection trajectories of the *P*-type LV when the 1st stage booster drops in point S_2^1 and the 3rd stage booster drops in point S_2^3 for various drop points of the 2nd stage booster

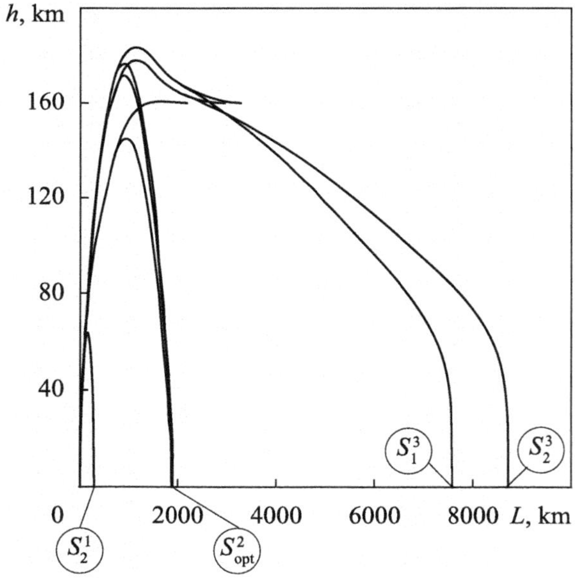

Fig. 6.73 Optimal branching injection trajectories of the *P*-type LV when the 1st stage booster drops in point S_2^1 and the 2nd stage booster drops in the optimal (free) point for two fixed drop points of the 3rd stage booster

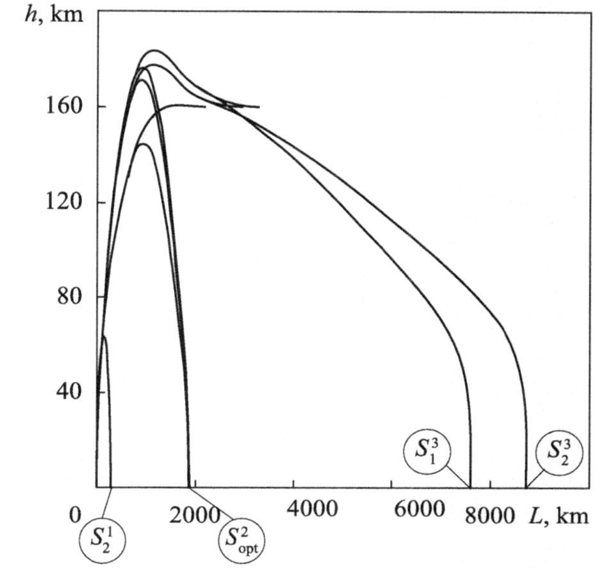

Fig. 6.74 Optimal branching injection trajectories of the *P*-type LV when the 1st stage booster drops in point S_2^1 and the 2nd stage booster drops in the optimal (free) drop points for two drop points of the 3rd stage booster

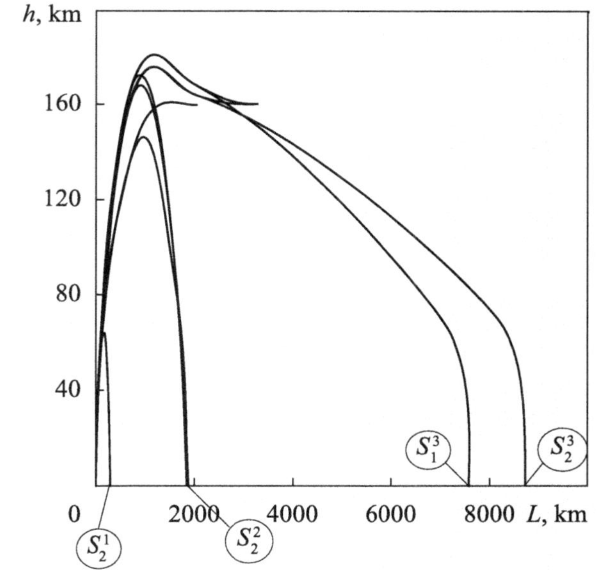

Fig. 6.75 Optimal branching injection trajectories of the *P*-type LV when the 1st stage booster drops in point S_2^1 and the 2nd stage booster drops in point S_2^2 for two drop points of the 3rd stage booster

dynamic pressure is reached in the fixed 1st stage segment, where the g-turn is carried out due to the current constraints on the angle of attack.

The influence of the admissible pitch angle rate and thrust increase rate on the maximum injected mass is shown in Figs. 6.79 and 6.80. At a maximum angular

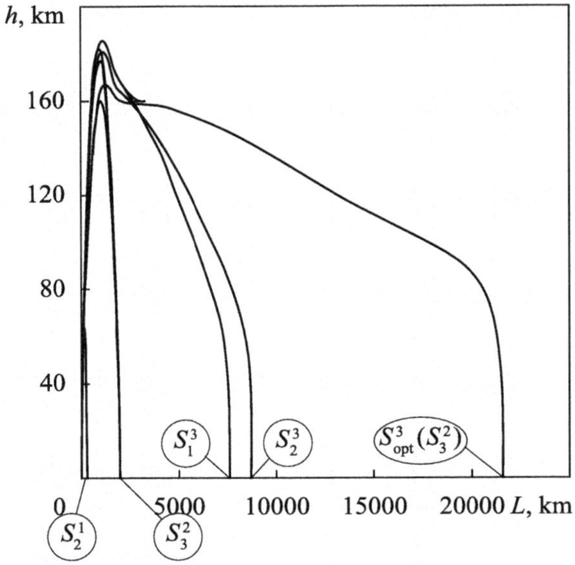

Fig. 6.76 Optimal branching injection trajectories of the *P*-type LV when the 1st stage booster drops in point S_2^1 and the 2nd stage booster drops in point S_3^2 for various drop points of the 3rd stage booster

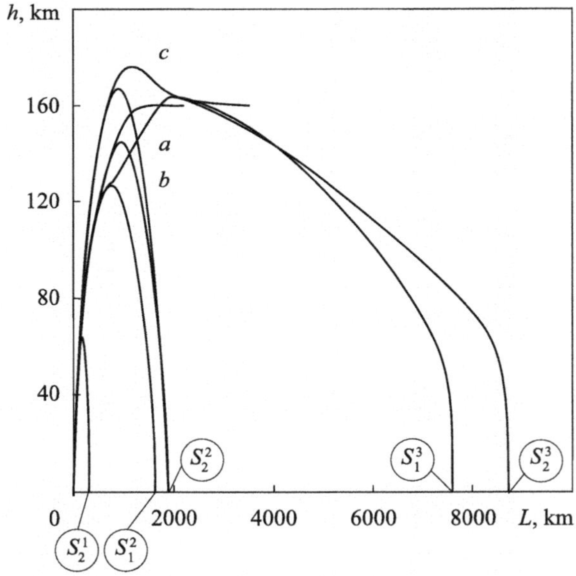

Fig. 6.77 Optimal branching trajectories of the *P*-type LV: **a** with optimal (free) SP drop points; **b** with drop points S_2^1,S_1^2 and S_1^3; **c** with SPs drop points S_2^1, S_2^2 and S_2^3

rate of RS over 5 deg/s, the loss of the injected mass does not exceed 25 kg. A further decrease in the angular rate leads to a sharp increase in the losses (Fig. 6.79). Reducing the duration of the thrust increase to reach the maximum mode leads to an almost linear increase in the injected mass with rate of 13 kg/s (Fig. 6.80).

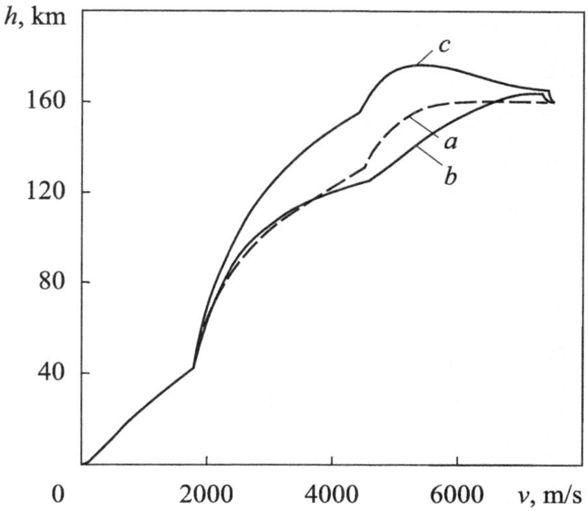

Fig. 6.78 Optimal injection trajectories of the *P*-type LV: **a**) with optimal (free) SP drop points; **b**) with drop points S_2^1,S_1^2 and S_1^3; **c**) with drop points S_2^1, S_2^2 and S_2^3

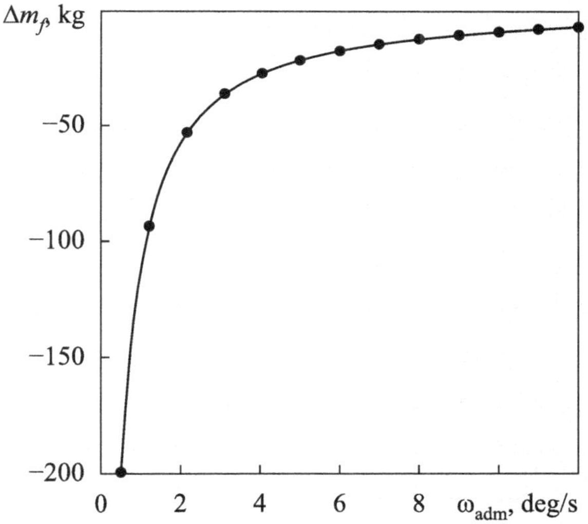

Fig. 6.79 The maximum injected mass change depending on the admissible pitch angle rate

6.4.5 Accuracy Verification of the Optimization Problem Solution

In order to control the accuracy of calculations, the constancy of the Hamiltonian was checked in LV trajectory except segments with constraints that contain explicitly the time (constraints on the pitch angular rate and thrust change rate). An example of the dependence of the Hamiltonian increment on the optimal *P*-type LV trajectory is shown in Fig. 6.81. The admissible error in solving the optimization problem (taking

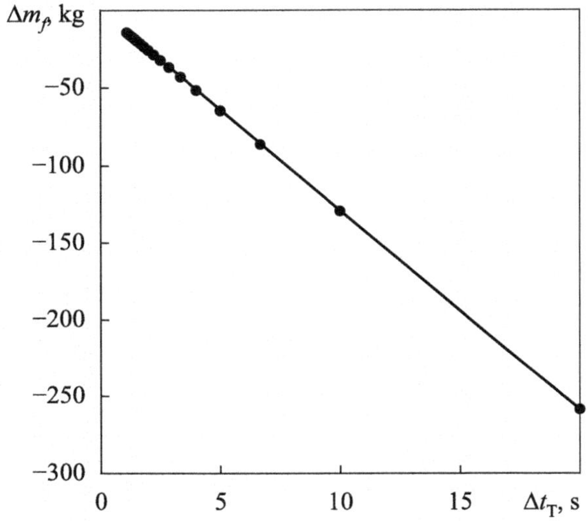

Fig. 6.80 Influence of the thrust increase duration on the maximum injected mass

into account the errors of numerical integration and BVP solution) corresponds to allowable injected mass uncertainty not more than $\Delta m_{\mathrm{orb}} = 0.1$ kg. Therefore, in accordance with the methodology described in Sect. 1.10, the maximum deviations of the Hamiltonian from a constant function were allowed within:

$$|\Delta H| < \frac{\Delta m_{\mathrm{orb}}}{m_i} \frac{\sqrt{R_0/g_0}}{t_f - t_i} \approx \frac{0.1}{7 \cdot 10^5} \cdot \frac{806}{760} \sim 10^{-7}.$$

The exception was the segments with constraints depending explicitly on time, where the Hamiltonian is not an invariant:

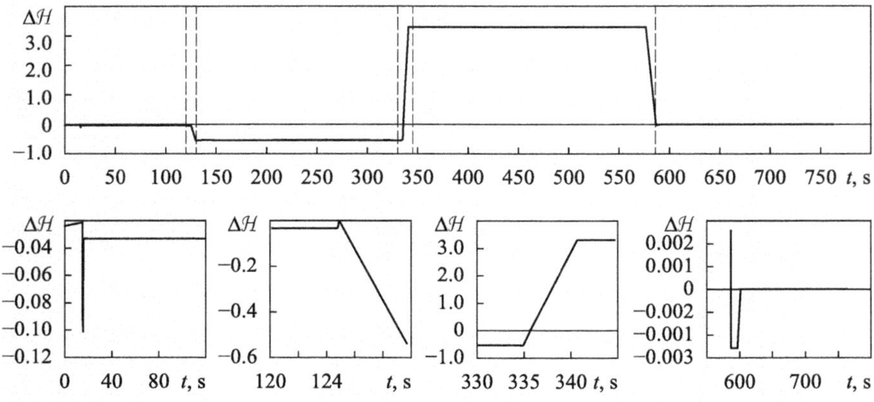

Fig. 6.81 Time-history of the Hamiltonian change on the *P*-type LV optimal trajectory

0 s ≤ t ≤ 15.00 s is the vertical ascent section;
15.00 s < t < 15.753 s is the pitch turn section with the given angle-of-attack;
15.753 s < t < 124.798 s is the g-turn section;
124.798 s < t < 124.949 s is the section, simulated of the aftereffect impulse when the 1st stage engines are turned off (linear thrust reduction);
t = 124.949 s is the 1st staging moment;
124.949 s < t < 129.670 s is linear increase in the 2nd stage thrust and its programmed pitch turn with limited angular rate;
334.920 s < t < 335.635 s is the section, simulated the aftereffect impulse when the 2nd stage engines are turned off (linear thrust reduction);
t = 335.635 s is the 2nd staging moment;
335.635 s < t < 340.660 s is the section of linear increase in the 3rd stage thrust;
335.685 s < t < 340.805 s is the program pitch angle change with the maximum admissible rate;
578.050 s < t < 586.576 s is the segment, simulated the aftereffect impulse when the 2nd stage engines are turned off (linear thrust reduction);
t = 586.576 s is the moment of the US separation;
586.576 s < t < 596.576 s is the passive segment of the US flight;
596.576 s < t < 601.376 s is the linear increase in the US thrust.

6.5 Through Optimization of ASTS Injection Trajectories with An-124 Carrier Aircraft

ASTS consists of An-124-100 carrier aircraft and two-stage LV, placed in a launching container (LC) inside the CA (Fig. 6.82). The LV is pushed nose forward from the LC through the CA tail section. A similar ASTS scheme was proposed by the AirLaunch Corporation [20]. Characteristics of a hypothetical LV taken into account in the calculations may not fully correspond to the specified project.

Optimization of trajectories and control of CA and LV was carried out taking into account constraints adopted for the CA and LV (see Sect. 6.5.1), the requirements for the safe separation of LV and CA, as well as the conditions for the safe CA flight in a failure of the "postponed launch" type. The requirement of safe separation in through optimization is taken into account in accordance with the methodology

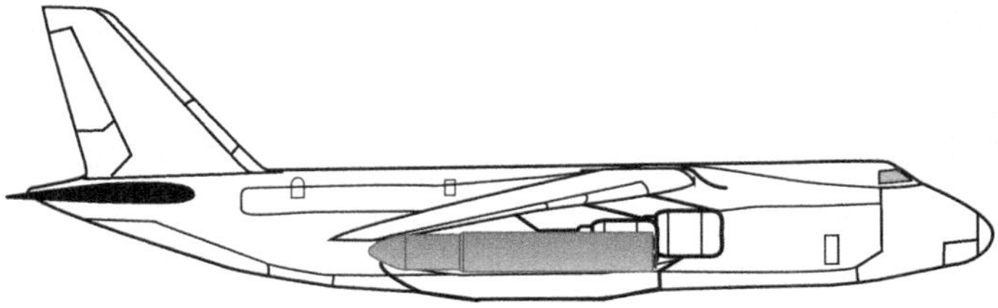

Fig. 6.82 Scheme of the AirLaunch ASTS with RS inside CA

described in Chap. 4. Therefore, the question of the feasibility of LV safe separation requires a special additional study, taking into account the features of CA control system, dynamic loading of the CA and LV structures, aerodynamic interference, etc. These aspects of the problem are not considered here.

The calculations were carried out using *ASTER* software. The relative error in calculating the optimal control and trajectory parameters did not exceed 10^{-8}, the absolute error in calculating the maximum injected mass, due to errors in the numerical integration of phase and conjugate ODE systems in all flight segments and the fulfillment of boundary conditions when solving the BVP, was no more than 0.1 kg.

The aerodynamic characteristics of LV and CA were set in accordance with the characteristics of the prototypes available in the data bank of *ASTER* program.

In the process of the through optimization, the branching trajectory was simulated, consisting of the following main segments (see Fig. 6.83):

the main branch

- CA departure from the optimal cruising flight mode;
- pre-launch maneuver of CA with LV;
- short-term pre-release maneuver to create conditions for safe LV separation;
- LV separation from CA and short-term (~ 5 s) passive flight;
- main engine ignition and intensive LV turn to reach the optimal pitch angle;
- LV 1st stage injection;
- LV 2nd stage insertion into the given polar orbit with HF separation in an intermediate point, where the dynamic pressure is rather small;

side branch 1 (*actual*)

- CA return to the cruising flight mode after LV separation;

side branch 2 (*Imaginary*)

- CA return after pre-launch and pre-release maneuvers without LV separation (failure of "postponed launch" type).

The functional is the injected mass to be maximized. The given operational limitations for CA and LV were taken into account.

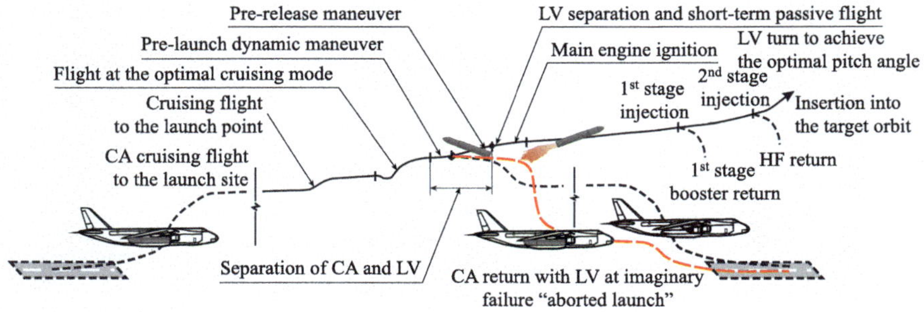

Fig. 6.83 Scheme of the AirLaunch ASTS branching trajectory

Two types of optimal injection trajectories are considered:

(1) Without a pre-launch maneuver of CA, when the pre-release maneuver begins with the optimal mode of cruising flight of CA with LV;
(2) With the optimal CA pre-launch maneuver of a limited duration ($\leq$40 s).

The advantage of the first type of trajectories is that they do not require the CA operation in "pull-up" flight modes that are not typical for heavy transport aircraft. The advantage of the second type trajectories is that they ensure the insertion of the maximum payload mass into the target orbit.

6.5.1 Source Data

ASTS consists of An-124–100 carrier aircraft and LV (Fig. 6.82). The initial CA mass is 300 t. Nominal LV data are given in Table 6.11.

Calculations of optimal spatial trajectories were carried out using *ASTER* program. The extended phase vectors had a dimension of 7. As the calculations show, the constraints on the imaginary aborted launch trajectories are not essential. Therefore, the side branch for CA is not taken into account. SP drop points are not limited. Therefore, the side branches corresponding to the return of the 1st stage SP and HF are not taken into account. The general dimension of the BVP is 8. Varied parameters of the BVP are the following: 7 initial values of conjugate variables (due to time constraints, the Hamiltonian at the initial point is not defined). Varied parameters of the inner BVP include the flight path angle and azimuth at the beginning of the g-turn of the 1st stage, 6 values of the conjugate variables at the initial points of two remaining side branches (the drop of SPs of 2nd and 3rd stages). The residuals are the conditions at the right ends of the main and three side branches.

Table 6.11 LV source data

Initial mass, t	80.0
Longitudinal load factor	$\leq$ 4.5
Normal load factor on the turn segment after main engine ignition	$\leq$ 1.0
Parameter $q\alpha$, kgf deg/m^2	$\leq$ 15,000
1st stage	
Initial thrust-to-weight ratio	2.25
Specific impulse, s	276/360
2nd stage	
Initial thrust-to-weight ratio	1.35
Specific impulse, s	370
Latitude of launch site	0
Orbit inclination, deg	90
Orbit altitude, km	200

6.5.2 *Influence of CA Steady Flight Modes at the Beginning the Pre-launch Maneuver*

The boundaries of the area of possible CA steady-state level flight (static selling) are shown in Fig. 6.84. The initial conditions correspond to the maximum cruising thrust of the D-18T engines. When performing a pre-release (generally, pre-separation) maneuver, the CA thrust increased to the nominal value, that is about 5% higher than the maximum cruising. The dots mark the initial modes of the CA flight considered in the parametric calculations. Nearby are the corresponding values of maximum injected mass m_f. The optimal initial cruise mode corresponds to the altitude and velocity slightly lower than the maximum achievable by CA in cruise flight. The time-changes in optimal trajectory variables for the best initial level flight conditions ($v_{i\ opt}$ ≈800 km/h, $h_{i\ opt}$ ≈11 km) are shown in Fig. 6.85.

Although the CA pre-launch dynamic maneuver is not performed, the parameters of LV motion after RS release are noticeably different from those given above, since the CA performs a short-term pre-release maneuver to ensure safe separation conditions, and a 5-s delay is introduced for firing the LV main engines after leaving the CA. The maximum injected mass for these initial conditions is m_f = 4,883 kg.

At the pre-launch maneuver, the CA engines operate at the nominal mode. A typical dependence of the maximum injected mass on the duration of the pre-launch dynamic maneuver is shown in Fig. 6.86. It is seen that the use of a short (~ 30–40 s) optimal pre-launch maneuver can gain in the injected mass up to 140 kg. The optimal CA pre-launch maneuver (including the pre-release maneuver) is shown in Fig. 6.87.

When performing the optimal CA pre-launch and pre-release maneuvers with a total duration of 35 s, the following optimal values of the phase variables are

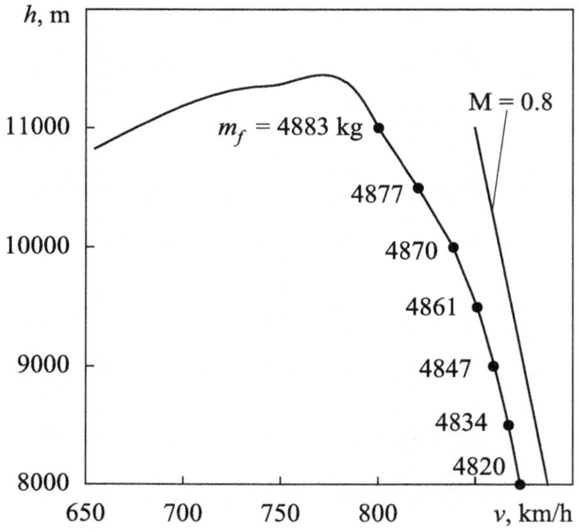

Fig. 6.84 The boundary of the area of possible steady-state level flight of CA with RS, and the maximum injected mass for various initial points of the pre-release maneuver

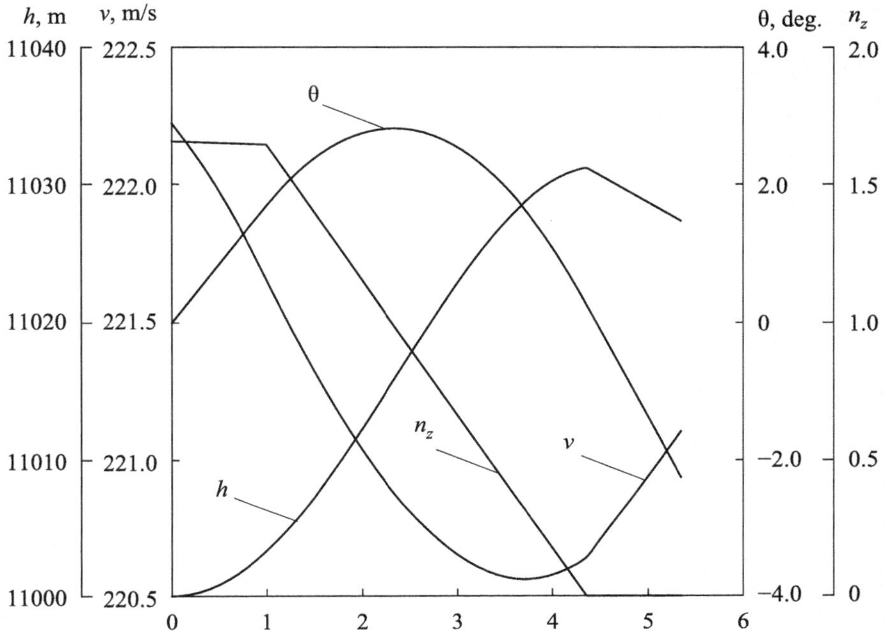

Fig. 6.85 The change of the altitude, velocity, flight path angle and normal load factor over time of the optimal CA pre-release maneuver for the best initial conditions of CA cruising flight

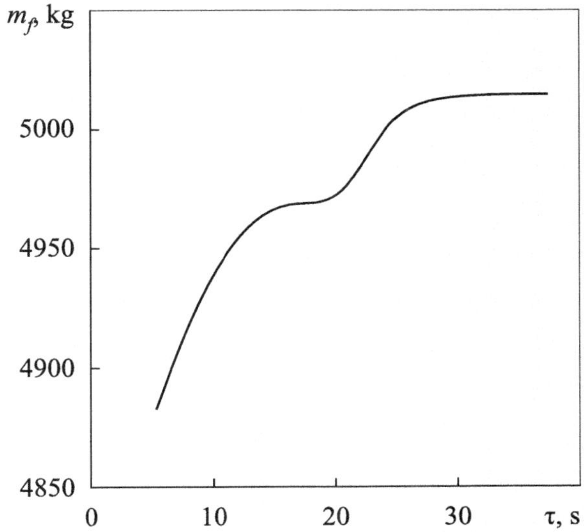

Fig. 6.86 Dependence of the maximum injected mass on the duration of CA pre-launch dynamic maneuver

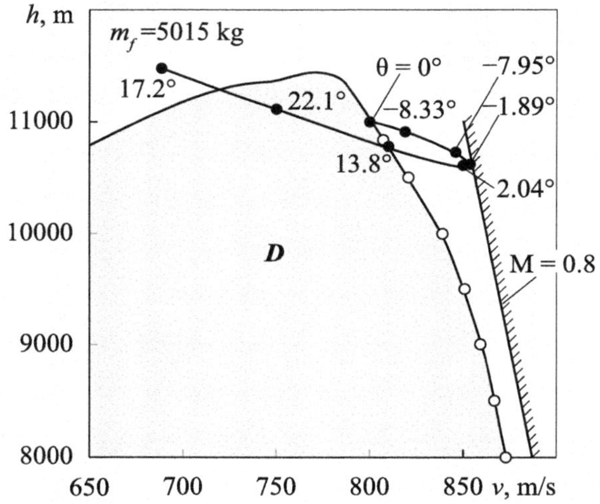

Fig. 6.87 The optimal CA pre-launch maneuver. Flight path angles are marked every 5 s of the flight

realized at the time of LV release: $v = 688$ km/h, $h = 11{,}480$ m, $\theta = 17.2°$. The corresponding maximum injected mass is $m_f = 5015$ kg. Close values of the injected mass are obtained at the beginning of the CA maneuver at lower altitudes, however, at this, the flight path angle at the end of the maneuver increases so, that can create additional difficulties in emergency case of the aborted launch.

It is important to note in Fig. 6.87 that the optimal path angle at the LV release moment is less than the maximum one on the optimal trajectory. The optimal CA pre-launch dynamic maneuver, generally, does not correspond to the traditional aircraft maneuvers of the "pull-up" type. Changing the trajectory variables on a traditional "pull-up" depends on maneuver duration τ. At $\tau < \tau^*$, the flight path angle and altitude increase monotonically, while the speed decreases. At $\tau > \tau^*$, the flight path angle begins to decrease. The optimal point on such the "pull-up", from which LV inserts the maximum payload, corresponds to the duration $\tau < \tau^*$. The optimal CA pre-launch maneuver is only at small τ coincides with the traditional "pull-up". With longer CA maneuvers, an increase in the injected mass can be ensured by optimizing the control program. The change in the maximum injected mass depending on the duration of the maneuver and a comparison it with the traditional "pull-up" maneuver is shown in Fig. 6.88.

6.5.3 *Influence of Safety Requirements for LV Airdrop*

Airdrop of LV, oriented aft to the oncoming flow, requires that after dropping and firing the main engines, LV turns intensively to reach the optimal pitch angles with minimum energy losses. In this case, quite large transverse load factors (0.5–0.7) inevitably arise and, most importantly, about 5–8 tons of LV fuel, which corresponds

Fig. 6.88 The maximum injected mass depending on the duration of the optimal CA pre-launch maneuver and a comparison with the typical aircraft pull-up maneuver

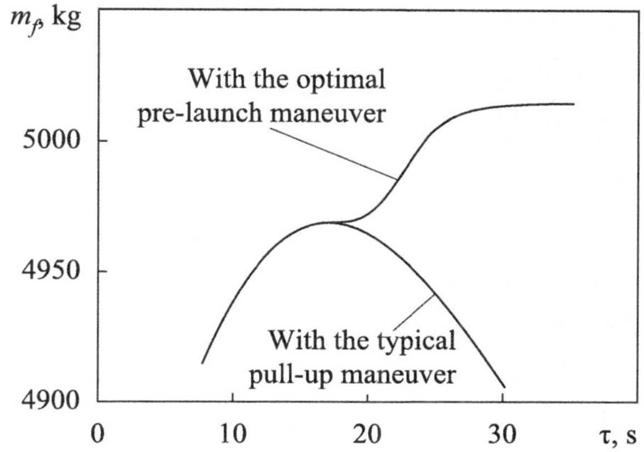

to a loss of 300–450 kg of the injected mass, are consumed inefficiently. These losses can be reduced by reversing the LV in LC and, accordingly, at airdropping. However, it induces problems of fuel intake for LV main engines, loading the LV aft at the exit from LC, etc.

Calculations of LV optimal branching trajectories for various pitch angles ϑ at time t_E of main engines ignition are carried out. The total duration of the optimal pre-launch and pre-release CA maneuvers was 35 s. The delay of main engine ignition after LV airdrop was fixed and amounted to 5 s.

The change in the payload mass with a variation of $\vartheta(t_E)$ in the range [− 72°, 180°] compared with the design case $\vartheta(t_E) = 180°$ is shown in Fig. 6.89. A qualitative picture of the influence of $\vartheta(t_E)$ on the maximum payload mass is shown in Fig. 6.90. Here, the hodograph of the vector is shown in the axes of the Earth's coordinate system. The direction of this vector coincides with the LV longitudinal axis at the engine ignition moment, and the module is equal to the corresponding increment of the maximum payload mass Δm_{pl}. It also shows LV velocity vector $\mathbf{v}(t_E)$ at the same time moment and the optimal LV orientation, to which LV tends during the turn, for various values of $\vartheta(t_E)$. That is, by the optimal orientation here we mean the LV attitude that it should have taken under optimal control at the current time moment, if there were no constraint on the angular rate of the LV's turn. The optimal LV orientation vectors form a sector with the maximum pitch angle $\vartheta_{opt} \approx 61°$ in case of $\vartheta(t_E) \approx 180° \div 160°$ and minimum $\vartheta_{opt} \approx 54°$ for $\vartheta(t_E) \approx -30°$ to $-40°$. The LV passive turn with the main engines ignition in a vicinity of the optimal pitch angles makes it possible to increase the payload mass by 17% compared to the case of ignition at a design pitch angle of 180°.

The dependence of the maximum injected mass on the admissible Mach number M_{max} for CA with LV is shown in Fig. 6.91. In the considered range [0.755, 0.800], this dependence is close to linear, and a decrease in M_{max} by 0.01 leads to a loss of ~ 10 kg of the maximum injected mass.

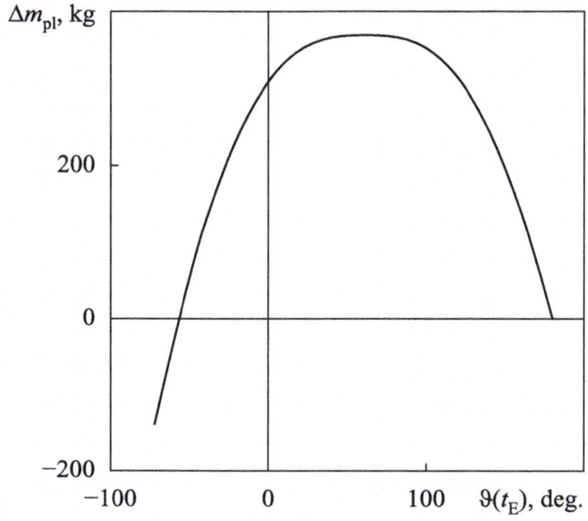

Fig. 6.89 The increment of the maximum payload mass depending on LV pitch angle $\vartheta(t_E)$ at the time of main engine ignition compared to the design case of $\vartheta(t_E) = 180°$

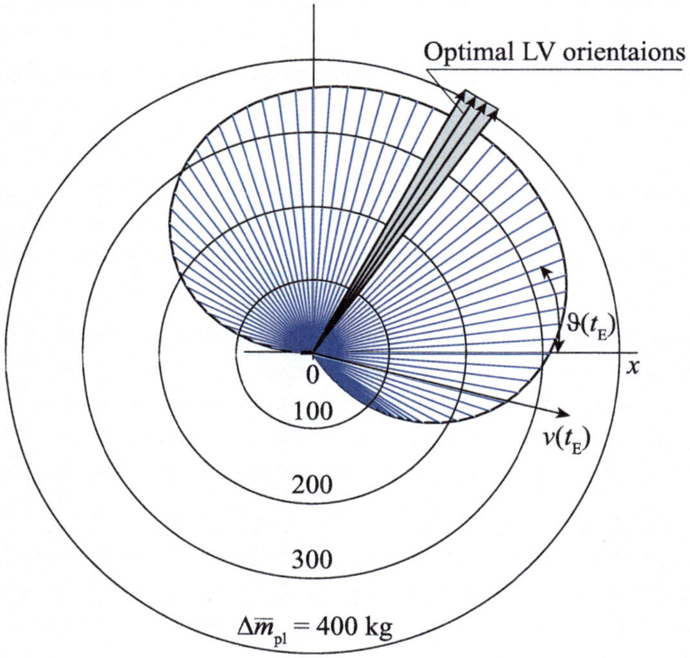

Fig. 6.90 Hodograph of the vector in the direction of the LV longitudinal axis at the moment of main engine ignition, the module of which is equal to the increment of maximum payload mass Δm_{pl}

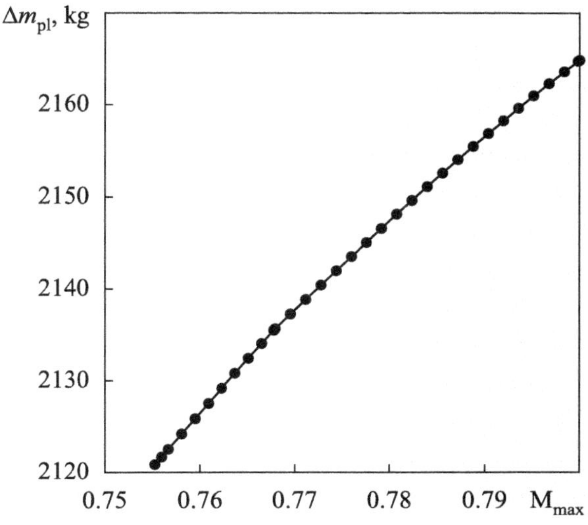

Fig. 6.91 Dependence of the maximum payload mass on the admissible Mach number for CA with LV

6.5.4 *Analysis of Optimal Solution Accuracy*

In order to control the accuracy of calculations, the constancy of the Hamiltonian was checked in flight sections without constraints containing the time (constraints on the pitch angle rate, thrust change rate, etc.). An example of a dependence of the Hamiltonian on the flight time on the optimal ASTS branching trajectory is shown in Fig. 6.92. The allowable error in the optimization problem solution (the errors of numerical integration and BVP solution) corresponded to 0.1 kg of the injected mass. Therefore, in accordance with the methodology described in Sect. 1.10, the maximum deviations of the Hamiltonian from a constant value were allowed within:

$$|\Delta H| < \frac{\Delta m_{\rm orb}}{m_i}\frac{\sqrt{R_0/g_0}}{t_f - t_i} \approx \frac{0.1}{300000}\cdot\frac{806}{360} \sim 10^{-6}.$$

In particular, as can be seen from Fig. 6.92, in time interval 13 s < t < 18 s, corresponding to the motion along the phase constraint on Mach number, the error of compliance with the considered integral of motion is several orders of magnitude less than the allowable one taking into account the rest flight time in relation to this segment duration (~ 360/5). In interval 19.2 s < t < 30.6 s, maintaining the constraint on the CA angle of attack, the error is about $6{\cdot}10^{-8}$ that is accordance the admissible value ($6{\cdot}10^{-8}{\cdot}360/11{\sim}2{\cdot}10^{-6}$) taking into account the rest flight time in relation to this segment duration (~ 360/11). In the atmospheric segment 50 s < t < 200 s of the LV optimal autonomous flight, the increments in the Hamiltonian lie within the allowable limit ($6{\cdot}10^{-7}{\cdot}310/150{\sim}1{\cdot}10^{-6}$). The error in the fulfillment of transversality conditions in the Hamiltonian at the right end of the trajectory is an order of magnitude lower than the allowable one.

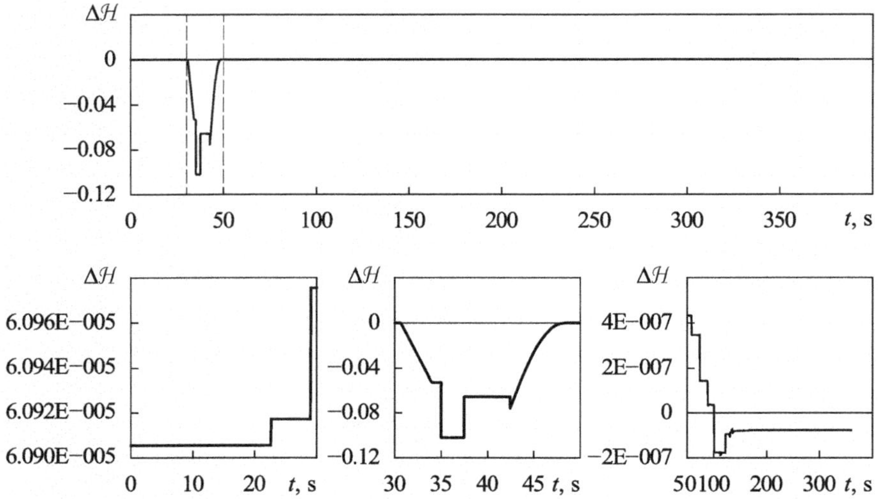

Fig. 6.92 Time-history of the Hamiltonian change on the optimal ASTS branching trajectory

Changes in the Hamiltonian between the 30th and 50th seconds are associated with the fulfillment of time-dependent constraints:

30.64 s < t < 33.99 s is the programmatic reduction of the angle of attack with a limited angular rate until reaching $n_z = 0$ (pre-release maneuver);
33.99 s < t < 34.99 s is the flight with $n_z = 0$;
$t = 34.99$ s is the beginning LV movement inside LC;
34.99 s < t < 37.46 S is LV acceleration in LC with $n_x = 1$;
$t = 37.46$ s is the LV exit from LC;
37.46 s < t < 42.46 s is the LV autonomous flight with the main engine turned off;
$t = 42.46$ s is the LV main engine ignition;
42.46 s < t < 47.46 s is the programmed LV pitch turn with a given angular rate;
$t = 47.46$ s is the time of attainment of the optimal LV attitude.

6.6 Through Trajectory Optimization of STS with Reusable Winged Boosters

LV with reusable winged 1st stage boosters (RWB) is considered. Two scenarios for the RWB return is investigated: basic and alternative. The Basic Concept was proposed by Khrunichev Space Center as the MRKS-1 project and provides for the return of the spent RWB to the launch site using special Air-Breathing Engines (ABE) in the cruising flight at subsonic speed [21]. The Alternative Concept is proposed in [22, 23], it allows a non-powered landing on one of the intermediate airfields, reached with the optimal spatial maneuvers of both the RWB and LV. The results of calculations of the optimal branching trajectories are presented, justifying the effectiveness of the proposed alternative solutions in terms of increasing the payload mass, reliability of launches and reducing their cost. Following RWB layouts are considered: with a delta and straight tilt wings, with and without subsonic ABE, and a special RWB-transformer.

6.6.1 Problem Overview

The future of LVs lies in fully reusable ones. In this sense, the LV with RWB returning to the launch site is one of the possible steps on this way. The RWB rescue in this project is primarily aimed at reducing SP drop fields. In addition, the possibility of reusing RWB makes it possible to reduce the cost of each launch in case of a sufficient frequency of launches. Additional opportunities for improving the RWB efficiency are opened up by the diversification of RWB and infrastructure use outside the project.

At the same time, the organization of RWB rescue will inevitably lead to an increase in the relative mass of the first stage structure and STS development costs. Although the elimination of drop areas undoubtedly remains dominant, every opportunity to reduce cost should be used in STS design. The developed methods let to evaluate the effectiveness of using the technological experience accumulated in aviation (aircraft with variable geometry, aerodynamic and structural optimization, optimal use of lifting capabilities, unpowered landing, etc.).

First of all, from these positions it is necessary to critically analyze the basic concept of the RWB return to the launch site with help of ABE, which are supposed to be started at a subsonic speed after passing through weightlessness and high overload conditions in dense layers of the atmosphere, which are atypical for such aircraft engines. Actually, the return of the RSB to the launch site does not raise any questions, unless it is understood literally "an urgent, non-stop return of the spent RWB to the launch site." Since in the basic-concept interpretation the ABE and additional fuel for them will first have to be "delivered into space" (separation of the 1st stage occurs at the Mach number ~ 7 at the altitude of ~ 50 km). This increases enormously the cost of each kilogram of ABE with its fuel, since it is determined by the specific cost of the payload delivered into orbit (more precisely, one tenth of this cost, taking into account the contribution of the 1st stage).

The alternative solution is to use not only the airfield in the launch site area, but also the existing network of airfields that RWB can reach in unpowered flight after staging. From such an intermediate airfield, the RWB can be delivered to the launch site by rail, water, road or air transport, or autonomously using attached ABE and fuel tank. In this case, it is possible to increase the injected payload and launch reliability due to:

- exclusion of ABE, its fuel and related infrastructure in the launched STS layout;
- reducing the RWB and 1st stage aerodynamic drag due to the best layout solutions without ABE;
- increasing the lifting capability of the 1st stage (see Chap. 3);
- reduction of RWB loads at the reentry in dense atmosphere layers, which can reduce the structure mass;
- elimination of the problem of ABE starting after flight in quasi-weightlessness, the impact of high load factors and heat fluxes at the reentry, etc.

6.6.2 Source Data

The MRKS-1 layout of two RWB and two Central Blocks (CB) of the 2nd stage (Fig. 6.93), presented in [21], is used as the Basic Concept here. The main characteristics of the STS main engines are given in Table 6.12 [24, 25]. At the STS start, all the engines of the RWB and CB are simultaneously ignited.

In the ascent phase, the longitudinal load factor was limited: $n_x \leq 4$, by the main engine throttling. It was assumed that the specific impulse does not depend on the throttle ratio; there is no fuel transferring between stages.

Four types of RWB are considered (Table 6.13 and Fig. 6.94). Types A and B correspond to [21]. Subscripts " + " and " – "indicate, respectively, the presence or absence of the ABE and its fuel tanks.

In the nose of the RWB-A$^+$, three ABEs are installed. On RWB-B$^+$, due to significantly better subsonic aerodynamic characteristics, two ABEs are installed.

The layout of the RWB-R [23, 26] differs significantly from A and B in terms of the form and manner of application, which provides for deep structural transformations to improve aerodynamic characteristics on different flight modes [27]:

- no ABE;
- straight swivel wing with foldable consoles; the wing aspect ratio is 1.3 times higher as compared to B$^+$;
- no vertical tail;
- before the return flight, RWB-R turns 180° (tail forward) simultaneously with the wing deployment;
- the tail is V-shaped (at angle of 45°) and moves out the body in return flight mode;
- main engines are closed with heat shielding doors after the RWB separation (the door mass equals that of the RWB B$^+$ fairing).

The RWB-R flight configurations are shown in Fig. 6.95: (a) as a part of STS; (b) in the autonomous return flight. After landing at an intermediate aerodrome, a pair of ABE of sufficient power can be installed on RWB, the wing tanks be filled with fuel, and other necessary operations be performed to prepare an unmanned flight to an airfield near the launch site. Such operations are not unique in aviation practice and, according to existing standards, take less than an hour [23].

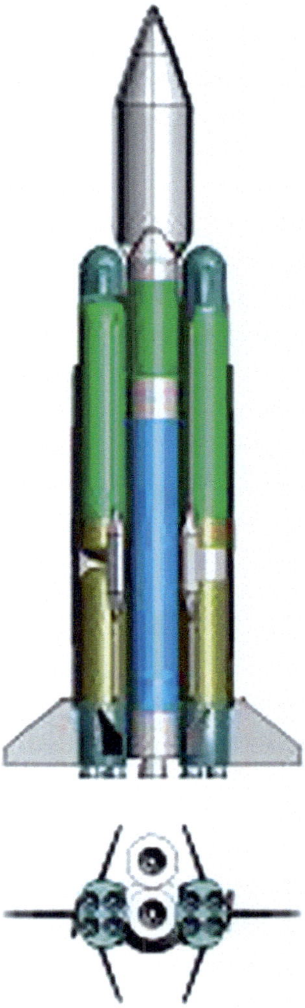

Fig. 6.93 The STS layout [21]

Table 6.12 The STS main engine characteristics

Name	RWB	CB
Number of engines on each unit	4	1
Engine's type	RD 191	RD0120
Engine thrust, tf	212/196	190/146
Specific impulse in vacuum, s	337	452

Table 6.13 RWB options

RWB type	ABE	Wing type	Return concept
A^+	Yes	Delta wing (like Buran ASV)	To the launch site
A^-	No		To an intermediate airfield
B^+	Yes	Straight swivel wing (like Baikal RWB)	To the launch site
B^-	No		To an intermediate airfield
R ("tRansformer")	No	Straight swivel wing with foldable consoles	To an intermediate airfield

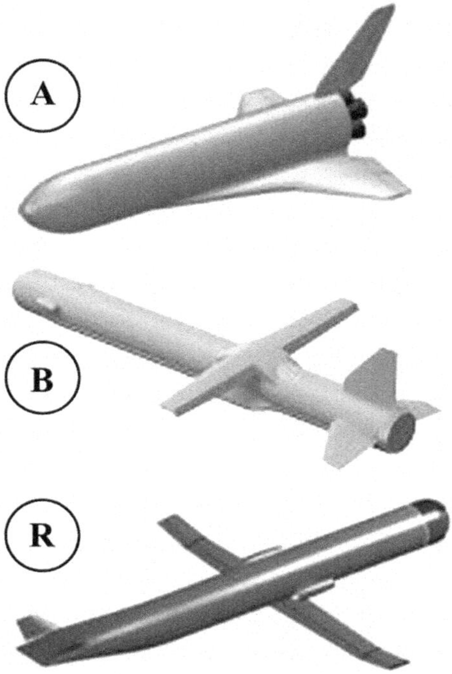

Fig. 6.94 The RWB layouts A, B [21] and R [23]

Schemes of the STS launch trajectories for the two concepts are shown in Fig. 6.96. After turning off main engines, the passive flight of the 2nd stage begins. The trajectory parameters at the end of the powered phase ensure the drop of the 2nd stage SP and HF at the Earth's antipodal point. After separation of the HF and 2nd stage SP, the payload with the booster block continues passive flight along the transfer orbit until the apogee at $h_{orb} = 200$ km, where they are given a velocity impulse to get the local circular speed. The specific thrust impulse during final ascent is equal to $J_\infty = 300$ s. The mass summary of STS with RWB-A^+ in a dimensionless form is given in Table 6.14. The relative masses of other types of RWB are presented in Table 6.15.

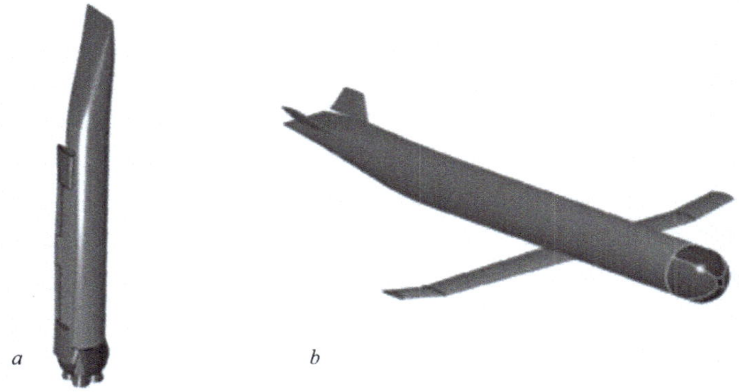

Fig. 6.95 The RWB-R flight configurations [23]: the launch configuration as part of STS (**a**) and the configuration in autonomous flight after separation (**b**)

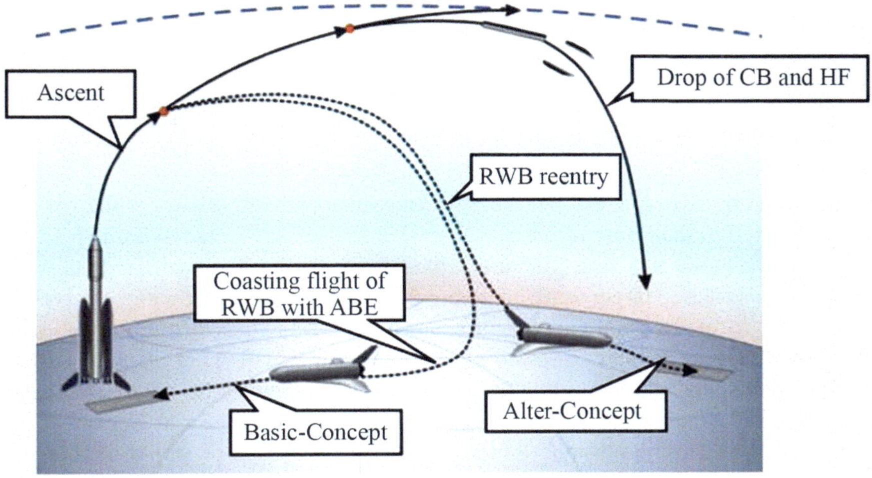

Fig. 6.96 Schemes of branching trajectories for two concepts of STS with RWB

To simplify the comparison of aerodynamic coefficients for different types of STS and RWB, the lift coefficient C_L was assumed to be proportional to angle of attack α with coefficient $C_{L\alpha}$, and the drag polar was quadratic:

$C_L = C_{L\alpha} \cdot \alpha$, $C_D = C_{D0} + A_D C_L^2$,

where C_D is the drag coefficient, C_{D0} is the drag coefficient at zero angle of attack, A_D is the induced drag coefficient. It was assumed that ABE does not affect lift, but a drag only (Table 6.16).

The aerodynamic characteristics of RWBs and corresponding types of the 1st stages are given in Tables 6.16 and 6.17 [22, 23]. The cross-sectional area of one CB is taken as the reference area at the STS ascent mode (RWB with a delta-shaped wing

Table 6.14 Relative mass distribution for STS with RWB-A$^+$

Characteristic	Value
Initial mass of STS	1
Initial mass of RWB-A$^+$, including masses of	2 × 0.3232
main engine fuel	2 × 0.2796
unprocessed fuel residues	2 × 0.002571
ABE fuel	6 × 0.001857
dry mass	2 × 0.03544
Initial mass of CB, including masses of	2 × 0.1558
main engine fuel	2 × 0.1397
unprocessed fuel residues and reserve	2 × 0.001664
dry mass	2 × 0.01444
HF mass	0.005714

Table 6.15 Relative initial masses of RWB types

RWB types				
A$^+$	A$^-$	B$^+$	B$^-$	R
1	1.1015	1.0155	1.091	1.1075

has the same body diameter). The relative reference areas of RWBs at the reentry mode (referred to the reference area of the 1st stage) are presented in Table 6.18.

According to the RWB aerodynamic characteristics, the RWB with a delta wing at high speeds (due to larger wing area) and the RWB with straight wing at subsonic speeds (due to greater elongation and span) have the greatest lifting capabilities. The greatest effect of increasing the aspect ratio of RWB-R wing reveals itself at low subsonic speeds and amounts to ~ 7%, which contributes to better conditions for unpowered landing. At subsonic speeds, the highest lift-to-drag ratio $K \approx 13$ has RWB-R (~ 30% higher than RWB-B$^-$) due to the lowest drag and the highest wing aspect ratio. The lifting capabilities of RWB with a delta wing is much lower, $K \approx 5.5$. At high supersonic speeds, the highest lift-to-drag ratio corresponds to delta-wing RWB ($K \approx 1.67$) and the smallest one corresponds to RWB with straight wing ($K \approx 1.1$).

6.6.3 *Statement of the Through Optimization Problem for STS with RWB*

Since the segments of the STS ascent and SP return to specified fields are interrelated, objective data on the optimal trajectories and influence functions of the problem parameters on the functional can only be obtained on the basis of the simultaneous

Table 6.16 Aerodynamic characteristics of different types of RWB

	A^+			A^-		
M	C_{D0}	A_D	$C_{L\alpha}$, deg^{-1}	C_{D0}	A_D	$C_{L\alpha}$, deg^{-1}
0	0.0518	0.159	0.0543	0.0518	0.159	0.0543
0.26	0.0518	0.159	0.0543	0.0518	0.159	0.0543
0.6	0.0548	0.205	0.0601	0.0548	0.205	0.0601
0.8	0.0631	0.245	0.0724	0.0631	0.245	0.0724
0.9	0.0942	0.280	0.0707	0.0942	0.280	0.0707
1	0.1393	0.261	0.0785	0.1300	0.261	0.0785
1.2	0.1556	0.317	0.0625	0.1380	0.317	0.0625
1.5	0.1476	0.384	0.0504	0.1300	0.384	0.0504
2	0.1347	0.498	0.0385	0.1150	0.498	0.0385
3	0.1200	0.680	0.0268	0.1000	0.680	0.0268
4	0.1082	0.859	0.0209	0.0905	0.859	0.0209
5	0.1036	1.018	0.0179	0.0850	1.018	0.0179
7	0.1033	1.075	0.0168	0.0835	1.075	0.0168
	B^+			R		
M	C_{D0}	A_D	$C_{L\alpha}$, deg^{-1}	C_{D0}	A_D	$C_{L\alpha}$, deg^{-1}
0	0.0635	0.044	0.2222	0.0444	0.034	0.2380
0.29	0.0635	0.044	0.2222	0.0444	0.034	0.2380
0.4	0.0637	0.047	0.2231	0.0449	0.036	0.2382
0.61	0.0925	0.075	0.2467	0.0673	0.059	0.2640
0.78	0.1375	0.114	0.2200	0.3448	0.100	0.2564
1	0.4000	0.100	0.2564	0.3560	0.229	0.0771
1.67	0.3922	0.229	0.0771	0.3450	0.405	0.0475
2.5	0.3654	0.405	0.0475	0.3378	0.528	0.0368
4	0.3462	0.528	0.0368	0.3311	0.614	0.0264
7	0.3364	0.614	0.0264	0.0444	0.622	0.2380

consideration of all flight segments, i.e., through optimization of branching trajectories. For this, *ASTER* complex [28] was used based on the rigorous application of PMP. The RWB return section was simulated taking into account the lifting capabilities and approximately optimal control of angle of attack α and bank angle γ in accordance with [1–5]:

$$\alpha_{\text{opt}} = \begin{cases} \alpha_{\text{adm}}, \psi_v \geq \pi/2, \\ \alpha_{K\max}, \psi_v < \pi/2, \end{cases} \quad \gamma_{\text{opt}} = \begin{cases} \gamma_{\max}, \psi_v \geq \gamma_{\max}, \\ \psi_v, \psi_v < \gamma_{\max}, \end{cases}$$

where $\psi_v = \arccos(\mathbf{e}_R \times \mathbf{e}_L, \mathbf{e}_R \times \mathbf{e}_v)$ is the sighting angle between the instantaneous motion plane and great circle plane passing through the center of mass and collinear

Table 6.17 Aerodynamic characteristics of the 1st stage with different types of RWB

	A^+			A^-			B^+			R		
M	C_{D0}	A_D	$C_{L\alpha}$, deg^{-1}	C_{D0}	A_D	$C_{L\alpha}$, deg^{-1}	C_{D0}	A_D	$C_{L\alpha}$, deg^{-1}	C_{D0}	A_D	$C_{L\alpha}$, deg^{-1}
0	1.244	0.0292	0.500	1.2444	0.0292	0.5	0.632	0.0448	0.400	0.612	0.0733	0.275
0.4	1.249	0.0296	0.530	1.2485	0.0296	0.53	0.632	0.0451	0.430	0.612	0.0737	0.272
0.6	1.294	0.0302	0.598	1.2937	0.0302	0.599	0.657	0.0448	0.478	0.635	0.0730	0.269
0.8	1.784	0.0287	0.700	1.784	0.0287	0.7	1.070	0.0395	0.571	1.070	0.0634	0.268
1	3.656	0.0239	0.750	3.091	0.0239	0.75	2.472	0.0297	0.650	2.301	0.0581	0.272
1.2	4.387	0.0246	0.718	3.6	0.0246	0.718	3.167	0.0302	0.633	3.029	0.0578	0.280
1.5	4.145	0.0276	0.668	3.5	0.0276	0.668	3.158	0.0336	0.599	2.850	0.0591	0.299
2	3.681	0.0321	0.596	3.2	0.0321	0.596	2.880	0.0387	0.547	2.434	0.0615	0.331
3	3.051	0.0435	0.490	2.7	0.0435	0.49	2.463	0.0507	0.455	1.898	0.0700	0.351
4	2.616	0.0621	0.405	2.33	0.0621	0.405	2.136	0.0678	0.380	1.570	0.0809	0.345
5	2.396	0.0830	0.340	2.1	0.0830	0.34	1.957	0.0877	0.329	1.399	0.0957	0.315
7	2.180	0.1126	0.272	1.855	0.1126	0.272	1.7244	0.1152	0.265	1.177	0.1184	0.260

Table 6.18 Relative reference areas of RWB types

	RWB type				
	A+	A−	B+	B−	R
1st stage	1	1	1	1	1
RWB	8.327	8.327	3.512	3.512	3.512

to given vector $\mathbf{e}_L$; $\alpha_{K_{\max}}(\mathbf{x})$ is the angle of attack at which the maximum lift-to-drag ratio $K_{\max}$ is reached; α_{adm} is the maximum admissible angle of attack in accordance with given constraints:

$$\alpha_{\text{adm}} = \begin{cases} \alpha_{\max}, \, n \leq n_{\text{adm}}, \\ \alpha|_{n=n_{\text{adm}}}, \, n > n_{\text{adm}}. \end{cases}$$

In this case, RWB attitude control vector $\mathbf{e}_\tau$ is given as

$$\mathbf{e}_\tau = a\mathbf{e}_\tau + b\mathbf{e}_v + c(\mathbf{e}_v \times \mathbf{e}_R), a = \cos\alpha_{\text{opt}} - b\sin\gamma_{\text{opt}}, b = \frac{\sin\alpha_{\text{opt}} \cdot \cos\gamma_{\text{opt}}}{\cos\theta},$$
$$c = \frac{\sin\alpha_{\text{opt}} \cdot \sin\gamma_{\text{opt}}}{\cos\theta},$$

where $\cos\theta = |\mathbf{e}_v \times \mathbf{e}_R|$.

Three following problems are considered.

Problem 1 Find STS optimal control program $\mathbf{e}_\tau$ (t) (on the ascent phase), so that, taking into account the constraints on the control and trajectory, but without constraints on the RWB landing point $\mathbf{x}_f^s$, in order to maximize the final mass inserted into the target orbit:

$$m_f \Rightarrow \max \tag{6.21}$$

Problem 2 Find optimal control program $\mathbf{e}_\tau$ (t) of STS (on the ascent phase) and RWB (on the return phase), in order to maximize the final injected mass (6.21), taking into account the constraints on control and trajectory and on a given RWB landing point.

Problem 3 Find optimal spatial attitude control program $\mathbf{e}_\tau(t)$ of STS during ascent and RWB during return, in order to maximize RWB planning range ΔL, taking into account the constraints on control and trajectory and a given injected mass m_f. Range ΔL is defined in the direction of unit vector $\mathbf{e}_L$ on the Earth's surface from the conditional center of the RWB feasible area (see Sects. 4.5, 5.5 and 6.4.2). The direction of $\mathbf{e}_L$ defined by angle ψ between the launch plane and orthodromic plane passing through vector $\mathbf{e}_L$ (Fig. 6.97):

$$\Delta L(\mathbf{e}_L(\psi)) \Rightarrow \max. \tag{6.22}$$

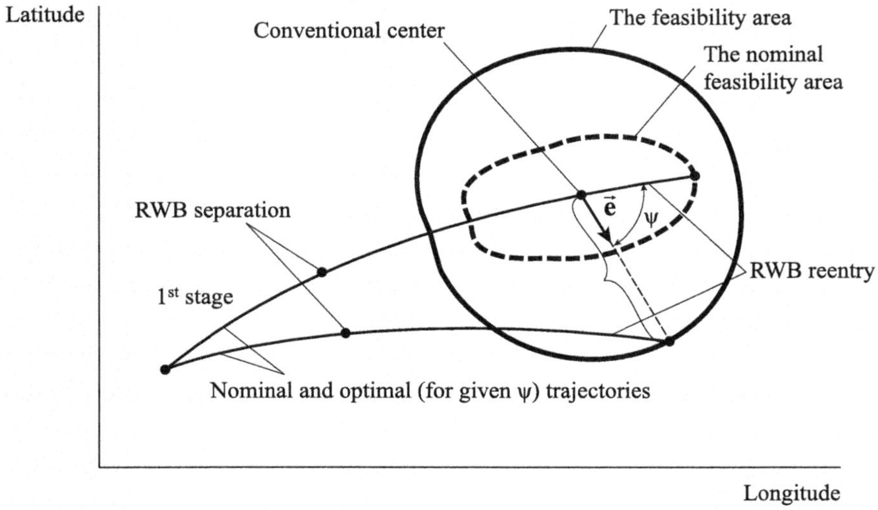

Fig. 6.97 The scheme for calculation of the RWB feasibility area boundary by continuing the Problem 3 optimal solution with respect to parameter ψ

The first and second problems are quite traditional; their solution allows determining the maximum STS injected mass with a free (Problem 1) and a given (Problem 2) RWB landing point.

The solution of the third problem (mutual to the second one) allows calculating the boundary of the Earth's surface area, feasible by RWB, with a given STS injected mass by continuously changing parameter ψ in the range from 0° to 360° (Fig. 6.97). In this case, a through optimization problem is solved, when the control of both the STS injection trajectory and RWB return maneuver is subjected to optimization. A similar formulation was used in [17] in the analytical synthesis of critical perturbations for STS SP drop trajectories (see Sect. 5.5).

The methodology for solving the optimization Problem 1 is described in Chap. 2.

When solving Problem 2, one should add the jump conditions for the conjugate vector and the Hamiltonian (transversality conditions) at the RWB separation point (5.73) and the transversality conditions at the right end of side branch (5.74) in accordance with the methodology described in Sect. 5.5.3.

When solving Problem 3 (see Sect. 5.5.4), the transversality conditions at the separation point are the same as in Problem 2, and at the right end of the side branch, determined by the given RWB landing altitude, they have the form (5.79).

6.6.4 Optimal STS Branching Trajectories with Optimal RWB Landing Points (Problem 1)

In calculations, it was assumed that the nominal target circular orbit has altitude $h_{\text{orb}} = 200$ km and inclination $i_{\text{orb}} = 51.7°$. As a result of solving Problem 1 for all RWB types, the maximum injected mass is obtained without constraints on the RWB landing point (Table 6.19). For RWBs with ABE, it is assumed that they return to the launch site. As a boundary condition at the right end of RWB return trajectory, altitude h_0 of ABE starting is set. The ABE fuel mass is estimated by the Breguet formula (not optimized) and is used when setting initial masses of RWB-A$^+$ and RWB-B$^+$.

Table 6.19 for STS, consisting of two RWB-A$^-$ and two CBs of the 2nd stage, shows that due to the denial of the ABE installation, the relative gain in the payload mass reaches

$$\Delta\overline{m}_{\text{pl}} = 0.1015. \tag{6.23}$$

The gain was obtained due to changes in several layout parameters. The influence of each of these parameters can be estimated using Bliss formula [29], which relates the variations in the right-hand sides $\Delta\mathbf{f}$ of the STS motion equation system with a change in the functional using conjugate vector $\mathbf{p}(t)$ known from the solution of the optimization problem:

$$\Delta m_f = \int_t F_{\text{inf}} dt, \tag{6.24}$$

where $F_{\text{inf}}(t) = (\mathbf{p}, \Delta\mathbf{f})$ defines the relative measure of the change in the functional with a local change in the vector of the right-hand sides caused by a variation in layout parameters.

A part of payload mass gain (6.23), determined by the improvement in aerodynamic properties due to ABE elimination, was estimated using (6.24), where Δf is caused by the replacement of RWB-A$^+$ by RWB-A$^-$. Figure 6.98 shows the local influence function F_{inf} caused by a change in aerodynamic characteristics, and the Mach number on the optimal injection trajectory of STS with RWB-A$^+$. It can be seen that the aerodynamic characteristics have the maximum influence in the transonic regime. The area under the curve $F(t)_{\text{inf}}$ determines the total increment of functional $\Delta m_f(\Delta\mathbf{f}_{\text{aer}})$ when the aerodynamic characteristics change along the entire trajectory. For RWB-A$^-$ without taking into account the change in RWB mass

$$\Delta\overline{m}_f(\Delta\mathbf{f}_{\text{aer}}) = 0.01624. \tag{6.25}$$

Table 6.19 Relative injected mass for various RWB types

	Types of RWB				
	A$^+$	A$^-$	B$^+$	B$^-$	R
$\overline{m}_f$	1	1.1015	1.0155	1.091	1.1075

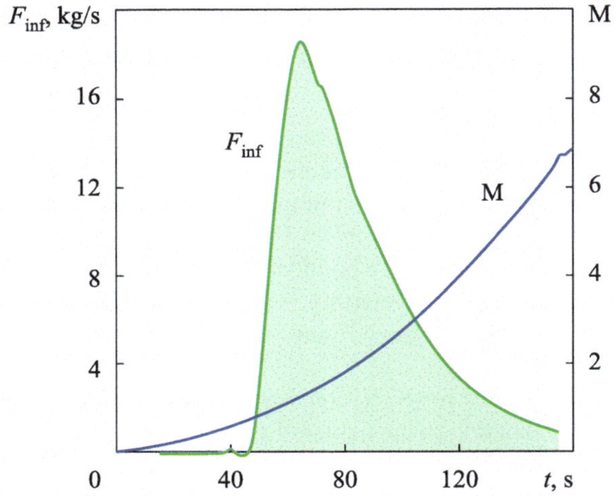

Fig. 6.98 Local influence function F_{inf} and Mach number on the optimal injection trajectory of STS with RWB-A+

In general, the rigorous optimization of branching STS trajectories provides a gain of at least 5–8% of the payload mass compared to the traditionally used g-turn, which implements the STS flight in dense atmosphere layers under nominal conditions with almost zero angle of attack [30] (see Sect. 2.2.2).

6.6.5 RWB Feasible Areas

To calculate RWB feasibility areas (FAs), a series of Problems 3 was solved for a successively changing angle ψ of deviation of unit vector $\mathbf{e}_L$. This vector sets a line on the Earth's surface on which RWB landing point could be located as far as possible from some FA conditional center (Fig. 6.97). The exact position of this center is not of great importance; the only requirement is that it lies inside the FA. In the calculations, the RWB landing point on the optimal trajectory in Problem 1 is used as this center. We will call the FA, obtained by the RWB quasi-optimal control programs with a fixed ascent trajectory, which corresponds to the optimal one in Problem 1, as the nominal (NFA).

All FAs were built taking into account the constraints on RWB angle of attack α, bank angle γ, admissible normal load factor n_z, and dynamic pressure q:

$$|\alpha| \leq \alpha_{max} = 50°,\ |\gamma| \leq \gamma_{max} = 65°,$$
$$n_z \leq n_{z\,adm} = 4,\ q \leq q_{adm} = 3000\ \text{kgf/m}^2. \quad (6.26)$$

It is obvious that the deviation of the landing point beyond the NFA will be accompanied by a decrease in maximum payload mass Δm_{pl}. Figure 6.99 shows FA boundaries with the optimal launch of STS with RWB-A$^-$ from Vostochny cosmodrome [31] into reference orbits with inclinations of 51.7° and 108° for various relative values of $\Delta\overline{m}_{\text{pl}}$: − 1%, − 2%, − 5% and − 10%. Symbols I, II mark the location of aerodromes of classes 1 and 2, potentially suitable for RWB landing [32, 33].

As follows from Fig. 6.99, payload insertion into orbits in inclination ranges

$$i_{\text{orb}} \in \left[51.7°, 61°\right] \text{ and } i_{\text{orb}} \in \left[77°, 108°\right] \tag{6.27}$$

can be performed with a loss of no more than 10% of the maximum payload mass obtained without the constraint on the landing point. This loss does not exceed the relative mass reduction in comparison with Basic Concept (cases of RWB-A$^+$ and RWB-B$^+$).

Orbital inclinations, upon insertion into which there are no available airfield for unpowered landing of RWB with a limit of 10% of the allowable loss of the maximum payload mass, lie in range:

$$i_{\text{orb}} \in (61°, 77°) \tag{6.28}$$

which is only a quarter of the considered inclination range (6.29).

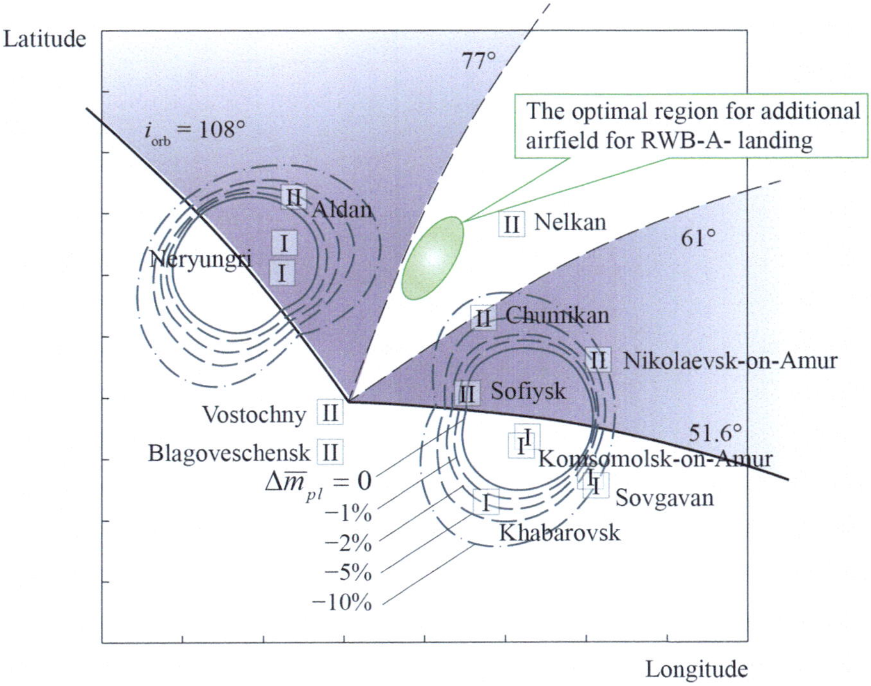

Fig. 6.99 Possible expansion of RWB-A$^+$ feasibility areas with the optimal control of LV and RWB at the cost of reducing maximum payload mass Δm_{pl}; and available airfields of the first I and second II classes

Similar calculations were carried out for STS with RWB-R. Fig. 6.100 shows the FAs of RWB-R when the payload is inserted into reference orbits with various inclinations. For comparison, NFA of RWB-A⁻ is given (without a payload mass loss). It can be seen that the best lift-to-drag ratio of the RWB-R makes it possible to implement almost all-azimuth launches in the considered range:

$$i_{\text{orb}} \in \left[51.7^\circ, 108^\circ\right]. \tag{6.29}$$

As follows from the above results, when STS is launched from Vostochny cosmodrome, airfields in Komsomolsk-on-Amur and Neryungri are located inside of FAs far from their boundaries, so that their feasibility does not require of RWB ultimate maneuverability, and there remains a significant margin to compensate for possible disturbances, including control losses.

An assessment of the influence of the target orbit inclination on the FA form was made. For this, RWB-R FAs were calculated in the coordinates of longitudinal x and lateral z ranges for orbit inclinations $i_{\text{orb}} = 51.6^\circ, 72^\circ, 108^\circ$ (Fig. 6.101). It can be seen that in these coordinates, FAs practically coincide for the entire range of inclinations. The change in maximum $\overline{x}_{\max}$, and minimum $\overline{x}_{\min}$ longitudinal ranges and maximum lateral range $\overline{z}_{\max}$ on the specified FA, related to ones for inclination $i_{\text{orb}} = 51.6^\circ$, is shown in Fig. 6.102. The maximum lateral range in FA practically does not depend on i_{orb} (6.29), and the changes in $x_{\max}$ and $x_{\min}$ do not exceed 3%.

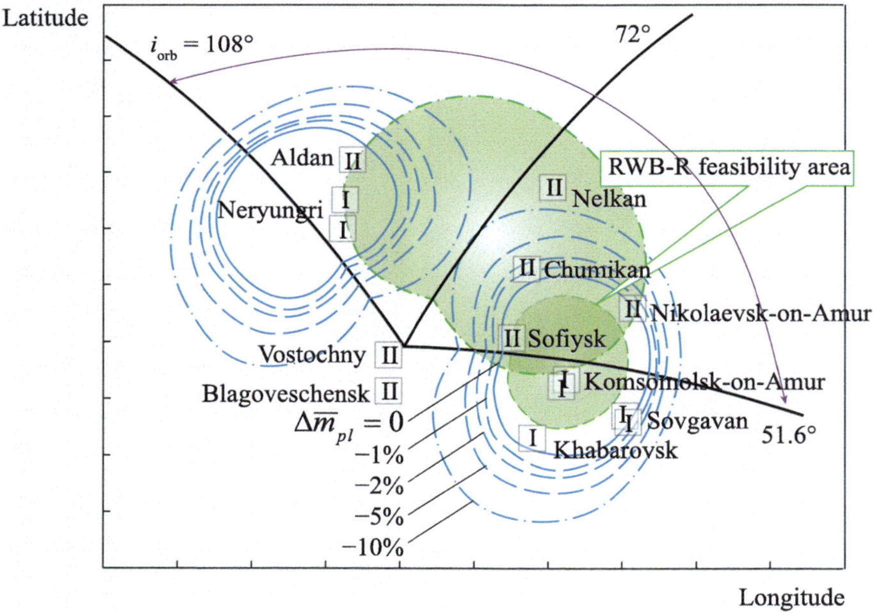

Fig. 6.100 Possible expansion of RWB-R feasibility areas with the optimal control of LV and RWB at the cost of reducing the maximum payload mass Δm_{pl}

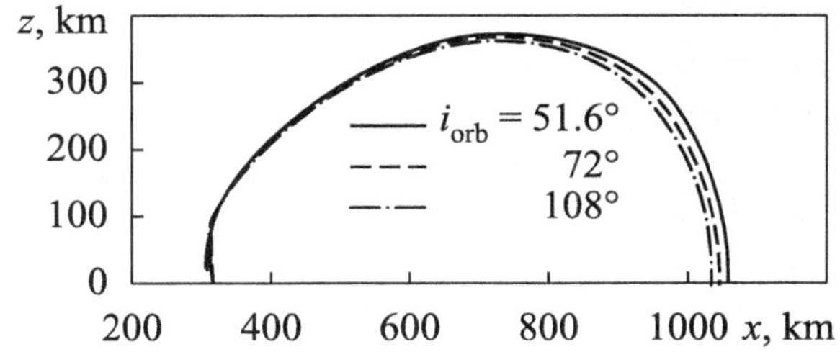

Fig. 6.101 Comparison of RWB-R feasibility area boundaries for three orbit inclinations in the plane of longitudinal x and lateral z ranges

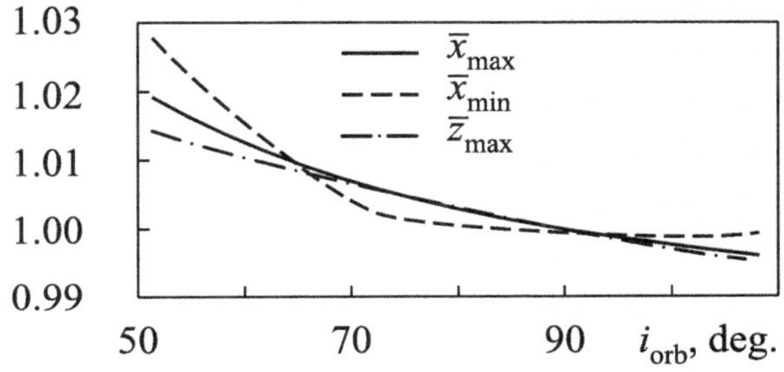

Fig. 6.102 Relative change in the maximum and minimum longitudinal ranges and the maximum lateral range of RWB-R feasibility areas with variation in the target orbit inclination

Since a small change in the maximum and minimum longitudinal ranges is associated with a different initial longitudinal velocity in the inertial coordinate system due to the Earth rotation, it is obvious, that if STS takes off from a launch site at a higher latitude (for example, from Plesetsk cosmodrome [34]), the FA in these coordinates has to be more stable with variations of the orbit inclination. For the same reason, FA in the coordinates of the longitudinal and lateral ranges will be invariant to STS launch site for target orbits close to the polar one.

Using the above simplifying considerations, FAs of RWB-B$^-$ were built for Vostochny and Plesetsk launch sites (Figs. 6.103 and 6.104).

6.6.6 Comparative Effectiveness of Various STS Injection Schemes

The results described above are summarized in Fig. 6.105, which shows the relative maximum payload mass of STSs with RWB-A$^-$ and RWB-R when injected into a reference orbit with various inclinations. The STS with RWB-R inserts the maximum payload mass into orbits with inclination range (6.29) without a loss, since due to the high maneuverability of RWB-R, the available intermediate landing airfields are feasible within the nominal FA, and the expansion of FA by changing the optimal launch trajectory of STS with RWB-R is not required (see Fig. 6.100). For the same reasons, there are no payload mass losses for inclinations $i_{\text{orb}} \in [51.7°, 55°]$ and $i_{\text{orb}} \in [86°, 108°]$, however, in range (6.28) the losses exceed 10%. If it is rational

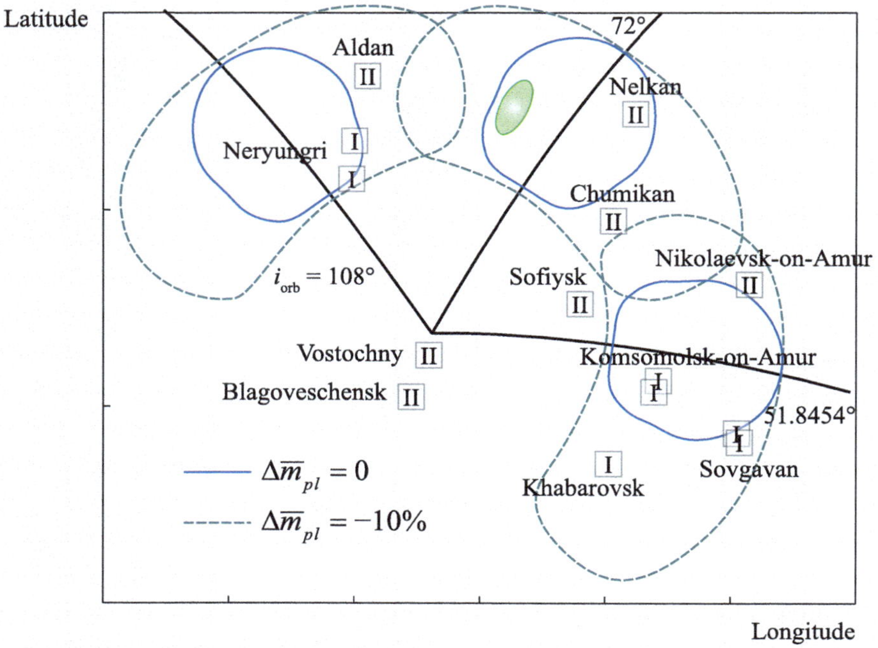

Fig. 6.103 RWB-B⁻ feasibility areas for the optimal STS injection from Vostochny launch site. The dashed line corresponds to the expansion of the area due to the LV optimal spatial maneuvering with a loss of 10% of the payload mass

to place one more airfield in addition to the existing ones for landing RWB-A⁻, then a payload mass loss can be avoided at any orbit inclination from considered range (6.29). The optimal region for such airfield is shown in Fig. 6.99. Thus, even for an RWB layout with a delta wing that is not the best in terms of the aerodynamic lift-to-drag ratio, an alternative concept with an intermediate unpowered landing makes it possible to increase the maximum payload mass by ~ 10% compared to the basic concept. At the same time, for any launch azimuth from 0° to 108°, no more than two or three airfields are required for the intermediate landing of the RWB. When launching from Vostochny launch site, airfields near Neryungri and Komsomolsk-on-Amur can be used as two airfields that serve the most launches with low and high inclinations. To serve orbits with medium inclinations, the same airfields can be used, but at the cost of some reduction in the maximum injected mass. At the same time, it still turns out to be no lower than with the basic injection scheme ($\Delta m_{\text{pl}} \geq -10\%$). Creation of an additional airfield for "unfavourable" inclinations (6.28) is a radical solution to the RWB return problem with increasing STS effectiveness in the entire considered range of orbit inclinations. From an economic point of view, such decision does not look burdensome for the STS project due to the obvious possibility of diversifying the use of the additional airfield in the interests of implementing the State

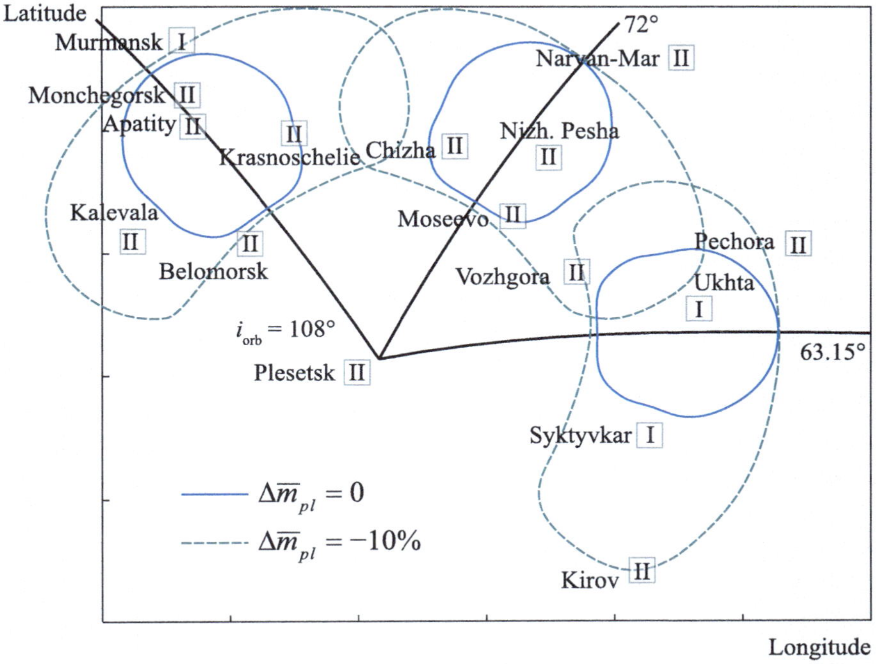

Fig. 6.104 RWB-B⁻ feasibility areas with optimal STS injection from Plesetsk launch site. The dashed lines correspond to the expansion of the area due to the LV optimal spatial maneuvering with a loss of 10% of the payload mass

Program for the development of transport infrastructure in the promising Eastern Region of Russia.

The presented analysis of the concepts of a partly reusable STS with 1st stage winged boosters confirms the effectiveness of such LV, which differs, first of all, in the absence of alienated SP drop areas. The results obtained for the proposed alternative concept with returning the reusable boosters with their unpowered landing on an intermediate airfield, feasible from the launch route, indicate that there are additional reserves for increasing the STS carrying capacity by almost 10% without reducing the reliability of operations for delivering payload to orbits. At the same time, location of such airfields inside and far from boundaries of RWB feasibility areas does not require the ultimate maneuverability to reach them, leaving a margin to compensate for possible disturbances and control losses.

All the calculations described above for through optimization of STS branching trajectories were performed using *ASTER* software. The convergence of the corresponding TPBVPs, the number of which amounted to thousands, was provided automatically, without an intervention of a researcher.

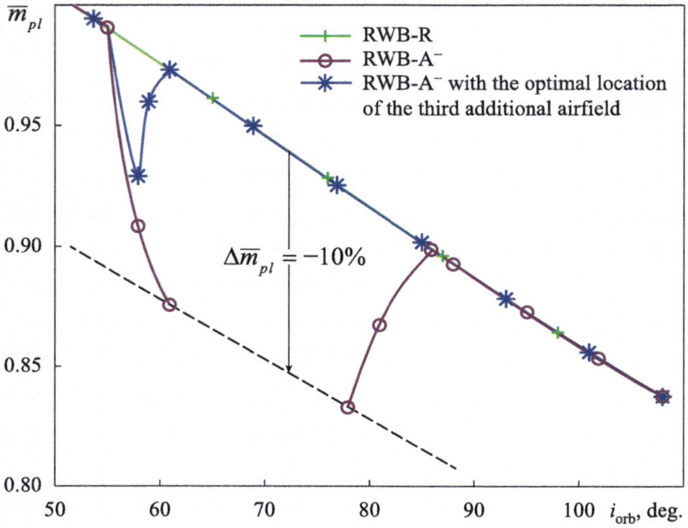

Fig. 6.105 The relative maximum payload mass injected by STS with RWB-A⁻ and RWB-R, depending on the orbit inclination, using no more than two airfields for landing and no more than three airfields for RWB-A⁻ landing with the optimal location of the third additional airfield

References

1. Filatyev, A.S.: Reduction of maximum dynamic and thermal loads during passive motion of hypersonic vehicles with the help of small control actions. Uchenye Zapiski TsAGI **9**(2) (in Russian), (1978)
2. Filatyev, A.S.: Approximate analytical synthesis of optimal control of hypersonic aircraft when moving in the atmosphere with a sub-circular speed. Parts 1 and 2. Uchenye Zapiski TsAGI **11**, (1 and 2) (in Russian), (1980)
3. Filatyev, A.S.: Space vehicle safety problem: reentry with subcircular speeds. IAA. 6.1-93-731. In: Heath, G.W. (eds.) Space Safety and Rescue 1993. Science and Technology Series, vol. 87, pp. 31–40. American Astronautical Society Publication (1993)
4. Filatyev, A.S.: Critical reentry conditions. In: The 19th International Symposium on Space Technology and Science, Yokohama, Japan, May 15–24, ISTS 94-c-32 (1994)
5. Filatyev, A.S., Yanova, O.V.: Nonequilibrium reentry. ICAS-2.2-1, 21st ICAS Congress, Sorrento, Italy, September 7–14, 1996
6. Ilyin, V.A., Filatyev, A.S.: Synthesis of optimal trajectories for launching into orbit, from any point of which a descent in the atmosphere is possible with the fulfillment of specified constraints. Cosm. Res. **23**(1), 37 (1985)
7. Filatyev, A.S., Yanova, O.V.: The optimization of ascent trajectories with continuum of branches, IAF-94-A.2.019. In: The 45th Congress of the International Astronautical Federation, October 9–14, Jerusalem, Israel, 1994
8. GOST 4401-64.: The standard atmosphere. Parameters. Moscow, Publishing House of Standards, 1964 (in Russian)
9. Parkinson, R.: The An-225/Interim Hotol launch vehicle. In: AIAA 3rd International Aerospace Planes Conference (1991). https://doi.org/10.2514/6.1991-5006

10. Lozino-Lozinsky, G.E., Dudar, E.N., Chelkis F.Yu. The MAKS project and perspectives of development. In: AIAA 7th International Space Planes and Hypersonic Systems and Technologies Conference, 18–22 November 1996, Norfolk, Virginia, USA
11. GOST 4401-81. The standard atmosphere. Parameters. Moscow, Publishing House of Standards, 1981 (in Russian)
12. Leutin, A.P., Sergeev, I.V., Demeshkina, V.V., Ilyin, V.A., Kirpishchikov, V.P. Optimal programmed control of CA and RV in the process of their separation. In: Proceedings of the 1st International Aerospace Conference "Prospects for Space Exploration", Moscow, Russia, Sept 28–Oct 2, 1992 (in Russian)
13. Filatyev, A.S., Yanova, O.V.: Optimization of space launching with limitations on fall zones for spent components. IAF-96-A.3.01. In: The 47th IAF Congress, Beijing, China (1996)
14. Kobelev, V.N., Milovanov, A.G.: Spacecraft Launch Vehicles, p. 528. RESTART, Moscow (2009). ISBN 978-5-904348-01-4 (in Russian)
15. Filatyev, A.S., Golikov, A.A., Demeshkina, V.V., Leutin, A.P.: Peculiarities of unguided reentry of space transportation system parts. Dokl. Phys. **55**(12), 597–601 (2010)
16. Filatyev, A.S., Golikov, A.A.: New topology of debris impact points dispersion. IAC-16-D1.IP.6. In: Proceedings of the 67th International Astronautical Congress, Guadalajara, Mexico, 26–30 Sept 2016
17. Filatyev, A.S., Yanova, O.V.: Through optimization of branched trajectories with random disturbances by the Pontryagin maximum principle. ICAS 2006-6.6.3, 2006. In: The 25th International Congress of the Aeronautical Sciences, Hamburg, 3–8 Sept 2006
18. Filatyev, A.S., Yanova, O.V.: Through optimization of branching injection trajectories by the Pontryagin maximum principle using stochastic models. Acta Astronaut. **68**(7–8), 1042–1050 (2011)
19. Filatyev, A.S., Yanova, O.V.: Through optimization of branching trajectories in view of random disturbances. VESTNIK Samara University. Aerosp. Mech. Eng. **1**(21), pp. 244–249 (2010)
20. Karpov, A., Ivanov, R., Kovalevsky, M.: Air launch aerospace international project. In: Proceedings of the 27th International Congress of The Aeronautical Sciences, 2010
21. Nesterov, V.E., Kuzin, A.I., Ivanov, V.L., Lozin, S.N., Lechov, P.A., Semenov, A.I., Panitchkin, N.G., Romashkin, A.M.: Re-usable space-rocket system. Innovations on development of Russian means of access to outer space. IAC-11-D2.4.4. In: The 62nd International Astronautical Congress, Cape Town, SA, 2011
22. Filatyev, A.S., Yanova, O.V.: The alternative concept of use of launch vehicles with recoverable winged boosters. IAC-11-D2.4.6. In: The 62nd International Astronautical Congress, Cape Town, SA (2011)
23. Filatyev, A.S., Buzuluk, V.I., Yanova, O.V., Ryabukha, N.N., Petrov, A.N.: Through optimization of branching injection trajectories by the Pontryagin maximum principle using stochastic models. Acta Astronaut. **100**(1), 11–21 (2014)
24. https://ru.wikipedia.org/wiki/RD-191.
25. https://ru.wikipedia.org/wiki/RD-0120#:~:text=RD-0120%20—%20liquid%20rocket %20engine%2C,onthe%20secondstage%20%20launchvehicle%20"Energy"
26. Ryabukha, N.N.: Reusable rocket-aviation module and the method of its return to the cosmodrome. Patent RU 2442727, 20.02.2012. FIPS. Bulletin of the invention, utility models. No. 5, 2012
27. Petrov, K.P.: Aerodynamics of Space Transport Systems, p. 368. Editorial URSS, Moscow (2000)
28. Filatyev, A.S., Yanova, O.V.: ASTER program package for the thorough trajectory optimization. In: 41st AIAA "Guidance, Navigation & Control" Conference, 6–9 Aug 2001, Montreal, Canada, AIAA-2001-4391
29. Bliss, G.A.: Mathematics for Exterior Ballistics. NY (1944)
30. Filatyev, A.S., Golikov, A.A., Shanygin, A.N., Voityshen, V.S.: The local distributed criteria method for multidisciplinary optimization of launcher parameters, Acta Astronaut. **65** (3–4), August-September 2009, pp. 584–590
31. https://ru.wikipedia.org/wiki/Vostochny_(cosmodrome)

32. https://ru.wikipedia.org/wiki/List_of_airports_of_Russia
33. https://ru.wikipedia.org/wiki/Airfield
34. https://ru.wikipedia.org/wiki/Plesetsk_(cosmodrome)

Chapter 7
Mathematical Models of a Spacecraft with an Electric Propulsion for Trajectory Optimization Problems

7.1 Introduction

The mathematical model of the spacecraft motion is a key factor that determines both the reliability of the calculation results and the efficiency of trajectory calculations. In particular, the choice of coordinate system and state vector is of great importance. In this work, the geocentric non-rotating coordinate system J2000, associated with the mean earth equator and ecliptic for the standard epoch January 1.5, 2000, is used to analyze the near-Earth trajectories of a spacecraft with an electric propulsion system. To analyze the selenocentric sections of the trajectory, a non-rotating selenocentric coordinate system is used, which is related to the mean equator of the Earth or the Moon at the J2000 epoch, and to analyze the heliocentric sections of the trajectory, a non-rotating heliocentric coordinate system is used, which is related to the ecliptic plane at the J2000 epoch. As the state coordinates of the spacecraft, its coordinates and velocity components in the Cartesian coordinate system (for a relatively small angular flight range) or osculating equinoctial elements (for multi-orbit flights) are used.

In a significant part of the problems considered in this work, the unperturbed motion of the spacecraft in the central Newtonian gravitational field under the action of the thrust of the electric propulsion system is analyzed. In many cases, however, external disturbing accelerations affect the spacecraft trajectory so strongly that they cannot be ignored even when conducting a preliminary analysis of the trajectories. In these cases, taking into account external disturbing accelerations becomes necessary. Depending on the task, it may be necessary to take into account disturbing accelerations from geopotential harmonics of the second and higher orders, the gravity of the Moon, the Sun and other distant celestial bodies, aerodynamic forces, and solar radiation pressure.

Boundary conditions impose constraints on the state vector of the spacecraft at the beginning and end of the flight and, in some cases, at some intermediate points. The simplest form of boundary conditions is for a point-to-point flight. In this case, the

A. S. Filatyev and V. G. Petukhov, *Through Optimization of Aerospace Vehicle Trajectories by the Pontryagin Maximum Principle*, Studies in Systems, Decision and Control 579, https://doi.org/10.1007/978-3-031-80756-5_7

initial and final state vectors are completely specified, except for the value of the final (or initial) mass of the spacecraft. Options with partially specified state vectors at the beginning and end of the flight are possible. For example, if the goal of the flight is to reach a given orbit at an arbitrary point, the final true anomaly (or true longitude) remains free. For low Earth orbits, the acceleration from the second zonal harmonic is several orders of magnitude greater than the typical thrust acceleration when using electric propulsion. In this case, it is often advisable to consider flights with free values of the final or initial right ascension of the ascending node and perigee arguments. Specific boundary conditions arise when considering interplanetary transfers in the context of the zero sphere of influence method or transfers in the Earth-Moon system in the context of a restricted three- or four-body problem.

In practice, it is often necessary to compute a program to control the thrust vector of an electric propulsion system so that a minimum of some functionality, such as flight time or mass of propellant, is ensured. The mathematical formulation of the problem of computing such an optimal control program is reduced to the optimal control problem. In this study, the Pontryagin maximum principle is used to solve optimal control problems. The maximum principle reduces the optimal control problem to a boundary value problem for a system of ordinary differential equations. The boundary value problem of the maximum principle is reduced in this study to the Cauchy problem using the continuation method (homotopy method).

One of the well-known disadvantages of the approach to solving problems of optimization of spacecraft trajectories based on the application of the maximum principle is the necessity to choose a good initial guess for the initial values of the conjugate variables. To overcome this drawback, a sequential solution of optimization problems for limited power (LP) and limited thrust (LT) trajectories is used. To optimize the LP trajectories, stable numerical methods have been developed that use a zero initial guess for the initial values of the conjugate variables, i.e., the coasting motion of the spacecraft along the initial orbit. To solve the LT problem, the continuation method of the LP problem is used, which allows us to solve the problem of choosing an initial approximation.

This chapter discusses mathematical models of the motion of a spacecraft with an electric propulsion system (Sect. 7.2), typical problems of optimizing the trajectories of spacecraft with an electric propulsion system are given (Sect. 7.3), an indirect approach to optimizing trajectories using the maximum principle is described (Sect. 7.4), mathematical models of the optimal trajectory are given for a spacecraft with an ideally controlled limited-power engine (Sect. 7.5) and with a limited-thrust propulsion system (Sect. 7.6) are presented, as well as a design model of a spacecraft for the problem of joint optimization of the trajectory and the main design parameters (Sect. 7.7).

7.2 Mathematical Models of the Motion of a Spacecraft with an Electric Propulsion System

The differential equations of the unperturbed motion of a spacecraft with an electric propulsion system in a central Newtonian gravitational field, written in an inertial Cartesian coordinate system with the origin at the gravity center, have the form:

$$\left.\begin{aligned} \frac{d^2\mathbf{r}}{dt^2} &= -\frac{\mu}{r^3}\mathbf{r} + \frac{\delta T}{m}\mathbf{e}, \\ \frac{dm}{dt} &= -\frac{\delta T}{c}, \end{aligned}\right\} \tag{7.1.1}$$

where $\mathbf{r}$ is the spacecraft position vector, t is the time, μ is the gravitational parameter of the primary, $r = |\mathbf{r}|$ is the distance of the spacecraft from the gravity center,δ is the throttling function ($\delta = 1$ when the EPS is operating and $\delta = 0$ when the EPS is OFF), T is the thrust, m is the mass of the spacecraft, $\mathbf{e}$ is the unit vector along the EPS thrust vector, and c is the exhaust velocity of the EPS. The inertial Cartesian coordinate system is usually the J2000 coordinate system or the international geoinertial coordinate system ICRS as defined in the International Earth Rotation and Referencing Service (IERS) convention [1]. (The ICRS coordinate system practically coincides with J2000, referred to the mean equator and the equinox on January 1.5, 2000, the ICRS rotation relative to J2000 does not exceed 0.023 arc seconds). When considering the geocentric motion of the spacecraft, the main plane of the inertial coordinate system coincides with the plane of the mean earth equator at the epoch of January 1.5, 2000, and when considering the heliocentric motion - with the ecliptic plane in the same epoch. The main axis of both coordinate systems is directed to the mean equinox of the epoch, so the transition between the ecliptic and equatorial coordinate systems is determined by a rotation around the x axis by the angle ε (the inclination of the plane of the mean Earth equator to the ecliptic at the epoch). The system of equations of spacecraft motion in the form (7.1.1) is used in this study mainly for the analysis of heliocentric low-thrust trajectories within the framework of the method of zero spheres of influence for a relatively small angular flight range, although for a number of test problems when using this system, optimal trajectories with an angular range of 50–100 revolutions were computed.

In practice, to reduce the computational error in the numerical integration of the spacecraft motion equations, motion equations in dimensionless variables are usually used. For heliocentric trajectories, the unit of length is usually 1 AU, and the unit of velocity is the circular velocity of orbital motion at a heliocentric distance of 1 AU. For geocentric trajectories, the radius of the initial or final orbit of the spacecraft is usually used as a unit of length (in particular, in this study, when analyzing flights to GEO, the GEO radius was used as a length unit), and as a unit of velocity, the velocity of orbital motion in a circular orbit with a radius equal to the length unit. With the units of length L* and velocity V* chosen in this way, the unit of time becomes equal to T* = L*/V*, and the unit of acceleration—A* = V*/T*. The unit of mass is usually the initial mass of the spacecraft m_0: M* $= m_0$. In the variables introduced in this way, the gravitational parameter of the gravity center is equal to 1 and the equations of spacecraft motion (7.1.1) take the form:

$$\left.\begin{aligned} \frac{d^2\mathbf{r}}{dt^2} &= -\frac{1}{r^3}\mathbf{r} + \frac{\delta T}{m}\mathbf{e}, \\ \frac{dm}{dt} &= -\frac{\delta T}{c}, \end{aligned}\right\} \tag{7.1.2}$$

To analyze the perturbed geocentric motion of a spacecraft with an electric propulsion system, differential equations of motion are used in the geocentric inertial Cartesian coordinate system J2000 or ICRS of the form:

$$\left.\begin{aligned} \frac{d^2\mathbf{r}}{dt^2} &= -\frac{\mu}{r^3}\mathbf{r} + \mathbf{Q}\frac{\partial U}{\partial \mathbf{r}} + \mathbf{a} + \frac{\delta T}{m}\mathbf{e}, \\ \frac{dm}{dt} &= -\frac{\delta T}{c}, \end{aligned}\right\} \tag{7.1.3}$$

where $\mathbf{Q}$ is the rotation matrix to account for the rotation of the Earth and the movement of the Earth's equatorial plane, which is the product of the precession, nutation matrices, the matrix of the motion of the celestial intermediate pole and the matrix of the Earth's own daily rotation; U is a perturbing function due to the non-centrality of the Earth's gravitational field, $\mathbf{a}$ is other perturbing accelerations (acceleration due to the gravity of the Moon, Sun and planets; acceleration due to solar radiation pressure; aerodynamic accelerations). Equations of motion in the form (7.1.3) are used in this study when calculating geocentric trajectories of interorbital transfer, flight trajectories to the Moon and flight trajectories to libration points with suboptimal feedback control.

The IERS and International Astronomical Union (IAU) definitions of coordinate and time systems were used in this study. The IAU SOFA (Standards of Fundamental Astronomy [2]) library was used to perform the coordinate system transformation calculations.

Of course, when analyzing multi-revolution trajectories, higher accuracy and computational performance are provided by the equations of motion in the osculating elements. To analyze multi-revolution trajectories of spacecraft with electric propulsion, equations of motion in modified equinox elements (p, e_x, e_y, i_x, i_y, L, [3]) are used:

$$\frac{dp}{dt} = \sqrt{\frac{p}{\mu}}\frac{2p}{q}a_t,$$
$$\frac{de_x}{dt} = \sqrt{\frac{p}{\mu}}\left[\sin L \cdot a_r + \frac{(q+1)\cos L + e_x}{q}a_t - \frac{e_y\xi}{q}a_n\right],$$
$$\frac{de_y}{dt} = \sqrt{\frac{p}{\mu}}\left[-\cos L \cdot a_r + \frac{(q+1)\sin L + e_y}{q}a_t + \frac{e_x\xi}{q}a_n\right], \tag{7.1.4}$$
$$\frac{di_x}{dt} = \sqrt{\frac{p}{\mu}}\frac{s^2}{2q}\cos L \cdot a_n, \quad \frac{di_y}{dt} = \sqrt{\frac{p}{\mu}}\frac{s^2}{2q}\sin L \cdot a_n,$$
$$\frac{dL}{dt} = \sqrt{\mu p}\left(\frac{q}{p}\right)^2 + \sqrt{\frac{p}{\mu}}\frac{\xi}{q} \cdot a_n, \quad \frac{dm}{dt} = -\delta\frac{T}{c},$$

where p is the semi-latus rectum, $e_x = e\cos(\omega + \Omega)$ and $e_y = e\sin(\omega + \Omega)$ are the elements of the eccentricity vector, $i_x = \mathrm{tg}\frac{i}{2}\cos\Omega$ and $i_y = \mathrm{tg}\frac{i}{2}\sin\Omega$ are the elements of the inclination vector, $L = \nu + \omega + \Omega$ is the true longitude, e is the eccentricity, ω is the argument of pericenter, Ω is the right ascension of the ascending node, i is the inclination, $s^2 = 1 + i_x^2 + i_y^2$, $q = 1 + e_x\cos L + e_y\sin L$, $\xi = i_x\sin L - i_y\cos L$, m is the spacecraft mass, T is the thrust, c is the exhaust velocity, δ is the throttling function ($\delta = 1$ when the propulsion system is ON and $\delta = 0$ when it is OFF),

$$a_t = \delta\frac{T}{m}\cos\vartheta\cos\psi + a_{pt},\ a_r = \delta\frac{T}{m}\sin\vartheta\cos\psi + a_{pr},\ a_n = \delta\frac{T}{m}\sin\psi + a_{pn}, \tag{7.1.5}$$

ϑ is the pitch angle (angle between the projection of the thrust vector onto the orbital plane and the circumferential direction),ψ is the yaw angle (the angle between the thrust vector and the orbital plane), and a_{pt}, a_{pr}, a_{pn} are the transversal, radial and binormal components of the perturbing acceleration, respectively.

In many studies (for example, [4]), the value $h = \sqrt{p/\mu}$ is used instead of the semi-latus rectum. When it is used, the first equation of system (7.1.4) takes the form $dh/dt = (h^2/q)a_t$, and in the remaining equations of (7.1.4) the expression μh^2 is substituted in place of the semi-latus rectum.

In many cases, it turns out to be advisable to use an angular independent variable instead of time in differential equations of motion. In this work, the auxiliary longitude K is used as such an independent variable, obeying the differential equation

$$\frac{dK}{dt} = \sqrt{\mu p}\left(\frac{q}{p}\right)^2 \tag{7.1.6}$$

Let us introduce into consideration the deviation of true longitude from the auxiliary longitude $L_K = L - K$, which, in accordance with (7.1.4) and (7.1.6), obeys the differential equation

$$\frac{dL_K}{dt} = \sqrt{\frac{p}{\mu}}\frac{\xi}{q}a_n \tag{7.1.7}$$

If we replace the sixth equation in the system (7.1.4) by (7.1.6) and (7.1.7), we obtain the following system of differential equations:

$$\begin{aligned}
&\frac{dp}{dt} = \sqrt{\frac{p}{\mu}}\frac{2p}{q}a_t,\\
&\frac{de_x}{dt} = \sqrt{\frac{p}{\mu}}\left[\sin L \cdot a_r + \frac{(q+1)\cos L + e_x}{q}a_t - \frac{e_y\xi}{q}a_n\right],\\
&\frac{de_y}{dt} = \sqrt{\frac{p}{\mu}}\left[-\cos L \cdot a_r + \frac{(q+1)\sin L + e_y}{q}a_t + \frac{e_x\xi}{q}a_n\right],\\
&\frac{di_x}{dt} = \sqrt{\frac{p}{\mu}}\frac{s^2}{2q}\cos L \cdot a_n, \quad \frac{di_y}{dt} = \sqrt{\frac{p}{\mu}}\frac{s^2}{2q}\sin L \cdot a_n,\\
&\frac{dL_K}{dt} = \sqrt{\frac{p}{\mu}}\frac{\xi}{q}a_n, \quad \frac{dK}{dt} = \sqrt{\mu p}\left(\frac{q}{p}\right)^2, \quad \frac{dm}{dt} = -\delta\frac{T}{c}.
\end{aligned} \tag{7.1.8}$$

In system (7.1.8), the true longitude L is replaced in all equations by its expression through the auxiliary longitude K and the deviation of the true longitude L_K: $L = K + L_K$. If the magnitudes of the thrust and perturbing accelerations are small, then the equations of motion (7.1.8) are reduced to a system of differential equations with 7 slow variables p, e_x, e_y, i_x, i_y, L_K, m and one fast variable K.

The right-hand side of the differential equation for K is positive for all values of the state coordinates for $p > 0$. Thus, if we exclude the case of degeneration of the osculating orbit into a straight line segment (case $p = 0$), we can replace the independent variable t by K using the relation

$$dK = \frac{dK}{dt}dt \Rightarrow \frac{d}{dK} = \frac{dt}{dK}\frac{d}{dt} = \frac{1}{\sqrt{\mu p}}\left(\frac{p}{q}\right)^2\frac{d}{dt} \tag{7.1.9}$$

As a result, we obtain the following system of differential equations:

$$
\begin{aligned}
\frac{dp}{dK} &= \frac{2p^3}{\mu q^3} a_t, \\
\frac{de_x}{dK} &= \frac{p^2}{\mu q^2}\left[\sin L \cdot a_r + \frac{(q+1)\cos L + e_x}{q} a_t - \frac{e_y \xi}{q} a_n\right], \\
\frac{de_y}{dK} &= \frac{p^2}{\mu q^2}\left[-\cos L \cdot a_r + \frac{(q+1)\sin L + e_y}{q} a_t + \frac{e_x \xi}{q} a_n\right], \\
\frac{di_x}{dK} &= \frac{p^2 s^2}{2\mu q^3}\cos L \cdot a_n, \quad \frac{di_y}{dK} = \frac{p^2 s^2}{2\mu q^3}\sin L \cdot a_n, \quad \frac{dL_K}{dK} = \frac{p^2}{\mu q^3}\xi \cdot a_n, \\
\frac{dm}{dK} &= -\frac{\delta}{\sqrt{\mu p}}\left(\frac{p}{q}\right)^2 \frac{T}{c}, \quad \frac{dt}{dK} = \frac{1}{\sqrt{\mu p}}\left(\frac{p}{q}\right)^2.
\end{aligned}
\tag{7.1.10}
$$

The use of the true longitude L as an independent variable in problems of spacecraft trajectory optimization has been considered previously in a number of publications [5–7]. It should be noted that the derivative dL/dt can change its sign at sufficiently high values of the binormal component of the thrust acceleration, therefore the use of true longitude as an independent variable requires control of the sign of its derivative and may turn out to be incorrect in a number of solutions. The use of auxiliary longitude K instead of true longitude L as an independent variable was proposed in [8]. Auxiliary longitude K, unlike L, has a positive time derivative on all non-rectilinear orbits, so using K as an independent variable is preferable to using L. In addition, it should be noted that in [5, 6] differential equations of motion with true longitude as an independent variable were used in the trajectory optimization procedure based on the direct approach, while this study considers the indirect approach based on the use of the maximum principle, allowing us to construct a more efficient and fast optimization method. When L is used as the independent variable, the thrust acceleration components appear in both the numerators and denominators of the right-hand sides of the differential equations, which leads to a very cumbersome expression for optimal control when the maximum principle is used. Using K instead of L as an independent variable allows us to greatly simplify the optimal control expressions that follow from the maximum principle, since in this case the right-hand sides of the differential equations of motion depend linearly on the components of thrust acceleration.

The parameters of the initial orbit are denoted by the subscript "0", and those of the final orbit by the subscript "f". Let us consider a flight from a given point of the initial orbit with true longitude L_0 with a given angular range ΔL. In this case, the final point of the trajectory will have a true longitude $L_f = L_0 + \Delta L$. Let us introduce the notation $\mathbf{x} = \left(p, e_x, e_y, i_x, i_y\right)^{\mathrm{T}}$. In this case, the boundary conditions for the free flight time have the form:

$$L_K(K_0) + K_0 = L_0, \mathbf{x}(K_0) = \mathbf{x}_0(L_0), m(K_0) = m_0, t(K_0) = t_0, \tag{7.1.11}$$

$$L_K\left(K_f\right) + K_f = L_f, \mathbf{x}\left(K_f\right) = \mathbf{x}_f\left(L_f\right), \tag{7.1.12}$$

where K_0 and K_f are the initial and final values of the auxiliary longitude, respectively, and t_0 is the initial flight time. Since L_K is included in the equations of motion (7.1.10) only in the form of the sum $L = L_K + K$, we can, without loss of generality, set $K_0 = 0$, and rewrite the initial conditions (7.1.11) in the form:

$$L_K(0) = L_0, \mathbf{x}(0) = \mathbf{x}_0(L_0), m(0) = m_0, t(0) = t_0. \quad (7.1.13)$$

If the transfer time Δt is given, the following condition is added to (7.1.12):

$$t(K_f) = t_f, \quad (7.1.14)$$

where $t_f = t_0 + \Delta t$ is the final transfer time.

The systems (7.1.4) and (7.1.10) are used in this work when optimizing multi-orbit geocentric trajectories, in particular to analyze the properties of optimal trajectories in the central gravitational field and in the presence of perturbing accelerations, to construct an algorithm for quickly assessing the main characteristics of optimal multi-orbit geocentric flights when carrying out space mission design and analysis to synthesize suboptimal feedback control.

Perturbing accelerations due to the non-centrality of the Earth's gravitational field were calculated in the Greenwich (rotating) geocentric coordinate system as derivatives of the perturbing function U, presented as a series expansion in spherical functions. To calculate these derivatives, the Cunningham algorithm [9] and the EGM96 model of the Earth's gravitational field [10] were used, and the IAU SOFA program library [2] was used to calculate the Earth's rotation parameters.

Perturbing accelerations from the gravity of distant celestial bodies were calculated as follows: if μ_p is the gravitational constant of a distant celestial body (Moon, Sun or planet), $\mathbf{r}$ is the spacecraft position vector, $\mathbf{r_p}$ is the position vector of the distant celestial body, then the vector of the perturbing acceleration caused by the gravity of this celestial body is calculated by the formula:

$$\mathbf{a_p} = \mu_p \cdot \left(\frac{\mathbf{r_p} - \mathbf{r}}{\left|\mathbf{r_p} - \mathbf{r}\right|^3} - \frac{\mathbf{r_p}}{\mathbf{r_p^3}} \right)$$

The JPL DE403/DE405 ephemeris software was used to calculate the coordinates of celestial bodies $\mathbf{r_p}$ [11]. The same software was used to determine the coordinates and velocity components of celestial bodies for their use in boundary conditions when calculating interplanetary trajectories, flight trajectories to the Moon and libration points, as well as when calculating the eclipsis effects on planetocentric trajectories.

When solving a number of problems on geocentric trajectories, the aerodynamic drag should be taken into account. The magnitude of acceleration due to aerodynamic drag force has been calculated using the following formula:

$$a_{aero} = c_d \frac{\rho v^2}{2} \frac{S_m}{m}$$

where c_d is the aerodynamic drag coefficient, ρ is the density of the atmosphere, v is the orbital velocity of the spacecraft, S_m is the cross-section area of the spacecraft, m is the mass of the spacecraft.

It is assumed that the aerodynamic drag force is directed against the orbital velocity vector of the spacecraft, i.e. the proper motion of the atmosphere is not taken into account.

To calculate the atmospheric density ρ, the dynamic upper atmosphere model MET (NASA Marshall Engineering Thermosphere model) [12] is used, which is based on the Yacchia model of the upper atmosphere.

In a number of cases, the spacecraft trajectory is significantly affected by the solar radiation pressure. The calculations used a simplified model for calculating the force of solar radiation pressure, in which the spacecraft is represented as a flat plate oriented with its normal to the Sun. In this case, the reflected and absorbed parts of the solar radiation produce force components acting in the same direction–along the internal normal to the plate, which coincides with the anti-solar direction (the diffuse component of the reflected flux was not taken into account). This model is quite reasonable, since most of the area of a spacecraft with a main solar electric propulsion system consists of solar panels, the outer normal of which is nominally directed toward the Sun. The spacecraft model in the form of a flat plate normally facing the Sun is dynamically equivalent to the spacecraft model in the form of a sphere of equal cross-section area and is traditionally used in ballistic calculation models.

Solar radiation pressure ($p = 4.56 \cdot 10^{-6}$ N/m^2) at a distance of 1 AU is calculated as $p = F/c$, where $F = 1366.1$ W/m^2 is the average power density of the solar radiation flux at a distance of 1 AU, and $c = 299{,}792{,}458$ m/s is the speed of light.

Within the accepted assumptions, acceleration from solar radiation pressure is calculated as:

$$\mathbf{a}_{SRP} = \frac{pS}{r^2 m}(1 + k_{SRP})\mathbf{s}$$

where S is the characteristic area of the reflective surface of the spacecraft (for a spacecraft with a solar electric propulsion system, this area is approximately equal to the area of solar panels), m is the spacecraft mass, r is the heliocentric distance of the spacecraft in AU, k_{SRP} is the reflection coefficient ($0 \leq k_{SRP} \leq 1$), and $\mathbf{s}$ is the unit vector in the Sun-SC direction.

When modeling perturbed controlled motion, the possibility of complete or partial shading of the spacecraft by the Earth and the Moon is taken into account. Both cylindrical and conical shadow models are used to calculate illumination. When using the conical model, the illumination coefficient is calculated as the ratio of the part of the solar disk area not covered by the Moon and the Earth to its total area. When calculating the solar radiation pressure force, the above formula for the acceleration vector from solar radiation pressure is multiplied by the illumination coefficient to account for the influence of umbra/penumbra. The shift of the geometric center of the unshaded part of the solar disk, caused by its partial shading, generally speaking, leads to a slight deviation of the direction of the light pressure vector from the Sun-SC direction. The influence of this deviation was considered negligible and was not taken into account in the calculations.

When calculating the umbra and penumbral areas, the spherical shape of the Sun, Earth and Moon was assumed. The coordinates of the Sun and the Moon were calculated using ephemeris software (JPL DE403/DE405 ephemeris), assuming a radius of 695,992 km for the Sun, 6378.137 km for the Earth, and 1738 km for the Moon. Atmospheric aberration was not taken into account when calculating the Earth's shadow.

A mathematical model of the shadow areas is used not only to calculate acceleration due to solar radiation pressure, but also to account for a more significant effect: the electric propulsion system cut-off when the spacecraft crosses the shadow zone if the power source for the EPS is a solar power supply system (PSS) and storage batteries are not used to provide the EPS operation of the in the shadow. In this case, it was assumed that the electric propulsion system cannot operate during periods of spacecraft motion within the penumbra cone.

Integration of the considered systems of differential equations of spacecraft motion was performed in most cases using the Dorman-Prince method of 7(8) order with an adaptive step or the Gregg-Bulirsch-Stoer (GBS) extrapolation method ODEX/ODEX2 with combined control of step length and order [13]. The integration was exited either when a specified point in time (or another independent variable) was reached, or when one of the specified exit functions turned to zero. To do this, after completing the calculation of each integration step, the change in sign of the exit functions and of the derivatives of the exit functions with respect to independent variable was checked. The algorithm for finding the zero of the exit function at the completed step was launched in 2 cases:

(1) if a change in the sign of the function is detected at the completed integration step and
(2) in the case of a change in the sign of the derivative of the exit function at the completed integration step, meaning the presence of a minimum or maximum of the exit function at this step, and provided that the exit function is positive at the beginning of the integration step, if the value of the derivative of this function at the step changes from a negative value to a positive value or with a negative initial value of the exit function when the sign of the derivative changes from plus to minus. In this case, the value of the independent variable t^* was determined at the integration step, at which the extremum of the exit function and the value of this function at the extreme point are reached. If the sign of the exit function at an extreme point differs from the sign of this function at the beginning of the integration step, then an algorithm is launched to find the zero of the exit function on the interval from the beginning of the integration step to t^*.

The search for the roots of the exit functions is performed by a combination of the Golden Section method and the modified Newton method. When determining the values of the exit functions to calculate the current state coordinates of the spacecraft an interpolation polynomial approximating the solution at the last performed integration step was used [13]. If there are several exit functions, the zeros of all functions are computed and the exit is performed at the moment the first zero is reached. It should be noted that the above algorithm, which makes it possible to process cases of double change of sign of the exit function at the integration step, is important for the problem of optimizing trajectories with thrust switchings in the case where the smoothing of the stepwise thrust function is not used. This is important because, as will be shown later, it is necessary to accurately determine the unknown parameters of the boundary value problem at which the structure of the trajectory changes, i.e. at the birth or disappearance of burn or coast trajectory sections.

7.3 Typical Problems of Optimizing the Trajectories of Spacecraft with an Electric Propulsion System

Typical problems of optimizing the trajectories of a spacecraft with an electric propulsion system include the problems of time minimization, fuel minimization, and final mass maximization. In a number of problems, it is advisable to consider the problems of thrust minimization (see Sect. 9.3) and maximization of the "useful" mass of the spacecraft (see Sects. 7.6, 9.4), that is, the part of the spacecraft mass that does not include the mass of the means that provide transfer to a given target orbit. The problem of thrust minimization is dual to the problem of minimizing the transfer time, in the sense that the solution to the problem of minimizing thrust at a fixed transfer time coincides with the solution to the problem of minimizing the transfer time if the given thrust value is equal to its minimum value. However, in the thrust minimization problem, the duration of the flight is usually fixed, which makes it possible to apply more efficient methods for its solution. The main application of the thrust minimization problem is to check the existence of a solution to a problem with given values of specific impulse and maximum thrust: if the minimum thrust is less than a given value of the maximum thrust, then a solution to this problem exists.

In the simplest case, the entire state vector is specified at the initial point of the trajectory, and the entire state vector except for the spacecraft mass is specified at its end point. For fixed departure time and flight duration, such boundary conditions correspond to the formulation of the rendezvous problem, i.e. of the problem of approaching the target (celestial body or other spacecraft) with zero relative velocity. If the velocity vector is not specified at the final moment of time, the final conditions correspond to the target flyby problem. Of course, in the general case, the initial and final transfer times, can be either fixed or free (chosen from the optimality conditions). If the final transfer time is free, then one should distinguish between the problems of flying to a fixed or moving end point. In this case, the flyby problem corresponds to the condition that the final vector of the spacecraft's position is equal to the position vector of the moving target at the final moment of time, while the rendezvous problem requires an additional condition that the vector of the final velocity of the spacecraft is equal to the velocity vector of the moving target at the final moment of time.

Optimization of heliocentric trajectories by the method of zero spheres of influence (ZSOI) leads to the need to select an optimal vector of the departure hyperbolic excess velocity. Depending on the formulation of the problem, the value of this vector can be fixed or free, but in any case the direction must be chosen from optimality conditions.

When considering flights along complex routes (for example, interplanetary trajectories with gravity assist maneuvers or with successive flybys or rendezvous with several celestial bodies), additional constraints are imposed on the trajectory. Within the framework of the ZSOI method, these are the conditions for the equality of the position vectors of the spacecraft and the celestial body at the moments of flyby. For the rendezvous problem, an additional constraint is imposed on the equality of the velocity vectors of the spacecraft and the celestial body at this moment, and when performing a gravity assist maneuver–on the equality of the values of the approaching and departing hyperbolic excess velocity and a limitation on the minimum flyby altitude, which is reduced to a limitation on the rotation angle of the hyperbolic excess velocity vector when performing a gravity maneuver. Of course, the time of flight, the gravity assist maneuver, the start and end of rendezvous of an intermediate celestial body can be specified or selected as optimal.

When optimizing inter-orbital trajectories, not all orbital elements that determine the initial and final vectors of the spacecraft position and velocity can be specified at the initial and final points, but only some. For example, a typical condition is a flight to a final orbit with a free final true longitude of the spacecraft, a free right ascension of the ascending node, or a free argument of the periapsis. The free values of the orbital elements of the initial and final orbits in these cases must satisfy the optimality conditions.

In this study, a solution to the trajectory optimization problem that satisfies the necessary optimality conditions is called a locally optimal solution. If the trajectory consists of several sections, the solution to the end-to-end optimization problem is required to satisfy the optimal conditions for connecting these sections. The problem of end-to-end optimization can also include the problem of simultaneous calculation of the optimal thrust vector control program and the optimal values of free trajectory parameters (initial and final positions of the spacecraft in given or partially specified orbits, free elements of the initial and final orbit). In many cases, there are many solutions that satisfy the necessary optimality conditions. In this case, the problem of global optimization is relevant, that is, calculating the best solution from the entire set of locally optimal solutions.

7.4 An Indirect Approach to Optimizing Trajectories by the Maximum Principle

To optimize the trajectories of a spacecraft with an electric propulsion system, this study uses the formulation of the optimal control problem in Lagrange form. In a typical case, it is required to calculate the control program $\mathbf{u}(t)$, the trajectory $\mathbf{x}(t)$ and (optionally) the initial t_0 and final t_f times of the flight, at which the functional

$$J = \int_{t_0}^{t_f} f_0(\mathbf{x}, \mathbf{u}, t)dt \tag{7.3.1}$$

reaches a minimum on admissible trajectories determined by a system of ordinary differential equations (equations of spacecraft motion)

$$\frac{d\mathbf{x}}{dt} = \varphi(\mathbf{x}, \mathbf{u}, t) \tag{7.3.2}$$

with boundary conditions

$$\boldsymbol{\psi}\left(t_0, \mathbf{x}(t_0); t_f, \mathbf{x}\left(t_f\right)\right) = 0 \tag{7.3.4}$$

We will seek a solution to the optimal control problem (7.3.1–7.3.3) from the condition of satisfying the necessary optimality conditions in the form of Pontryagin maximum principle [14]. Let us introduce the Pontryagin function

$$H = -f_0(\mathbf{x}, \mathbf{u}, t) + \mathbf{p}^{\mathrm{T}}\varphi(\mathbf{x}, \mathbf{u}, t) \tag{7.3.5}$$

where $\mathbf{p}$ is the vector of conjugate variables, and the Lagrange end-point function

$$l = \boldsymbol{\lambda}^{\mathrm{T}}\boldsymbol{\psi}\left(t_0, \mathbf{x}(t_0); t_f, \mathbf{x}\left(t_f\right)\right) \tag{7.3.6}$$

where $\boldsymbol{\lambda}$ is the vector of Lagrange multipliers.

It follows from the maximum principle that the control vector $\mathbf{u}$ must be selected from the maximum condition of the Pontryagin function in the domain of admissible controls $\mathbf{U}$:

$$\hat{\mathbf{u}}(\hat{\mathbf{x}}(t), \hat{\mathbf{p}}(t), t) = \arg\operatorname*{abs\,max}_{\mathbf{u}(t)\in\mathbf{U}(t)} H(\hat{\mathbf{x}}(t), \mathbf{u}(t), \hat{\mathbf{p}}(t), t) \tag{7.3.7}$$

In this section, the optimal values of variables will be denoted by the symbol ^ (in the subsequent sections this designation will be omitted). The Hamiltonian of the optimal control problem can be obtained by substituting the expression for optimal control (7.3.7) as a function of time, state and conjugate variables into the Pontryagin function (7.3.5):

$$\hat{H}(\hat{\mathbf{x}}(t), \hat{\mathbf{p}}(t), t) = H(\hat{\mathbf{x}}(t), \hat{\mathbf{u}}(\hat{\mathbf{x}}(t), \hat{\mathbf{p}}(t), t), \hat{\mathbf{p}}(t), t) \tag{7.3.8}$$

The system of differential equations for the optimal motion has the form:

$$\left.\begin{aligned} \frac{d\hat{\mathbf{x}}}{dt} &= \frac{\partial \hat{H}(\hat{\mathbf{x}}, \hat{\mathbf{p}}, t)}{\partial \hat{\mathbf{p}}}, \\ \frac{d\hat{\mathbf{p}}}{dt} &= -\frac{\partial \hat{H}(\hat{\mathbf{x}}, \hat{\mathbf{p}}, t)}{\partial \hat{\mathbf{x}}}. \end{aligned}\right\} \tag{7.3.9}$$

Boundary conditions (7.3.4) are supplemented by transversality conditions:

$$\left.\begin{aligned} \hat{\mathbf{p}}(\hat{t}_0) &= \frac{\partial l(\hat{t}_0, \hat{\mathbf{x}}(\hat{t}_0); \hat{t}_f, \hat{\mathbf{x}}(\hat{t}_f); \boldsymbol{\lambda})}{\partial \hat{\mathbf{x}}(\hat{t}_0)}, \\ \hat{\mathbf{p}}(\hat{t}_f) &= -\frac{\partial l(\hat{t}_0, \hat{\mathbf{x}}(\hat{t}_0); \hat{t}_f, \hat{\mathbf{x}}(\hat{t}_f); \boldsymbol{\lambda})}{\partial \hat{\mathbf{x}}(\hat{t}_f)} \end{aligned}\right\} \tag{7.3.10}$$

and, if optimization of the initial and (or) final time moments is necessary, by stationarity conditions:

$$\left.\begin{aligned} \hat{H}(\hat{t}_0) &= -\frac{\partial l(\hat{t}_0, \hat{\mathbf{x}}(\hat{t}_0); \hat{t}_f, \hat{\mathbf{x}}(\hat{t}_f))}{\partial \hat{t}_0}, \\ \hat{H}(\hat{t}_f) &= \frac{\partial l(\hat{t}_0, \hat{\mathbf{x}}(\hat{t}_0); \hat{t}_f, \hat{\mathbf{x}}(\hat{t}_f))}{\partial \hat{t}_f}. \end{aligned}\right\} \tag{7.3.11}$$

Let $\dim(\mathbf{x}) = \dim(\mathbf{p}) = n$, $\dim(\boldsymbol{\psi}) = \dim(\boldsymbol{\lambda}) = m \leq 2n$. Then it follows from the maximum principle that the optimal control problem (7.3.1–7.3.3) is reduced to a boundary value problem for the system of ordinary differential Eqs. (7.3.9), the solution of which requires the calculation of n initial values $\hat{\mathbf{x}}(\hat{t}_0)$, n initial values $\hat{\mathbf{p}}(\hat{t}_0)$, m constant Lagrange multipliers $\boldsymbol{\lambda}$ and, if necessary, initial $\hat{t}_0$ and final $\hat{t}_f$ times (in total $2n + m + 2$ variables) in such a way that m boundary conditions (7.3.4), $2n$ transversality conditions (7.3.10) and, if it is necessary to calculate the initial and final times, 2 stationarity conditions (7.3.11) are satisfied (in total $2n + m + 2$ conditions). In practice, the Lagrange multipliers can usually be eliminated from the system (7.3.10) and the dimension of the boundary value problem is reduced to $2n + 2$. If the initial state vector is given explicitly (as part of the boundary condition vector (7.3.4)), the number of unknown parameters and the number of conditions is reduced to $n + 2$. In addition, if the initial and final times of the flight are fixed, then the dimension of the boundary value problem is reduced to n.

In the special case when the boundary conditions (7.3.4) have the form

$$\mathbf{x}(t_0) - \mathbf{x}_0 = 0 \tag{7.3.12}$$

$$\mathbf{x}(t_f) - \mathbf{x}_f = 0, \tag{7.3.13}$$

and t_0 and t_f are fixed, to solve the boundary value problem it is only necessary to calculate the vector $\hat{\mathbf{p}}(t_0)$ (n scalar variables) under which condition (7.3.13) (n scalar variables) is satisfied.

The solution to the boundary value problem of the maximum principle is implemented in practice by numerical methods and, as a rule, it is the most difficult part of the problem of optimizing the trajectory of a spacecraft with an electric propulsion system. Typical methods for optimizing spacecraft trajectories based on the application of the maximum principle have the following disadvantages:

- limited area of convergence of the solution, which leads to the need to search for a good initial approximation before solving the optimization problem;
- high sensitivity of trajectories to variations of unknown parameters of the boundary value problem of the maximum principle;
- lack of reliable information about the existence and number of extreme solutions before starting to solve the problem;
- the existence of many solutions that satisfy optimality conditions, among which it is often necessary to find the best;
- jumps in the matrix of partial derivatives of the residuals of the boundary value problem for unknown parameters of the boundary value problem when changing the structure of the control program on the trajectory (sequences of coasting and thrusting sections in trajectory optimization problems with the thrust switching), which can lead to failures of numerical methods for solving the problem;
- the complexity of the mathematical model of the optimal motion of the spacecraft (a joint system of differential equations for state and conjugate variables) for perturbed optimal control problems, since it is necessary to calculate the partial derivatives of the perturbing accelerations with respect to the state variables, while the perturbing accelerations themselves often need to be calculated not by finite formulas, but by some iterative algorithms.

Recent advances in applied computational mathematics allow for a new approach to indirect methods for solving optimal control problems based on the maximum principle and, to a large extent, correct the above-mentioned shortcomings of numerical methods built on their basis, while maintaining high computational performance and accuracy of calculations.

For example, to expand the convergence region of the numerical method for solving the boundary value problem of the maximum principle, the method of continuous continuation (or the Newtonian homotopy method) is used, which makes it possible to formally reduce the boundary value problem to the Cauchy problem. A series of continuation methods have been developed that allow optimizing certain types of problems without choosing an initial approximation (more precisely, when using trivial initial approximations) for unknown parameters of the boundary value problem.

The essence of the continuation method lies in the formal reduction of the boundary value problem under consideration to the Cauchy problem. The boundary value problem for the dynamic system (7.3.9) with boundary conditions of the type (7.3.4), (7.3.10), (7.3.11) can be represented as an equation for residuals:

$$\mathbf{f}(\mathbf{z}) = 0 \tag{7.3.14}$$

where $\mathbf{z}$ is the vector of unknown parameters of the boundary value problem.

With an initial guess value for the decision vector of the boundary value problem $\mathbf{z}_0$, we calculate the vector of residuals (7.3.14):

$$\mathbf{f}(\mathbf{z}_0) = \mathbf{b} \tag{7.3.15}$$

Let's consider the immersion of Eq. (7.3.14) into the one-parameter family

$$\mathbf{f}(\mathbf{z}) = (1 - \tau)\mathbf{b} \tag{7.3.16}$$

where τ is the continuation parameter, and represent the vector $\mathbf{z}$ as a function of this parameter: $\mathbf{z} = \mathbf{z}(\tau)$, with $\mathbf{z}(0) = \mathbf{z}_0$ from Eq. (7.3.15). We require that equality (7.3.16) be satisfied for any $0 \le \tau \le 1$. It is obvious that when $\tau = 0$, the Eq. (7.3.16) coincides with (7.3.15), and when $\tau = 1$ – with the equation for residuals for the desired boundary value problem (7.3.14).

In fact, Eq. (7.3.16) represents a Newtonian homotopy between the system of equations $\mathbf{f}(\mathbf{z}) - \mathbf{b} = 0$ with a known solution $\mathbf{z} = \mathbf{z}_0$ and the original system of equations $\mathbf{f}(\mathbf{z}) = 0$.

Differentiating Eq. (7.3.16) with respect to the continuation parameter τ and resolving the resulting expression with respect to the derivative $d\mathbf{z}/d\tau$, we obtain a formal reduction of Eq. (7.3.14) to the Cauchy problem:

$$\begin{gathered} \frac{d\mathbf{z}}{d\tau} = -\left(\frac{\partial \mathbf{f}}{\partial \mathbf{z}}\right)^{-1} \mathbf{b}, \\ \mathbf{z}(0) = \mathbf{z}_0,\ 0 \le \tau \le 1. \end{gathered} \tag{7.3.17}$$

Obviously, by integrating (7.3.17) over τ from 0 to 1, due to (7.3.15) and (7.3.16), we can define the desired decision vector of the boundary value problem (7.3.14) in the form $\mathbf{z} = \mathbf{z}(1)$. In fact, for boundary value problems of the maximum principle (7.3.9), (7.3.4), (7.3.10), and (7.3.11), the Cauchy problem (7.3.17) implements the continuation over final conditions (CFC) from the initial solution, corresponding to the initial guess value for the decision vector of the boundary value problem, to the desired solution.

We will call the differential Eqs. (7.3.17) differential equations of the continuation method. Of course, for the continuation method to be applicable, the existence and non-degeneracy of the partial derivative matrix $\partial\mathbf{f}/\partial\mathbf{z}$ over the entire continuation interval $\tau \in [0;\ 1]$ is required.

In some cases, it may be advisable to introduce a continuation parameter τ to the right-hand sides of the differential equations of motion and to the boundary conditions. In this case, the residual function $\mathbf{f}$ explicitly depends on the continuation parameter τ, and the differential equations of the continuation method will take the form:

$$\frac{d\mathbf{z}}{dt} = -\mathbf{f}_{\mathbf{z}}^{-1}\left(\mathbf{b} + \frac{\partial \mathbf{f}}{\partial \tau}\right) \tag{7.3.18}$$

Calculation of the sensitivity matrix $\partial \mathbf{f}/\partial \mathbf{z}$ (and of the vector $\partial \mathbf{f}/\partial \tau$) can be done, for example, using finite differences, joint integration of differential equations for the elements of this matrix with the equations of optimal motion of the spacecraft, or by the complex step differentiation method for high-precision numerical differentiation.

In the version with finite-difference calculation of the sensitivity matrix, options with direct and central differences were practically used. Such variants of the continuation method turned out to be quite practicable and in many cases allow to obtain results with minimal effort for the preparation of a mathematical model of optimal movement, since in this case differential equations for the elements of the sensitivity matrix are not used. However, when using finite differences, the efficiency of the continuation method algorithm is reduced due to the relatively low accuracy of the derivative calculation. For example, when using forward differences with an optimal step, the number of valid significant digits in the calculated derivative is 2 times less than the number of valid significant digits in the function value due to the machine representation of real numbers with a limited number of digits. Another reason for the decrease in the efficiency of the method is the presence of "noise" in the calculated values of the state and conjugate vectors, which manifests itself during the numerical integration of trajectories with adaptive step size. Such effect arises due to the operation of the automatic step selection procedure in the numerical integration procedure and was noted in [13]. In the same publication, a way to combat this effect was proposed, which consists in "freezing" the sequence of steps of numerical integration. In this case, the automatic step selection procedure is used only when integrating the nominal trajectory, and when calculating neighboring trajectories, the same sequence of steps is used as for the nominal trajectory. Experience with a version of the continuation method using a finite-difference approximation of the sensitivity matrix has shown that it can be used for trajectories with a limited angular range (of the order of 3–4 orbits) in cases where the spacecraft does not come close to the gravity center.

A version of the continuation method with calculation of partial derivatives during joint integration of differential equations for the optimal system and matrix elements $\partial \mathbf{f}/\partial \mathbf{z}$ proved to be more efficient in terms of numerical stability and speed, despite a significant increase in the order of the system of differential equations. The explanation for this fact is the higher accuracy of calculating partial derivatives when using the second method. These advantages are especially noticeable on multi-orbit trajectories and in cases where the trajectory passes near the gravity center. The disadvantage of this option is the high complexity of developing a mathematical model of the optimal motion: it requires the derivation of differential equations and boundary conditions for the elements of the sensitivity matrix, and in the case of discontinuous control, the calculation of jumps in the elements of the sensitivity matrix at the control discontinuity points.

One of the most promising and universal variants of the continuation method turned out to be the variant using the complex step method for calculating derivatives.

The complex step differentiation method was proposed in [15–17], and in [18, 19] it was applied to the problem of optimizing the trajectories of low-thrust spacecraft for automatically generating the right-hand sides of the differential equations for the adjoint system by numerical differentiation of the optimal Hamiltonian for perturbed problems. The use of the complex step differentiation method in the procedure in the solution of a boundary value problem to determine the trajectory of a low-thrust spacecraft was probably first proposed in [20].

The formula for differentiating the function $f(x)$ using the complex step method is:

$$\frac{df(x)}{dx} = \frac{\mathrm{Im}[f(x+ih)]}{h} + O(h^2)$$

where h is the differentiation step, and $i = \sqrt{1}$. This formula can be obtained by expanding the value of a real function $f(x)$ into a Taylor series by varying the argument of the function along the imaginary axis.

The complex step method ensures that the accuracy of determining the derivative is equal to the accuracy of calculating the function ε at the differentiation step h, which is less than $\varepsilon^{1/2}$. Figure 7.1 shows, using a model example of numerical differentiation of an analytic function, the error in estimating the first derivative when using the forward difference method and the complex step method depending on the differentiation step for the model problem.

Application of the complex step method to solving the problem of optimizing the trajectories of low-thrust spacecraft made it possible to significantly improve the convergence of the continuation method with a relative simplicity of implementation, comparable to the simplicity of implementing a version of the continuation method with finite-difference calculation of derivatives.

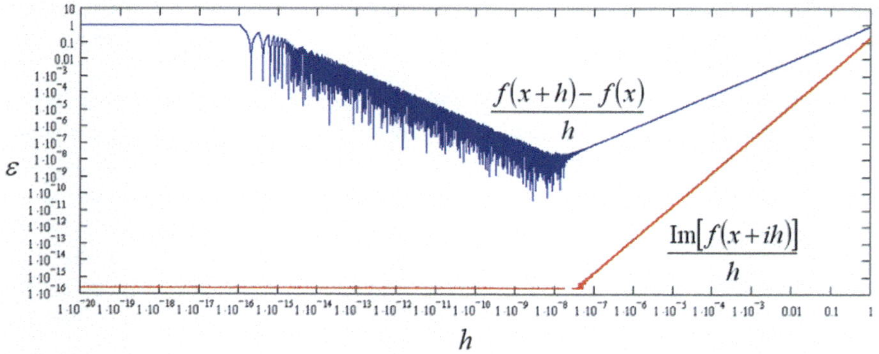

Fig. 7.1 Relative errors in calculating derivatives by the finite-difference method and the complex-step method depending on the differentiation step h

Of course, differentiation by the complex step method requires calculating the vector of residuals of the boundary value problem in the complex domain. Therefore, to implement this version of the continuation method, a method is required for numerically solving systems of ordinary differential equations with a state vector defined in the complex domain.

A specific implementation of the numerical method of continuous continuation, especially in the form (7.3.18), depends significantly on the optimization problem under consideration. Therefore, the details of the implementation of the continuation method for specific problems will be considered further when describing methods for solving them.

The solution of the of LP trajectory optimization problem, which exists for arbitrary nonsingular boundary conditions, in many cases provides sufficient information about the existence of LT trajectories. Therefore, the method proposed in this study for the continuation of an LP solution into an LT solution makes it possible in many cases to resolve the question of the existence of an LT solution after the calculation of the LP solution before starting the next continuation stage. If the maximum thrust acceleration on the optimal LP trajectory does not exceed the thrust acceleration in the LT problem, and the characteristic velocity on the LP trajectory does not exceed the maximum permissible increment of the characteristic velocity in the LT problem, then there is obviously a solution to the LT problem. However, a solution to the LT problem may exist even with some violation of these restrictions. Therefore, to diagnose the existence of an optimal LT trajectory, it is proposed to first solve the thrust minimization problem ($T_{\min}$-problem). The solution to this problem makes it possible to calculate the region of existence of LT trajectories with minimum fuel consumption for a fixed flight duration or angular range: such trajectories exist if the minimum thrust found does not exceed the thrust value specified in the LT problem.

To reduce the sensitivity of trajectories to variations in the unknown parameters of the maximum principle boundary value problem, a number of problems use the division of the trajectory into sections (for example, between gravity assist maneuvers) and propose a special scheme for selecting the composition of the unknown parameters of the boundary value problem.

The trajectory optimality conditions can be satisfied on a certain set of unknown parameters of the boundary value problem, and each set of unknown parameters from this set corresponds to a locally optimal trajectory, characterized by a certain cost function (flight duration, final mass of the spacecraft, etc.). For example, for multi-orbit spacecraft transfers with constant thrust, there are many local extrema corresponding to different angular flight ranges. To calculate trajectories with an optimal ratio of duration and angular range, either the angular range continuation method or the formulation of the problem with a fixed angular range and free transfer time are used.

Jumps in the elements of the sensitivity matrix (partial derivatives of the boundary conditions with respect to the parameters of the boundary value problem) due to the appearance or disappearance of active or coasting sections are one of the main reasons for the failure of traditional numerical methods in solving the problem of trajectory optimization with thrust switching. To overcome this difficulty, we use a method of smoothing the thrust relay function based on the properties of the switching function, which ensures the smoothing of jumps in the elements of the sensitivity matrix and thus allows the reliable determination of the optimal trajectory structure.

The problem of complexity in the derivation of differential equations of optimal motion in perturbed problems is solved using high-precision numerical or automatic differentiation methods (complex step method, use of algebra of dual or complex dual numbers, etc.).

The combination of the listed techniques makes it possible to effectively solve almost any problem of optimization of the trajectories of a spacecraft with an electric propulsion system using indirect methods based on the application of the maximum principle.

7.5 Mathematical Model of the Optimal Trajectory of a Spacecraft with an Ideally Controlled Limited-Power Thruster

Let us consider the problem of optimizing the trajectory of a spacecraft with an ideally controlled limited-power thruster of, following mainly the publications [20–25].

An ideally controlled (IC) thruster is a mathematical model of an electric propulsion (EP). In this model, it is assumed that the only limitation is imposed on the specific impulse and thrust of the electric propulsion—their product is considered given. Within the given constraint, the specific impulse and thrust of the IC thruster can vary arbitrarily. Optimization of trajectories of spacecraft with an IC thruster has been considered in many publications, for example [26–33]. With a certain degree of boldness, we can state that in the mechanics of space flight and in the theory of optimal control the problem of optimizing the trajectories of a spacecraft with an IC thruster plays a role that is similar to that of the restricted three-body problem in celestial mechanics and analytical dynamics. The differential equations of this problem, like the equations of the restricted three-body problem, have a beautiful symmetrical form and cannot be integrated by quadrature. The variety of particular solutions to both problems, along with their relative simplicity, symmetry and smoothness, has led to the fact that the problem of optimizing flights of a spacecraft with an IC thruster has become a kind of "touchstone" for new approximate methods for solving optimal control problems, like the restricted three-body problem for methods of analytical dynamics.

From a practical point of view, the IC thruster model, although in most cases not being sufficiently justified from a technical point of view due to the complexity of implementing the control of the electric propulsion in a wide range of specific impulse and mass flow rate, nevertheless allows one to determine a lower estimate of the power consumption required to solve the transportation problem under consideration. In addition, recently there have appeared projects of electric propulsion with characteristics close to those of the IC thruster. In particular, the VASIMR project

(Variable Specific Impulse Magnetoplasma Rocket) [34, 35] is being considered both for automatic interplanetary spacecraft and for a manned expedition to Mars.

Another reason for the interest in the problem of optimizing the trajectories of a spacecraft with an IC thruster is the possibility of using its solutions as an initial approximation for solving more complicated problems of optimizing the trajectories of spacecraft equipped with engines with more realistic control characteristics [20, 25, 36].

In what follows, the problem of optimizing the trajectory of a spacecraft with a limited-power IC engine will be called the LP problem.

The jet power P_b of the electric propulsion system (EPS) is equal to half of the product of the thrust T by the exhaust velocity c:

$$P_b = \frac{Tc}{2} \tag{7.4.1}$$

On the other hand, the EPS jet power can be expressed as the product of the EPS efficiency η_{EPS} and the electrical power P consumed by the EPS. Both η_{EPS} and P may depend on the spacecraft position vector $\mathbf{r}$ and time t. The dependence of the EPS electrical power on the state coordinates of the spacecraft and time is usually related to the type of onboard power supply system (PSS) of the spacecraft and with the effects of degradation of the characteristics of the power supply and propulsion systems. The most common types of onboard spacecraft PSSs are based on solar and nuclear power sources. The electrical power of a solar PSS depends on the heliocentric distance of the spacecraft, while the power of a nuclear PSS is usually assumed to be constant. Preliminary space mission design and analysis for a spacecraft with electric propulsion is usually performed under the assumption of constant efficiency. Therefore, the case of variable electric power corresponds to the use of a solar electric propulsion system (SEPS), and the case of constant electric power corresponds to the use of a nuclear electric propulsion system (NEPS). Another reason for changes in the electrical power of the electric propulsion system is the degradation of the characteristics of the onboard PSS and thrusters. Therefore, in the general case, P_b also depends on $\mathbf{r}$ and t. Hence,

$$P_b = \eta(\mathbf{r}, t) P_{b0}$$

where $\eta(\mathbf{r},\ t) = \eta_{EPS}(\mathbf{r},\ t)P(\mathbf{r},\ t)/P_{b0}$, $\eta(\mathbf{r}, t) = \eta_{EPS}(\mathbf{r}, t)P(\mathbf{r}, t)\big/ P_{b0}$, $P_{b0} = \eta_{EPS}(\mathbf{r}_0, t_0)P(\mathbf{r}_0, t_0)$ is the initial jet power, $\mathbf{r}_0$ is the initial spacecraft position vector and t_0 is the initial time. Given the dependency $T = -\dot{m}c$, where $\dot{m}$ is the spacecraft mass, the differential equation for the spacecraft mass will take the form:

$$\frac{dm}{dt} = -\frac{T^2}{2P_b} = -\frac{m^2 a^2}{2P_{b0}\eta(\mathbf{r}, t)} \tag{7.4.2}$$

where $a = T/m$ is the thrust acceleration.

Equation (7.4.2) has the following solution:

$$m(t) = \frac{m_0}{1 + \frac{m_0}{P_{b0}} J(t)} \tag{7.4.3}$$

where

$$J(t) = \frac{1}{2} \int_{t_0}^{t} \frac{a^2}{\eta(\mathbf{r}, t)} dt \tag{7.4.4}$$

J(t) is the cost function of the LP-problem.

If a spacecraft moves in a potential external field with a force function Ω, the position vector $\mathbf{r}$ and velocity vector $\mathbf{v}$ of the spacecraft satisfy the differential equations:

$$\frac{d\mathbf{r}}{dt} = \mathbf{v}, \quad \frac{d\mathbf{v}}{dt} = \Omega_{\mathbf{r}} + \mathbf{a} \tag{7.4.5}$$

Here Ω is the force function ($\Omega = 1/r$, $r = |\mathbf{r}|$) for the central Newtonian field in dimensionless coordinates and $\mathbf{a}$ ($a = |\mathbf{a}|$) is the thrust acceleration vector.

Applying the maximum principle to the optimal control problem (7.4.2), (7.4.4), (7.4.5), we have the following Pontryagin function:

$$H = -\frac{a^2}{2\eta}\left(1 + \frac{m^2 p_m}{P_{b0}}\right) + \mathbf{p}_{\mathbf{v}}^{\mathrm{T}} \Omega_{\mathbf{r}} + \mathbf{p}_{\mathbf{v}}^{\mathrm{T}} \mathbf{a} + \mathbf{p}_{\mathbf{r}}^{\mathrm{T}} \mathbf{v} \tag{7.4.6}$$

where p_m, $\mathbf{p_r}$, $\mathbf{p_v}$ are the variables conjugate to m, $\mathbf{r}$ and $\mathbf{v}$, respectively.

From the maximum condition (7.4.6) for $\mathbf{a}$, the optimal control has the form:

$$\mathbf{a} = \frac{\eta P_{b0}}{P_{b0} + m^2 p_m} \mathbf{p_v} \tag{7.4.7}$$

and the equations of optimal motion are of the form:

$$\left.\begin{aligned}
\frac{d^2\mathbf{r}}{dt^2} &= \Omega_{\mathbf{r}} + \frac{\eta P_{b0}}{P_{b0} + m^2 p_m} \mathbf{p_v}, \\
\frac{dm}{dt} &= -\frac{\eta P_{b0} m^2 p_v^2}{2\left(P_{b0} + m^2 p_m\right)^2}, \\
\frac{d^2\mathbf{p_v}}{dt^2} &= \Omega_{\mathbf{rr}} \mathbf{p_v} + \frac{P_{b0} p_v^2}{2\left(P_{b0} + m^2 p_m\right)} \frac{\partial \eta}{\partial \mathbf{r}}, \\
\frac{dp_m}{dt} &= \frac{\eta P_{b0} m p_m p_v^2}{\left(P_{b0} + m^2 p_m\right)^2},
\end{aligned}\right\} \tag{7.4.8}$$

where $p_v = |\mathbf{p_v}|$. This system has the first integral

$$\frac{d}{dt}\left(m^2 p_m\right) = 2mp_m\frac{dm}{dt} + m^2\frac{dp_m}{dt} = 0 \Rightarrow m^2 p_m = C_m \tag{7.4.9}$$

Consequently, if the spacecraft mass is free at any boundary (for example, at a final point), then $p_m = 0$ at that point due to the transversality conditions [14] and $C_m = 0$ (and therefore $p_m(t) \equiv 0$ due to (7.4.9)). In this case, the differential equations for $\mathbf{r}$ and $\mathbf{p_v}$ do not depend on m and p_m, and the differential equation for m does not depend on p_m. Thus, the system (7.4.8) can be decomposed into independent dynamic (equations for $\mathbf{r}$ and $\mathbf{p_v}$) and parametric (equation for the spacecraft mass) parts if m is free at least at one boundary point. This property of the LP problem was discovered by Irving [31] half a century ago. It should be noted that the decomposition of the optimal control problem into dynamic and parametric parts is impossible in a rather exotic case, when there are constraints on the spacecraft mass at both ends of the trajectory.

Taking into account the specified decomposition of the problem, the equations for the dynamic part can be written as:

$$\left.\begin{aligned} \frac{d^2\mathbf{r}}{dt} &= \Omega_{\mathbf{r}} + \eta\mathbf{p_v}, \\ \frac{d^2\mathbf{p_v}}{dt^2} &= \Omega_{\mathbf{rr}}\mathbf{p_v} + \frac{p_v^2}{2}\frac{\partial\eta}{d\mathbf{r}}, \end{aligned}\right\} \tag{7.4.10}$$

The equation for mass is:

$$\frac{dm}{dt} = -\frac{\eta m^2 p_v^2}{2P_{b0}} \tag{7.4.11}$$

The spacecraft mass can be calculated from (7.4.3) using (7.4.4), (7.4.7) after solving (7.4.10). Most authors use such approach. However, consideration of the homotopy between the LP-problem and the limited thrust problem (LT-problem) requires the use of the complete system (7.4.8).

Equations (7.4.8) should be supplemented with boundary conditions. Within the assumptions of the ZSOI method, typical initial conditions for the heliocentric section of an interplanetary trajectory have the form:

$$\mathbf{r}(t_0) = \mathbf{r}_0(t_0), \mathbf{v}(t_0) = \mathbf{v}_0(t_0), m(t_0) = m_0 \tag{7.4.12}$$

if the hyperbolic excess velocity of the planet of departure is zero, or:

$$\mathbf{r}(t_0) = \mathbf{r}_0(t_0), \mathbf{v}(t_0) = \mathbf{v}_0(t_0) + V_{\infty 0}\mathbf{p}_{\mathbf{v}0}/p_{v0}, m(t_0) = m_0 \tag{7.4.13}$$

otherwise. Here $\mathbf{r}_0$, $\mathbf{v}_0$ are the position and velocity vectors of the departure planet, respectively, t_0 is the specified departure time, m_0 is the initial mass of the spacecraft,

$\mathbf{p}_{v0} = \mathbf{p}_v(t_0)$, $p_{v0} = |\mathbf{p}_{v0}|$. The second equation of (7.4.13) is derived from the well-known transversality condition [27] (this condition will be analyzed in more detail in Sect. 9.2).

Typical terminal conditions correspond to the rendezvous or flyby problems. For the rendezvous problem, these conditions have the form:

$$\mathbf{r}(t_f) - \mathbf{r}_f(t_f) = 0, \; \mathbf{v}(t_f) - \mathbf{v}_f(t_f) = 0, \tag{7.4.14}$$

and for the flyby the boundary conditions and transversality conditions have the form:

$$\mathbf{r}(t_f) - \mathbf{r}_f(t_f) = 0, \; \mathbf{p}_v(t_f) = 0. \tag{7.4.15}$$

Here $\mathbf{r}_f$, $\mathbf{v}_f$ are the position and velocity vectors of the planet of arrival, respectively, t_f is the specified arrival time.

Of course, the mathematical model of the optimal trajectory of a spacecraft with a limited-power propulsion system can (and, in some cases, it makes sense to) be formulated using a different form of writing the equations of motion of the spacecraft, for example, in equinoctial elements, as in (7.1.10).

In the remainder of this section we will limit ourselves to considering the problem of optimizing an unperturbed trajectory with a fixed angular range in true longitude ΔL. The parameters of the initial orbit are denoted by the subscript "0", and those of the final orbit - by the subscript "f". Let us consider a flight from the initial orbit point with true longitude L_0 with angular distance ΔL. In this case, the end point of the trajectory will have a true longitude $L_f = L_0 + \Delta L$. Let us introduce the notation $\mathbf{x} = (p, e_x, e_y, i_x, i_y)^{\mathrm{T}}$. In this case, the boundary conditions for the arbitrary flight time take the form:

$$L_K(K_0) + K_0 = L_0, \mathbf{x}(K_0) = \mathbf{x}_0(L_0), m(K_0) = m_0, t(K_0) = t_0, \tag{7.4.16}$$

$$L_K(K_f) + K_f = L_f, \mathbf{x}(K_f) = \mathbf{x}_f(L_f), \tag{7.4.17}$$

where K_0 and K_f are the initial and final values of the auxiliary longitude, respectively, t_0 is the initial flight time. Since L_K is included in the equations of motion (7.1.10) only in the form of the sum $L = L_K + K$, without loss of generality we can set $K_0 = 0$, and the initial conditions (7.4.16) can be rewritten in the form:

$$L_K(0) = L_0, \mathbf{x}(0) = \mathbf{x}_0(L_0), m(0) = m_0, t(0) = t_0. \tag{7.4.18}$$

In case the flight time Δt given, the following condition is added to the final conditions (7.4.17):

$$t(K_f) = t_f, \tag{7.4.19}$$

where $t_f = t_0 + \Delta t$—final flight time.

Consider the case of constant power ($\eta(\mathbf{r}, t) \equiv 1$). Then the problem of minimizing the propellant consumption corresponds to the problem of optimizing the cost function $J_{LP}(t_f) = \frac{1}{2}\int_{t_0}^{t_f} a_{LP}^2(t)dt$, where $a_{LP}(t) = \sqrt{a_{LPr}^2(t) + a_{LPt}^2(t) + a_{LPn}^2(t)}$ is the thrust acceleration, and a_{LPr}, a_{LPt}, a_{LPn} are the radial, transversal and binormal components of thrust acceleration, respectively. If K is used instead of t as an independent variable, it is necessary to replace the variable in the last integral, so the minimized cost function will take the form:

$$J_1 = \frac{1}{2}\int_{t_0}^{t_f} a_{LP}^2 dt = \frac{1}{2}\int_{K_0}^{K_f} \frac{a_{LP}^2}{q^2}\sqrt{\frac{p^3}{\mu}}dK \quad (7.4.20)$$

The dynamic equations of unperturbed motion of a spacecraft with an ideally controlled thruster have the following form:

$$\begin{aligned}
\frac{dp}{dK} &= \frac{2p^3}{\mu q^3}a_{LPt},\\
\frac{de_x}{dK} &= \frac{p^2}{\mu q^2}\left[\sin L \cdot a_{LPr} + \frac{(q+1)\cos L + e_x}{q}a_{LPt} - \frac{e_y \xi}{q}a_{LPn}\right],\\
\frac{de_y}{dK} &= \frac{p^2}{\mu q^2}\left[-\cos L \cdot a_{LPr} + \frac{(q+1)\sin L + e_y}{q}a_{LPt} + \frac{e_x \xi}{q}a_{LPn}\right],\\
\frac{di_x}{dK} &= \frac{p^2 s^2}{2\mu q^3}\cos L \cdot a_{LPn},\ \frac{di_y}{dK} = \frac{p^2 s^2}{2\mu q^3}\sin L \cdot a_{LPn},\ \frac{dL_K}{dK} = \frac{p^2}{\mu q^3}\xi \cdot a_{LPn}.
\end{aligned} \quad (7.4.21)$$

Since the right-hand sides of these equations do not depend on the spacecraft mass and flight time, differential equations for m and t are not included in this system. Let us consider the problem of optimizing the unperturbed trajectory of a spacecraft with an ideally controlled thruster between two given points with a fixed angular range and free flight time. The boundary conditions for such a problem have the following form:

$$L_K(0) = L_0, \mathbf{x}(0) = \mathbf{x}_0(L_0), \quad (7.4.22)$$

$$L_K(K_f) + K_f = L_f, \mathbf{x}(K_f) = \mathbf{x}_f(L_f), \quad (7.4.23)$$

Thus, it is required to find a thrust acceleration program $\mathbf{a}_{LP}(K) = [a_{LPt}(K), a_{LPr}(K), a_{LPn}(K)]^{\mathrm{T}}$ that would minimize the cost function (7.4.20) by solving the system of differential Eqs. (7.4.21) satisfying the initial conditions (7.4.22) and final conditions (7.4.23).

The Pontryagin function of the optimal control problem (7.4.20) - (7.4.23) has the form:

$$H_{LP} = -\frac{1}{2}\frac{a_{LP}^2}{q^2}\sqrt{\frac{p^3}{\mu}} + \frac{p^2}{q^3\mu}\mathbf{A}^T\mathbf{a}_{LP} \tag{7.4.24}$$

where $\mathbf{A}^{\mathrm{T}} = (A_t, A_r, A_n)$,

$$\begin{aligned}
A_t &= 2p \cdot p_p + \left[(q+1)\cos L + e_x\right]p_{ex} + \left[(q+1)\sin L + e_y\right]p_{ey},\\
A_r &= q \cdot \left(\sin L \cdot p_{ex} - \cos L \cdot p_{ey}\right),\\
A_n &= \xi \cdot \left(-e_y p_{ex} + e_x p_{ey} + p_{LK}\right) + \frac{s^2}{2}\left(\cos L \cdot p_{ix} + \sin L \cdot p_{iy}\right),
\end{aligned}$$

$p_p, p_{ex}, p_{ey}, p_{ix}, p_{iy}, p_{LK}$ are the variables conjugate to the corresponding state variables of the system p, e_x, e_y, i_x, i_y, L_K.

From the condition for the maximum of the Pontryagin function (28) in the control $\mathbf{a}_{LP}$, it is easy to obtain an expression for the optimal control:

$$\mathbf{a}_{LP} = \frac{1}{q}\sqrt{\frac{p}{\mu}}\mathbf{A} \tag{7.4.25}$$

By substituting it into (7.4.24), we obtain the expression for the Hamiltonian of the optimal control problem under consideration:

$$H_{LP} = k\mathbf{A}^T\mathbf{A} \tag{7.4.26}$$

Here $k = \frac{1}{2q^4}\sqrt{\frac{p^5}{\mu^3}}$. The optimal motion equations will obviously take the form:

$$\frac{d\mathbf{x}}{dK} = \frac{\partial H_{LP}}{\partial \mathbf{p_x}}, \frac{dL_K}{dK} = \frac{\partial H_{LP}}{\partial p_{LK}}, \frac{d\mathbf{p_x}}{dK} = -\frac{\partial H_{LP}}{\partial \mathbf{x}}, \frac{dp_{LK}}{dK} = -\frac{\partial H_{LP}}{\partial L_K}. \tag{7.4.27}$$

To solve the LP problem under consideration, it is necessary to determine 5 initial values of the conjugate variables $\mathbf{p_x} = (p_p, p_{ex}, p_{ey}, p_{ix}, p_{iy})^{\mathrm{T}}$, the initial value p_{LK} and the final value of the auxiliary longitude K_f (7 unknown variables in total). Boundary conditions (7.4.22) and (7.4.23) define only 6 equations for determining these 7 unknown variables and must be supplemented with a transversality condition. To derive it, we write the Lagrange end function in the following form:

$$\begin{aligned}
l = {} & (L_K(K_0) + K_0 - L_0)\lambda_L^0 + \left(L_K\left(K_f\right) + K_f - L_f\right)\lambda_L^f + \\
& + (\mathbf{x}(K_0) - \mathbf{x}_0(L_0))^{\mathrm{T}}\boldsymbol{\lambda}_{\mathbf{x}}^0 + \left(\mathbf{x}\left(K_f\right) - \mathbf{x}_f\left(L_f\right)\right)^{\mathrm{T}}\boldsymbol{\lambda}_{\mathbf{x}}^f,
\end{aligned}$$

where λ_L^0, λ_L^f, $\boldsymbol{\lambda}_{\mathbf{x}}^0$, $\boldsymbol{\lambda}_{\mathbf{x}}^f$ are the Lagrange multipliers. The conditions of transversality will take the form:

$$\mathbf{p_x}(K_0) = \frac{\partial l}{\partial \mathbf{x}(K_0)} = \boldsymbol{\lambda}_{\mathbf{x}}^0,\ \mathbf{p_x}\left(K_f\right) = -\frac{\partial l}{\partial \mathbf{x}\left(K_f\right)} = -\boldsymbol{\lambda}_{\mathbf{x}}^f,$$
$$p_{LK}(K_0) = \frac{\partial l}{\partial L_K(K_0)} = \lambda_L^0,\ p_{LK}\left(K_f\right) = -\frac{\partial l}{\partial L_K\left(K_f\right)} = -\lambda_L^f,$$
$$H(K_0) = -\frac{\partial l}{\partial K_0} = -\lambda_L^0,\ H\left(K_f\right) = \frac{\partial l}{\partial K_f} = \lambda_L^f.$$

These conditions imply the relations $p_{LK}(K_0) + H(K_0) = 0$ and $p_{LK}\left(K_f\right) + H\left(K_f\right) = 0$. Let us show that $p_{LK} + H$ is the first integral of the system (7.4.27). Indeed, $\frac{dp_{LK}}{dK} = -\frac{\partial H}{\partial L_K} = -\frac{\partial H}{\partial K} = -\frac{dH}{dK}$, since K and L_K are included in the Hamiltonian (7.4.26) only in the form of the sum $K + L_K$ and taking into account the known properties of non-autonomous Hamiltonian systems. It follows that $d(p_{LK} + H)/dK = 0$, that is, $p_{LK} + H$ is the first integral. Consequently, the equality of $p_{LK} + H$ to zero for any value of K implies the identical equality of this expression to zero along the entire trajectory. Taking into account the accepted equality $K_0 = 0$, the additional boundary condition is reduced to the form $p_{LK}(0) + H(0) = 0$.

Thus, for the considered boundary value problem of the maximum principle, the initial conditions take the form:

$$L_K(0) = L_0,\ \mathbf{x}(0) = \mathbf{x}_0(L_0),\ p_{LK}(0) + H_{LP}(0) = 0. \tag{7.4.28}$$

The latter condition is reduced to a quadratic equation for $p_{LK}(0)$:

$$a_0 p_{LK}^2(0) + b_0 p_{LK} + c_0 = 0,$$

where $a_0 = k\xi^2, b_0 = 1 + 2kA_{n0}\xi, c_0 = k \cdot \left(A_t^2 + A_r^2 + A_{n0}^2\right)$. Obviously, if the starting point of the trajectory coincides with a descending or ascending node, then $\xi = 0$ and $p_{LK}(0) = -c_0$. Otherwise:

$$p_{LK}(0) = \frac{-b_0 \pm \sqrt{b_0^2 - 4a_0c_0}}{2a_0}. \tag{7.4.29}$$

It should be noted that in solving the boundary value problem (7.4.27), (7.4.28), (7.4.23) the necessary conditions for the optimality of the flight time $p_t(K_f) = 0$ are satisfied, since only if this condition is fulfilled, after having added time to the state variables, the Hamiltonian retains the form (7.4.26). Moreover, $p_t(K) \equiv 0$ due to the independence of the right-hand sides of the optimal motion equations on time.

7.6 Mathematical Model of the Optimal Trajectory of a Spacecraft with a Limited-Thrust Propulsion System

The LP problem uses an idealized mathematical model of an electric propulsion system. As noted above, the problem of spacecraft transfer with a limited-thrust electric propulsion system (hereinafter referred to as the LT problem) is much more realistic. For its description, we will follow [20, 36]. In the LT problem, it is assumed that the jet power P_b is a function of the spacecraft position and time, while the thrust T and the exhaust velocity c are functions of the jet power:

$$P_b = \eta(\mathbf{r}, t)P_{b0} \tag{7.5.1}$$

$$T = T[\eta(\mathbf{r}, t)] \tag{7.5.2}$$

$$c = c[\eta(\mathbf{r}, t)] \tag{7.5.3}$$

The expressions (7.5.1), (7.5.2) and (7.5.3) define the mathematical model of the control characteristics of the power supply and propulsion system, that is, the dependence of the power available for the electric propulsion system, its thrust and exhaust velocity on the spacecraft flight conditions (state coordinates and time). A special case of the LT problem is, for example, the problem of flight with given constant values of thrust and exhaust velocity.

Equations of motion of a spacecraft with an electric propulsion system in an external potential field with a force function Ω in the context of the LT problem have the form:

$$\frac{d\mathbf{r}}{dt} = \mathbf{v},\ \frac{d\mathbf{v}}{dt} = \Omega_{\mathbf{r}} + \delta\frac{T}{m}\mathbf{e},\ \frac{dm}{dt} = -\delta\frac{T}{c} \tag{7.5.4}$$

where δ is the throttling function ($\delta = 1$ during the thruster operation, and $\delta = 0$ during coasting), $\mathbf{e}$ is a unit vector directed along the thrust vector.

Let us consider the problem of fuel minimization, that is, the problem of minimizing the cost function:

$$J = \int_{t_0}^{t_f} \delta\frac{T}{c}dt \tag{7.5.5}$$

The Pontryagin function of the optimal control problem has the form:

$$H = -\delta\frac{T}{c} + \mathbf{p}_{\mathbf{r}}^{\mathrm{T}}\mathbf{v} + \mathbf{p}_{\mathbf{v}}^{\mathrm{T}}\Omega_{\mathbf{r}} + \delta\frac{T}{m}\mathbf{p}_{\mathbf{v}}^{\mathrm{T}}\mathbf{e} - \delta p_m\frac{T}{c} \tag{7.5.6}$$

where $\mathbf{p}_{\mathbf{r}} = -d\mathbf{p}_{\mathbf{v}}/dt$, $\mathbf{p}_{\mathbf{v}}$, and p_m are the variables conjugate to $\mathbf{r}$, $\mathbf{v}$ and m, respectively.

Optimal control can be obtained from maximizing the Pontryagin function (7.5.6) for controls **e** and δ:

$$\mathbf{e} = \mathbf{p_v}/p_v, \tag{7.5.7}$$

$$\delta = \begin{cases} 1, S > 0 \\ 0, S < 0 \end{cases} \tag{7.5.8}$$

where S is the switching function,

$$S = \frac{p_v}{m} - \frac{1 + p_m}{c} \tag{7.5.9}$$

The complete system of differential equations of optimal motion for the LT problem, in accordance with the maximum principle, has the form:

$$\left.\begin{aligned} \frac{d^2\mathbf{r}}{dt^2} &= \Omega_\mathbf{r} + \delta T \frac{\partial S}{\partial \mathbf{p_v}}, \\ \frac{dm}{dt} &= \delta T \frac{\partial S}{\partial p_m}, \\ \frac{d^2\mathbf{p_v}}{dt^2} &= \Omega_{\mathbf{rr}}\mathbf{p_v} + \delta\left[\frac{\partial T}{\partial \eta} S + T \frac{\partial S}{\partial c}\frac{\partial c}{\partial \eta}\right]\frac{\partial \eta}{\partial \mathbf{r}}, \\ \frac{dp_m}{dt} &= -\delta T \frac{\partial S}{\partial m}. \end{aligned}\right\} \tag{7.5.10}$$

The boundary conditions for standard LT problems coincide with the variants of boundary conditions for the LP problem (7.4.12–7.4.15).

Consider the problem of optimizing a trajectory with a fixed angular range in true longitude ΔL using equinoctial elements. The initial and final conditions of the problem under consideration are the same as for the previously considered case of the LP trajectory optimization ((7.4.16) - (7.4.19)).

Let us consider the problem of propellant minimization for a flight with a limited-thrust thruster with a constant exhaust velocity, a fixed angular distance and free flight time, that is, the problem of minimizing the cost function

$$J = \int_{t_0}^{t_f} \delta \frac{T}{m} dt = \int_{K_0}^{K_f} \delta \frac{T}{q^2 m}\sqrt{\frac{p^3}{\mu}} dK \tag{7.5.11}$$

for a dynamic system (7.1.10), (7.4.18), (7.4.17). In the general case, the optimal LT trajectory contains a number of the thruster firings and shutdowns that is unknown in advance.

The Pontryagin function of the LT problem has the form:

$$H = H_T + H_t, \tag{7.5.12}$$

where

$$H_T = \delta T k_1 \cdot \left[\frac{k_2}{m} \cdot (A_r \sin\vartheta \cos\psi + A_t \cos\vartheta \cos\psi + A_n \sin\psi) - \frac{p_m + 1}{c}\right], \tag{7.5.13}$$

$$H_t = \frac{1}{\sqrt{\mu p}}\left(\frac{p}{q}\right)^2 p_t, \tag{7.5.14}$$

p_m, p_t are the variables conjugate to the corresponding state variables of the system m, t, $k_1 = \frac{1}{q^2}\sqrt{\frac{p^3}{\mu}}$, $k_2 = \frac{1}{q}\sqrt{\frac{p}{\mu}}$.

It is obvious that only the term H_T (7.5.13) of the Pontryagin function (7.5.12) explicitly depends on the control δ,ϑ,ψ. Maximizing (7.5.13) by δ,ϑ,ψ, we obtain expressions for optimal control:

$$\sin\vartheta = A_r/A_{rt}, \cos\vartheta = A_t/A_{rt}, \sin\psi = A_n/A, \cos\psi = A_{rt}/A, \tag{7.5.15}$$

$$\delta = (1 + \mathrm{sign}S)/2, S \neq 0;\ \delta \in [0;\ 1], S = 0, \tag{7.5.16}$$

where $A_{rt} = \sqrt{A_r^2 + A_t^2}$, $A = \sqrt{A_r^2 + A_t^2 + A_n^2}$, $S = \frac{k_2 A}{m} - \frac{p_m+1}{c}$ is the switching function. In the future, we will not consider the possibility that the switching function is equal to 0 at a finite interval of change in the auxiliary longitude, i.e. we will exclude from consideration trajectories with sections of singular control.

Substituting expressions for optimal control (7.5.15), (7.5.16) into (7.5.13), we obtain the following expression for H_T:

$$H_T = \delta T k_1 \cdot \left[\frac{k_2 A}{m} - \frac{p_m + 1}{c}\right] \tag{7.5.17}$$

Substituting it in (7.5.12) instead of (7.5.13), we obtain an expression for the optimal Hamiltonian H.

The optimal motion equations will take the form:

$$\frac{d\mathbf{x}_1}{dK} = \frac{\partial H}{\partial \mathbf{p}_{\mathbf{x}1}}, \frac{d\mathbf{p}_{\mathbf{x}1}}{dK} = -\frac{\partial H}{\partial \mathbf{x}_1} \tag{7.5.18}$$

where $\mathbf{x}_1{}^{\mathrm{T}} = (p,\ e_x,\ e_y,\ i_x,\ i_y,\ L_K,\ m,\ t)$ is the state vector, and $\mathbf{p}_{\mathbf{x}1}{}^{\mathrm{T}} = (p_p, p_{ex}, p_{ey}, p_{ix}, p_{iy}, p_{LK}, p_m, p_t)$ is the vector of conjugate variables.

To solve the LT problem, it is necessary to determine 8 initial values of the conjugate variables $\mathbf{p}_{\mathbf{x}1}$ and the final value of the auxiliary longitude K_f (9 unknown variables in total). The boundary conditions (7.4.17), (7.4.18) define only 6 equations to determine these 9 unknown variables and must be supplemented with three transversality conditions. To derive them, we write the Lagrange end function in the form:

$$\begin{aligned} l = {} & (L_K(K_0) + K_0 - L_0)\lambda_L^0 + \left(L_K\left(K_f\right) + K_f - L_f\right)\lambda_L^f + \\ & +(\mathbf{x}(K_0) - \mathbf{x}_0(L_0))^{\mathrm{T}}\boldsymbol{\lambda}_{\mathbf{x}}^0 + \left(\mathbf{x}\left(K_f\right) - \mathbf{x}_f\left(L_f\right)\right)^{\mathrm{T}}\boldsymbol{\lambda}_{\mathbf{x}}^f + \\ & (m(K_0) - m_0)\lambda_m^0 + (t(K_0) - t_0)\lambda_t^0, \end{aligned}$$

where $\lambda_L^0, \lambda_L^f, \boldsymbol{\lambda}_{\mathbf{x}}^0, \boldsymbol{\lambda}_{\mathbf{x}}^f, \lambda_m^0, \lambda_t^0$ are the undetermined Lagrange multipliers. The transversality conditions will take the form:

$$\begin{aligned} \mathbf{p}_{\mathbf{x}}(K_0) &= \tfrac{\partial l}{\partial \mathbf{x}(K_0)} = \boldsymbol{\lambda}_{\mathbf{x}}^0,\ \mathbf{p}_{\mathbf{x}}\left(K_f\right) = -\tfrac{\partial l}{\partial \mathbf{x}\left(K_f\right)} = -\boldsymbol{\lambda}_{\mathbf{x}}^f, \\ p_{LK}(K_0) &= \tfrac{\partial l}{\partial L_K(K_0)} = \lambda_L^0,\ p_{LK}\left(K_f\right) = -\tfrac{\partial l}{\partial L_K\left(K_f\right)} = -\lambda_L^f, \\ p_m(K_0) &= \frac{\partial l}{\partial m(K_0)} = \lambda_m^0,\ p_m\left(K_f\right) = -\frac{\partial l}{\partial m\left(K_f\right)} = 0, \\ p_t(K_0) &= \frac{\partial l}{\partial t(K_0)} = \lambda_t^0,\ p_t\left(K_f\right) = -\frac{\partial l}{\partial t\left(K_f\right)} = 0, \\ H(K_0) &= -\tfrac{\partial l}{\partial K_0} = -\lambda_L^0,\ H\left(K_f\right) = \tfrac{\partial l}{\partial K_f} = \lambda_L^f, \end{aligned}$$

where $\mathbf{p}_{\mathbf{x}} = (p_p, p_{ex}, p_{ey}, p_{ix}, p_{iy})^{\mathrm{T}}$.

The relations $p_m(K_f) = 0$, $p_t(K_f) = 0$, $p_{LK}(K_0) + H(K_0) = 0$ and $p_{LK}\left(K_f\right) + H\left(K_f\right) = 0$ follow from these conditions. As in the LP problem, it is easy to show that $p_{LK} + H$ is the first integral of the system (20). Consequently, the equality of $p_{LK} + H$ to zero for any K value entails the identical equality of this expression to zero along the entire trajectory. Taking into account the assumed equality $K_0 = 0$, the additional boundary condition is reduced to the form $p_{LK}(0) + H(0) = 0$.

This equation can be used to calculate the unknown value $p_{LK}(0)$, which allows us to reduce the number of unknown parameters of the boundary value problem to 8. Finally, the boundary conditions for the problem under consideration can be written as:

$$L_K(0) = L_0, \mathbf{x}(0) = \mathbf{x}_0(L_0), m(0) = m_0, t(0) = t_0, p_{LK}(0) + H(0) = 0, \quad (7.5.19)$$

$$L_K\left(K_f\right) + K_f = L_f, \mathbf{x}\left(K_f\right) = \mathbf{x}_f\left(L_f\right), p_m\left(K_f\right) = 0, p_t\left(K_f\right) = 0, \quad (7.5.20)$$

Moreover, to solve the boundary value problem of the maximum principle (7.5.18), (7.5.19), (7.5.20), it is necessary to determine 8 components of the vector $\mathbf{p}_{\mathbf{x}1}(0)$ and the final auxiliary longitude K_f, at which the latter equation in (7.5.19) is satisfied, and 8 equations of (7.5.20).

Due to the Hamiltonian dependence on p_{LK}, the latter equation in (7.5.19) is reduced to a quadratic equation for $p_{LK}(0)$, the solution of which has the following form:

$$p_{LK}(0) = \frac{-b_0 \pm \sqrt{b_0^2 - 4a_0c_0}}{2a_0}, \tag{7.5.21}$$

where

$$a_0 = a_1^2\xi^2 - 1,\ b_0 = 2\left[a_1^2 A_{n0}\xi - H_{00}\right],\ c_0 = a_1^2\left(A_r^2 + A_t^2 + A_{n0}^2\right) - H_{00}^2,\ a_1 = \delta T\frac{k_1k_2}{m},$$

$$H_{00} = -\delta T k_1 \frac{p_m + 1}{c} + H_t,\ A_{n0} = \xi \cdot \left(-e_y p_{ex} + e_x p_{ey}\right) + \frac{s^2}{2}\left(\cos L \cdot p_{ix} + \sin L \cdot p_{iy}\right).$$

The main advantage of the considered formulation of the problem of trajectory optimization with a fixed angular distance and free flight time is that for the similar families of trajectories, judging by the numerical analysis carried out earlier (for example, in [8, 37]), there is only one solution that satisfies the necessary optimality conditions. When using a more traditional formulation–trajectory optimization with a fixed flight time and free angular range–there are many optimal solutions with different numbers of orbits [38–40], which limits the range of convergence of numerical methods and makes it difficult to calculate the globally optimal solutions.

It should be noted that when optimizing unperturbed trajectories or in the case when the perturbing accelerations do not depend explicitly on time, the differential equation for time can be excluded from the system of differential equations of optimal motion along with the boundary conditions for time and the time-conjugate variable. In this case, the Hamiltonian does not explicitly depend on time, so $dp_t/dK \equiv 0$ and $p_t(K) \equiv 0$ due to the last equation of (7.5.20).

7.7 Design Model of a Spacecraft for the Problem of Joint Optimization of Trajectory and Main Design Parameters

The main reason for using electric propulsion systems is their high specific impulse, which makes it possible to achieve high values of characteristic velocity with low propellant consumption. The main disadvantage of the electric propulsion system is the low thrust that is proportional to the electric power available on board the spacecraft for the electric propulsion system and inversely proportional to the specific impulse. Low thrust results in the need for a long-term operation of the electric propulsion system to provide the required increment in the characteristic velocity and in the long flight duration. To increase the thrust of the electric propulsion system, it is necessary either to decrease its specific impulse or increase the electric power. A decrease in specific impulse leads to an increase in the propellant consumption, and an increase in electric power leads to an increase in the mass of the electric propulsion system and onboard systems that provide power supply to the electric propulsion system. Obviously, for each specific space transportation mission of limited duration, there must be optimal values of the specific impulse and electric power (or thrust) of the electric propulsion system that ensure the maximum useful mass of the spacecraft, by which we mean the final mass of the spacecraft minus the mass of EPS with the propellant storage and feed system (PSFS), and mass other onboard systems required for EPS operation only.

In the traditional formulation of the problems of optimizing the trajectories of a spacecraft with an electric propulsion system, the main design parameters–initial mass, power of the power supply system (PSS), thrust and specific impulse of the electric propulsion system–are considered as given. This is certainly justified when the design of the spacecraft has been formed and the means of its launching have been selected. However, at the initial stages of spacecraft design, it is necessary to obtain an idea of what values of the design parameters of the spacecraft and electric propulsion system will provide the best performance during the space transportation operation under consideration. Despite the fact that many of these parameters can only be selected to a limited extent in practice, the knowledge of their optimal values and of the achievable values of the space mission performance indicators with optimal design parameters allows one to make reasonable technical decisions in the early stages of spacecraft development.

The main design parameters of a spacecraft with an electric propulsion system can be divided into two categories. The first category includes parameters that define the level of technology used. To ensure the maximum efficiency of the space transportation operation, their values should be maximum or minimum, but in fact they are limited by the technologies used and cannot be improved within the framework of the specific mission. Such parameters include, for example, the specific mass of the solar panels and other elements of the spacecraft power supply system, the specific mass and thrust efficiency of the EPS, and the tank coefficient (the ratio of the final mass of PSFS to the mass of propellant consumed to generate thrust). The second category includes parameters that can be selected when designing a spacecraft and electric propulsion system. Such parameters may include the electric power of solar panels, the maximum and minimum power consumption of the electric propulsion system, and the specific impulse of the electric propulsion system. The optimal values of these parameters, ensuring the maximum payload mass of the spacecraft, are associated with the trajectory parameters. Therefore, the joint optimization of the trajectory and main design parameters of a second category spacecraft is of great practical interest. Such joint optimization makes it possible to evaluate the maximum achievable efficiency indicators, to justify the main characteristics of the power supply system of the spacecraft and the electric propulsion system, to select the most suitable electric propulsion thrusters (EPT) for the implementation of a specific mission, and also to justify the main parameters of the newly designed EPT.

In publications, the problem of optimization of the main design parameters is often called parametric, in contrast to the dynamic problem–the problem of trajectory optimization [27]. It is well known that for an ideally controlled engine (in the case when only the power proportional to the product of the thrust and the exhaust velocity of the electric propulsion system is limited, and the values of the thrust itself and of the exhaust velocity can vary arbitrarily within this limitation), the dynamic problem is completely separated from the parametric one and can be is solved independently, and the solution to the parametric problem is defined by simple relations using the results of solving the dynamic problem (Ref. to Sect. 7.4).

For a problem with limited thrust and exhaust velocity of the electric propulsion system, such a division of the optimization problem into dynamic and parametric parts is not known. Therefore, until now, the typical approach to the joint optimization of the trajectory and main design parameters of a spacecraft with electric propulsion consists in solving a sequence of problems with different values of design parameters and analyzing the dependences of the efficiency indicators of the resulting solutions (for example, the payload mass of the spacecraft) on the values of varied design parameters [27, 41, 42]. This approach requires large computational costs, is characterized by low accuracy of optimization of the main design parameters, and may have problems with convergence when the properties of the optimal trajectory change with changes in the values of the varied parameters.

Therefore, we will consider an approach based on the use of the necessary optimality conditions of the maximum principle for the simultaneous optimization of the trajectory and the main design parameters, which can significantly reduce the computational cost of solving the problem and, using the continuation method, ensure good convergence to the optimal solution. The ideas of joint optimization used here were presented in [43, 44]. A similar approach was used in [45] in the analysis of a manned mission to Mars.

This section presents the necessary conditions for the optimality of the main design parameters for cases when the electric power of the electric propulsion system is constant or depends on the heliocentric distance of the spacecraft. The first case corresponds to a single-mode electric propulsion system with a nuclear or solar PSS on board. In this case, if a solar PSS is used, its power must be sufficient to ensure the operation of the electric propulsion system at the maximum required heliocentric distance. The second case corresponds to the electric propulsion system, the power and thrust of which are continuously adjusted as the available power of the on-board solar PSS changes.

A combined scheme of a flight between two planets with a fixed duration is considered. In this scheme, a spacecraft with an electric propulsion system is launched onto a departure trajectory from the planet of departure by a launch vehicle equipped with high-thrust propulsion system. Either the initial mass of the spacecraft is given, or the dependence of the initial mass of the spacecraft in the departure trajectory on the value of the departure hyperbolic excess velocity is given (in what follows, approximations of such dependencies are used, which are presented in tabular form in the user manuals for launch vehicles [46, 47], using the 6th-order polynomials). The heliocentric trajectory of the spacecraft is formed by the electric propulsion system. The problem of optimizing the trajectories of direct heliocentric transfers between two planets is considered within the framework of the assumptions of the ZSOI method [48]; the section of motion within the sphere of influence of the destination planet is not analyzed. The optimization of the trajectory, in addition to the calculation of the optimal control (program for controlling the thrust vector of the electric propulsion system), may include optimization of a number of trajectory parameters (departure date, magnitude and direction of departure hyperbolic excess velocity). Simultaneous optimization of control, trajectory and design parameters allows us to get an idea of the best option for the designed space mission and to draw some conclusions concerning the properties of such optimal solutions and possibility for their use.

For the problem of joint optimization of the trajectory of the main design parameters, which will be presented in detail in Sect. 9.4, we consider a simple spacecraft design model. In this design model, we assume that the spacecraft consists of the "useful" part, the power supply and propulsion unit (PSPU) and the PSFS. The considered scheme of spacecraft division into components is presented in Fig. 7.2. The optimization goal is to maximize the mass of the useful part of a spacecraft, which includes the payload (the target instruments) and part of the PSS and other service systems of a spacecraft, which ensure the payload operation. The PSPU includes a part of the PSS that is required to provide EPS operation only, the power processing unit (PPU) of the EPS, and the EPT itself. The PSFS comprises the tanks for storing the EPS propellant with pipes and fittings to ensure propellant feed to the EPT. It should be noted that the accepted scheme of spacecraft division into components is not quite traditional. It is only necessary for the convenience of presenting a spacecraft design model for solving the optimization problem under consideration.

The power supply system consists of a primary power source (the nuclear power unit or the solar arrays with uniaxial drives), storage batteries, and a power control unit to stabilize the voltage of the spacecraft power supply bus. When solar arrays are used, their axis of rotation is perpendicular to the EPS thrust vector, so that by using the three-axis spacecraft attitude control and rotating the plane of solar arrays, one can independently orient the EPS thrust vector in the desired direction and the normal to the working plane of the solar arrays toward the Sun to ensure maximum power generation at any given time.

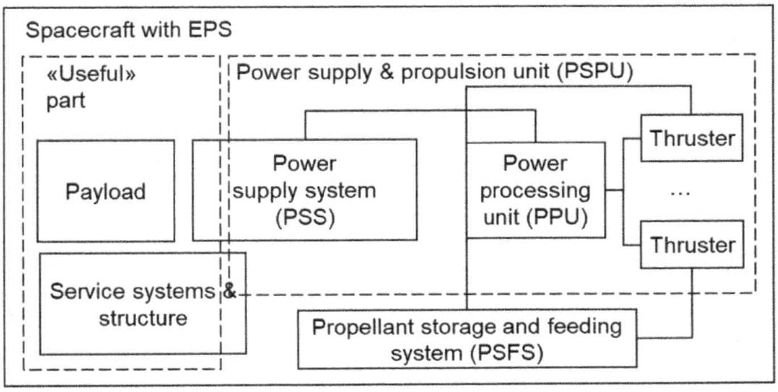

Fig. 7.2 A simple design model of a spacecraft with an EPS

At the beginning of flight with the heliocentric distance of 1 AU, the EPS electric power P_{BOL} is one of the selectable spacecraft design parameters. The dependence of the electric power of the electric propulsion system P on the heliocentric distance r in the general case is presented in the form:

$$P = P_{BOL}\rho(r, r_{\min}) \tag{7.6.1}$$

where, if the solar panels and the power-controlled electric propulsion system are used, $\rho(r, r_{\min}) = \min(r^{-k}, r_{\min}^{-k})$, $r_{\min}$ is the spacecraft heliocentric distance at which the maximum power of the EPS is reached (r and $r_{\min}$ are expressed in astronomical units), and k is the parameter that depends on the type and design characteristics of the solar arrays and has a value ranging from 1.7 to 2.0. The limitation of the maximum power of solar arrays is related to the limitation of the power that can be transmitted through the drives of the solar panels, the limited power of the voltage regulation equipment of the spacecraft power bus and the need to ensure the thermal conditions of the solar panels when the spacecraft approaches the Sun, which requires turning the normal to the working plane of the solar arrays away from the Sun. In the case of using a single-mode (unregulated) electric propulsion system and a nuclear power source (or solar panels with the power sufficient to ensure the operation of the electric propulsion system and spacecraft service systems at the maximum heliocentric distance), $\rho(r, r_{\min}) \equiv 1$.

The mass of solar arrays providing EPS operation is proportional to their initial power P_{BOL} at the distance of 1 AU from the Sun, while the mass of the solar array drives (through which the electric power generated by the solar arrays is transmitted), the mass of the power regulator unit of a spacecraft PSS and the mass of EPS (thrusters, pivoting mechanisms, EPS power processing unit) is proportional to the maximum electric power generated by the solar panels at the heliocentric distance equals or less than $r_{\min}$–($P_{BOL}\ \rho(r_{\min}, r_{\min})$, which, is equal to the maximum power consumption of the controlled EPS. The PSFS wet mass is proportional to the mass of the EPS propellant. Thus, after the SC separation from the LV, the initial mass m_0 of the spacecraft with controlled-in-power EPS, can be represented as follows:

$$m_0 = m_u + m_{\text{PSPU}} + m_{\text{PSFS}} = m_u + [\gamma_1 + \gamma_2\rho(r_{\min}, r_{\min})]P_{BOL} + (1 + a_t)m_p, \tag{7.6.2}$$

where m_u is the spacecraft payload mass, m_{PSPU} is the mass of the PSPU, m_{PSFS} is the mass of the PSFS, γ_1 is the specific mass of solar arrays (in units of mass per unit of electric power generated at a heliocentric distance of 1 AU), γ_2 is the specific mass of the power regulator unit and solar array drives of the PSS for the EPS power supply and of the EPS itself (in units of mass per unit of a maximum estimated electric power consumption of the EPS), a_t is the tank coefficient that is equal to the ratio of the dry mass of PSFS and unconsumed propellant remnants at the end of flight to the mass of consumed propellant, m_p is the mass of EPS propellant consumed by the end of a transfer. It is obvious that, if a single-mode EPS is used, $\rho(r_{\min}, r_{\min}) = 1$; thus simplifying the expression for the spacecraft initial mass:

$$m_0 = m_u + m_{\mathrm{PSPU}} + m_{\mathrm{PSFS}} = m_u + \gamma_3 P_{BOL} + (1 + a_t) m_p, \tag{7.6.3}$$

where $\gamma_3 = \gamma_1 + \gamma_2$.

Among the spacecraft design parameters to be optimized, in addition to the electric power P_{BOL}, there is the EPS effective exhaust velocity c. The thrust T, exhaust velocity c, total thrust efficiency η, and power consumption P of the electric propulsion unit are related by the well-known relationship:

$$\frac{Tc}{2\eta} = P \tag{7.6.4}$$

In the general case, the total thrust efficiency may depend on c and P, and the exhaust velocity c–on P. However, these dependencies are largely determined by the specific implementation of the EPS, in particular, by the number of thrusters in the EPS, and by the power of a single thruster. For this reason, at the preliminary design stage of spacecraft, when choosing the main EPS parameters, we will suppose the total thrust efficiency η to be given and constant throughout the trajectory, and the chosen value of exhaust velocity c to be constant and independent of the current power of the EPS. Obviously, with respect to (7.6.4), the thrust can be calculated by the following expression:

$$T = \frac{2\eta P}{c} \tag{7.6.5}$$

Thus, the chosen (optimized) design parameters are: the initial EPS electric power P_{BOL}, the EPS exhaust velocity c, and, in the general case, the heliocentric distance $r_{\min}$, at which the maximum calculated power of a spacecraft's PSS and EPS is reached. The initial spacecraft mass is determined by characteristics of the launch vehicle and by the departure hyperbolic velocity excess $V_{\infty 0}$, which can also be optimized. The other design parameters, listed in this Section (η, a_t, γ_1, γ_2, γ_3, k) belong to the first group of design parameters that determine the level of technologies used. They should be specified as input data for solving the problem of joint optimization of the trajectory and the main design parameters P_{BOL}, c, and $r_{\min}$.

7.8 Conclusion

This section presents mathematical models of the motion of a spacecraft with an electric propulsion system. The equations of motion of a spacecraft with an electric propulsion system are given in the inertial Cartesian coordinate system and in the modified equinoctial elements; the auxiliary longitude is introduced as an angular independent variable. A description of the main perturbing accelerations is given, which will be considered later in the monograph when calculating trajectories using quasi-optimal feedback control.

Typical problems of optimizing the trajectories of spacecraft with an electric propulsion system are considered, including the problems of maximizing the final mass of the spacecraft, and minimizing fuel consumption, thrust or flight time.

A description is given of the indirect approach to optimizing trajectories using the maximum principle to reduce the optimal control problem to a boundary value problem, the continuation method to reduce the boundary value problem to the Cauchy problem, and the complex step method for high precision calculation of the required derivatives.

Mathematical models of optimal trajectories of a spacecraft with an ideally controlled thruster with the limited power and limited thrust are presented, including systems of ordinary differential equations for state and conjugate variables, equations for optimal control and boundary conditions with transversality conditions for typical flights.

The design model of the spacecraft is considered, which is subsequently used in the problem of joint optimization of the trajectory and the main design parameters of the spacecraft.

References

1. McCarthy D.D., Petit G. (eds.).: IERS technical note No. 32. IERS Conventions (2003). IERS Conventions Centre, Frankfurt am Main, p 127 (2004)
2. The SOFA software libraries. International Astronomical Union. Division 1: Fundamental Astronomy, Commission 19: Rotation of the Earth, Standards of Fundamental Astronomy Board. Release 9 p 252 (2012)
3. Walker M.J.H., Ireland B., Owens J.: A set of modified equinoctial elements, Celestial Mechanics. **36**, 409–419 (1985), **38**, 391 (1986)
4. Sackett, L.L., Malchow, H.L., Edelbaum, T.N.: Solar electric geocentric transfer with attitude constraints: analysis. NASA CR-134927 (1975)
5. Graham, K.F., Rao, A.V.: Minimum-time trajectory optimization of low-thrust earth-orbit transfers with eclipsing. J. Spacecr. Rocket. **53**(2), 289–303 (2016)
6. Betts, J.T.: Optimal low-thrust orbit transfers with eclipsing. Optim. Control. Appl. Methods **36**(2), 218–240 (2015)
7. Schaff S.: Low-thrust multi-revolution orbit transfers. In: Fasano, G., Pinter, J.D. (eds.): Space engineering, Springer Optimization and Its Applications 114, Springer International Publishing Switzerland, 337–367 (2016)
8. Petukhov, V.G.: Application of the angular independent variable and its regularizing transformation in the problems of optimizing low-thrust trajectories. Cosm. Res. **57**(5), 351–363 (2019)
9. Cunningham, L.E.: On the computation of the special harmonic terms needed during numerical integration of artificial satellite. Celest. Mech. **2**, 207–216 (1970)
10. Lemoine, F.G., Kenyon, S.C., Factor, J.K. et al.: The development of the joint NASA GSFC and NIMA geopotential model EGM96, NASA Goddard Space Flight Center, Greenbelt, Maryland, 20771 USA (1998)
11. Standish, E.M., Newhall, X.X., Williams, J.G., Folkner, W.F.: JPL planetary and lunar ephemerides, DE403/LE403, JPL IOM 314.10–127
12. Owens, J.K.: NASA Marshall engineering thermosphere model–version 2.0. NASA/TM-2002–211786, June 2002, Marshall Space Flight Center, Alabama, p 41

13. Hairer, F., Norsett, S.P., Wanner, G.: Solving ordinary differential equations I. Non-Stiff Problems. Spinger, Verlag, Berlin (1987)
14. Pontryagin, L.S., Boltyansky, V.G., Gamkrelidze, R.V., Mishchenko, E.F.: Mathematical theory of optimal processes. M.: Nauka (1976)
15. Lyness, J.N., Moller, C.B.: Numerical differentiation of analytical functions. SIAM J. Numer. Anal. **4**, 202–210 (1967)
16. Lyness, J.N.: Numerical algorithms based on the theory of complex variables. Proc. ACM 22nd Nat. Conf. Thompson Book Co. Washington, DC, pp. 124–134 (1967)
17. Squire, W., Trapp, G.: Using complex variables to estimate derivatives of real functions. SIAM Rev. **40**(1), 110–112 (1998)
18. Dargent T.: Automatic minimum principle formulation for low thrust optimal control in orbit transfers using complex numbers. Int. Symp. Space Flight Dyn. Toulouse, France, p 9 (2009)
19. Lantoine, G.A., Russel, R.P., Dargent, T.: Using multicomplex variables for automatic computation of high-order derivatives. AAS 10–218 p 18 (2010)
20. Petukhov, V.G.: Homotopic approach to low-thrust trajectory optimization: numerical technique and tools. In: 4th International conference on astrodynamics tools and techniques, 3–6 May 2010, ESA/ESAC, Madrid, Spain. ESA Proceedings WPP-308, p 8
21. Petukhov, V.G.: Optimization of interplanetary trajectories of spacecraft with an ideally controlled engine by the continuation method. Space Res. **46**(3), 224–237 (2008)
22. Eneev, T.M., Efimov, G.B., Konstantinov, M.S., Petukhov, V.G., Akhmetshin, R.Z., Fedotov, G.G.: Mercury-to-pluto range missions using solar-nuclear electric propulsion. Keldysh Inst. Appl. Math., Russ. Acad. Sci. preprint 111 (1996)
23. Eneev, T.M., Egorov, V.A., Efimov, G.B., Konstantinov, M.S., Petukhov, V.G., Akhmetshin, R.Z., Fedotov, G.G.: Some methodical problems of low-thrust trajectory optimization. Keldysh Inst. Appl. Math., Russ. Acad. Sci. preprint 110, pp. 1–24 (1996)
24. Petukhov, V.G.: One numerical method to calculate optimal power-limited trajectories. International Electric Propulsion Conference. IEPC-95–221, Russia, Moscow, p 8 (1995)
25. Petukhov, V.G., Konstantinov, M.S., Fedotov, G.G.: 1st ACT Global Trajectory optimization competition: results found at Moscow aviation institute and Khrunichev state research and production space center. Acta Astronaut. **61**(9), 775–785 (2007)
26. Beletsky, V.V., Egorov, V.A.: Interplanetary flights with constant power engines. Cosm. Res. **2**(3) (1964)
27. Grodzovsky, G.L., Ivanov, Y.N., Tokare, V.V.: Mechanics of space flight. Optim. Probl. M., Nauka (1975)
28. Sukhanov, A.A., Prado, A.F.B. de A.: Interorbital flights with low thrust in a arbitrary field of forces. Space Res. **51**(2), pp. 159–170 (2013)
29. Edelbaum, T.N.: Optimum power-limited orbit transfer in strong gravity fields. AIAA J. **3**(5), pp. 921–925 (1965)
30. Haissig, C.M., Mease, K.D., Vinh, N.X.: Minimum-fuel power-limited transfers between coplanar elliptical orbits. Paper IAF-91–346, Montreal (1991)
31. Irving, J.H.: Low thrust flight: variable exhaust velocity in gravitational fields. Space Techn. **10**(4), pp. 1959
32. Kluever, C.A.: Geostationary orbit transfers using solar electric propulsion with specific impulse modulation. J. Spacecr. Rocket. **41**(3), pp. 461–466 (2004)
33. Melbourne, W.G., Sauer, C.G.: Optimum interplanetary rendezvous with power-limited vehicles. AIAA J. **1**(1) (1963)
34. Chang Diaz, F., Hsu, M., Braden, E. et al.: Rapid mars transits with exhaust-modulated plasma propulsion. NASA TP-1995–3539. Johnson Space Center, Houston, Texas (1995)
35. Chang Diaz, F., Squire, J., Bengston R. et al.: The physics and engineering of the VASIMR engine. AIAA-00–3756 (2000)
36. Petukhov, V.G.: A continuation method for optimizing low-thrust interplanetary trajectories. Space Res. **50**(3), 258–270 (2012)
37. Akhmetshin, R.Z.: Planar problem of an optimal transfer of a low-thrust spacecraft from high-elliptic to geosynchronous orbit. Cosm. Res. **42**(3), 238–249 (2004)

38. Haberkorn, T., Martinon, P., Gergaud, J.: Low thrust minimum-fuel orbital transfer: a homotopic approach. J. Guid. Control. Dyn. **27**(6), 1046–1060 (2004)
39. Taheri, E., Junkins, J.L.: How many impulses redux. J. Astronaut. Sci. **67**(2), 257–334 (2020)
40. Oberle, H.J., Taubert, K.: Existence and multiple solutions of the minimum-fuel orbit transfer problem. J. Optim. Theory Appl. **95**(2), 243–262 (1997)
41. Vavrina, M.A., Englander, J.A., Ghosh, A.R.: Coupled low-thrust trajectory and system optimization via multi-objective hybrid optimal control. AAS/AIAA Space Flight Mechanics Meeting, American Astronautical Soc. Paper 2015–397 (2015)
42. Kwon, K., Lantoine, G., Russell, R.P., Mavris, D.N.: A study on simultaneous design of a hall effect thruster and its low-thrust trajectory. Acta Astronaut. **119**, 34–47 (2016)
43. Petukhov, V.G.: Joint optimization of the low-thrust trajectory and the main design parameters of electric propulsion system. IAC-15-C1.1.5, Jerusalem, p 9 (2015)
44. Petukhov V.G., Ivanyukhin A.V. Joint optimization of the main design parameters of the electric rocket propulsion system and the trajectory of the spacecraft. News of the Russian Academy of Sciences. Energy, No. 2, 2016, pp. 92–101.
45. Klimov, S.S.: Joint optimization of trajectories and energy-power parameters of Martian expeditionary complexes with electric rocket propulsion systems. Cosmonaut. Rocket. Sci. **4**(97), pp. 71–83 (2017)
46. Proton launch system mission planner's guide. Revision 7. International Launch Services, p 395 (2009)
47. Soyuz user's manual. Arianespace, Issue 2, Revision 0, p 244 March 2012
48. Bate, R.R., Mueller, D.D., White, J.E.: Fundamentals of astrodynamics. Dover Publications, New York (1971)

Chapter 8
Optimization of Limited-Power Interplanetary Trajectories

8.1 Introduction

Let us consider a problem of optimizing the heliocentric trajectory between two planets for a spacecraft (SC) with an ideally controlled limited-power (LP) propulsion. Optimization techniques for the LP interplanetary trajectories considered in this chapter are described in [1–11]. For the non-singular boundary conditions (i.e. when the SC coordinates do not coincide with the singular points of the right-hand sides of the differential equations of SC motion), the dynamic part of the LP problem always has a solution, since the thrust acceleration is not limited. In fact, it is a set of LP solutions for arbitrary non-singular boundary conditions. The main difficulty in solving the LP-trajectory optimization problem by indirect methods is the problem of choosing the initial guess values of decision vector that ensures convergence of the numerical solution. In particular, the problem of choosing the initial guess values is related to the bifurcation of optimal solutions and to the high sensitivity of the residuals of the corresponding boundary value problem to variations in its decision vector.

Therefore, the aim of the study, the results of which are presented in this chapter, was to develop a numerical technique that does not require the user to choose an initial guess to optimize an LP-trajectory. This problem was solved through various modifications of the continuous numerical technique based on the Newtonian homotopy. In particular, numerical algorithms have been developed to compute the optimal trajectory of interplanetary transfer using the spacecraft motion along the initial orbit without thrust as an initial approximation, and the method of continuation by the gravitational parameter of the center of attraction, which makes it possible to calculate optimal trajectories with a given angular distance with a trivial (zero) initial guess values for the initial value of the co-state variables on an almost regular basis. Such an initial guess corresponds to the SC coasting motion along the initial orbit.

A. S. Filatyev and V. G. Petukhov, *Through Optimization of Aerospace Vehicle Trajectories by the Pontryagin Maximum Principle*, Studies in Systems, Decision and Control 579, https://doi.org/10.1007/978-3-031-80756-5_8

It should be noted that the numerical techniques developed are quite demanding in terms of computational productivity. In particular, their implementation requires integration of embedded systems of nonlinear ordinary differential equations. However, modern computers enable a good performance for the typical problems of optimizing LP-trajectories using the developed numerical techniques due to their stability.

The statement of the LP-trajectory optimization problem was given in Sect. 7.5. Using the maximum principle, the problem of optimizing a direct interplanetary LP-trajectory is reduced to a boundary value problem for a system of differential Eqs. (7.4.8) (or (7.4.10), (7.4.11)) with initial conditions (7.4.12) in the case of "zero" junction of the planetocentric and heliocentric sections of the trajectory (i.e. at departure from the planet with zero asymptotic velocity) or (7.4.13) with a non-zero departing hyperbolic excess velocity, and with final conditions (7.4.14) for the rendezvous problem (approach to planet-target with alignment of orbital velocities) or (7.4.15) for the flyby problem with an arbitrary vector of final hyperbolic excess velocity.

8.2 Numerical Implementation of the Continuation by Boundary Conditions

The efficiency of the Cauchy problem solution procedure for the continuation method (7.3.17) depends on the properties of the function $\mathbf{f}(\mathbf{z})$, the choice of the initial guess value for the decision vector of the boundary value problem $\mathbf{z}$, the method of calculation of the matrix of partial derivatives $\partial\mathbf{f}/\partial\mathbf{z}$ and the characteristics of the numerical integration method.

In the dynamic part of the LP-problem with the fixed departure date t_0 and arrival date t_f, the components of vector $\mathbf{f}$ are defined by the expressions (7.4.14) for the rendezvous problem and (7.4.15) for the flyby problem (6 scalar components in total). The components of the decision vector of the LP-problem $\mathbf{z}$ are the initial values of vector $\mathbf{p}_\mathbf{v}$ and its time derivative (also 6 scalar values).

In the case of LP-problem for the motion in the central Newtonian field, the right-hand sides of the optimal motion differential equations have a singularity at a point coinciding with the center of gravity (for typical dependencies of EP power on the state variables and time). The sensitivity matrix $\partial\mathbf{f}/\partial\mathbf{z}$ is nondegenerate everywhere, except for the boundary hypersurfaces in the space of unknown parameters of the boundary value problem, which separate the domains of attraction of different extremals. Implementation of the smooth continuation (7.3.17) is thus limited by the requirements of non-collision with the gravity center when integrating the equations of optimal motion (7.4.8) or (7.4.10) and non-intersection of boundary hypersurfaces in the process of integrating the problem (7.3.17).

The advantage of the continuation method (7.3.17) in comparison with the previously used numerical techniques was most clearly manifested in the traditional problem of choosing an initial guess value $\mathbf{z}_0$. It turned out, that at zero value of the initial hyperbolic excess velocity, the zero initial guess value $\mathbf{z}_0 = 0$ ensures convergence of the method (7.3.17) to some extremal solution in most problems. Let's consider the LP-problem with a zero departing hyperbolic excess velocity as the basic LP-problem. The zero initial guess value in the basic LP-problem corresponds to the coasting motion along the departure planet orbit, and is thus natural for the user of the method. Hence, there is no problem of choosing an initial guess value of the decision vector for the basic LP-problem.

Once the solution to the basic LP-problem is calculated, the resulting decision vector $\mathbf{z}$ can be used as the initial guess for the LP-problem with modified values for the departure date, flight duration, and departing hyperbolic excess velocity.

With a large difference in the mean motions of the departure planet and the target planet on the optimal trajectories obtained by the method (7.3.17) with a zero initial guess value $\mathbf{z}_0$, when the flight duration increases, there is a tendency for the maximum heliocentric distance to increase in the case of flight to the inner planet and for the minimum heliocentric distance to decrease in the case of a flight to the outer planet. The reason for such tendencies is the existence of a sequence of local extrema of the optimal control problem corresponding to a different number of complete revolutions around the Sun. For a fixed flight duration and with zero $\mathbf{z}_0$, a trajectory is calculated that corresponds to the coasting motion (coinciding with the orbital motion of the departure planet at zero departing hyperbolic excess velocity). Such a trajectory has a certain angular distance and is located in the attraction region of a locally optimal trajectory with a close angular distance. Therefore, the solution of the problem (7.3.17) results in convergence of the continuation process with this local extremum. The proximity of the angular distances of the trajectories at the beginning and at the end of the continuation process is ensured by changing the minimum or maximum heliocentric distance of the spacecraft during the continuation process (see Fig. 8.1), which is achieved in the LP-problem by increasing the thrust with a corresponding decrease in the specific impulse. Such solution satisfies all necessary optimality conditions, but it is quite obvious that there may be other local extremals of the problem of flight within a fixed time with an angular distance that differs from that obtained for the whole number of revolutions. It is also obvious that the cost function values of the extremals will differ for different angular distances.

This noted multi-extremality of the optimal control problem leads to the need to develop a method that allows us to obtain a locally optimal (satisfying the necessary optimality conditions arising from the maximum principle) solution to the LP-problem with a given number of complete revolutions, which, for a given departure date and flight duration, corresponds to a given angular distance. This problem will be considered in the next section.

The differential Eqs. (7.4.10) can be rewritten in the following form:

$$\left.\begin{aligned} \ddot{\mathbf{r}} &= \Omega_{\mathbf{r}} + \eta \mathbf{p}_{\mathrm{v}}, \\ \ddot{\mathbf{p}}_{\mathrm{v}} &= \Omega_{\mathbf{rr}} \mathbf{p}_{\mathrm{v}} + \mathbf{h} \mathbf{p}_{\mathrm{v}}^{\mathrm{T}} \mathbf{p}_{\mathrm{v}}, \end{aligned}\right\} \tag{8.1.1}$$

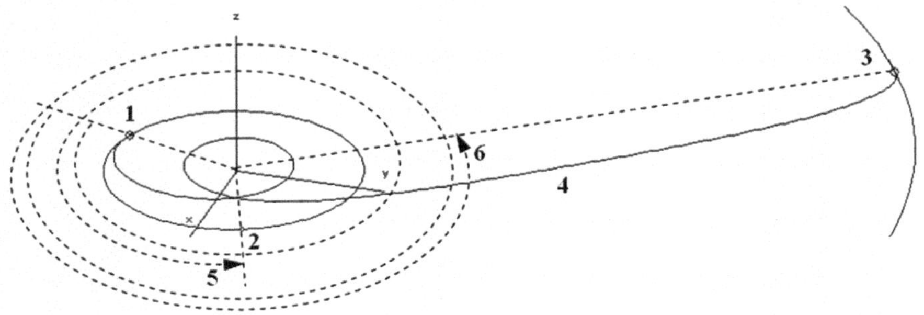

Fig. 8.1 Calculation of the optimal LP-trajectory for the Jupiter rendezvous problem by the basic continuation method. Departure on 1.6.2001, flight duration 500 days. 1—the position of the Earth at time of departure; 2—the position of the Earth at time of spacecraft arrival at Jupiter; 3—the position of Jupiter at the final moment of time; 4—the trajectory; 5—the angular range of the Earth's motion during the flight; 6—the angular range of the trajectory

where $\mathbf{h} = [\partial\eta(\mathbf{r}, t)/\partial\mathbf{r}]/2$. We will use the method of joint integration of differential equations of optimal motion with differential equations for the elements of the sensitivity matrix. The differential equations for the sensitivity matrix elements are:

$$\left.\begin{aligned} \frac{d^2}{dt^2}\left(\frac{\partial r_i}{\partial e}\right) &= \sum_k \Omega_{r_i r_k}\frac{\partial r_k}{\partial e} + \eta\frac{\partial p_{vi}}{\partial e} + \left(\sum_k \frac{\partial\eta}{\partial r_k}\frac{\partial r_k}{\partial e}\right)p_{vi}, \\ \frac{d^2}{dt^2}\left(\frac{\partial p_{vi}}{\partial e}\right) &= \sum_m\sum_k \Omega_{r_i r_k r_m} p_{vk}\frac{\partial r_m}{\partial e} + \sum_k \Omega_{r_i r_k}\frac{\partial p_{vk}}{\partial e} + 2h_i\sum_k p_{vk}\frac{\partial p_{vk}}{\partial e} + \frac{\partial h_i}{\partial e}p_v^2, \end{aligned}\right\} \tag{8.1.2}$$

where i, m, k vary from 1 to 3, while the variable e stands for some element of the decision vector of the boundary value problem. Note that if the EP power depends on the radius-vector and time only:

$$\begin{aligned} \mathbf{h} &= \frac{1}{2r}\frac{\partial\eta}{\partial r}\mathbf{r}, \\ \frac{\partial\mathbf{h}}{\partial e} &= \frac{1}{2r^2}\left(\frac{\partial^2\eta}{\partial r^2} - \frac{1}{r}\frac{\partial\eta}{\partial r}\right)\mathbf{r}^{\mathrm{T}}\frac{\partial\mathbf{r}}{\partial e}\mathbf{r} + \frac{1}{2r}\frac{\partial\eta}{\partial r}\frac{\partial\mathbf{r}}{\partial e}. \end{aligned} \tag{8.1.3}$$

For example, if $\eta = 1/r^2$ (that is typical for solar electric propulsion system):

$$\mathbf{h} = -\frac{1}{r^4}\mathbf{r},\ \frac{\partial\mathbf{h}}{\partial e} = \frac{1}{r^4}\left(\frac{4}{r^2}\mathbf{r}^{\mathrm{T}}\frac{\partial\mathbf{r}}{\partial e}\mathbf{r} - \frac{\partial\mathbf{r}}{\partial e}\right) \tag{8.1.4}$$

The system (8.1.1), (8.1.2), (7.4.12) or (7.4.13) and (7.4.14) or (7.4.15) is supplemented by additional initial conditions for the sensitivity matrix elements:

$$\frac{\partial \mathbf{r}(0)}{\partial \mathbf{p}_{\mathrm{v}o}} = \frac{\partial \mathbf{r}(0)}{\partial \dot{\mathbf{p}}_{\mathrm{v}o}} = \frac{\partial \mathbf{p}_{\mathrm{v}}(0)}{\partial \dot{\mathbf{p}}_{\mathrm{v}o}} = \frac{d}{dt}\left(\frac{\partial \mathbf{p}_{\mathrm{v}}(0)}{\partial \mathbf{p}_{\mathrm{v}o}}\right) = 0,$$
$$\frac{d}{dt}\left(\frac{\partial \mathbf{r}(0)}{\partial \mathbf{p}_{\mathrm{v}o}}\right) = \frac{V_\infty}{p_{vo}}\left(\mathbf{E} - \frac{1}{p_{vo}^2}\mathbf{p}_{\mathrm{v}o}\mathbf{p}_{\mathrm{v}o}^{\mathrm{T}}\right), \quad \frac{\partial \mathbf{p}_{\mathrm{v}}(0)}{\partial \mathbf{p}_{\mathrm{v}o}} = \frac{d}{dt}\left(\frac{\partial \mathbf{p}_{\mathrm{v}}(0)}{\partial \dot{\mathbf{p}}_{\mathrm{v}o}}\right) = \mathbf{E}, \tag{8.1.5}$$

where $\mathbf{E}$ is the unity matrix.

Integration of the continuation differential Eqs. (7.3.17) with the Euler method is equivalent to the use of the modified Newton method for the original boundary value problem. The advantages of the continuation method are particularly evident in the case of the higher-order numerical integration. In the limit, with the "ideal" method of numerical integration for a smooth dynamical system, the convergence region of the continuation method to some optimal solution coincides with the domain of attraction of this solution in the space of decision variables of the boundary value problem.

In practice, the system (7.3.17) has been successfully integrated by the 15th-order Everhart method [12], the 7(8)-th order Dormand-Prince method and the Gregg-Bulirsch-Stoer method [13]. In these algorithms, the adaptive step-size control procedure plays a role similar to the one-dimensional search for the minimum of a function in the direction of its gradient in the modified Newton methods.

8.3 Branching of Solutions and Gravitational Parameter Continuation

It was noted earlier that for the given boundary conditions and flight time, there are many optimal trajectories that differ in the number of entire revolutions around the gravity center. The domains of attraction of the corresponding solutions to the LP-problem in the space of decision variables $\mathbf{z}$ are separated by hypersurfaces, on which the sensitivity matrix $\partial \mathbf{f}/\partial \mathbf{z}$ is degenerate.

To regularize the control problem optimization, it is necessary to construct an algorithm that ensures convergence to a solution with the given angular flight distance for the given initial guess $\mathbf{z}_0$.

Let us consider optimization of the basic interplanetary trajectory, i.e. the rendezvous problem with zero initial hyperbolic excess velocity and zero initial guess value ($\mathbf{z}_0 = 0$). The departure date and flight time T are assumed to be fixed. The initial position of the spacecraft coincides with the position of the Earth on the date of departure: $\mathbf{r}(0) = \mathbf{r}_0(0)$, and its final position coincides with the position of the target planet on the date of arrival $t = T$: $\mathbf{r}(T) = \mathbf{r}_k(T)$ (Fig. 8.2).

If $\mathbf{z} = 0$, the thrust acceleration is equal to 0 along the entire trajectory and at the time moment T the spacecraft will arrive at the same point as the departure planet at the moment T, i.e., at a point with the position vector $\mathbf{r}_0(T)$. By varying the gravitational parameter of the central celestial body, at a time t = T it is possible to bring the spacecraft moving along the initial orbit to the point of intersection of this

Fig. 8.2 Determining the initial value of the gravitational parameter. 1—the initial orbit; 2—the final orbit

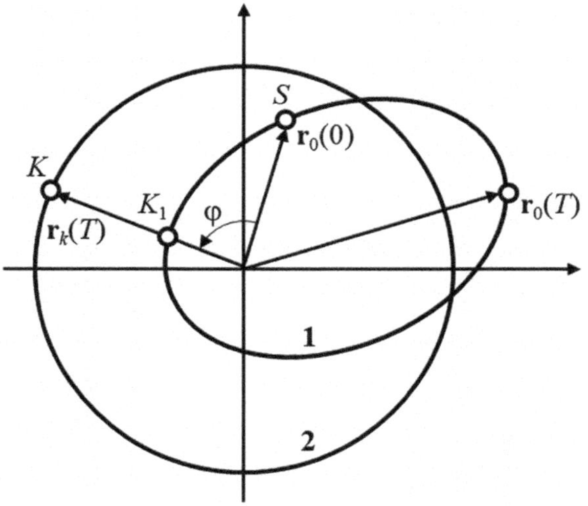

orbit with the projection of the given final position vector $\mathbf{r}_k(T)$ onto the orbit plane (point K_1 in Fig. 8.2). This initial value of the gravitational parameter μ_0 is found as follows.

Let the SC true anomaly in the initial orbit at the initial time moment at the point S be ν_0, and at the point K_1 – $\nu_k = \nu_0 + \varphi$, where φ is the angle between vector $\mathbf{r}_0(0)$ and the projection of the vector $\mathbf{r}_k(T)$ onto the initial orbit plane. Given the eccentricity e of the initial orbit, the known values of the true anomaly ν_0 and ν_k can be used to determine the corresponding values of the mean anomalies M_0 and M_k ($M = E - e\sin E$, where $E = 2\ \mathrm{arctg}\{[(1 - e)/(1 + e)]^{0.5}\mathrm{tg}(\nu/2)\}$ is the eccentric anomaly). Along the Keplerian orbit, however, the mean anomaly changes linearly in time: $M = M_0 + n(t - t_0)$, where $n = (\mu_0/a^3)^{0.5}$ is the mean motion in the initial orbit, and a is the semi-major axis of the initial orbit. Therefore, in order to transfer the spacecraft to the point K_1 by the final time moment T, it is necessary that the equality $M_k - M_0 + 2\pi N_r = nT$ be satisfied, where N_r is the integer number of revolutions around the central body that the spacecraft should make during the time T before arriving at the point K_1 at the final time moment. Taking into account the above expression for the mean motion n, it follows from this expression that:

$$\mu_0 = [(M_k + 2\pi N_r - M_0)/T]^2 a^3. \tag{8.2.1}$$

To implement a smooth continuation with respect to the gravitational parameter of the central body μ, the following dependence is used: $\mu(\tau) = \mu_0 + (1 - \mu_0)\tau$ (with the dimensionless form of the SC motion equations, the final value of the gravitational parameter μ corresponding to the physical formulation of the problem is equal to 1). For the initial and final orbits to be invariant under the transformation $\mu(\tau)$, it suffices to assume that the velocity $\mathbf{v}(t, \tau)$ of a point moving along any of these orbits is equal to $\mathbf{v}(t, \tau) = \mu(\tau)^{0.5}\ \mathbf{v}(t, 1)$, where $\mathbf{v}(t, 1)$ is the velocity of the corresponding point at τ

$= 1$ ($\mu = 1$). For the central Newtonian gravitational field Ω the above relations are exactly fulfilled, while for other fields, which differ only slightly from the central Newtonian one, they are only approximately fulfilled. The presented variant of the continuation method along the gravitational parameter works in the latter case also, if the main term of the force function Ω (i.e. μ/r) has a predominant effect on the motion of the spacecraft.

For the considered case, the system of ordinary differential equations of the optimal spacecraft motion has the following form:

$$\left.\begin{aligned} \frac{d^2\mathbf{r}}{dt^2} &= \mu(\tau)\Omega_{\mathbf{r}} + \eta\mathbf{p_v}, \\ \frac{d^2\mathbf{p_v}}{dt^2} &= \mu(\tau)\Omega_{\mathbf{rr}}\mathbf{p_v} + \frac{1}{2}p_v^2\frac{\partial\eta}{\partial\mathbf{r}}, \end{aligned}\right\} \tag{8.2.2}$$

where $\mu(\tau)$ is the function of the continuation parameter. Considering the above, the boundary conditions for the rendezvous problem take the following form:

$$\begin{aligned} &\mathbf{r}(0) = \mathbf{r_0}, \dot{\mathbf{r}}(0) = \mu^{1/2}(\tau)\mathbf{v_0}, \\ &\mathbf{r}(T) = \mathbf{r_k}, \dot{\mathbf{r}}(T) = \mu^{1/2}(\tau)\mathbf{v_k}, \end{aligned} \tag{8.2.3}$$

and for the flyby problem they take the form:

$$\begin{aligned} &\mathbf{r}(0) = \mathbf{x_0}, \dot{\mathbf{r}}(0) = \mu^{1/2}(\tau)\mathbf{v_0}, \\ &\mathbf{r}(T) = \mathbf{x_k}, \mathbf{p_v}(T) = 0. \end{aligned} \tag{8.2.4}$$

If the sensitivity matrix is computed by joint integration with the system (8.2.2), system (8.2.2) is supplemented by the equations:

$$\left.\begin{aligned} \frac{d^2}{dt^2}\left(\frac{\partial r_i}{\partial e}\right) &= \mu(\tau)\sum_k \Omega_{r_ir_k}\frac{\partial r_k}{\partial e} + \eta\frac{\partial p_{vi}}{\partial e} + \left(\sum_k \frac{\partial\eta}{\partial r_k}\frac{\partial r_k}{\partial e}\right)p_{vi}, \\ \frac{d^2}{dt^2}\left(\frac{\partial p_{vi}}{\partial e}\right) &= \mu(\tau)\sum_m\sum_k \Omega_{r_ir_kr_m}p_{vk}\frac{\partial r_m}{\partial e} + \mu(\tau)\sum_k \Omega_{r_ir_k}\frac{\partial p_{vk}}{\partial e} + 2h_i\sum_k p_{vk}\frac{\partial p_{vk}}{\partial e} + \frac{\partial h_i}{\partial e}p_v^2, \end{aligned}\right\} \tag{8.2.5}$$

where e is an element of the vector $\mathbf{p_{v0}}$ or vector $\dot{\mathbf{p}}_{\mathbf{v0}}$, and

$$\left.\begin{aligned} \frac{d^2}{dt^2}\left(\frac{\partial r_i}{\partial \tau}\right) &= \frac{\partial\mu}{\partial\tau}\Omega_{r_i} + \mu(\tau)\sum_k \Omega_{r_ir_k}\frac{\partial r_k}{\partial\tau} + \eta\frac{\partial p_{vi}}{\partial\tau} + \left(\sum_k \frac{\partial\eta}{\partial r_k}\frac{\partial r_k}{\partial\tau}\right)p_{vi}, \\ \frac{d^2}{dt^2}\left(\frac{\partial p_{vi}}{\partial\tau}\right) &= \frac{\partial\mu}{\partial\tau}\sum_k \Omega_{r_ir_k}p_{vk} + \mu(\tau)\sum_m\sum_k \Omega_{r_ir_kr_m}p_{vk}\frac{\partial r_m}{\partial\tau} + \mu(\tau)\sum_k \Omega_{r_ir_k}\frac{\partial p_{vk}}{\partial\tau} \\ &\quad +2h_i\sum_k p_{vk}\frac{\partial p_{vk}}{\partial\tau} + \frac{\partial h_i}{\partial\tau}p_v^2 \end{aligned}\right\} \tag{8.2.6}$$

with the additional initial conditions:

$$\frac{\partial \mathbf{r}(0)}{\partial \mathbf{p}_{vo}} = \frac{\partial \mathbf{r}(0)}{\partial \dot{\mathbf{p}}_{vo}} = \frac{\partial \mathbf{p}_v(0)}{\partial \dot{\mathbf{p}}_{vo}} = \frac{d}{dt}\left(\frac{\partial \mathbf{p}_v(0)}{\partial \mathbf{p}_{vo}}\right) = \frac{d}{dt}\left(\frac{\partial \mathbf{r}(0)}{\partial \mathbf{p}_{vo}}\right) = 0,$$
$$\frac{\partial \mathbf{p}_v(0)}{\partial \mathbf{p}_{vo}} = \frac{d}{dt}\left(\frac{\partial \mathbf{p}_v(0)}{\partial \dot{\mathbf{p}}_{vo}}\right) = \mathbf{E},$$
$$\frac{\partial \mathbf{r}(0)}{\partial \tau} = \frac{\partial \mathbf{p}_v(0)}{\partial \tau} = 0,$$
$$\frac{d}{dt}\left(\frac{\partial \mathbf{r}(0)}{\partial \tau}\right) = \frac{1}{2\mu^{1/2}}\frac{\partial \mu}{\partial \tau}\mathbf{v}_0, \quad \frac{d}{dt}\left(\frac{\partial \mathbf{p}_v(0)}{\partial \tau}\right) = 0. \tag{8.2.7}$$

In this case, the residual function **f** explicitly depends on the continuation parameter τ, so the differential equation of the continuation method has the form (7. 3.18).

The difference between boundary condition continuation and gravitational parameter continuation is illustrated in Fig. 8.3 using the example of an optimal LP trajectory to Mars. The figure shows that the trajectories calculated with the intermediate values of the continuation parameter τ have various angular distances when using the continuation method along the boundary conditions and the constant angular distance when using continuation along the gravitational parameter. This helps to avoid a situation of intersection in the process of solving the boundary value problem of hypersurfaces $\partial \mathbf{f}/\partial \mathbf{z} = 0$, separating extremals with different numbers of entire revolutions. As a rule, the gravitational parameter continuation method converges faster than the boundary condition continuation method: Fig. 8.4 makes it clear that the first method requires 7 steps of integrating the differential equations to obtain the optimal trajectory, while the gravitational parameter continuation drives to the solution in only 4 steps.

8.4 Numerical Examples

The continuation method has shown high efficiency in LP-trajectory optimization problems (see, e.g., [1–11]). Here are some numerical examples to demonstrate its application.

Let us consider the optimal trajectories of the Jupiter flyby for cases when the EPS power is constant or proportional to $r^{-1.6}$. The first case corresponds to the use of nuclear EPS (NEPS), and the second case—to the solar EPS (SEPS). The departure dates and flight durations in both cases are assumed to be equal, and the departure hyperbolic velocity excess is assumed to be 0. Fig. 8.4 shows both trajectories. The strokes on the trajectories here and below show the thrust acceleration vector at equal time intervals. From Fig. 8.4, it can be seen that a relatively small perihelion radius is maintained with SEPS to ensure that the thrust acceleration is large enough to increase the aphelion radius as quickly as possible. Figure 8.5 shows the time dependencies of thrust acceleration, pitch (the angle between the circumferential direction and the projection of the thrust vector onto the osculating orbital plane), and yaw (the

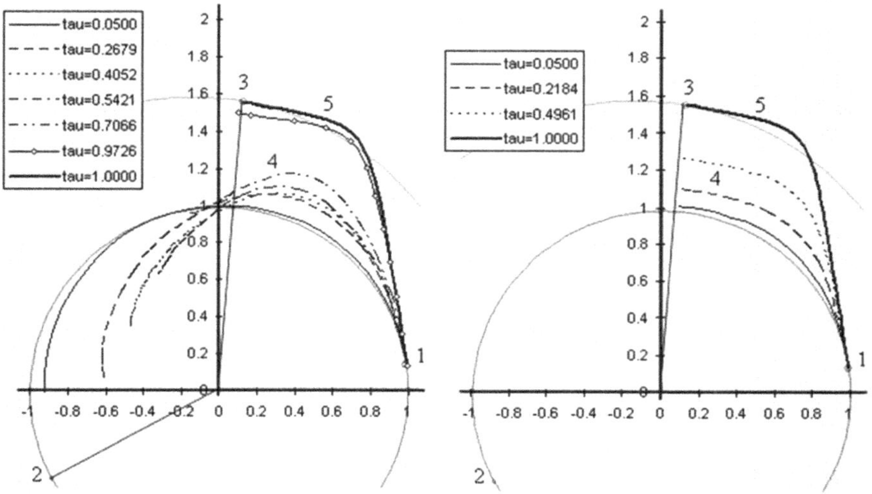

Fig. 8.3 Calculation of the Earth-Mars transfer (departure date 1.10.2001, duration 200 days). On the left is the basic continuation method, on the right is the continuation with respect to the gravity parameter, 0 full revolutions. 1—Earth at time of departure; 2—Earth at time of arrival at Mars; 3—Mars at time of arrival; 4—trajectories at $\tau < 1$; 5—optimal trajectory ($\tau = 1$)

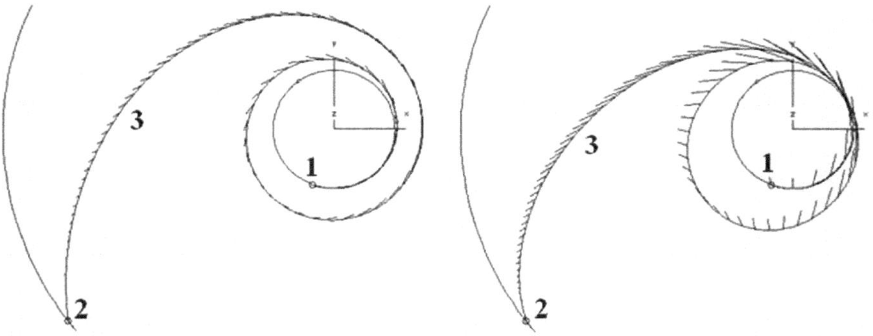

Fig. 8.4 Comparison of the Jupiter flyby trajectories for a spacecraft using NEP (left) and SEP (right). The departure date is June 1, 2001, the departure hyperbolic excess of velocity is zero, and the flight duration is 1700 days.1—the position of the Earth at time of departure, 2—the position of Jupiter at time of the flyby, 3—the transfer trajectory

angle between the osculating orbital plane and the thrust vector). The maximum value of thrust acceleration on the trajectory of a spacecraft with SEPS is reached at perihelion and exceeds the maximum value of thrust acceleration on the trajectory of a spacecraft with NEPS. The pitch angle on the trajectory of the spacecraft with SEPS is close to 180° to keep the perihelion radius small on the second revolution.

On the trajectory of a spacecraft with NEPS, the pitch angle always oscillates around zero to maximize the semi-latus rectum.

Figure 8.6 shows the dependence of the cost function and the required characteristic velocity to realize the Mars flyby on the value of the departure hyperbolic excess velocity. The date of departure from Earth and the duration of the transfer are assumed to be fixed, and the EPS power is assumed to be constant and its value is also fixed. When the hyperbolic excess velocity is zero, $J = 0.8834\ \mathrm{m^2/s^3}$, and the characteristic velocity is 3423 m/s. When the hyperbolic excess velocity is 3265 m/s, $J = 0.0\ \mathrm{m^2/s^3}$, i.e. the trajectory becomes fully coasting.

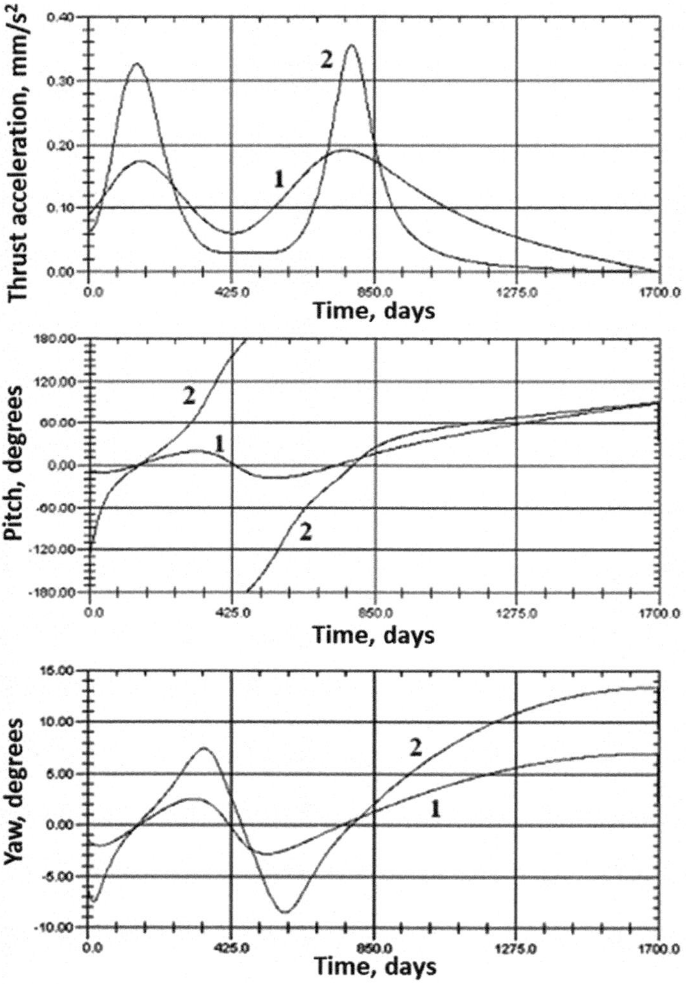

Fig. 8.5 Comparison of the control functions on the Jupiter flyby trajectories for a spacecraft using for NEP (1) and SEP (2)

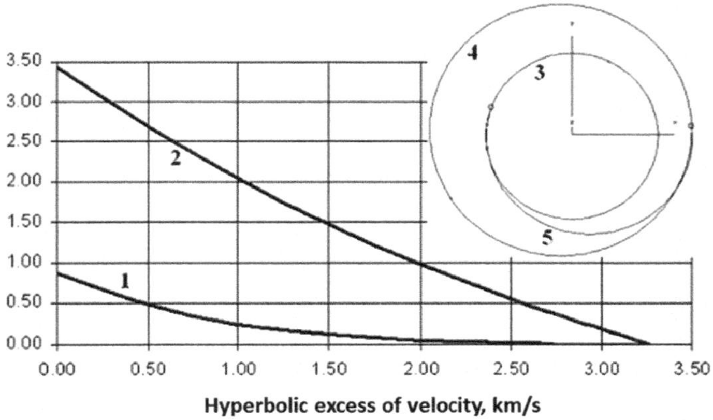

Fig. 8.6 The cost function J [m^2/s^3] and the characteristic velocity [km/s] as a function of the departure hyperbolic velocity excess for the trajectory of Mars flyby by a spacecraft with NEP. The departure date is March 1, 2001, the duration of the flight is 270 days. 1—cost function, 2—characteristic velocity, 3—Earth orbit, 4—Mars orbit, 5—transfer trajectory

Figure 8.7 shows a set of LP-optimal trajectories for Mars rendezvous. The duration of all trajectories is 300 days. The EPS power was assumed to be constant. The departure hyperbolic velocity excess was assumed to be 0. The dates of departure from Earth of different trajectories differ from each other by one month. The left column corresponds to the departure dates in 2015, the middle column—in 2016, and the bottom column—in 2017. Optimal trajectories with angular range of flight up to 720° were calculated. The top row shows trajectories with angular range up to 360°, and the bottom row shows trajectories with angular range from 360° to 720°. All calculations were performed using the method of continuation by gravitational parameter. The figure shows that at too small angular range trajectories go beyond the orbit of Mars, acquiring in many cases a loop-shaped form, including sections of retrograde motion. In contrast, trajectories with too large an angular range include regions of motion inside the Earth's orbit to ensure that this angular range is reached in a limited specified time.

The dependence of the cost function (7.4.4) on the departure date for the considered family of trajectories is represented by the upper curve in Fig. 8.7. The lower curve corresponds to a similar family of trajectories, but with a duration of 400 days. Trajectories containing less than one revolution are marked with thick lines.

To test the possibilities of calculating trajectories with a large number of revolutions, we considered the problem of calculating the LP-optimal Mercury rendezvous trajectory with 100 whole orbits of 14,000 days duration and zero departure hyperbolic velocity excess. The problem was solved by the gravitational parameter continuation method; partial derivatives were calculated by joint integration of the corresponding differential equations with the equations of motion of the spacecraft. Of course, for practical applications such a flight duration (more than 38 years) is too long. This numerical example is intended only to demonstrate the capabilities of

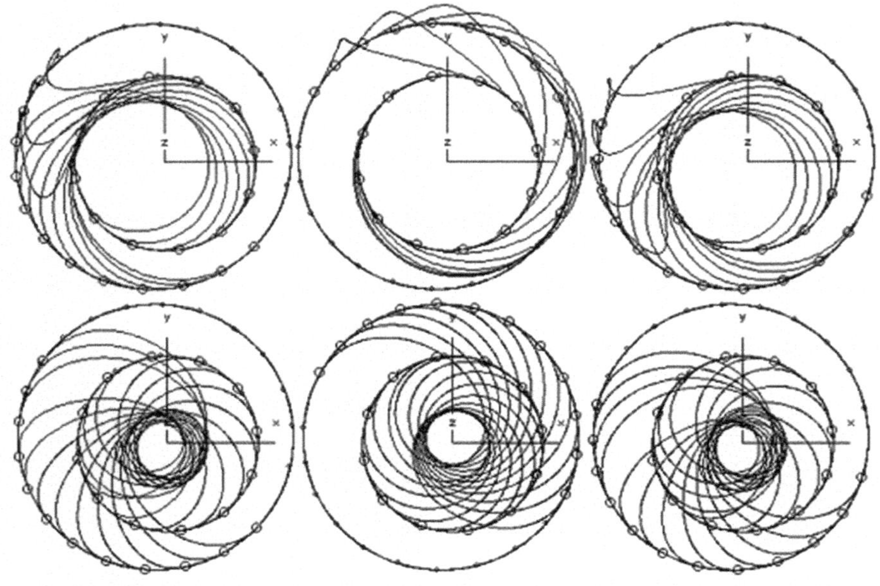

Fig. 8.7 Sets of LP-optimal trajectories to the Mars. The flight time is 300 days, the hyperbolic velocity excess is zero, and the departure dates of the neighboring trajectories differ by 1 month. The upper row corresponds to an angular range of less than 360°, the lower row corresponds to an angular range of 360–720°. The left column shows the departure dates in 2015, the middle one—in 2016, and the right one—in 2017

the method. The optimal LP-trajectory and control are presented in Fig. 8.8. It can be seen that the pitch angle oscillates in the vicinity of 180°, and the amplitude of yaw angle oscillation grows from 30° at the beginning of the flight to 50° when approaching Mercury.

In spite of the fact that in this numerical example the problem of optimization of a 100-orbit LP-trajectory using differential equations of motion of the spacecraft in Cartesian coordinates has been successfully solved, it is more efficient to use equations of motion of the spacecraft in osculating orbital elements to optimize multi-revolution trajectories. Trajectories with a very large number of orbits are characteristic of planetocentric trajectories, where the ratio of thrust acceleration to gravitational acceleration is small. Optimization of multi-turn interorbital flights around the Earth using equations of motion in osculating orbital elements will be considered in Chaps. 11 and 12.

Let us consider a numerical example demonstrating branching of optimal LP-trajectories. For this purpose, let us solve the problem of optimization of the Pluto rendezvous trajectory at the date of departure from the Earth on January 1, 2001 with zero departure hyperbolic velocity excess. We will vary the flight duration in the range from 2000 to 8000 days, and the number of integer orbits of the trajectory from 0 to 3. The main results of calculations are presented in Fig. 8.9. It can be seen

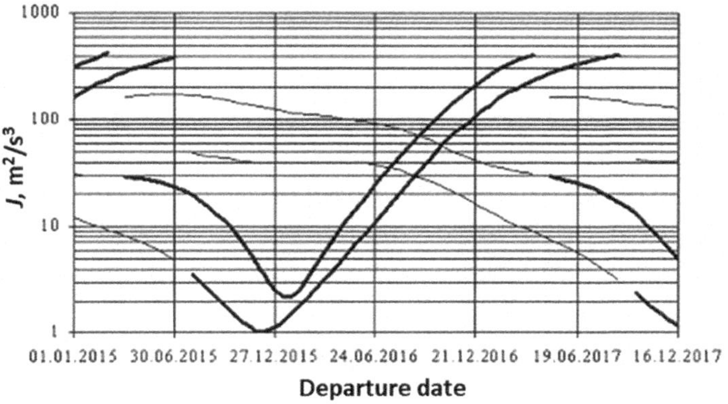

Fig. 8.8 The dependence of the cost function for the LP-optimal trajectories of the Earth-Mars transfer (rendezvous, NEP) on the departure date. The hyperbolic velocity excess is zero, the flight time is 300 and 400 days. The bold line corresponds to the angular range of the trajectory less than 1 revolution, the thin line corresponds to the angular range of the trajectory from 1 to 2 revolutions

that at flight time from 2000 to 2600 days the minimum value of the cost function is provided by trajectories with angular range less than 360°. At flight duration from 2600 to 5300 days, trajectories with angular range from 360° to 720° become optimal. In the range of flight duration from 5300 to 7200 days the minimum of cost function is achieved on trajectories with angular range from 720° to 1080°, and with further increase of duration—with angular range more than 1080°. Figure 8.10 shows that at flight durations of about 2600, 5300, and 7200 days, the same value of the cost function is achieved on trajectories with different numbers of revolutions around the Sun.

The developed methodology was successfully applied to optimize LP-trajectories within the framework of the ephemeris model of the restricted three-body problem. Figure 8.11 shows the results of calculating the flight trajectory from the initial high elliptical Earth orbit to the L_1 libration point of the Earth-Moon system. The inclination of the initial orbit was chosen to be 63°, and its apogee is located in the vicinity of the lunar orbit. The calculations were performed using the continuation method by the Earth's gravitational parameter. The gravitational parameter of the Moon was also varied from zero value at $\tau = 0$ to the actual value at $\tau = 1$. The obtained optimal LP-trajectory approaches close to the Moon in the middle of the flight, which can be seen by the kink in the trajectory at point 7. In fact, at this point the lunar gravity assist maneuver is realized, which was found automatically, without introducing additional boundary conditions to ensure numerical convergence of the method.

Flights with large changes in inclination and eccentricity are typical in the analysis of many missions to small celestial bodies of the Solar System. Let us consider the optimization of an asteroid rendezvous trajectory with the following Keplerian orbital elements in the heliocentric ecliptic coordinate system: semi-major axis

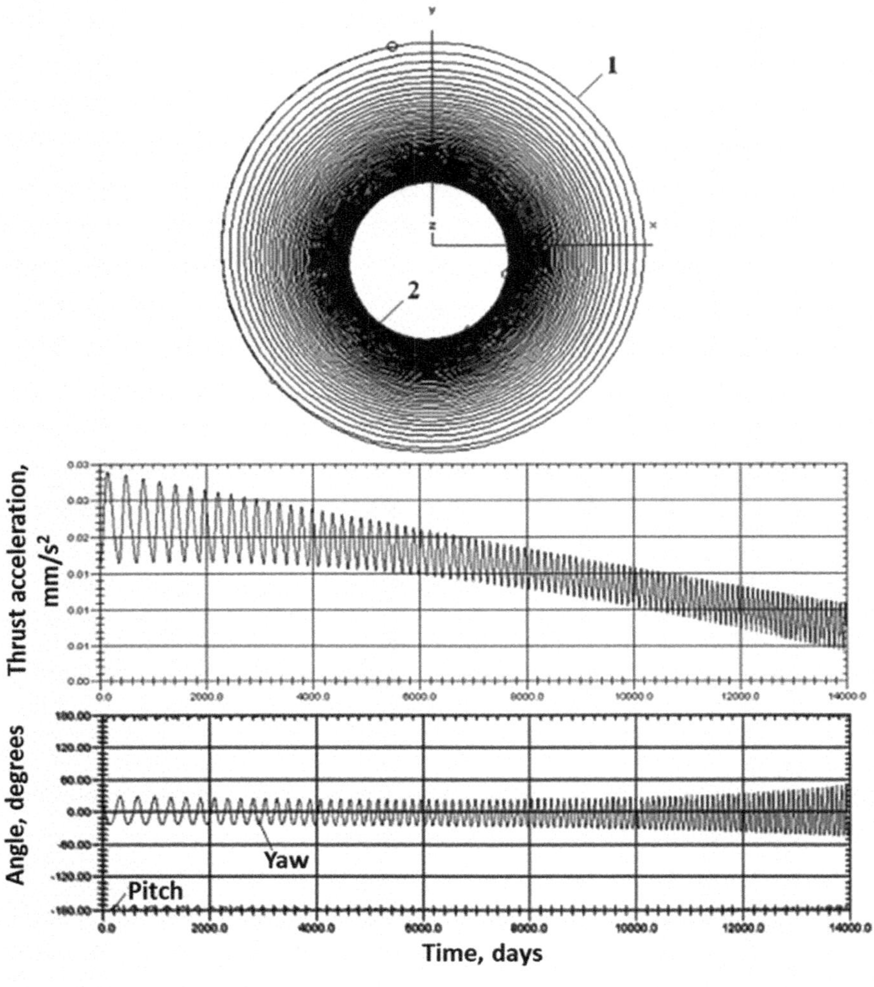

Fig. 8.9 Trajectory (top) and control (by the thrust acceleration, pitch and yaw angles) for transfer to Mercury (1—Earth orbit, 2—Mercury orbit)

1.0780195 A.U., eccentricity 0.8267806, argument of pericenter 31.2143°, inclination 122.88504°, right ascension of the ascending node 88.16747°, and mean anomaly 121.45003° at the epoch JD 2448700.5. This orbit corresponds to that of the asteroid #1566 (Icarus) with an inclination increased by 100°. Figure 8.12 shows the optimal LP-trajectory of the flight with zero departing hyperbolic excess velocity and three complete revolutions around the Sun for the departure date of January 1, 2001. The method of continuation by gravitational parameter was used to solve the problem.

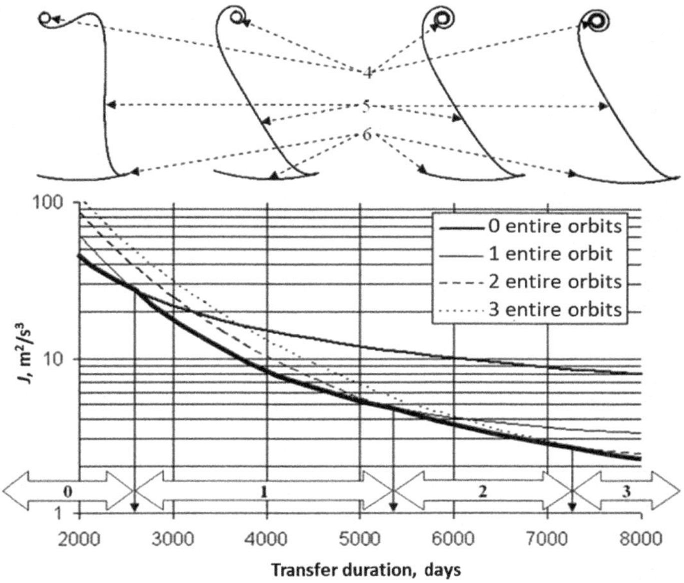

Fig. 8.10 Optimal trajectories to Pluto with different duration and angular range, departure date January 1, 2001. 0, 1, 2, 3—domains of optimality for trajectories with 0, 1, 2 or 3 full revolutions, respectively, 4—Earth's orbit, 5—trajectory, 6—Pluto's orbit

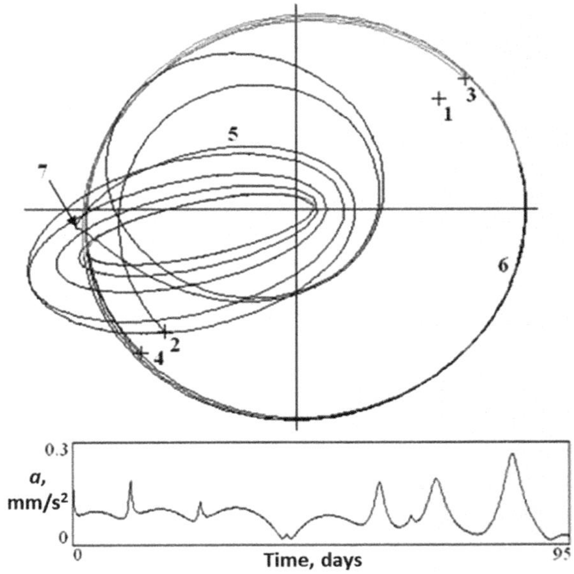

Fig. 8.11 The optimal LP-trajectory of the Earth-Moon L1 rendezvous with a gravity assisted maneuver (top) and the variation of thrust acceleration with time (bottom). 1, 2—the position of L1 at the initial and final moments of time; 3, 4—the position of the Moon at the initial and final moments of time; 5—transfer trajectory; 6—lunar orbit; 7—gravity assisted maneuver

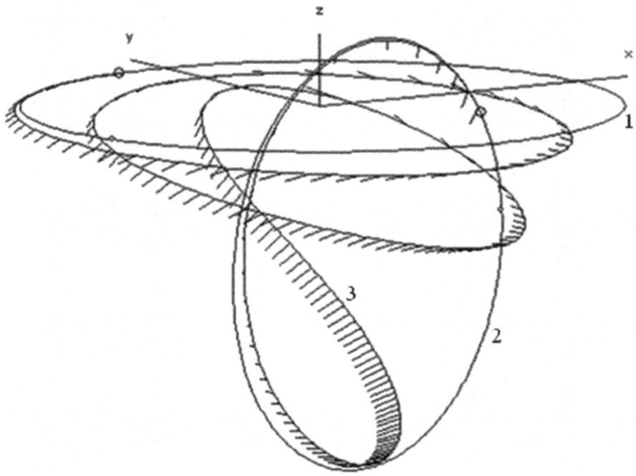

Fig. 8.12 An example of a small celestial body rendezvous trajectory with the retrograde orbital motion. 1—Earth orbit, 2—asteroid orbit, 3—trajectory

8.5 Conclusion

The variants of the parameter continuation method presented in this section have proven effective for optimizing LP-trajectories. The boundary condition continuation method and the gravitational parameter continuation method mutually complement each other, allowing, on the one hand, to vary the main mission parameters (departure date, initial hyperbolic excess velocity, duration and angular distance), and, on the other hand, to use trivial initial guess values (e.g., corresponding to the coasting motion along the orbit of the departure planet).

The numerical continuation methods described above have been implemented in a number of software tools (see, e.g. Fig. 8.13) that support the optimization of LP-trajectories with rendezvous and flyby of major planets of the Solar system, asteroids, comets, and other user-defined celestial bodies.

A typical optimization procedure with the developed software tools starts with the gravitational parameter continuation in a wide range of departure dates, flight durations, and angular distances. The most interesting trajectories thus obtained are then treated by the boundary condition continuation to analyze the trajectory dependence on the initial hyperbolic excess velocity, t_0 and t_f. One of the most important functions of the developed software is the data preparation for the LP-problem continuation to the limited-thrust problem.

The developed methods and software have been successfully applied to the optimization of the trajectories of the NEP spacecraft (constant EP power) and those of the SEP spacecraft (variable EP power), for the flyby and rendezvous problems, for the central gravity field and a restricted three-body problem. The continuation methods implemented in the software tools are effective for the optimization of both relatively simple trajectories with a low angular distance and for multi-revolution trajectories; for the optimization of trajectories with retrograde orbits and gravitational maneuvers along the trajectory; and for the optimization of interplanetary trajectories with high eccentricity and inclination values.

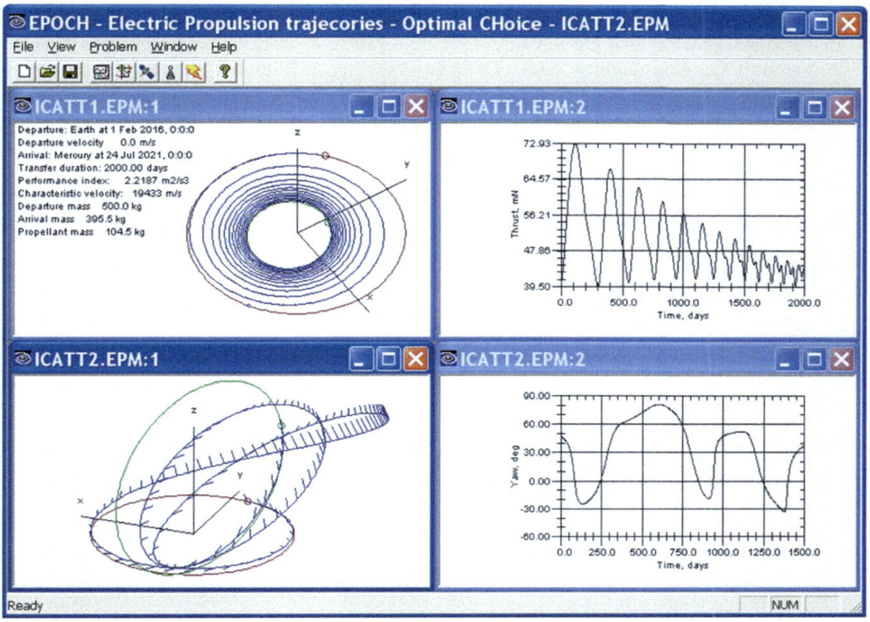

Fig. 8.13 The main window of the software package for optimizing LP-trajectories

References

1. Petukhov, V.G.: Continuation method for interplanetary low-thrust trajectory optimization. Space Sci **50**(3), 258–270 (2012)
2. Petukhov, V.G.: Continuation method for low-thrust trajectory optimization. In: Abstracts of the Fifth International Aeronautical Congress IAC'06. Moscow, 27–31 Aug 2006
3. Petukhov, V.G.: Optimization of interplanetary trajectories for spacecraft with ideally-controlled thrust by continuation method. Space Sci **46**(3), 224–237 (2008)
4. Petukhov, V.G.: Trajectory optimization and motion evolution of low-thrust spacecraft. Ph.D. thesis in Engineering Science, specialty 05.07.09 «Dynamics, ballistics, and control of flying vehicles». MAI, M., 1996, p. 132
5. Petukhov, V.G.: Low-thrust spacecraft trajectory optimization. A Workshop of SRI RAS on dynamics and control, Moscow, SRI RAS, 2000. http://arc.iki.rssi.ru/seminar/200006/OLTTR2.ppt
6. Eneev, T.M., Efimov, G.B., Konstantinov, M.S., Petukhov, V.G., Akhmetshin, R.Z., Fedotov, G.G.: Mercury-to-Pluto range missions using solar-nuclear electric propulsion. Keldysh Inst. Appl. Math. Russ. Acad. Sci. Prepr. **111** (1996)
7. Eneev, T.M., Egorov, V.A., Efimov, G.B., Konstantinov, M.S., Petukhov, V.G., Akhmetshin, R.Z., Fedotov, G.G.: Some methodical problems of low-thrust trajectory optimization. Keldysh Inst. Appl. Math. Russ. Acad. Sci. Prepr. **110**, 1–24 (1996)
8. Konstantinov, M.S., Petukhov, V.G.: Methodical aspects of optimization of complex interplanetary trajectories (global trajectory optimization). IAC-06-C1.4.03 (2006)
9. Petukhov, V.G.: Homotopic approach to low-thrust trajectory optimization: numerical technique and tools. In: 4th International Conference on Astrodynamics Tools and Techniques, 3–6 May 2010, ESA/ESAC, Madrid, Spain, p. 8. ESA Proceedings WPP-308

10. Petukhov, V.G.: One numerical method to calculate optimal power-limited trajectories. In: International Electric Propulsion Conference, p. 8. IEPC-95-221, Russia, Moscow (1995)
11. Petukhov, V.G., Konstantinov, M.S., Fedotov, G.G.: 1st ACT global trajectory optimization competition: results found at Moscow Aviation Institute and Khrunichev State Research and Production Space Center. Acta Astronaut. **61**(9), 775–785 (2007)
12. Everhart, E.: Implicit single sequence methods for integrating orbits. Celest. Mech. **10**, 35 (1974)
13. Hairer, F., Norsett, S.P., Wanner, G.: Solving Ordinary Differential Equations I. Non-Stiff Problems. Spinger, Berlin (1987)

Chapter 9
Optimization of Interplanetary Trajectories for a Limited-Thrust Spacecraft

9.1 Introduction

Let us consider the problem of optimizing the heliocentric section of the interplanetary trajectory between two planets for a spacecraft equipped with a limited-thrust engine. The mathematical model of the LT problem given in Sect. 7.6 adequately represents the main features of operation of modern and promising electric propulsion systems, so the optimization of LT trajectories is certainly of significant practical interest. However, from a methodological point of view, the optimization of LT trajectories is much more difficult than the optimization of LP trajectories.

The main reasons for the complexity of LT-trajectory optimization are the lack of a priori information about the existence and number of solutions and the violation of continuity in the dependence of the derivatives of the residuals of the boundary value problem with respect to the unknown parameters of the boundary value problem when changing the structure of the trajectory (changing the number and sequence of thrusting and coasting trajectory sections) in the process of solving the problem with the thrust switching. In many cases, the optimization process is complicated by the high sensitivity of LT trajectories to variations in unknown parameters of the boundary value problem.

The absence of a theorem on the existence of an optimal LT trajectory does not allow us to construct a regular method for solving the problem; however, it is possible to implement a numerical method that determines the existence of a solution under the given boundary conditions and characteristics of the electric propulsion system and calculates the optimal solution, if it exists.

The possibility of implementing such a method is based on the following considerations. Let us consider the problem of minimizing the thrust required for the transfer between fixed points in a given time. This problem is dual to the minimum-time problem. In particular, all the necessary optimality conditions for both problems are identical, except that in the thrust minimization problem the transversality condition is used to calculate the thrust instead of calculating the flight duration in the

A. S. Filatyev and V. G. Petukhov, *Through Optimization of Aerospace Vehicle Trajectories by the Pontryagin Maximum Principle*, Studies in Systems, Decision and Control 579, https://doi.org/10.1007/978-3-031-80756-5_9

minimum-time problem. The solution to the thrust minimization problem of corresponds to the minimum-time trajectory for the minimum thrust that ensures the flight in a given time. It is easy to show that the problem of thrust minimization is smooth due to the absence of thrust switching and fixed boundary conditions, even for the case of the moving start and end points (for example, planets), since the start and end times are fixed. This problem can be used to calculate the region of existence of the optimal trajectory with the thrust switching. Obviously, a solution to the latter problem exists, if the minimum thrust value of the corresponding thrust minimization problem is less than or equal to the value of the available thrust. Unfortunately, the thrust minimization problem itself has a limited domain of existence due to the limitation on the required propellant mass. To avoid numerical degeneration, we can consider the problem of minimizing thrust acceleration. The thrust acceleration minimization problem is a limiting case of the thrust minimization problem corresponding to zero mass flow rate (or infinite specific impulse). A solution to the thrust acceleration minimization problem exists for arbitrary non-degenerate boundary conditions. This problem can be solved using a LP-trajectory or the solution to the Lambert problem as an initial approximation and a numerical continuation method. Therefore, it is possible to construct a method for calculating optimal trajectories with finite thrust, starting with the solution of the corresponding problem of thrust acceleration minimization, followed by numerical continuation through the mass flow rate to the thrust minimization problem and then to the trajectory optimization problem with thrust switching. This method makes it possible to detect the degeneracy associated with a lack of thrust or specific impulse. In fact, this method allows to calculate the boundaries of the region of existence of trajectories with thrust switching. This method allows, first, to design a very stable numerical algorithm, and, second, to reliably find departure windows.

However, it should be noted that the LP problem solution provides quite a lot of information about the existence of an LT solution. In fact, for the existence of an LT solution, it is sufficient if the thrust value of on the LP solution does not exceed the given thrust value of the LT problem, and the characteristic velocity on the LP solution does not exceed the maximum permissible increment of the characteristic velocity in the LT problem based on the available propellant store. In this context, it is possible to construct effective methods for continuing the LP solution into an LT solution with the thrust switching. To do this, it is necessary to overcome the difficulties associated with jumps in the partial derivatives of the residual vector with respect to the unknown parameters of the boundary value problem.

This section describes the method for the continuation of the optimal LP trajectory into the optimal LT trajectory and gives numerical examples of the application of the developed method to practical problems. The presentation is basically the same as in [1–6].

9.2 Homotopy Between the Limited Power and Limited Thrust Problems

Jumps in the elements of the sensitivity matrix (partial derivatives of the boundary conditions with respect to the parameters of the boundary value problem) due to the appearance or disappearance of active or passive sections, as noted above, are one of the main reasons for the failure of traditional numerical methods in solving the LT problem. The high sensitivity and instability of the corresponding systems of ordinary differential equations are not a significant limitation for modern numerical methods.

To explain the mechanism of the formation of jumps in partial derivatives of boundary conditions with respect to the parameters of the boundary value problem, let us consider one of the equations of the system (7.5.10), which has the following general form:

$$\frac{dx}{dt} = f_0 + \delta(S) f_1 \tag{9.1.1}$$

where f_0, f_1 and S are the continuous twice differentiable functions, and x is a certain state coordinate of the spacecraft.

At a regular point (in the section where the switching function S has a definite sign), the differential equation for the partial derivative of the state coordinate x with respect to the unknown parameter of the boundary value problem z has the form:

$$\frac{d}{dt}\left(\frac{\partial x}{\partial z}\right) = \frac{\partial f_0}{\partial z} + \delta(S)\frac{\partial f_1}{\partial z}, \tag{9.1.2}$$

and at the switching point (at $t = t_s$: $S = 0$) it has the following form:

$$\frac{d}{dt}\left(\frac{\partial x}{\partial z}\right) = \frac{\partial f_0}{\partial z} + \delta(S)\frac{\partial f_1}{\partial z} + \frac{\partial \delta}{\partial S}\frac{\partial S}{\partial z} f_1. \tag{9.1.3}$$

The jump in the partial derivative at the switching point t_s can be calculated by integrating (9.1.3) over the infinitesimal interval containing this point:

$$\Delta\left(\frac{\partial x}{\partial z}\right) = \int_{t_s^-}^{t_s^+}\left(\frac{\partial f_0}{\partial z} + \delta\frac{\partial f_1}{\partial z}\right)dt + \int_{t_s^-}^{t_s^+}\frac{\partial \delta}{\partial S}\frac{\partial S}{\partial z} f_1 dt = 0 + \left.\frac{\partial S}{\partial z} f_1\right|_{t=t_s}\int_{t_s^-}^{t_s^+}\frac{\partial \delta}{\partial S}dt$$
$$= \left.\frac{\partial S}{\partial z} f_1\right|_{t=t_s}\int_{0^-}^{0^+}\frac{\partial \delta}{\partial S}\frac{dt}{dS}dS = \left.\frac{1}{dS/dt}\frac{\partial S}{\partial z} f_1\right|_{t=t_s}. \tag{9.1.4}$$

Let us consider the case when at $t = t_s$ the following conditions are met: $S = 0$, $dS/dt = \varepsilon$, $|\varepsilon| \ll 1$, $d^2S/dt^2 \cdot dS/dt < 0$, which correspond to the starting point of the small thrusting section ($\varepsilon > 0$) or the coasting section ($\varepsilon < 0$).

Then in the t_s neighborhood the following expansion is valid:

$$S = \dot{S}\Delta t + \frac{1}{2}\ddot{S}\Delta t^2 + O(\Delta t^3) \tag{9.1.5}$$

where $\Delta t = t - t_s$. The end point t_{s1} and the duration $\varepsilon_t = t_{s1} - t_s$ of the small thrusting or coasting section starting at the point t_s can be found by equating (4.1.5) to zero:

$$\varepsilon_t = t_{s1} - t_s = -\frac{2}{\ddot{S}}\dot{S} = -\frac{2}{\ddot{S}}\varepsilon. \tag{9.1.6}$$

In accordance with (9.1.5), the derivative dS/dt at the point $t = t_{s1}$ is equal to the derivative dS/dt at the point $t = t_s$, with the opposite sign, up to values of second order of smallness. Therefore, taking (9.1.4) into account, we can obtain an expression for the increment $\partial x/\partial z$ as a result of the total impact of two jumps at adjacent switching points t_s and t_{s1}:

$$\Delta\left(\frac{\partial x}{\partial z}\right)\bigg|_{t=t_s} + \Delta\left(\frac{\partial x}{\partial z}\right)\bigg|_{t=t_{s1}} = \frac{1}{\varepsilon}\left\{\frac{\partial S}{\partial z}f_1\bigg|_{t=t_s} - \frac{\partial S}{\partial z}f_1\bigg|_{t=t_{s1}}\right\}. \tag{9.1.7}$$

But the second term in curly brackets in expression (9.1.7) can be calculated through the following expansion in the vicinity of t_s:

$$\frac{\partial S}{\partial z}f_1\bigg|_{t=t_{s1}} = \frac{\partial S}{\partial z}f_1\bigg|_{t=t_s} + \left\{f_1\frac{d}{dt}\left(\frac{\partial S}{\partial z}\right) + \frac{df_1}{dt}\frac{\partial S}{\partial z}\right\}\bigg|_{t=t_s}\varepsilon_t + O(\varepsilon_t^2). \tag{9.1.8}$$

Substituting expression (9.1.6) into (9.1.8) for the duration of the section ε_t we get:

$$\Delta\left(\frac{\partial x}{\partial z}\right)\bigg|_{t=t_s} + \Delta\left(\frac{\partial x}{\partial z}\right)\bigg|_{t=t_{s1}} = -\frac{2}{\ddot{S}}\left\{f_1\frac{d}{dt}\left(\frac{\partial S}{\partial z}\right) + \frac{df_1}{dt}\frac{\partial S}{\partial z}\right\}\bigg|_{t=t_s} + O(\varepsilon_t). \tag{9.1.9}$$

Going to the limit $\varepsilon_t \to 0$ ($S = 0$, $dS/dt \to 0$), we obtain the condition for a jump in the partial derivative of the phase coordinate with respect to the parameter of the boundary value problem at the point t_s*, where the switching function touches the x-axis (that is, at the point where the thrusting or coasting section of the trajectory appears or disappears):

$$\Delta\left(\frac{\partial x}{\partial z}\right)\bigg|_{t=t_s^*} = -\frac{2}{\ddot{S}}\left\{f_1\frac{d}{dt}\left(\frac{\partial S}{\partial z}\right) + \frac{df_1}{dt}\frac{\partial S}{\partial z}\right\}\bigg|_{t=t_s^*}. \tag{9.1.10}$$

It is obvious that the jump (9.1.10) translates into a jump in the final value of the partial derivative of the phase coordinate x, which leads to a jump in the sensitivity matrix and to the corresponding computational complications in all numerical methods for solving the LT problem that use such matrix.

So, the main idea of this section is to smoothly continue the LP problem into the LT problem while avoiding discontinuities in the sensitivity matrix. Smooth continuation requires a smooth approximation of the throttling function δ by t and τ. Moreover, the smoothed thrust function should be close to the step function $\delta(S)$ at $\tau = 1$ and the sign of the smoothed throttling function must match the sign of the switching function. The following function is equal to $\delta(S)$ at any point where S $\neq$ 0:

$$\delta_1(S) = \frac{1}{2}\left[\frac{S}{|S|} + 1\right] \tag{9.1.11}$$

Regularization and smoothing of (9.1.11) can be carried out using a small regularizing term $\varepsilon > 0$ in the denominator of the first term:

$$\delta_2(S) = \frac{1}{2}\left[\frac{S}{|S| + \varepsilon} + 1\right]. \tag{9.1.12}$$

This function is smooth and positive at any point, but its time derivative may be too large for small ε. Otherwise, if ε is large, δ_2 is substantially different from δ. To facilitate the continuation process, we introduce the following dependence of the thrust function on τ:

$$\delta_3(S, \tau) = \frac{1}{2}\left[\frac{S}{1 - \tau + \tau \cdot |S| + \varepsilon(\tau)} + 1\right], \tag{9.1.13}$$

where

$$\varepsilon(\tau) = (1 - \tau) \cdot \varepsilon_0 + \tau \cdot \varepsilon_k. \tag{9.1.14}$$

At $\tau = 0$, $\delta_3 = 0.5 \cdot [S/(1 + \varepsilon_0) + 1]$, and when $\tau = 1$, $\delta_3 = 0.5 \cdot [S/(|S| + \varepsilon_k) + 1] = \delta_2$. The options ε_0 and ε_k control the convergence of the function δ_3 to the step thrust function δ (more precisely, to its smooth approximation δ_2) when increasing the continuation parameter τ from 0 to 1. The time derivative of δ_3 at the switching point increases with τ in such a way that the differential equations for the optimal spacecraft motion become truly stiff only when τ is very close to 1, and ε is small.

Finally, the following system of ordinary differential equations of spacecraft motion is proposed, realizing the homotopy between the LP and LT problems:

$$\left.\begin{aligned}
\frac{d^2\mathbf{r}}{dt^2} &= \Omega_{\mathbf{r}} + \frac{(1-\tau)\cdot\eta P_{b0}+\tau\cdot\delta_3 T}{(1-\tau)\cdot\left(P_{b0}+m^2 p_m\right)+\tau\cdot m p_v}\mathbf{p}_{\mathbf{v}}, \\
\frac{dm}{dt} &= -\frac{(1-\tau)\cdot\eta P_{b0} m^2 p_v^2+\tau\cdot\delta_3 T}{(1-\tau)\cdot 2\left(P_{b0}+m^2 p_m\right)^2+\tau\cdot c}, \\
\frac{d^2\mathbf{p}_{\mathbf{v}}}{dt^2} &= \Omega_{\mathbf{rr}}\mathbf{p}_{\mathbf{v}} + \frac{(1-\tau)\cdot P_{b0} p_v^2+\tau\cdot\delta_3\left(\frac{\partial T}{\partial\eta}S+T\frac{1+p_m}{c^2}\frac{\partial c}{\partial\eta}\right)}{(1-\tau)\cdot\left(P_{b0}+m^2 p_m\right)+\tau}\frac{\partial\eta}{\partial\mathbf{r}}, \\
\frac{dp_m}{dt} &= \frac{(1-\tau)\cdot\eta P_{b0} m p_m p_v^2+\tau\cdot\delta_3 T p_v}{(1-\tau)\cdot\left(P_{b0}+m^2 p_m\right)+\tau\cdot m^2}.
\end{aligned}\right\} \quad (9.1.15)$$

The above system corresponds to the LP problem when $\tau = 0$ and to the LT problem (except for the difference between δ and δ_2) when $\tau = 1$.

Figure 9.1 shows δ_3 as a function of time and τ for an example of the switching function (lower curve in Fig. 9.1) with $\varepsilon = 10^{-5}$. When τ is small, $\delta_3(t, \tau)$ is a smooth function, and it approaches a step function as τ approaches 1.

Of course, the time derivative of $\delta_3(t, \tau)$ approaches δ-function at $\tau \to 1$ and $\varepsilon \to 0$ (see Fig. 9.2). Therefore, the system of differential Eqs. (4.1.15) becomes stiff and an appropriate method is required to solve it.

An important note concerns the way the sensitivity matrix $\partial\mathbf{f}/\partial\mathbf{z}$ is calculated. A simultaneous system of ordinary differential equations, including (9.1.15) and differential equations for the elements $\partial\mathbf{f}/\partial\mathbf{z}$, may be too stiff due to the properties of δ_3. Moreover, complex dependencies (7.5.1)–(7.5.3) or a complex model of external forces can become a significant obstacle to the integration of this system of differential equations. Therefore, the finite differences were initially used to calculate $\partial\mathbf{f}/\partial\mathbf{z}$. The limited accuracy of the finite-difference representation of partial derivatives limited the convergence of the homotopy method. For example, a version of numerical homotopy with finite-difference calculation of derivatives made it possible to calculate optimal LT trajectories with no more than 3–4 orbits around the Sun.

To overcome this limitation, the complex step differentiation method was used to calculate derivatives with high accuracy (see Sect. 7.3).

So, the computational scheme for optimizing the LT trajectory is as follows:

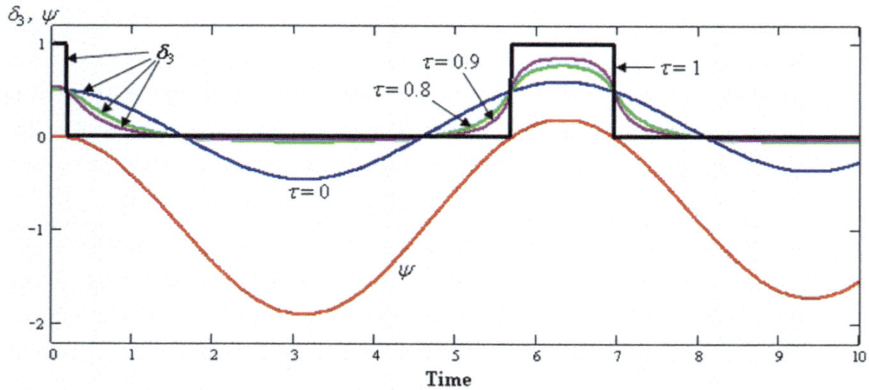

Fig. 9.1 Modified thrust function δ_3 as a function of time and continuation parameter τ (ψ is an example of a switching function)

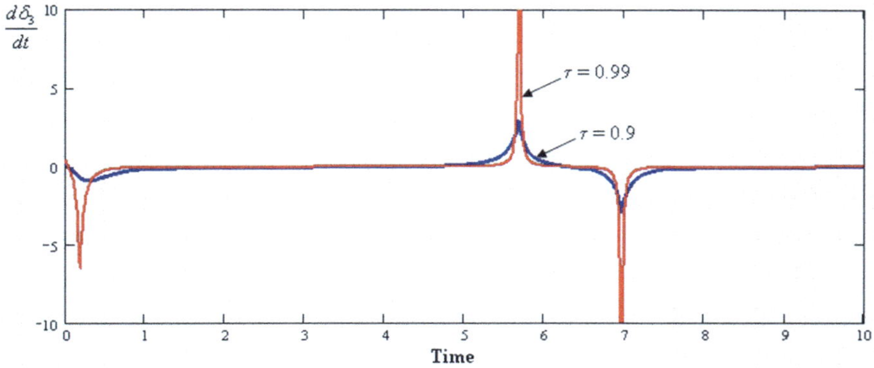

Fig. 9.2 The time derivative of δ_3 as a function of t and τ

(1) The LP trajectory is optimized using, for example, the method of continuation with respect to the gravitational parameter.
(2) The continuation method (7.3.18) is applied to the system of ordinary differential equations (9.1.15) with the appropriate boundary conditions. As a result, the optimal LT trajectory is calculated, which has a smooth approximation of the thrust function δ_2. The values from the LP solution are used as an initial approximation for the initial values of the conjugate variables.
(3) The initial values of the conjugate variables obtained as a result of step 2 are used as an initial approximation when optimizing the LT trajectory with a true (step) throttling function δ.

The latter step can be skipped if δ_2 is close enough to δ.

Such LT trajectory optimization scheme has been implemented in software. The developed software can use the finite difference method or the complex step method for numerical differentiation. It supports rendezvous and flyby missions, including flybys with the fixed arrival hyperbolic excess velocity. A scheme for continuing the LP trajectory with zero initial hyperbolic excess velocity into the LT trajectory with a given initial hyperbolic excess velocity has been implemented, for which the initial conditions (7.4.13) have been rewritten as:

$$\mathbf{r}(t_0) = \mathbf{r}_0(t_0),\ \mathbf{v}(t_0) = \mathbf{v}_0(t_0) + \tau V_\infty \mathbf{p}_{v0}/p_{v0},\ m(t_0) = m_0. \tag{9.1.16}$$

In addition to (9.1.12), there are many smoothed representations of the relay function of thrust. For example, the following expressions are used in various papers:

$$\delta_2(S) = \frac{1}{2}\left(1 + \frac{2}{\pi}\mathrm{arctg}\frac{S}{\varepsilon(\tau)}\right), \tag{9.1.17}$$

$$\delta_2(S) = \left[1 + \exp\left(-\frac{S}{\varepsilon(\tau)}\right)\right]^{-1}, \tag{9.1.18}$$

$$\delta_2(S) = \frac{1}{2}\left(1 + \text{th}\frac{S}{\varepsilon(\tau)}\right). \tag{9.1.19}$$

9.3 Critical Value of the Departure Hyperbolic Excess Velocity and Homotopy Between the Solution of the Lambert Problem and the Optimal Trajectory with Limited Thrust

The form of the transversality condition for interplanetary transfer with a given departure hyperbolic excess velocity $V_\infty > 0$, used e.g. in [1–7] and many other publications, is correct only for the sufficiently small values of this hyperbolic excess and is not entirely correct in the general case. To derive the general form of this transversality condition, consider a direct flight between two planets.

At the initial point of the trajectory, in accordance with the boundary conditions, the following equality must be satisfied:

$$g = \left(\mathbf{v} - \mathbf{v}_{pl}\right)^T \left(\mathbf{v} - \mathbf{v}_{pl}\right) - V_\infty^2 = 0 \tag{9.2.1}$$

Therefore, the transversality condition has the form:

$$\mathbf{p_v} = \frac{1}{2}\nu\frac{\partial g}{\partial \mathbf{v}} = \nu\left(\mathbf{v} - \mathbf{v}_{pl}\right), \tag{9.2.2}$$

whereν is the Lagrange multiplier.

Previously, in the known studies, this condition, after the exclusion of the multiplier ν, was interpreted in the form:

$$\mathbf{v} - \mathbf{v}_{pl} = V_\infty \frac{\mathbf{p_v}}{p_v}, \tag{9.2.3}$$

where $V_\infty = |\mathbf{v}(t_0) - \mathbf{v}_{pl}(t_0)|$ is the specified value of the departure hyperbolic excess velocity.

However, condition (9.2.3) corresponds to the assumption that ν in (9.2.2) is positive, i.e. to the assumption that the basis vector and the vector of hyperbolic excess velocity are co-directional. This assumption is valid only for the sufficiently small values of the hyperbolic excess velocity. The experience gained in the numerical solution of many boundary value problems of the Pontryagin maximum principle for heliocentric trajectories confirmed that the condition (9.2.3) is valid at the beginning of the heliocentric section of the trajectory for $V_\infty \in \left[0;\ V_\infty^*\right)$, where $V_\infty{}^*$ is a certain critical value of the departure hyperbolic excess velocity. For the flyby problem (when the value of the hyperbolic excess velocity at the end point of the trajectory is not limited), the value of V_∞^*corresponds to a completely coasting trajectory and can be obtained from the solution of the Lambert problem.

As an illustration confirming the change of ν sign when passing through V_∞^*, let us consider solutions to the problem of the optimal flyby of Mars with the following parameters: date of departure from Earth 06/01/2035 (JD 2464479.5), flight duration 210 days, initial spacecraft mass on a heliocentric trajectory 500 kg, the electric propulsion system thrust 0.083 N, and the EPS specific impulse 1500 s. The solution to the Lambert problem under the specified conditions leads to a departure hyperbolic excess velocity $V_\infty^* = 3534.9$ m/s, while the heliocentric trajectory of the flight to Mars will be coasting. The results of the flight trajectory optimization with initial hyperbolic excess velocity $V_\infty = 3400$ m/s (less than V_∞^*) lead to $\nu = 0.249769 > 0$ ($\mathbf{V}_\infty = (3.343847073; 0.56776024; 0.237336398)^T$, $\mathbf{p_v}(t_0) = (1.6703781; 0.28361756; 0.11855735)^T$, $\mathbf{V}_\infty \uparrow\uparrow \mathbf{p_v}(t_0)$), and with the initial hyperbolic excess of velocity $V_\infty = 4000$ m/s (more than V_∞^*) they lead to $\nu = -0.15766 < 0$ ($\mathbf{V}_\infty = (3.677402073; 0.81978824; 1.343374168)^T$, $\mathbf{p_v}(t_0) = (-1.1603111; -0.25866326; -0.42386679)^T$, $V_\infty \uparrow\downarrow \mathbf{p_v}(t_0)$). Both trajectories have one active section in the middle part of the trajectory; the spacecraft mass delivered to Mars is 496.159 kg for the trajectory with $V_\infty = 3400$ m/s and 487.825 kg for the trajectory with $V_\infty = 4000$ m/s. Figure 9.3 shows the time dependence of the angle of attack (the angle between the vector $\mathbf{p_v}$ and the heliocentric velocity vector of the spacecraft) for a trajectory with a departure hyperbolic excess of 3600 m/s and 4000 m/s. It can be seen from the figure that the thrust vector, collinear with $\mathbf{p_v}$, increases the orbital velocity of the spacecraft on a trajectory with a subcritical hyperbolic excess (the angle of attack in the active section changes from 12.2° to 11.4°), and decreases it on a trajectory with a supercritical hyperbolic excess (the angle of attack on the active section varies from 135.2° to 128.6°).

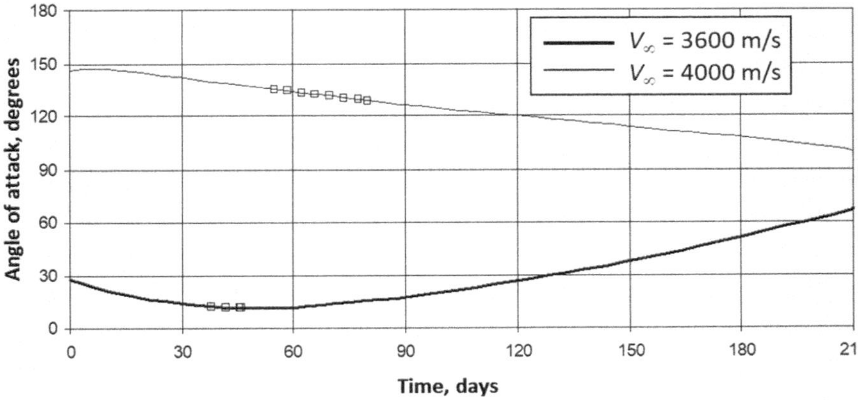

Fig. 9.3 The angle of attack (the angle between the $\mathbf{p_v}$ vector and the heliocentric velocity vector of the spacecraft) for a trajectory with a departure hyperbolic excess of 3600 m/s (thick line) and 4000 m/s (thin line). The trajectory sections with EPS in operation are marked with squares

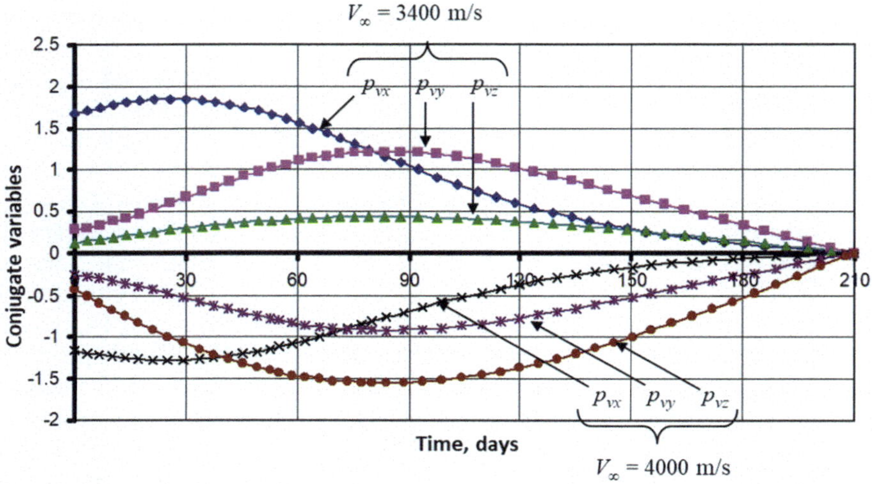

Fig. 9.4 Time dependence of the components of the vector conjugate to the spacecraft velocity on trajectories with subcritical and supercritical values of the departure hyperbolic velocity excess

Figure 9.4 shows the time dependence of the vector components conjugate to the spacecraft velocity on trajectories with subcritical and supercritical values of the departure hyperbolic excess velocity. In this case, these components are positively semidefinite for a subcritical hyperbolic excess of velocity and negatively semidefinite for a supercritical hyperbolic excess. At a critical value of the hyperbolic excess velocity (that is, on the Keplerian arc), these components are identically equal to 0.

It would seem that the case $V_\infty < V_\infty^*$ is typical, and the case $V_\infty > V_\infty^*$ is exotic and is not of great practical interest. However, when designing interplanetary trajectories with low thrust and gravity assist maneuvers, the case of supercritical (high V_∞^*) values of the departure hyperbolic excess velocity on the trajectory section after the ordinary gravity assist maneuver is quite typical. As an example of such cases, one can mention the trajectories of the launch of a spacecraft with an electric propulsion system to Jupiter with a gravity assist maneuver near the Earth and the trajectories of a solar probe with the electric propulsion system with gravity assist maneuvers near the Earth and Venus. On such trajectories, the hyperbolic excess velocity during a gravity assist maneuver near the Earth can reach 8–12 km/s, and near the Venus—16–18 km/s or more. Often such an asymptotic velocity is excessive to reach the next point on the flight route (the destination planet or the planet at which the next gravity assist maneuver will take place). In this case (with an unlimited flyby altitude in the current gravity assist maneuver), the thrust vector must be directed against the vector of the hyperbolic excess velocity, that is, the multiplier ν becomes negative, and the expression (9.2.3) becomes incorrect; its correct form is:

$$\mathbf{v} - \mathbf{v}_p = -V_\infty \frac{\mathbf{p_v}}{p_v}. \tag{9.2.4}$$

However, the use of the transversality condition of the form (9.2.4), is complicated by the extremely high sensitivity of the residuals of the boundary value problem to variations in the direction $\mathbf{V}_\infty$ at high values of $V_\infty > V_\infty{}^*$. In addition, the change of sign of ν at the point $V_\infty = V_\infty{}^*$ does not allow to use the transversality condition of the form (9.2.3), (9.2.4) for a smooth continuation of solutions with respect to the value V_∞ beyond $V_\infty{}^*$.

Therefore, for the problems in which V_∞ may exceed $V_\infty{}^*$, it is logical to use the boundary conditions and transversality conditions in their original form (9.2.1), (9.2.2) without excluding the Lagrange multiplier ν from this system. This increases the total number of boundary conditions by one (two Eqs. (9.2.1) and (9.2.2) are used instead of one (9.2.3) or (9.2.4)), as well as the number of unknown parameters of the boundary value problem (an unknown value of the Lagrange multiplier ν is added). Moreover, this allows to organize a regular continuation process by the value of V_∞ up to values exceeding $V_\infty{}^*$.

As a starting point for the continuation process, one can use the solution obtained when using the method of continuation by the gravitational parameter with $V_\infty = 0$. In this case, however, there is a singularity at the initial point of continuation, when using relations (9.2.1) and (9.2.2). Indeed, it follows from the comparison of (9.2.1) and (9.2.2) that $\nu = p_v/V_\infty$ at $V_\infty < V_\infty{}^*$, hence ν tends to infinity as V_∞ tends to zero (in general $p_v \neq 0$). Therefore, at the initial continuation interval it is more convenient to use the quantity $\lambda = 1/\nu = V_\infty/p_v$ instead of ν; this quantity being equal to 0 at $V_\infty = 0$ (the qualitative dependence of the multipliers ν and λ on V_∞ is shown in Fig. 9.5). In this case, the continuation scheme at the initial continuation interval ($V_\infty \in [0;\ V_{\infty 1}]$) has the form:

$$\begin{aligned}
&g = \left(\mathbf{v} - \mathbf{v}_{pl}\right)^T \left(\mathbf{v} - \mathbf{v}_{pl}\right) - V_\infty^2(\tau) = 0,\\
&\lambda \mathbf{p_v} = \mathbf{v} - \mathbf{v}_{pl} \Rightarrow \mathbf{v} = \mathbf{v}_{pl} + \lambda \mathbf{p_v},\\
&V_\infty(\tau) = \tau \cdot V_{\infty 1},\\
&\tau \in [0;\ 1],\\
&\lambda|_{\tau=0} = 0,\\
&\mathbf{p_v}|_{\tau=0} = \mathbf{p}_{\mathbf{v}0},
\end{aligned} \tag{9.2.5}$$

where $\mathbf{p}_{\mathbf{v}0}$ is the initial guess for $\mathbf{p_v}$ at $V_\infty = 0$, e.g. obtained by using the continuation with respect to the gravitational parameter.

After calculating the solution for $V_\infty = V_{\infty 1} < V_\infty{}^*$ using the continuation scheme (9.2.5), it is necessary to proceed to another continuation scheme using the multiplier $\nu = 1/\lambda$:

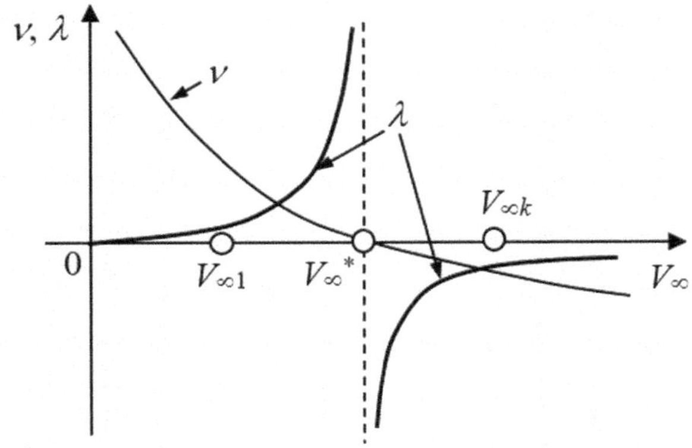

Fig. 9.5 Dependence of the multipliers ν and λ on V_∞

$$
\begin{aligned}
& g = \left(\mathbf{v} - \mathbf{v}_{pl}\right)^T \left(\mathbf{v} - \mathbf{v}_{pl}\right) - V_\infty^2(\tau) = 0, \\
& V_\infty(\tau) = (1 - \tau) \cdot V_{\infty 1} + \tau \cdot V_{\infty k}, \\
& \mathbf{p}_\mathbf{v} = \nu\left(\mathbf{v} - \mathbf{v}_{pl}\right), \\
& \tau \in [0;\, 1], \\
& \nu|_{\tau=0} = \left.\frac{1}{\lambda}\right|_{V_\infty = V_{\infty 1}}, \\
& \mathbf{v}|_{\tau=0} = \mathbf{v}|_{V_\infty = V_{\infty 1}},
\end{aligned}
\tag{9.2.6}
$$

where $V_{\infty k}$ is the given hyperbolic excess velocity. Condition $\mathbf{p}_\mathbf{v}|_{\tau=0} = \mathbf{p}_\mathbf{v}|_{V_\infty = V_{\infty 1}}$ is provided automatically if (9.2.6) is satisfied.

Thus, on the first segment of continuation according to the scheme (9.2.5), the unknown parameters of the boundary value problem are the initial values of $\mathbf{p}_\mathbf{r}$, $\mathbf{p}_\mathbf{v}$, p_m and λ, and on the second segment—according to the scheme (9.2.6)—the initial values of $\mathbf{p}_\mathbf{r}$, $\mathbf{v}$, p_m and ν.

Alternatively, on the first segment of continuation $0 \le V_\infty \le V_{\infty 1} < V_\infty{}^*$ it is possible to use the transversality conditions in the form (9.2.3). In this case, to ensure the transition from the first segment to the second one, it is necessary to use the following expression for ν:

$$\nu = p_v / V_\infty \tag{9.2.7}$$

at the end point of the first continuation segment, that is, at $V_\infty = V_{\infty 1}$, so that at the beginning of the second continuation section the following condition is satisfied:

$$\nu|_{\tau=0} = \frac{p_v|_{V_\infty = V_{\infty 1}}}{V_{\infty 1}}. \tag{9.2.8}$$

In this case, the unknown parameters of the boundary value problem at the initial section of the continuation are the initial values of $\mathbf{p}_\mathbf{x}$, $\mathbf{p}_\mathbf{v}$, p_m, and the continuation scheme at the initial section has the form:

$$\begin{aligned} & g = (\mathbf{v} - \mathbf{v}_{pl})^T (\mathbf{v} - \mathbf{v}_{pl}) - V_\infty^2(\tau) = 0, \\ & \mathbf{v} = \mathbf{v}_{pl} + \frac{V_\infty(\tau)}{p_v} \mathbf{p_v}, \\ & V_\infty(\tau) = \tau \cdot V_{\infty 1}, \\ & \tau \in [0;\, 1], \\ & \mathbf{p_v}|_{\tau=0} = \mathbf{p}_{v0}. \end{aligned} \tag{9.2.9}$$

In the case of our interest, the initial approximation is the solution of the Lambert problem. Consequently, the value of the departure hyperbolic excess velocity on the initial approach trajectory will be non-zero, so it is rational to use the initial conditions of the form (9.2.6).

Having, for convenience, re-designated $\lambda_0 = \nu$, we obtain the final form of the initial conditions:

$$\begin{aligned} & \mathbf{r}(t_0) = \mathbf{r}_0(t_0), \\ & \mathbf{v}(t_0) = \mathbf{v}_0(t_0) + \mathbf{V}_{\infty 0}, \\ & m(t_0) = m_0, \\ & \mathbf{p_r}(t_0) = \mathbf{p}_{r0}, \\ & \mathbf{p_v}(t_0) = \lambda_0 \mathbf{V}_{\infty 0}, \\ & p_m(t_0) = p_{m0}. \end{aligned} \tag{9.2.10}$$

When using this form of initial conditions, the unknown parameters of the boundary value problem are:

$$\mathbf{V}_{\infty 0},\ \lambda_0,\ \mathbf{p}_{r0},\ \text{and}\ p_{m0}, \tag{9.2.11}$$

in total, 8 scalar parameters. When using continuation from the initial Keplerian arc into a low-thrust trajectory with a given value of the departure hyperbolic excess velocity $V_{\infty 0}^*$ at the starting point of the trajectory, the following equation must be satisfied:

$$\mathbf{V}_{\infty 0}^T \mathbf{V}_{\infty 0} - V_{\infty 0}^2(\tau) = 0 \tag{9.2.12}$$

where

$$V_{\infty 0}^2(\tau) = (1 - \tau)\left(V_{\infty 0}^{lamb}\right)^2 + \tau\left(V_{\infty 0}^*\right)^2, \tag{9.2.13}$$

$V_{\infty 0}^{lamb}$ is the value of the departure hyperbolic excess velocity obtained from the solution of the Lambert problem.

Thus, when using the proposed continuation scheme, 8 unknown parameters of the boundary value problem (9.2.11) and 1 boundary condition (9.2.12) are specified at the initial point.

9.4 The Problem of Thrust Minimization, the Domain of Solution Existence in the Plane Specific Impulse—Thrust and Automation of Solving the Problem of Calculating the Optimal Transfer

One of the fundamental problems in solving trajectory optimization problems for a spacecraft with finite thrust is the lack of a theorem on the solution existence. This problem is one of the reasons for the difficulty in constructing stable and efficient numerical optimization methods. In fact, if the numerical optimization method does not converge to the desired solution, there is no way to determine whether this is due to the failure of the numerical method itself or to the lack of a solution to the optimal control problem. Therefore, the problem of numerical determination of the solution existence domain is relevant, since it opens the way to the creation of practically regular numerical methods for optimizing the trajectories of spacecraft with finite thrust.

Let us consider a typical problem of trajectory optimization for a finite-thrust spacecraft, namely the problem of optimizing a heliocentric trajectory between two planets in a central Newtonian gravitational field for a fixed time Δt. Let us rewrite the spacecraft motion Eq. (7.5.4) in the form:

$$d\mathbf{r}/dt = \mathbf{v}, d\mathbf{v}/dt = \Omega_{\mathbf{r}} + \delta T/m \cdot \mathbf{e}, dm/dt = -\delta\beta. \quad (9.3.1)$$

where $\beta = T/c$ is the mass flow rate. It is assumed that the thrust and mass flow are not controlled, i.e. they can take either a given maximum value or a zero value.

In the problem under consideration, the initial time moment t_0 and the final time moment $t_f = t_0 + \Delta t$ are given. The initial conditions are of the form (7.4.12), and the final conditions are of the form (7.4.14).

The minimum propellant problem (LT-problem) requires the choice of control functions $\mathbf{e}(t)$ and $\delta(t)$ of the dynamic system (9.3.1), which should ensure the satisfaction of the boundary conditions (7.4.12), (7.4.14) and minimization of the cost function (7.5.5), which can be rewritten as:

$$J_0 = \int_{t_0}^{t_f} \delta\beta dt \to \min_{\mathbf{e}(t),\delta(t)}. \quad (9.3.2)$$

The domain of existence of a solution to this problem is limited by the minimum values of thrust and exhaust velocity. In fact, to perform such a transfer, a certain finite increment in the characteristic velocity is required, for which, in order to be realized in a fixed time $\Delta t = t_f - t_0$, a sufficient magnitude of thrust acceleration and, therefore, that of the thrust is required,. Since the permissible propellant mass flow rate is usually limited from above by a certain value $m_{p\max}$:

$$\int_{t_0}^{t_f} \delta\beta dt \leq m_{p\max} < m_0, \tag{9.3.3}$$

then the mass flow rate β must be sufficiently low, and therefore the exhaust velocity must be sufficiently high.

Moreover, for each mass flow rate $\beta \in [0;\beta_{\max}]$ there is a minimum thrust value $T_{\min}$ at which there is a solution to the problem (9.3.1), (9.3.2), (7.4.12), (7.4.14); besides:

$$\beta_{\max} = m_{p\max}/\Delta t \tag{9.3.4}$$

if $\delta(t) \equiv 1$. Obviously, this particular case of continuously acting thrust occurs during the flight with minimum thrust, since in the presence of the coasting sections on the optimal trajectory, the thrust can decrease until they disappear.

Thus, the minimum thrust depends of on the mass flow rate:

$$T_{\min}(\beta),\ \beta \in [0;\ \beta_{\max}], \tag{9.3.5}$$

and the solution to the LT problem (9.3.1), (9.3.2), (7.4.12), (7.4.14) exists for

$$T(\beta) \geq T_{\min}(\beta),\ \beta \in [0;\ \beta_{\max}]. \tag{9.3.6}$$

Condition (9.3.6) thus determines the domain of existence of a solution to problem (9.3.1), (9.3.2), (7.4.12), (7.4.14). The goal of the problem considered in this section is to construct a stable (regular) numerical algorithm for determining the boundary of the domain of existence of the solution (9.3.5). After its calculation, we can determine the corresponding parametric dependence of the minimum thrust $T_{\min}$ on the available exhaust velocity c, taking into account the relation:

$$c = T_{\min}(\beta)/\beta,\ c \in [c_{\min};\ \infty]. \tag{9.3.7}$$

The proposed approach to calculating the boundary of the domain of existence of a solution to the problem (9.3.1), (9.3.2), (7.4.12), (7.4.14) was first proposed in [3]. This section contains a detailed description of the methods for constructing the boundary of the domain of existence of solutions on the "exhaust velocity—thrust" plane and the contour plots of minimum thrust and characteristic velocity on the "launch date—flight duration" plane.

Thus, the problem of minimizing the amount of thrust for a flight in a given time between fixed points is considered. This problem is dual to the minimum-time problem, but the transversality condition is used to calculate the thrust instead of calculating the flight duration. The solution to the thrust minimization problem corresponds to the minimum-time trajectory for the thrust magnitude that ensures the transfer in a given time. It is easy to show that the thrust minimization problem is smooth due to the absence of thrust switching and fixed boundary conditions, even in the case of the moving start and end points (for example, planets), since the start and end times are fixed. This problem can be used to calculate the domain of existence of the optimal trajectory with thrust switching (9.3.1), (9.3.2), (7.4.12), (7.4.14) in the space of design-ballistic parameters $\{m_0, T, c, t_0, \Delta t\}$.

9.4.1 Thrust Acceleration Minimization Problem

Unfortunately, the thrust minimization problem itself has a bounded existence domain that is limited by the required propellant mass. Therefore, we will start constructing the boundary of the solution existence domain the with the case $\beta = 0$ (zero mass flow rate, infinite specific impulse). This case corresponds to a flight with a constant spacecraft mass m_0, and, consequently, with a constant thrust acceleration $a = T/m_0$. Let us consider the problem of minimizing the magnitude of such thrust acceleration. It is obvious that a solution to this problem exists for any non-degenerate boundary conditions.

Due to the constancy of thrust acceleration in the thrusting sections of the trajectory, there is a certain freedom in choosing the functional of the optimal control problem that minimizes the value of such constant thrust acceleration. Since the flight time is fixed and the thrust acceleration a is constant, the condition for the minimum of a coincides with the condition for the minimum of the cost function:

$$J_1 = \frac{1}{2}\int_{t_0}^{t_f} a^2 dt. \tag{9.3.8}$$

The equations of spacecraft motion with constant thrust acceleration have the form:

$$d\mathbf{r}/dt = \mathbf{v}, d\mathbf{v}/dt = \Omega_{\mathbf{x}} + \delta a\mathbf{e}. \tag{9.3.9}$$

The initial conditions have the form:

$$\mathbf{r}(t_0) = \mathbf{r}_0(t_0), \mathbf{v}(t_0) = \mathbf{v}_0(t_0), \tag{9.3.10}$$

and the final ones are of the form:

$$\mathbf{r}(t_f) = \mathbf{r}_f(t_f), \mathbf{v}(t_f) = \mathbf{v}_f(t_f), \tag{9.3.11}$$

By applying the maximum principle to the problem (9.3.8)–(9.3.11), we define the Pontryagin function:

$$H = -a^2/2 + \mathbf{p}_\mathbf{r}^\mathrm{T}\mathbf{v} + \mathbf{p}_\mathbf{v}^\mathrm{T}\Omega_\mathbf{r} + \delta a\mathbf{p}_\mathbf{v}^\mathrm{T}\mathbf{e}. \tag{9.3.12}$$

From the condition for the maximum of the Pontryagin function (9.3.12) with respect to **e** and δ, we obtain the optimal control:

$$\mathbf{e} = \mathbf{p}_\mathrm{v}/p_v, \delta \equiv 1. \tag{9.3.13}$$

Substituting equation (9.3.13) into (9.3.12) we obtain an expression for the Hamiltonian:

$$H = -a^2/2 + \mathbf{p}_\mathbf{r}^\mathrm{T}\mathbf{v} + \mathbf{p}_\mathbf{v}^\mathrm{T}\Omega_\mathbf{r} + ap_v, \tag{9.3.14}$$

which leads to the following system of differential equations for the optimal spacecraft motion:

$$d\mathbf{r}/dt = \mathbf{v}, d\mathbf{v}/dt = \Omega_\mathbf{r} + a\frac{\mathbf{p}_\mathbf{v}}{p_v}, d\mathbf{p}_\mathbf{r}/dt = -\Omega_\mathbf{rr}\mathbf{p}_\mathbf{v}, d\mathbf{p}_\mathbf{v}/dt = -\mathbf{p}_\mathbf{r}. \tag{9.3.15}$$

To obtain the condition for determining the minimum value of thrust acceleration, consider the extended functional:

$$J_1^* = \int_{t_0}^{t_f} \left[a^2/2 + \mathbf{p}_\mathbf{r}^\mathrm{T}(d\mathbf{r}/dt - \mathbf{v}) + \mathbf{p}_\mathbf{v}^\mathrm{T}(d\mathbf{v}/dt - \Omega_\mathbf{r} - \delta a\mathbf{e})\right]dt. \tag{9.3.16}$$

The minimum value of thrust acceleration can be obtained from the condition of zero variation of this functional with the variation of a:

$$\delta_a J_1^* = \partial J_1^*/\partial a \cdot \delta a = 0 \Rightarrow \partial J_1^*/\partial a = 0 \tag{9.3.17}$$

or

$$\partial J_1^*/\partial a = \int_{t_0}^{t_f} (a - \delta\mathbf{p}_\mathbf{v}^\mathrm{T}\mathbf{e})dt = \int_{t_0}^{t_f} (a - p_v)dt = 0.$$

Thus, the minimum thrust acceleration is achieved under the following condition:

$$\int_{t_0}^{t_f} p_v dt = \int_{t_0}^{t_f} a dt = a \cdot \left(t_f - t_0\right) = V_{ch},$$

or, finally,

$$\int_{t_0}^{t_f} p_v dt = a \cdot \Delta t. \tag{9.3.18}$$

Thus, to solve the considered problem of minimizing the constant thrust acceleration, it is necessary to find such initial values of the conjugate variables $\mathbf{p_r}(t_0)$, $\mathbf{p_v}(t_0)$ and such thrust acceleration a that the final conditions (9.3.11) and the condition (9.3.18) are satisfied.

In practice, to solve the problem of optimal control (9.3.15), (9.3.10), (9.3.11), (9.3.18), the method of continuous continuation (Newtonian homotopy) between the LP-problem and the minimum thrust acceleration problem was used.

The Newtonian homotopy between these two problems was realized by immersing the system of Eq. (9.3.15) in a one-parameter family:

$$d\mathbf{r}/dt = \mathbf{v},\, d\mathbf{v}/dt = \Omega_\mathbf{r} + \frac{1 - \tau + \tau a}{1 - \tau + \tau p_v}\mathbf{p_v},\, d\mathbf{p_r}/dt = -\Omega_{\mathbf{rr}}\mathbf{p_v},\, d\mathbf{p_v}/dt = -\mathbf{p_r}. \tag{9.3.19}$$

When $\tau = 0$, the Eq. (9.3.19) coincide with the equations of optimal motion of a spacecraft with an ideally controlled engine, and when $\tau = 1$ they coincide with the equations of optimal motion for the problem of minimizing thrust acceleration. To ensure continuity of the continuation, condition (9.3.18) is also immersed in the one-parameter family

$$\int_{t_0}^{t_f} p_v dt = (1 - \tau) \int_{t_0}^{t_f} p_v^{LP} dt + a\Delta t. \tag{9.3.20}$$

with the same parameterτ, where ${p_v}^{LP}$ is the absolute value of the primer vector for the LP-problem. To ensure that condition (9.3.20) is satisfied throughout the entire continuation interval $\tau \in [0; 1]$, the initial value of thrust acceleration a must be zero at $\tau = 0$.

The boundary value problem (9.3.19), (9.3.10), (9.3.11), (9.3.20) is solved by the continuation method (7.3.18).

9.4.2 Thrust Minimization Problem

After having calculated optimal trajectory with the minimum thrust acceleration (i.e. with $\beta = 0$), one can use the parameter continuation method to construct the desired dependence of the minimum thrust on the mass flow (9.3.5).

Similar to the case of calculating the minimum thrust acceleration, it can be noted that the minimum thrust is reached at the minimum value of the cost function:

$$J_2 = \frac{1}{2}\int_{t_0}^{t_f} T^2 dt, \tag{9.3.21}$$

since the flight time $\Delta t = t_f - t_0$ is fixed, and the thrust T is constant when the engine is operating.

The equations of motion have the form (9.3.1), and the boundary conditions have the form (7.4.12), (7.4.14).

Applying the Pontryagin maximum principle to the optimal control problem (9.3.1), (7.4.12), (7.4.14), (9.3.21), we write the Hamiltonian in the form:

$$H = -T^2/2 + \mathbf{p}_\mathbf{r}^\mathrm{T}\mathbf{v} + \mathbf{p}_\mathbf{v}^\mathrm{T}\boldsymbol{\Omega}_\mathbf{r} + \delta\big((T/m)\cdot\mathbf{p}_\mathbf{v}^\mathrm{T}\mathbf{e} - p_m\beta\big) \tag{9.3.22}$$

Maximization of the Hamiltonian (9.3.22) in the thrust direction **e** leads to the optimal control:

$$\mathbf{e} = \mathbf{p}_\mathrm{v}/p_v, \tag{9.3.23}$$

and substitution of (9.3.23) into (9.3.22) leads to the expression:

$$H = -T^2/2 + \mathbf{p}_\mathbf{r}^\mathrm{T}\mathbf{v} + \mathbf{p}_\mathbf{v}^\mathrm{T}\boldsymbol{\Omega}_\mathbf{r} + \delta((T/m)\cdot p_v - p_m\beta). \tag{9.3.24}$$

The Hamiltonian (9.3.24) leads to the following system of optimal motion equations:

$$\begin{aligned} &d\mathbf{r}/dt = \mathbf{v},\, d\mathbf{v}/dt = \boldsymbol{\Omega}_\mathbf{r} + (\delta T/m)\cdot\mathbf{p}_\mathbf{v}/p_v,\, dm/dt = -\delta\beta, \\ &d\mathbf{p}_\mathbf{r}/dt = -\boldsymbol{\Omega}_{\mathbf{rr}}\mathbf{p}_\mathbf{v},\, d\mathbf{p}_\mathbf{v}/dt = -\mathbf{p}_\mathbf{r},\, dp_m/dt = \left(\delta T/m^2\right)\cdot p_v. \end{aligned} \tag{9.3.25}$$

Since the spacecraft mass at the end point of the trajectory is not specified, it follows from the transversality condition that at this point the variable p_m conjugate to the mass is equal to 0:

$$p_m\left(t_f\right) = 0 \tag{9.3.26}$$

In addition, it follows from the last equation of the system (9.3.25) that the derivative dp_m/dt is non-negative, therefore, $p_m(t) \le 0$ on the optimal trajectory. It follows that if there is at least one thrusting section, the factor multiplying δ in the expression for the Hamiltonian (9.3.24) is positive. Thus, the condition of the Hamiltonian maximization (9.3.24) in δ leads to the optimal thrust function being identically equal to 1:

$$\delta(t) = 1. \tag{9.3.27}$$

Thus, as in the minimum thrust acceleration problem, the spacecraft engine operates constantly in the minimum thrust problem.

Substituting (9.3.27) into (9.3.25), we obtain the following equations of optimal motion:

$$\begin{aligned} &d\mathbf{r}/dt = \mathbf{v}, d\mathbf{v}/dt = \Omega_{\mathbf{r}} + (T/m) \cdot \mathbf{p}_{\mathbf{v}}/p_v, dm/dt = -\beta, \\ &d\mathbf{p}_{\mathbf{r}}/dt = -\Omega_{\mathbf{rr}}\mathbf{p}_{\mathbf{v}}, d\mathbf{p}_{\mathbf{v}}/dt = -\mathbf{p}_{\mathbf{r}}, dp_m/dt = \left(T/m^2\right) \cdot p_v. \end{aligned} \tag{9.3.28}$$

To derive the condition for thrust minimization, consider the extended functional of the original optimal control problem (9.3.1), (7.4.12), (7.4.14), (9.3.21):

$$\begin{aligned} J_2^* =& \frac{1}{2}\int_{t_0}^{t_f} T^2 dt + \int_{t_0}^{t_f} \mathbf{p}_{\mathbf{r}}^{\mathrm{T}}(d\mathbf{r}/dt - \mathbf{v})dt \\ &+ \int_{t_0}^{t_f} \mathbf{p}_{\mathbf{v}}^{\mathrm{T}}(d\mathbf{v}/dt - \Omega_{\mathbf{r}} - (\delta T/m) \cdot \mathbf{e})dt \\ &+ \int_{t_0}^{t_f} p_m \cdot (dm/dt + \delta\beta)dt. \end{aligned} \tag{9.3.29}$$

The minimum thrust value can be obtained from the condition that the variation of this functional is equal to zero with a variation of T:

$$\delta_T J_2^* = \partial J_2^*/\partial T \cdot \delta T = 0 \Rightarrow \partial J_2^*/\partial T = 0 \tag{9.3.30}$$

or

$$\partial J_2^*/\partial T = \int_{t_0}^{t_f} \left(T - \delta \mathbf{p}_{\mathbf{v}}^{\mathrm{T}}\mathbf{e}/m\right)dt = \int_{t_0}^{t_f} (T - p_v/m)dt = 0.$$

Thus, the minimum thrust is reached under the condition:

$$\int_{t_0}^{t_f} (p_v/m)dt = \int_{t_0}^{t_f} Tdt = T \cdot \left(t_f - t_0\right)$$

or, finally,

$$\int_{t_0}^{t_f} (p_v/m)dt = T \cdot \Delta t. \tag{9.3.31}$$

Consequently, the thrust minimization problem is reduced to the solution of a boundary value problem for a system of ordinary differential Eq. (9.3.28) with boundary conditions (7.4.12), (7.4.14), (9.3.26), (9.3.31). The unknown parameters of this boundary value problem are the initial values of the conjugate variables $\mathbf{p_r}(t_0)$, $\mathbf{p_v}(t_0)$, $p_m(t_0)$ and the minimum thrust value T (8 unknowns in total), which must take such values that the equalities (7.4.14), (9.3.26), (9.3.31) (8 equations in total) are satisfied.

The boundary value problem (9.3.28), (7.4.12), (7.4.14), (9.3.26), (9.3.31) is also solved by the method of continuation with respect to a parameter, which is similar to the method used to calculate trajectory with the minimum thrust acceleration. The difference is in the parameterization used and in the initial approximation.

The following expression is used in the system (9.3.28) instead of β:

$$\beta = \tau\beta_{\max}, \tag{9.3.32}$$

and the values of $\mathbf{p_r}(t_0)$, $\mathbf{p_v}(t_0)$ from the solution to the problem of the thrust acceleration minimization are used as an initial approximation,

$$T = m(t_0)a,$$

and

$$p_m(t_0) = -\frac{a}{m(t_0)} \int_{t_0}^{t_f} p_v dt,$$

where the value of a and the function $p_v(t)$ are also taken from the solution to the problem of minimizing thrust acceleration.

The given parameterization allows to guarantee the existence of a solution with minimum thrust, since for $\beta \in [0;\beta_{\max}]$ the final mass of the spacecraft, due to (9.3.3), (9.3.4), is greater than zero: $m(t_f) \geq m(t_0) - m_{p\max} > 0 \; \forall \; \tau \in [0;\, 1]$. Together with the use of the specified initial approximation, this parameterization ensures that all boundary conditions (7.4.12), (7.4.14), (9.3.26), (9.3.31) are satisfied in the process of variation of the continuation parameterτ from 0 to 1. Thus, in the continuation process the continuous dependence of the minimum thrust T on the mass flow$\beta = \tau\beta_{max}$: $0 \le \beta \le \beta_{max}$ is calculated for each $0 \le \tau \le 1$.

9.4.3 Algorithm for Calculating the Boundary of the Solution Domain

Thus, the following algorithm is proposed for calculating the boundary of the domain of existence of a solution to the problem of flight between two planets in a fixed time with thrust switching (9.3.1), (7.4.12), (7.4.14), (9.3.21):

(1) For a spacecraft with an ideally controlled engine, the optimal LP-trajectory for the given time is calculated with the given boundary conditions. To solve this problem, practically regular methods presented in Sect. 10 and in [7] can be used.
(2) The problem of the constant thrust acceleration minimization (9.3.8)–(9.3.11) is solved by the method of continuation from the optimal LP-trajectory presented in Sect. 9.3.1.
(3) The trajectory with the minimum thrust acceleration is continued into the trajectory with the minimum thrust and maximum permissible mass flow β_{max} using the method presented in Sect. 9.3.2. In the process of this continuation, the minimum thrust dependence on the mass flow (9.3.5) is calculated, which is the desired boundary of the domain of existence of a solution to the problem with the thrust switching.

It should be noted that the calculation of the optimal LP-trajectory with an ideally controlled engine is, generally speaking, not necessary. It is possible to construct a stable scheme of continuation into a trajectory with the minimum thrust acceleration by the continuation from the coasting motion (for example, from the motion along the orbit of the departure planet or from the solution of the Lambert problem).

In addition, this method is applicable to other problems with different boundary conditions, such as the flyby problem, the problem with a non-zero departure hyperbolic excess velocity, the problems with an explicit dependence of thrust and exhaust velocity on state coordinates and time, and a number of other problems. For example, to solve the problem with the non-zero departure hyperbolic excess velocity $V_{\infty 0}$ with a value less than the critical one, it is sufficient to replace the initial conditions (7. 4.12) by (7.4.13).

9.5 Joint Optimization of Control, Main Trajectory and Design Parameters of a Spacecraft with a Limited Thrust Engine

This section discusses the problem of joint optimization of the heliocentric trajectory of a spacecraft with an electric propulsion system and the main design parameters of the spacecraft. A combined scheme of interplanetary transfer with a fixed duration between two planets is considered. In this scheme, a spacecraft with an electric propulsion system is launched onto a departure trajectory from the planet of departure by a launch vehicle with the high-thrust propulsion systems. Either the initial mass of the spacecraft or the dependence of the initial mass of the spacecraft on the departure trajectory on the departure hyperbolic excess velocity is given (such dependencies, which are presented in tabular form in the user manuals for launchers [8, 9], are approximated using the 6th-order polynomials). The heliocentric trajectory of the spacecraft is formed by the electric propulsion system. The problem of optimizing the trajectories of direct heliocentric flights between two planets is considered under the assumptions of the ZSOI method [10–12]; the motion within the sphere of action of the destination planet is not analyzed. Trajectory optimization, in addition to the calculation of the optimal control (program for controlling the thrust vector of the electric propulsion system), may include optimization of a number of trajectory parameters (departure date, magnitude and direction of the departure hyperbolic excess velocity). Simultaneous optimization of control, trajectory and design parameters allows us to get an idea of the best option for the designed space mission and draw some conclusions about the properties of such optimal solutions and the possibility of their use.

The necessary conditions for the optimality of the main design parameters are presented for cases where the electric power of the electric propulsion system is constant or depends on the heliocentric distance of the spacecraft. The first case corresponds to a single-mode electric propulsion system with a nuclear or solar power supply system (PSS). In this case, if a solar PSS is used, its power must be sufficient to ensure operation of the electric propulsion system at the maximum required heliocentric distance. The second case corresponds to an electric propulsion system, the power and thrust of which are continuously adjusted as the available power of the on-board solar PSS changes.

9.5.1 *Cost Function, Motion Equations and Control*

Let us consider the problem of joint optimization of the main design parameters of a spacecraft and the trajectory of a direct heliocentric flight between two given planets in a fixed time using a simple design model of a spacecraft with an electric propulsion system, presented in Sect. 7.6. The task of the joint trajectory optimization is to maximize the useful mass of the spacecraft m_u, which is equivalent to minimizing the cost function:

$$J_1 = -m_u = -m_0(V_{\infty 0}) + \left[\gamma_1 + \gamma_2 \rho(r_{\min}, r_{\min})\right] P_{BOL} + (1 + a_t) m_p. \quad (9.4.1)$$

Since, with the accepted assumptions, all specific masses included in this formula are given and constant, and the equation for the spacecraft mass m variation can be written in the form

$$\frac{dm}{dt} = -\delta \frac{T}{c} = -\delta \frac{2\eta P(P_{BOL}, r, r_{\min})}{c^2}, \quad (9.4.2)$$

the functional J_1 minimization is equivalent to the minimization of the cost function

$$J_2 = -m_0(V_{\infty 0}) + \gamma_3 P_{BOL} + (1 + a_t) \int_{t_0}^{t_f} \delta \frac{2\eta P(P_{BOL}, r, r_{\min})}{c^2} dt, \quad (9.4.3)$$

where, in general, $\gamma_3 = \gamma_1 + \gamma_2 \rho(r_{\min}, r_{\min})$, $P(P_{BOL}, r, r_{\min}) = P_{BOL} \rho(r, r_{\min})$.

Let us reduce the optimal control problem to the Lagrange form, for which we introduce the non-integral term of the functional J_2, which is constant on the trajectory, under the integral sign, having previously divided all terms on the right side of the expression for J_2 by a positive constant $(1 + a_t)$. As a result, we obtain a cost function equivalent to the J_2, which will be used in further calculations:

$$J = \int_{t_0}^{t_f} \left[-\frac{m_0(V_{\infty 0})}{\Delta t \cdot (1 + a_t)} + \gamma P_{BOL} + \delta \frac{2\eta P(P_{BOL}, r, r_{\min})}{c^2} \right] dt, \quad (9.4.4)$$

where $\gamma = \gamma_3 / [\Delta t \cdot (1 + a_t)] = \left[\gamma_1 + \gamma_2 \rho(r_{\min}, r_{\min})\right] / [\Delta t \cdot (1 + a_t)]$,$\Delta t = t_f - t_0$ is the fixed flight time.

The equations of motion of a spacecraft with an electric propulsion system on the heliocentric section of the trajectory have the form:

$$\frac{d\mathbf{r}}{dt} = \mathbf{v}, \; \frac{d\mathbf{v}}{dt} = \Omega_{\mathbf{r}} + \delta \frac{T}{m} \mathbf{e}, \; \frac{dm}{dt} = -\delta \frac{T}{c}. \quad (9.4.5)$$

The spacecraft motion can be controlled by selecting the thrust vector orientation program $\mathbf{e}(t)$ and the electric propulsion on/off program $\delta(t)$. The goal of control is to achieve a given finite manifold in space of state variables for a given flight time with a minimum value of the cost function (9.4.4).

9.5.2 Conditions for Optimal Power and Exhaust Velocity

In the general case, the initial electric power of the propulsion system P_{BOL}, the heliocentric distance of the spacecraft $r_{\min}$, at which the maximum power of the solar power supply and the propulsion systems of the spacecraft is reached, and the EPS exhaust velocity c are the design parameters to be optimized. To derive optimality conditions for these quantities, we introduce the corresponding dummy phase variables. Since these optimized quantities are constant along the trajectory, the differential equations describing their variation have a trivial form.

After replacing thrust by its expression in terms of the electric power consumed, exhaust velocity and thrust efficiency, we obtain the following complete system of differential equations of the considered dynamic system in the heliocentric inertial coordinate system:

$$\begin{aligned} &\frac{d\mathbf{r}}{dt} = \mathbf{v},\ \frac{d\mathbf{v}}{dt} = \Omega_{\mathbf{r}} + \delta\frac{2\eta P}{mc}\mathbf{e},\ \frac{dm}{dt} = -\delta\frac{2\eta P}{c^2}, \\ &\frac{dP_{BOL}}{dt} = 0,\ \frac{dr_{\min}}{dt} = 0,\ \frac{dc}{dt} = 0, \end{aligned} \tag{9.4.7}$$

where $P = P(P_{BOL}, r, r_{\min}) = P_{BOL}\rho(r, r_{\min})$.

The Pontryagin function of the optimal control problem under consideration has the form:

$$H = \frac{m_0(V_{\infty 0})}{\Delta t \cdot (1 + a_t)} - \gamma P_{BOL} - \delta\frac{2\eta P}{c^2} + \mathbf{p}_{\mathbf{r}}^{\mathrm{T}}\mathbf{v} + \mathbf{p}_{\mathbf{v}}^{\mathrm{T}}\Omega_{\mathbf{r}} + \delta\frac{2\eta P}{mc}\mathbf{p}_{\mathbf{v}}^{\mathrm{T}}\mathbf{e} - \delta\frac{2\eta P}{c^2}p_m, \tag{9.4.8}$$

where $\mathbf{p_r}$, $\mathbf{p_v}$, p_m are the variables conjugate to $\mathbf{r}$, $\mathbf{v}$ and m, respectively. Maximizing the Pontryagin function in the δ and $\mathbf{e}$ control, we obtain the Hamiltonian of the optimal control problem:

$$H = \frac{m_0(V_{\infty 0})}{\Delta t \cdot (1 + a_t)} - \gamma P_{BOL} + \mathbf{p}_{\mathbf{r}}^{\mathrm{T}}\mathbf{v} + \mathbf{p}_{\mathbf{v}}^{\mathrm{T}}\Omega_{\mathbf{r}} + \delta\frac{2\eta P}{c}S, \tag{9.4.9}$$

and the optimal control itself:

$$\mathbf{e} = \frac{\mathbf{p_v}}{p_v},\ \delta = \begin{cases} 1,\ S > 0, \\ 0,\ S < 0, \end{cases} \tag{9.4.10}$$

where

$$S = \frac{p_v}{m} - \frac{p_m + 1}{c} \tag{9.4.11}$$

is the switching function.

The optimal motion equations have the form:

$$\begin{aligned}
&\frac{d\mathbf{r}}{dt} = \mathbf{v},\ \frac{d\mathbf{v}}{dt} = \boldsymbol{\Omega}_{\mathbf{r}} + \delta\frac{2\eta P}{mc}\frac{\mathbf{p_v}}{p_v},\ \frac{dm}{dt} = -\delta\frac{2\eta P}{c^2},\\
&\frac{dP_{BOL}}{dt} = 0,\ \frac{dr_{\min}}{dt} = 0,\ \frac{dc}{dt} = 0,\\
&\frac{d\mathbf{p_r}}{dt} = -\boldsymbol{\Omega}_{\mathbf{rr}}\mathbf{p_v} - \delta\frac{2\eta}{c}P_{BOL}\frac{\partial\rho(r, r_{\min})}{\partial r}\frac{\mathbf{r}}{r}S,\ \frac{d\mathbf{p_v}}{dt} = -\mathbf{p_r},\ \frac{dp_m}{dt} = \delta\frac{2\eta P}{c}\frac{p_v}{m^2},\\
&\frac{dp_{P\,BOL}}{dt} = \gamma(r_{\min}) - \delta\frac{2\eta}{c}\rho(r, r_{\min})S,\ \frac{dp_c}{dt} = \delta\frac{2\eta}{c^2}P\cdot\left(\frac{p_v}{m} - 2\frac{p_m+1}{c}\right),\\
&\frac{dp_{r\min}}{dt} = \frac{\partial\gamma(r_{\min})}{\partial r_{\min}}P_{BOL} - \delta\frac{2\eta}{c}P_{BOL}\frac{\partial\rho(r, r_{\min})}{\partial r_{\min}}S,
\end{aligned} \tag{9.4.12}$$

where $p_{P\,BOL}, p_c, p_{r\min}$ are variables that are conjugated to P_{BOL}, c and $r_{\min}$, respectively. Here, taking into account the expression for γ, $\frac{\partial\gamma(r_{\min})}{\partial r_{\min}} = \frac{\gamma_2}{\Delta t\cdot(1+a_t)}\frac{\partial\rho(r_{\min}, r_{\min})}{\partial r_{\min}}$.

Since the values of P_{BOL}, $r_{\min}$ and c are optimized, they are free at both ends of the trajectory. Therefore, due to the transversality conditions, the variables conjugated to them must take zero values at the left and right ends of the trajectory:

$$\begin{aligned}
&p_{P\,BOL}(t_0) = 0, p_c(t_0) = 0, p_{r\min}(t_0) = 0,\\
&p_{P\,BOL}(t_f) = 0, p_c(t_f) = 0, p_{r\min}(t_f) = 0.
\end{aligned} \tag{9.4.13}$$

These conditions must be added to the initial and final conditions of the maximum principle boundary value problem in the case of optimizing P_{BOL}, $r_{\min}$ and c together with the trajectory. Obviously, the above optimality conditions for P_{BOL}, $r_{\min}$ and c can be rewritten in terms of the equality of the following integrals to zero:

$$\begin{aligned}
&\int_{t_0}^{t_f}\left[\gamma(r_{\min}) - \delta\frac{2\eta}{c}\rho(r, r_{\min})S\right]dt = 0,\\
&\int_{t_0}^{t_f}\left[\delta\frac{2\eta}{c^2}P\cdot\left(\frac{p_v}{m} - 2\frac{p_m+1}{c}\right)\right]dt = 0,\\
&\int_{t_0}^{t_f}\left[\frac{\partial\gamma(r_{\min})}{\partial r_{\min}}P_{SA\,BOL} - \delta\frac{2\eta}{c}P_{SA\,BOL}\frac{\partial\rho(r, r_{\min})}{\partial r_{\min}}S\right]dt = 0.
\end{aligned} \tag{9.4.14}$$

Obviously, when considering a single-mode electric propulsion system, the necessary optimality conditions for $r_{\min}$ should be excluded from consideration.

9.5.3 Boundary Conditions and Optimality Conditions for Trajectory Parameters

Within the framework of the ZSOI method, the durations of planetocentric sections are considered to be zero, the position vector of the spacecraft at the moment of the beginning or end of the heliocentric section of the trajectory coincides with the heliocentric radius vector of the planet, and the spacecraft velocity vector at these moments is equal to the sum of the vectors of the heliocentric velocity of the planet and the hyperbolic velocity excess. Therefore, at the initial time moment t_0 the boundary conditions can be written as:

$$\mathbf{r}(t_0) = \mathbf{r}_0(t_0),\ \mathbf{v}(t_0) = \mathbf{v}_0(t_0) + \mathbf{V}_{\infty 0},\ m(t_0) = m_0(V_{\infty 0}) \tag{9.4.15}$$

where $\mathbf{r}_0$ and $\mathbf{v}_0$ are the position and velocity vectors of the planet of departure, respectively, $\mathbf{V}_{\infty 0}$ is the vector of the departure hyperbolic excess velocity.

If the value of the departure hyperbolic excess velocity $V_{\infty 0} = |\mathbf{V}_{\infty 0}|$ at the initial moment t_0 is given and is not equal to 0, then the condition for the initial vector of the heliocentric velocity of the spacecraft from (9.4.15) can be rewritten as:

$$\psi = \frac{1}{2}\left[\mathbf{v}(t_0) - \mathbf{v}_{pl0}(t_0)\right]^2 - \frac{1}{2}V_{\infty 0}^2. \tag{9.4.16}$$

The boundary condition (9.4.16) leads to the transversality condition:

$$\mathbf{p}_{\mathbf{v}}(t_0) = \mu_{V0} \cdot \left(\frac{\partial \psi}{\partial \mathbf{v}(t_0)}\right)^T = \mu_{V0} \cdot \left[\mathbf{v}(t_0) - \mathbf{v}_{pl0}(t_0)\right] = \mu_{V0}\mathbf{V}_{\infty 0} \tag{9.4.17}$$

where μ_{V0} is the Lagrange multiplier.

For sufficiently small values of $\mathrm{V}_{\infty 0}$ (more precisely, for $V_{\infty 0}$ less than some critical value), the Lagrange multiplier μ_{V0} is positive; so, the Eq. (9.4.17) can be rewritten as:

$$\mathbf{V}_{\infty 0} = V_{\infty 0}\frac{\mathbf{p}_v(t_0)}{p_v(t_0)}, \tag{9.4.18}$$

that is, the vector of the departure hyperbolic excess velocity must be collinear with the vector $\mathbf{p}_{\mathbf{v}}(t_0)$. The system of initial conditions (9.4.15) then has the form:

$$\mathbf{r}(t_0) = \mathbf{r}_0(t_0),\ \mathbf{v}(t_0) = \mathbf{v}_0(t_0) + V_{\infty 0}\frac{\mathbf{p}_{\mathbf{v}}(t_0)}{p_v(t_0)},\ m(t_0) = m_0(V_{\infty 0}). \tag{9.4.19}$$

In the case of zero departure hyperbolic excess velocity, the initial conditions (9.4.19) are simplified:

$$\mathbf{r}(t_0) = \mathbf{r}_0(t_0),\ \mathbf{v}(t_0) = \mathbf{v}_0(t_0),\ m(t_0) = m_0(0). \tag{9.4.20}$$

In general, for an arbitrary (possibly large enough) value of $V_{\infty 0}$, the initial conditions (9.4.19) should be rewritten in the form:

$$\mathbf{r}(t_0) = \mathbf{r}_0(t_0),\ \mathbf{v}(t_0) = \mathbf{v}_0(t_0) + \mathbf{V}_{\infty 0},\ m(t_0) = m_0(V_{\infty 0}),\ \mathbf{p}_\mathbf{v}(t_0) = \mu_{V0}\mathbf{V}_{\infty 0},\ \mathbf{V}_{\infty 0}^T\mathbf{V}_{\infty 0} - V_{\infty 0}^2 = 0, \tag{9.4.21}$$

In this case, the Lagrange multiplier μ_{V0} is added to the decision vector of the boundary value problem, and $V_{\infty 0}$ can be used instead of $\mathbf{p}_\mathbf{v}(t_0)$.

Let us derive the conditions necessary for the optimality of the departure hyperbolic excess velocity vector, considering the known dependence of the initial spacecraft mass m_0 in the departure trajectory on the departure hyperbolic excess velocity $V_{\infty 0}$:

$$m(t_0) = m_0(V_{\infty 0}). \tag{9.4.22}$$

Let us introduce a dummy state variable $V_{\infty 0}$:

$$\frac{dV_{\infty 0}}{dt} = 0. \tag{9.4.23}$$

The differential equation for a variable conjugated to $V_{\infty 0}$ has the form:

$$\frac{dp_{V\infty 0}}{dt} = -\frac{\partial H}{\partial V_{\infty 0}} = -\frac{1}{\Delta t \cdot (1 + a_t)}\frac{\partial m_0}{\partial V_{\infty 0}}. \tag{9.4.24}$$

At the final moment t_f, the value of the variable $V_{\infty 0}$ is not fixed, therefore:

$$p_{V\infty 0}(t_f) = 0. \tag{9.4.25}$$

Since the right-hand side of (9.4.24) is constant, then

$$p_{V\infty 0}(t_0) = \frac{1}{1 + a_t}\frac{\partial m_0}{\partial V_{\infty 0}}. \tag{9.4.26}$$

At the initial time moment t_0, the following conditions must be satisfied with respect to the departure hyperbolic excess velocity:

$$\boldsymbol{\psi} = \begin{pmatrix} (\mathbf{v}(t_0) - \mathbf{v}_0(t_0))^T(\mathbf{v}(t_0) - \mathbf{v}_0(t_0)) - V_{\infty 0}^2 \\ m(t_0) - m_0(V_{\infty 0}) \end{pmatrix} = 0. \tag{9.4.27}$$

It follows that at $t = t_0$ the values of the variables conjugated to velocity, departure hyperbolic excess velocity and mass must satisfy the following conditions:

$$\mathbf{p}_v(t_0) = 2\mu_1(\mathbf{v}(t_0) - \mathbf{v}_0), p_{V\infty0}(t_0) = -2\mu_1 V_{\infty0} - \mu_2\frac{\partial m_0}{\partial V_{\infty0}}, p_m(t_0) = \mu_2, \tag{9.4.28}$$

where μ_1, μ_2 are the Lagrange multipliers.

Taking (9.4.26) into account, the second and the third equations of (9.4.28) allow us to obtain the following expression for the Lagrange multiplier μ_1:

$$\mu_1 = -\frac{1}{2V_{\infty0}}\left[p_m(t_0) + \frac{1}{1+a_t}\right]\frac{\partial m_0}{\partial V_{\infty0}}. \tag{9.4.29}$$

Substituting this expression into the first equation of (9.4.28), we obtain the necessary optimality condition for the optimal vector of the departure hyperbolic excess velocity $\mathbf{V}_{\infty0} = \mathbf{v}(t_0) - \mathbf{v}_0$:

$$\mathbf{p}_v(t_0) = -\frac{1}{V_{\infty0}}\left[p_m(t_0) + \frac{1}{1+a_t}\right]\frac{\partial m_0}{\partial V_{\infty0}}\mathbf{V}_{\infty0}. \tag{9.4.30}$$

As in the case of the given value of the departure hyperbolic velocity vector, the vector conjugate to the velocity on the optimal trajectory must be parallel to the vector of the initial (departure) hyperbolic excess velocity. The optimal value of the departure hyperbolic excess velocity is determined using the condition resulting from (9.4.30) in the form:

$$p_v(t_0) = -\left[p_m(t_0) + \frac{1}{1+a_t}\right]\frac{\partial m_0}{\partial V_{\infty0}}. \tag{9.4.31}$$

The boundary conditions at the final time moment t_f for the rendezvous problem (approaching the destination planet with zero hyperbolic excess velocity) have the form:

$$\mathbf{r}(t_f) = \mathbf{r}_f(t_f), \mathbf{v}(t_f) = \mathbf{v}_f(t_f) \tag{9.4.32}$$

where $\mathbf{r}_f$, $\mathbf{v}_f$ are the position and velocity vectors of the destination planet, while for the flyby problem (approaching the destination planet with an arbitrary hyperbolic excess of velocity) such boundary conditions have the following form:

$$\mathbf{r}(t_f) = \mathbf{r}_f(t_f). \tag{9.4.33}$$

In the latter case, the final conditions (9.4.33) must be supplemented by the transversality conditions:

$$\mathbf{p}_\mathbf{v}(t_f) = 0. \tag{9.4.34}$$

If the moment t_0 of the departure from the planet is not specified, but the flight time Δt is fixed, the transversality condition for computing the optimal t_0 value has the following form:

$$\mathbf{p}_{\mathbf{r}0}^{\mathrm{T}}\mathbf{V}_{\infty 0} + \mathbf{p}_{\mathbf{v}0}^{\mathrm{T}}(\boldsymbol{\Omega}_{\mathbf{r}0} - \mathbf{a}_0) - \mathbf{p}_{\mathbf{v}f}^{\mathrm{T}}(\boldsymbol{\Omega}_{\mathbf{r}f} - \mathbf{a}_f) + \delta_0 T_0 S_0 - \delta_f T_f S_f = 0 \qquad (9.4.35)$$

for the rendezvous problem, and

$$\mathbf{p}_{\mathbf{r}0}^{\mathrm{T}}\mathbf{V}_{\infty 0} + \mathbf{p}_{\mathbf{v}0}^{\mathrm{T}}(\boldsymbol{\Omega}_{\mathbf{r}0} - \mathbf{a}_0) - \mathbf{p}_{\mathbf{r}f}^{\mathrm{T}}\mathbf{V}_{\infty f} - \mathbf{p}_{\mathbf{v}f}^{\mathrm{T}}\boldsymbol{\Omega}_{\mathbf{r}f} + \delta_0 T_0 S_0 - \delta_f T_f S_f = 0, \qquad (9.4.36)$$

for the flyby problem. Here the values with the subscript "0" are calculated at $t = t_0$, and those with the subscript "f" are calculated at $t = t_f = t_0 + \Delta t$, $\mathbf{a}_0$ is the acceleration vector of the departure planet at the moment of departure t_0, $\mathbf{a}_f$ is the acceleration vector of the target planet at the moment of arrival t_f, $\mathbf{V}_{\infty f} = \mathbf{v}(t_f) - \mathbf{v}_f(t_f)$ (for the flyby problem). If the mathematical models of the gravitational field, in which the planets and spacecraft move, are the same, then the expressions (9.4.35), (9.4.36) can be simplified taking into account the equalities $\boldsymbol{\Omega}_{\mathbf{r}0} - \mathbf{a}_0 = \boldsymbol{\Omega}_{\mathbf{r}f} - \mathbf{a}_f \equiv 0$.

In all considered options, the final value of the spacecraft mass $m(t_f)$ is not specified, therefore, the condition of transversality on the variable conjugate to the mass must be added to the final conditions:

$$p_m(t_f) = 0. \qquad (9.4.37)$$

9.5.4 Boundary Value Problem and Method for Its Solution

The optimization problem under consideration has been reduced to a boundary value problem for a system of ordinary differential Eq. (9.4.12) with the initial conditions (9.4.19), (9.4.20) or (9.4.21), to which the necessary additional initial conditions from the first three equations of (9.4.13) are added, if the optimization of P_{BOL}, c and $r_{\min}$ is necessary. If it is necessary to optimize $V_{\infty 0}$, (9.4.31) is added to the initial conditions. The final conditions are (9.4.32) or (9.4.33), (9.4.34), supplemented by the transversality conditions (9.4.37), and, if necessary, by the additional transversality conditions at the final moment of time for the variables conjugated to P_{BOL}, c and $r_{\min}$ from the last three equations of (9.4.13). If it is necessary to optimize t_0, the mixed conditions (9.4.35) or (9.4.36) containing the values of the state variables at both ends of the trajectory are added to the listed initial and final conditions.

The unknown parameters of the boundary value problem are the values of the conjugate variables $\mathbf{p}_\mathbf{r}$, $\mathbf{p}_\mathbf{v}$ and p_m at the initial time t_0 and, if necessary, the values of the main design parameters P_{BOL}, c and $r_{\min}$ and of the trajectory parameters $V_{\infty 0}$ and t_0. To satisfy the necessary optimality conditions, it is necessary to calculate such values of these parameters that six final conditions (9.4.32) for the rendezvous problem or six boundary conditions (9.4.33), (9.4.34) for the flyby problem, one transversality condition (9.4.37) and, if necessary, three transversality conditions for P_{BOL}, c and $r_{\min}$ at time t_f (the last three equations of (9.4.13)), the $V_{\infty 0}$ optimality condition (9.4.31) and the t_0 optimality condition (9.4.35) or (9.4.36) are satisfied. Thus, the order of the boundary value problem (the number of unknown variables and the number of boundary conditions that must be satisfied) can vary from 7, if only the control is optimized, to 12, if the control and all considered design and trajectory parameters are jointly optimized.

To solve the boundary value problem, the method of continuation with respect to the parameter (7.3.18) is used. As an initial guess values for the unknown parameters of the boundary value problem, the solution of the LP-problem is used (see Sect. 7. 4 and Chapter 8); if necessary, the method of continuation with respect to the gravitational parameter is used to calculate the extremal with a given number of integer revolutions around the Sun (see Sect. 8.2). As an initial guess values for the exhaust velocity, its average value for a LP-solution with the given power can be taken. The initial guess for $V_{\infty 0}$ can be chosen to be zero, and for t_0—an arbitrary value within the range of variation of the departure date of interest. The question of choosing the initial guess for $r_{\min}$ turned out to be quite difficult; in practice, this initial guess was chosen manually to ensure the convergence of the solution to the boundary value problem. On the right-hand sides of the system of differential equations for optimal motion, a homotopy was implemented between the right-hand sides of differential equations for LP- and LT-problems in the form (9.1.15). At the initial value of the continuation parameter $\tau = 0$ the right-hand sides of these differential equations coincide with the right-hand sides of the differential equations for a spacecraft with an ideally controlled engine, and at the final value of the continuation parameter $\tau =$ 1—with the right-hand sides of the system (9.4.12).

To regularize the continuation process when solving a boundary value problem with discontinuous control, the thrust function smoothing of the form (9.1.12) is used:

$$\delta_\varepsilon(S, \varepsilon) = \frac{1}{2}\left(\frac{S}{|S| + \varepsilon} + 1\right)$$

Here ε is the regularizing (smoothing) term. The function $\delta_\varepsilon(x, \varepsilon)$ is smooth for any $> \varepsilon\ 0$ and tends to the Heaviside function with a discontinuity at the point $x = 0$ for $\varepsilon \to 0$. In turn, ε is represented as a linear function of τ: $\varepsilon(\tau) = (1 - \tau)\varepsilon_0 + \tau\varepsilon_f$, where ε_0, ε_f are the parameters regulating the degree of smoothing at the beginning and at the end of the continuation process.

The right-hand sides of the optimal motion equations include the function $\rho(r, r_{\min})$, which has a corner point at $r = r_{\min}$, and its derivative, which has a discontinuity at this point. To facilitate the solution of the boundary value problem of the maximum principle, a smooth approximation of this function of the following form is used:

$$\rho(r, r_{\min}) = r_{\min}^{-k}\delta_\varepsilon(r_{\min} - r, \varepsilon) + r^{-k}\delta_\varepsilon(r - r_{\min}, \varepsilon).$$

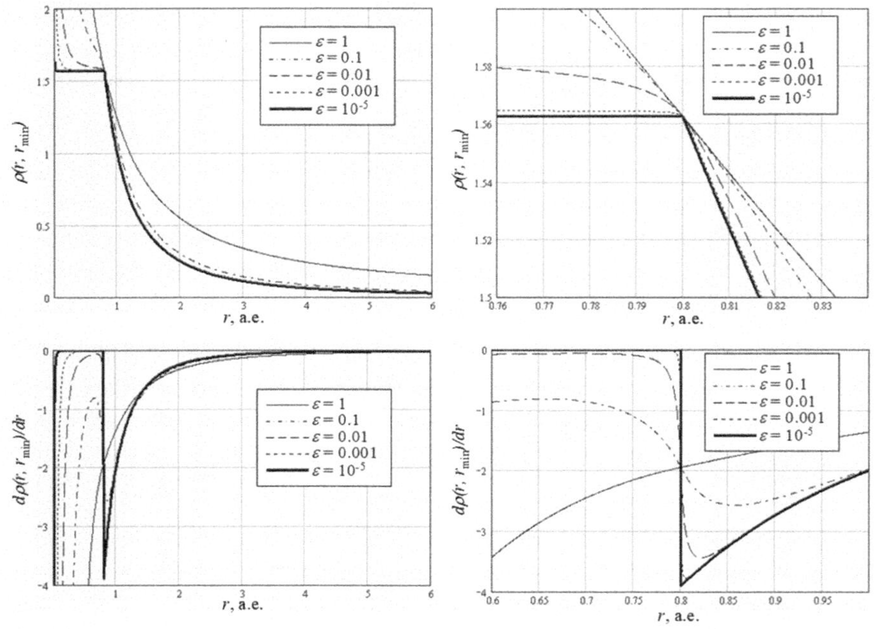

Fig. 9.6 Dependence of the function $\rho(r, r_{\min})$ and of its r-derivative on r with $k = 2$, $r_{\min} = 0.8$

The dependencies of the smoothed representation of the function $\rho(r, r_{\min})$ and of its derivative on r are shown in Fig. 9.6. One can see the smoothness of these dependencies and the practical indistinguishability of the smoothed function from the original one at $\varepsilon = 10^{-5}$.

9.6 Numerical Examples

Let's consider examples of optimization performed using the described methods and calculation tools. The first set of numerical examples demonstrates the continuation of optimal LT trajectories from LP trajectories using the method described in Sect. 9.1.

The first example is the rendezvous with the asteroid 2000 WO107 (semimajor axis $a = 0.9114029$ AU, eccentricity $e = 0.78065556$, inclination $i = 7.7829°$). The mission under consideration has the following parameters: launch date February 1, 2017, flight time 1300 days, asymptotic departure velocity 0 m/s; initial mass of the spacecraft 500 kg. In the case of the LT problem, the thrust is 80 mN, the specific impulse is 4000 s. Figure 9.7 shows the projection of the trajectory onto the ecliptic plane (for LP and LT problems). The corresponding time dependencies of the thrust and switching function are shown in Fig. 9.9. The trajectory has 7 thrusting and 7 coasting sections; the first and the sixth thrusting sections and the last coasting

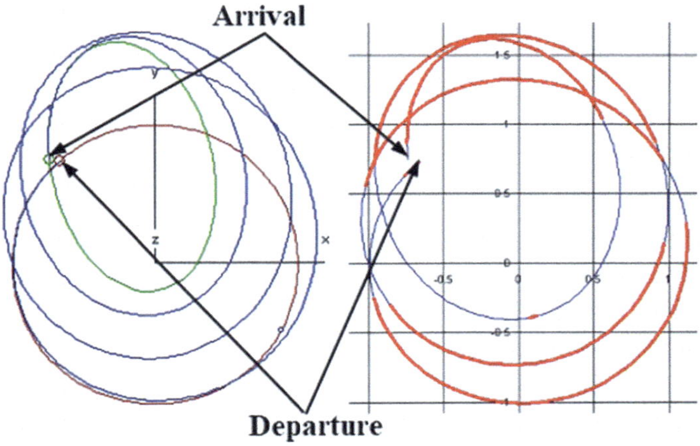

Fig. 9.7 Asteroid 2000 WO107 rendezvous—a projection of the trajectory onto the plane of the ecliptic. On the left—LP-trajectory, on the right—LT-trajectory, thick lines correspond to the operating EPS

section are very small. The evolution of the thrust versus time as the LP trajectory continues into the LT trajectory is shown in Fig. 9.8. With the τ growth, the thrust becomes a step function.

Figures 9.10, 9.11, 9.12 and 9.13 show the Mercury rendezvous trajectories with 10 full revolutions. Figures 9.10 and 9.11 correspond to the constant jet power case, and Figs. 9.12 and 9.13 show trajectories for the $P_b = P_{b0}r^{-1.7}$ case. Testing of the programs included the calculation of very long flights to Mercury with 50 full orbits (Figs. 9.14 and 9.15) and the Jupiter rendezvous with 10 full orbits (Figs. 9.16 and 9.17). It is worth paying attention to the significant difference in the shapes of the LP and LT trajectories in the last example.

As the next example demonstrating the multi-extremal nature of optimal trajectories that is associated with a different number of orbits, let us consider the problem of the direct flight of a spacecraft with a NEPS (electric power of 200 kW) from Earth to Jupiter. The initial hyperbolic excess velocity is assumed to be 0 m/s, the initial spacecraft mass on the trajectory of departure from the Earth is 16150 kg, the thrust of the electric propulsion system is 6.4527757 N and the specific impulse is 4500 s. The departure date is assumed to be 02/05/2017, and the flight duration is 1200 days. The Jupiter is approached with zero hyperbolic excess velocity (the rendezvous problem).

For the formulated problem, 5 local extrema were found that satisfy the necessary conditions of the maximum principle with the number of full orbits around the Sun ranging from 0 to 4. The projections of the found trajectories onto the ecliptic plane are presented in Fig. 9.18. Figure 9.19 shows the time dependences of the switching and thrust functions. It can be seen that in all cases, except for the solution with 1 full orbit, the flight duration is 1200 days. For a solution with 1 full orbit, there is an initial coasting section with a duration of 221.602279 days, which, with zero

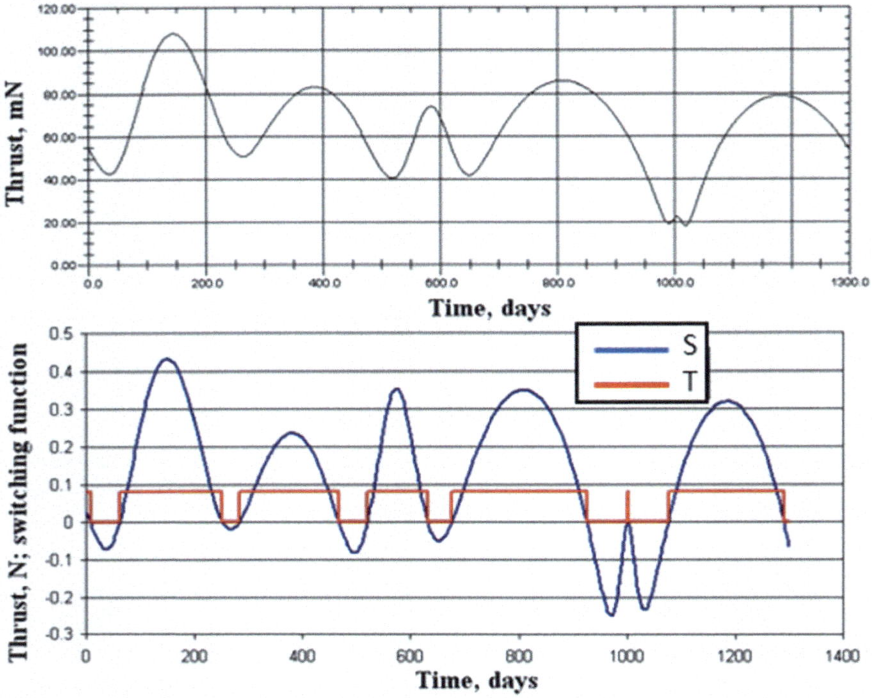

Fig. 9.8 Asteroid 2000 WO107 rendezvous problem—the thrust versus time for the LP-trajectory (top) and the thrust and the switching function versus time for the LT-trajectory (bottom)

departure hyperbolic excess, is equivalent to a reduction of the flight duration to 978.397721 days.

It is obvious from Figs. 9.19 and 9.20 that the number of thrusting sections varies from 2 to 5 on different trajectories. The final spacecraft mass is 2268.38464 kg for a trajectory with 0 full orbits, 10,693.675636 kg for a trajectory with 1 full orbit, 7661.817453 kg for a trajectory with 2 full orbits, 5429.070282 kg for a trajectory with 3 full orbits and 3903.505554 kg for a trajectory with 4 full orbits. It was not possible to find trajectories with a larger angular distance and, apparently, they do not exist. It is interesting to note that on the trajectory with the minimum angular distance there is a double change in the spacecraft motion direction (direct-retrograde-direct), and such trajectory differs substantially from the corresponding trajectory with an ideally controlled thruster (Fig. 9.20); the only globally optimal trajectory is the trajectory in which the actual duration of the flight is less than the specified one due to the presence of a coasting section at the beginning of the flight, in which the spacecraft moves along the Earth's orbit. Although formally the globally optimal trajectory contains one full orbit, the reduction of the flight time due to the long initial coasting segment leads to the fact that its actual angular distance is also less

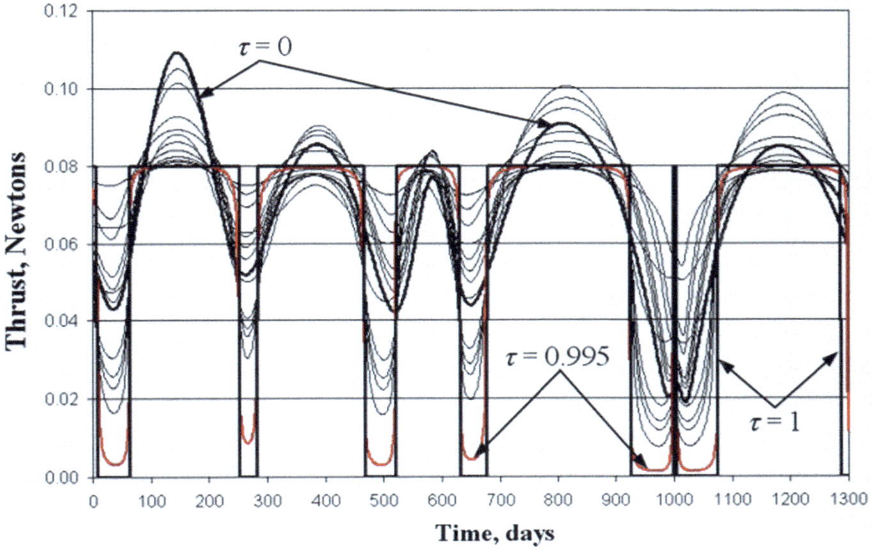

Fig. 9.9 The evolution of the thrust dependence on time for the LP-trajectory continuation into the LT-trajectory

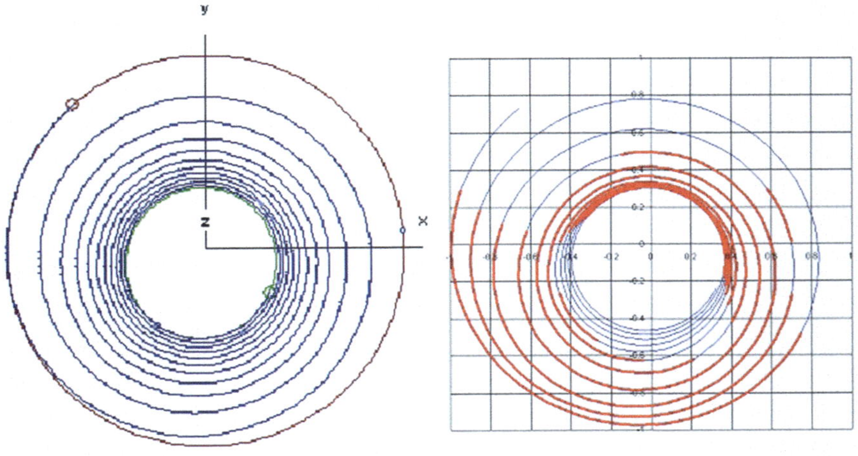

Fig. 9.10 The Mercury rendezvous problem, constant jet power, 10 full revolutions. On the left—LP-trajectory, on the right—LT-trajectory

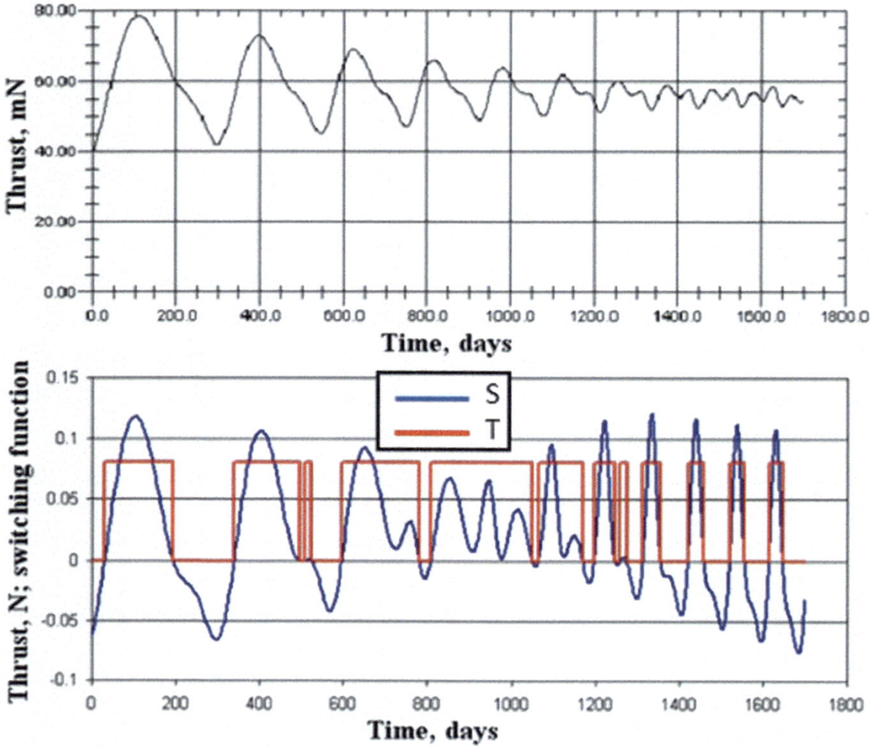

Fig. 9.11 The Mercury rendezvous problem, constant jet power, 10 entire orbits. At the top is the dependence of thrust on time for the LP-trajectory, at the bottom is the dependence of thrust and the switching function on time for the LT-trajectory

than 360°, since in fact the exit from the Earth's sphere of influence can be performed at the moment of the beginning of the first thrusting section.

The second set of numerical examples demonstrates the calculation of minimum thrust trajectories using the method described in Sect. 9.3.

Let us consider the problem of a flight to Mercury with a departure date from Earth of May 10, 2020 and a flight duration of 2000 days. According to the analysis of optimal LP-trajectories, performed using the methods of Sect. 10, the minimum values of the cost function and characteristic velocity (global minimum) are achieved on a trajectory consisting of 12 full orbits.

Then the algorithm for calculating the boundary of the domain of existence of the solution from Sect. 9.3 was applied to the problem. As a result of solving the problem of calculating the trajectory with the minimum thrust acceleration, it was found that this acceleration is 0.1173986 mm/s^2. This thrust acceleration is minimal at infinite exhaust velocity (zero mass flow rate).

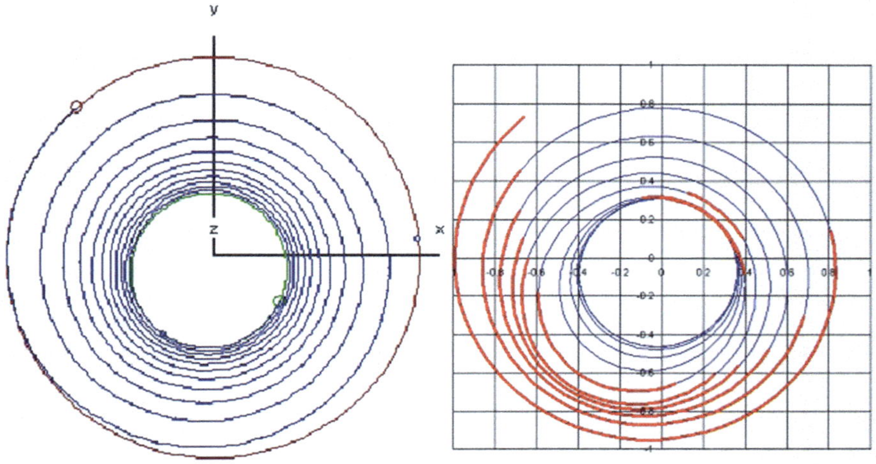

Fig. 9.12 Mercury rendezvous problem, $P_b = P_{b0}r^{-1.7}$, 10 full revolutions. On the left—LP-trajectory, on the right—LT-trajectory

The minimum thrust acceleration trajectory was further continued into the minimum thrust trajectory. The permissible propellant consumption was assumed to be 600 kg, and the initial mass of the spacecraft was assumed to be 1000 kg. As a result of such continuation, the dependence of the minimum thrust $T_{\min}$ on the specific impulse $I_{sp} = c/g_0$, where $g_0 = 9.80665$ m/s^2, was obtained (Fig. 9.21). The domain of existence of solutions to the problem with thrust switching on the specific impulse—thrust plane, located above the minimum thrust curve is highlighted in this figure. On the left side, this domain is limited by the maximum allowable propellant consumption (600 kg in the problem under consideration). Such left boundary is represented by a dotted line; it corresponds to the specific impulse of 2347.078 s.

It is obvious that if the available spacecraft thrust is $T = T_{\min}(I_{sp})$, there are no coasting sections on the trajectory, i.e. for such thrust the flight time of 2000 days is minimal, and the flight trajectory with minimum thrust for a given time is also a solution to the problem of minimizing the spacecraft flight time with $T = T_{\min}(I_{sp})$.

Figure 9.22 shows the dependence of the relative final mass (ratio of final to initial mass) of the spacecraft on the specific impulse for the minimum thrust trajectories. Obviously, this curve represents the lower limit of the achievable relative final mass of the spacecraft, since if the thrust increases, then the characteristic velocity decreases and the relative final mass increases also.

Figure 9.23 shows the dependence of the EPS jet power $P_b = Tc/2$ on the specific impulse for the minimum thrust trajectories. Obviously, this curve is a lower estimate of the required jet power, since it corresponds to the minimum thrust for any given value of specific impulse. If the EPS efficiency η is known, then the electric power required to provide it can be calculated as $P_e = P_b/\eta$.

Thus, the implementation of an algorithm for calculating the solution existence domain provides a lot of information, including:

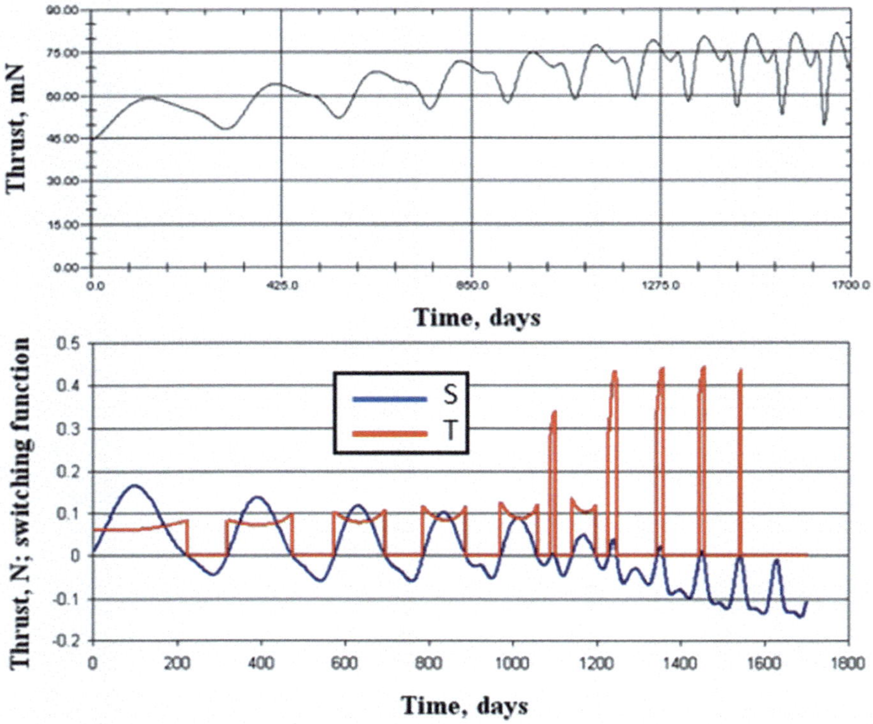

Fig. 9.13 The Mercury rendezvous problem, $P_j = P_{b0}r^{-1.7}$, 10 full revolutions. At the top is the time dependence of the thrust for the LP-trajectory, at the bottom is the time dependence of the thrust and the switching function for the LT-trajectory

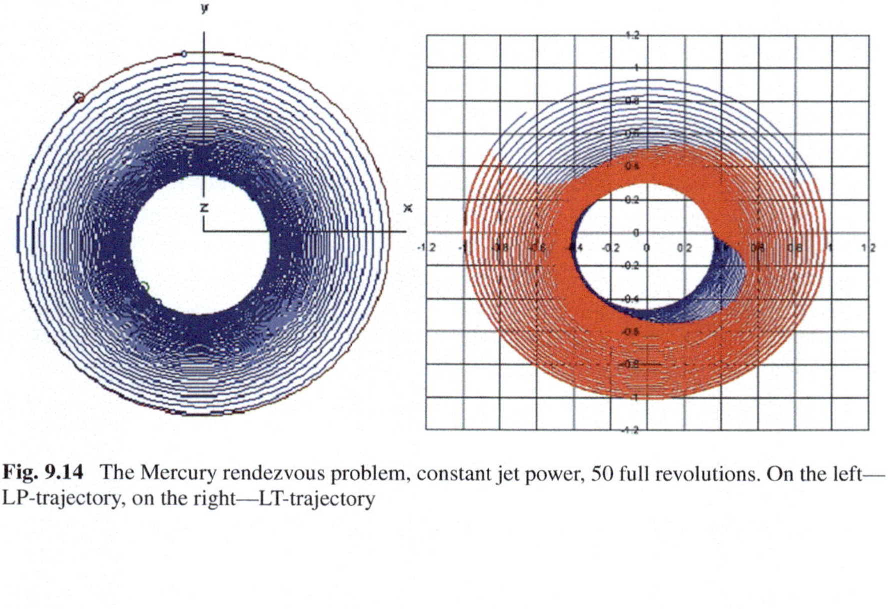

Fig. 9.14 The Mercury rendezvous problem, constant jet power, 50 full revolutions. On the left—LP-trajectory, on the right—LT-trajectory

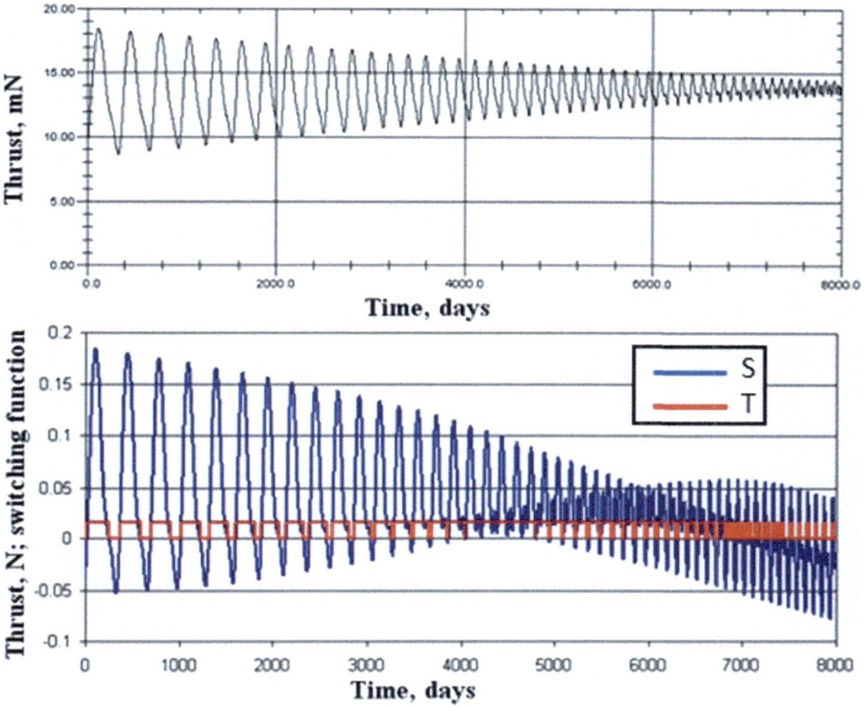

Fig. 9.15 The Mercury rendezvous problem, constant jet power, 50 full revolutions. At the top is the time dependence of the thrust for the LP-trajectory, at the bottom is the time dependence of the thrust and the switching function for the LT-trajectory

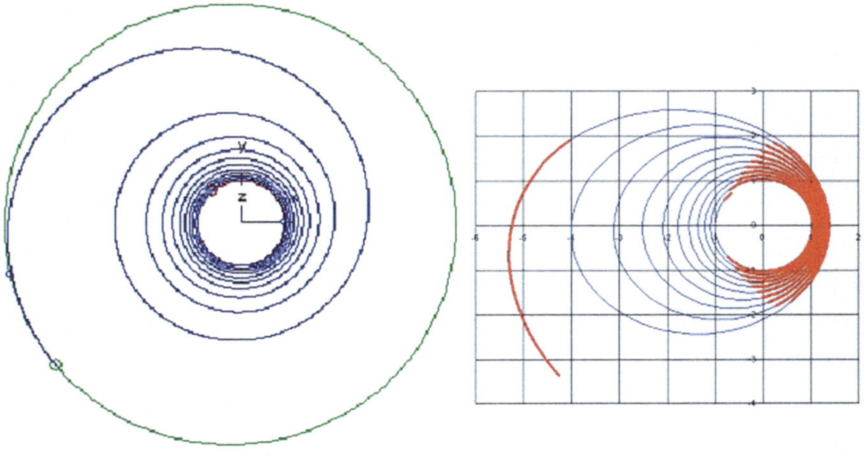

Fig. 9.16 The Jupiter rendezvous problem, constant jet power, 10 full revolutions. On the left—LP-trajectory, on the right—LT-trajectory

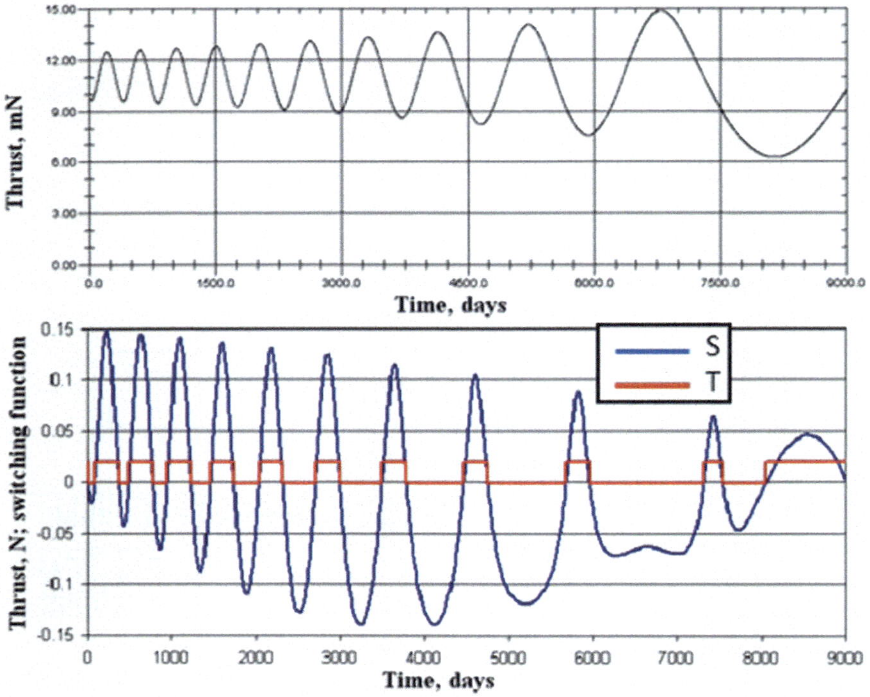

Fig. 9.17 The Jupiter rendezvous problem, constant jet power, 10 full revolutions. At the top is the time dependence of the thrust for the LP-trajectory, at the bottom is the time dependence of the thrust and the switching function for the LT-trajectory

(1) Dependence of minimum thrust on the available specific impulse;
(2) Minimum permissible specific impulse;
(3) The lower limit of the final spacecraft mass as a function of the available specific impulse;
(4) The lower limit of the EPS jet power as a function of the available specific impulse.

Let us consider the flight trajectories to Mercury with the specific impulse of 4500 s. The point with such a specific impulse is marked with a circle in Figs. 9.21 and 9.22, namely on the curve corresponding to the minimum thrust trajectory.

At the specific impulse of 4500 s, the minimum thrust is $T_{\min} = 93.547$ mN. On the minimum thrust trajectory, the electric propulsion system operates continuously, and there are no coasting sections. Figure 9.24 shows trajectories to Mercury with different thrust values—from the minimum value of 93.547 mN (Fig. 9.24a) to 120 mN (Fig. 9.24e). When the thrust exceeds the minimum value, coasting sections appear on the trajectory, the number and duration of which increase as the thrust increases. Coasting sections of the trajectory in Fig. 9.24 are marked with a dotted line. The location of the such trajectories in the domain of solution existence on

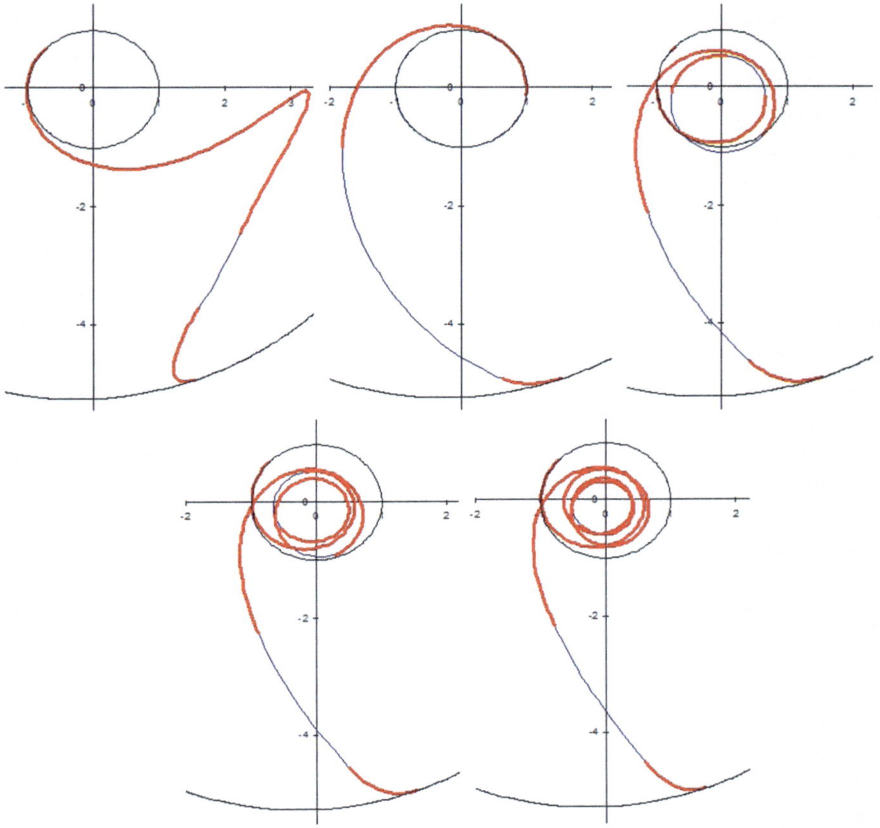

Fig. 9.18 A set of local extremums in the problem of optimizing a direct LT-trajectory (constant jet power) to Jupiter under the same boundary conditions and spacecraft design parameters

the specific impulse–thrust plane is marked with triangles in Fig. 9.21. With the thrust of 93.6 mN (Fig. 9.24b), a single coasting section appears on the last revolution of the trajectory. As the thrust increases to 95 mN (Fig. 9.24c), the duration of this coasting section increases, with additional coasting sections appearing on two previous revolutions and on the first revolution. As the thrust continues to increase, coasting sections appear on every revolution (Fig. 9.30e). The time dependencies of thrust and switching function for these trajectories are shown in Fig. 9.25.

All presented trajectories were obtained by the method of continuation from an LP- trajectory with smoothing of the throttling function in the form (9.1.12). Preliminary calculation of the boundary of the solution existence domain guaranteed the convergence of this method to the optimal solution when the thrust exceeded the minimum value.

As the next example, consider an Earth-Mars flight with a minimum thrust and duration of 350 days. Let us assume that the initial mass of the spacecraft is 1000 kg.

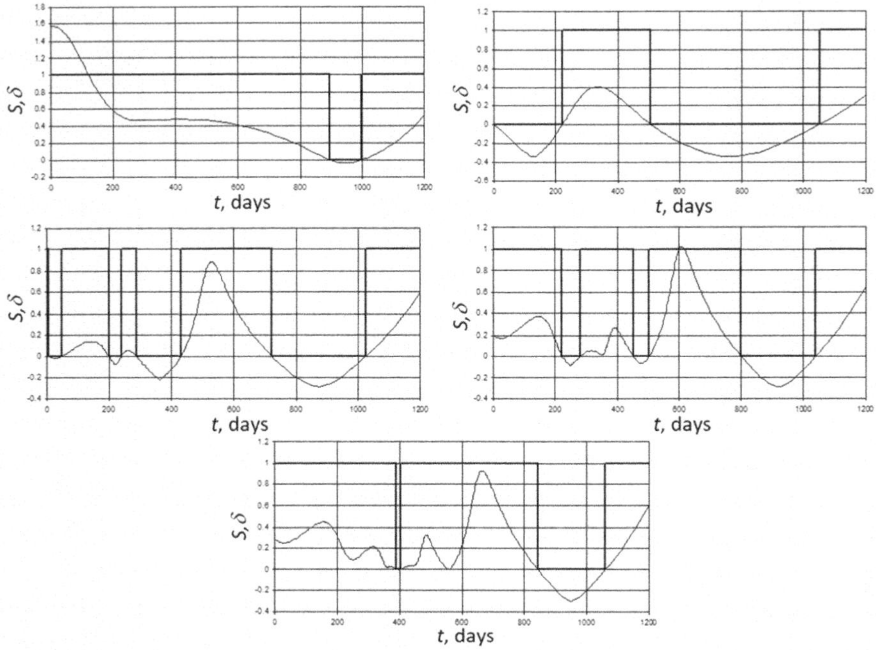

Fig. 9.19 Switching function ψ (thin line) and thrust function δ (thick line) at various local extremums in the problem of optimizing af direct LT-trajectory to Jupiter

Figure 9.26 shows the calculation results obtained by the algorithm given in Sect. 9.3 for the boundary of the domain of existence of the solution $T_{\min}(I_{sp})$ at an allowable propellant consumption of up to 99% of the initial spacecraft mass for an angular flight range of less than 360° and a date of departure from the Earth of 1.1.2035. The minimum value of the specific impulse (268.482 s) occurs when the maximum allowable propellant mass is consumed ($0.99m(t_0)$).

Some results of the calculation of trajectories to Mars with a minimum thrust and duration of 350 days for the launch dates from 1.1.2035 to 5.1.2036 are shown in Figs 9.27, 9.28 and 9.29. Data for the range of specific impulse from a certain minimum value corresponding to the propellant consumption of $0.99m(t_0)$ to infinity were obtained for each launch date. In this case, the minimum specific impulse varies from 268.482 s to 2281.416 s, depending on the start date.

Figures 9.27, 9.28 and 9.29 show the results interpolated from the calculated dependencies for the specific impulse of the electric propulsion system of 7000 s. In Fig. 9.28, on the left are the dependences of the minimum thrust on the launch date for the trajectories with angular range in the ranges 0°–360° and 360°–720°, and on the right—the launch window with the available thrust of 0.25 N. Figure 9.28 shows the dependence of the minimum jet power on the departure date, and Fig. 9.29—the dependence of the minimum relative final mass of the spacecraft on the launch date.

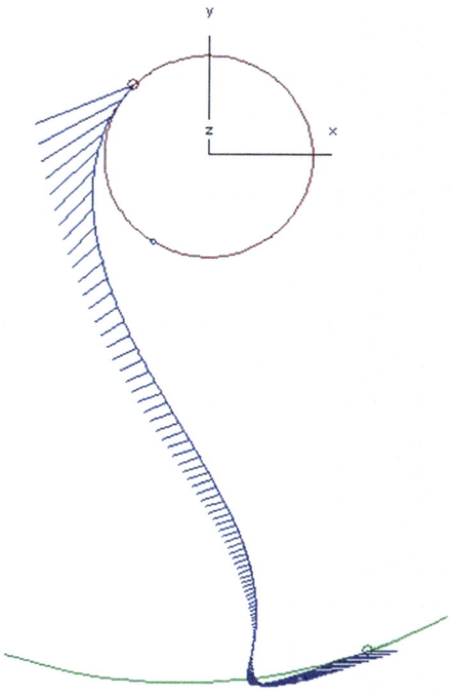

Fig. 9.20 The LP-trajectory to Jupiter (the departure date is 02/05/2017, the flight duration is 1200 days, the departure hyperbolic velocity is zero, the problem is the rendezvous problem, 0 full revolutions)

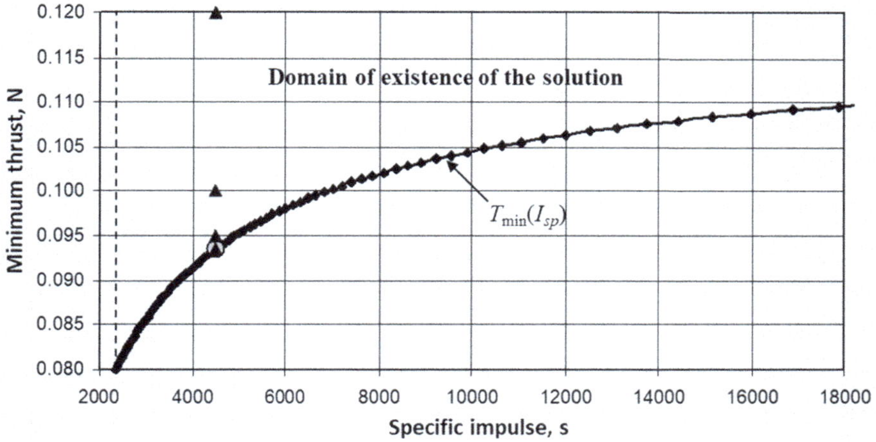

Fig. 9.21 The domain of existence of the solution to the problem of flight to Mercury

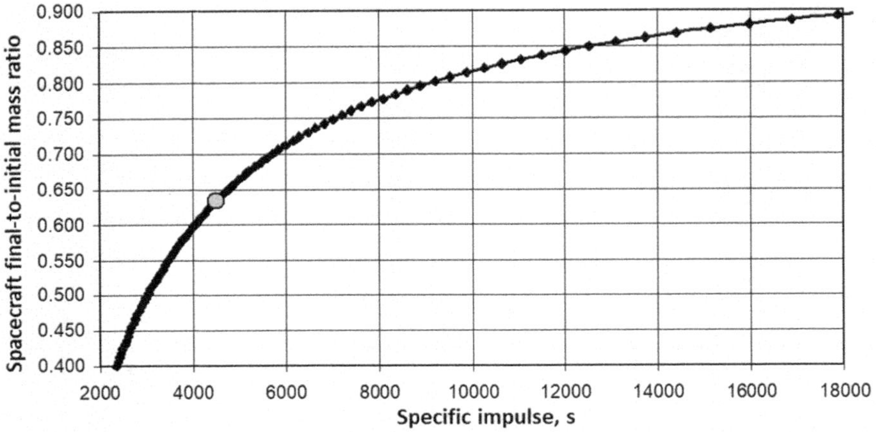

Fig. 9.22 The ratio of the spacecraft's final to initial mass as it approaches Mercury as a function of the specific impulse for minimum-thrust trajectories

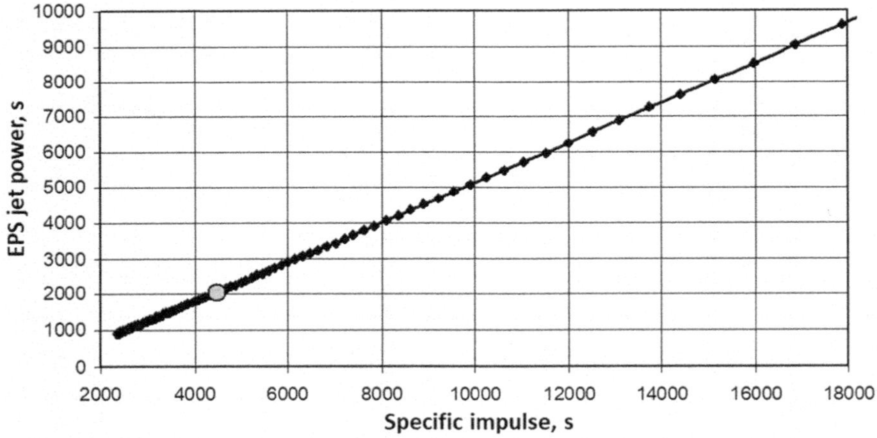

Fig. 9.23 EPS jet power versus specific impulse on the minimum thrust trajectories to Mercury

Visual and fairly complete information about the influence of the launch date and duration of the flight with minimum thrust can be presented in the form of contour plots of the main characteristics of such flights (minimum thrust, characteristic velocity, minimum jet power of the electric propulsion system, etc.) on the plane launch date t_0—transfer duration Δt. To plot such isolines, it is sufficient to calculate the minimum-thrust trajectories on a narrow enough mesh by t_0 and Δt. A direct approach to solving this problem leads to high computational costs: if the calculations are performed on a uniform mesh by t_0 and Δt, which contains n_1 nodes

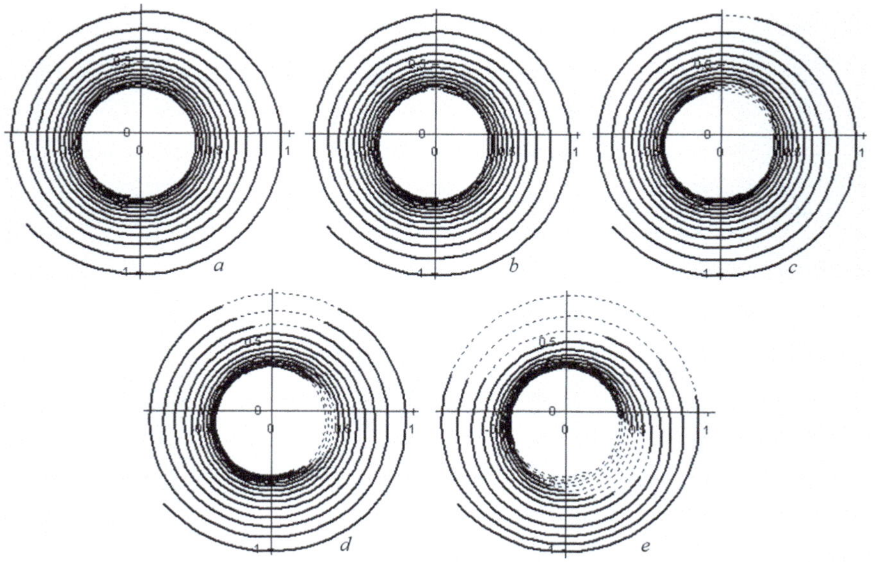

Fig. 9.24 Projection of trajectories to Mercury on the ecliptic plane in the heliocentric ecliptic coordinate system J2000 (thick solid lines—thrusting sections, dotted lines—coasting sections): *a*—trajectory with minimum thrust $T = 93.547$ mN; *b*—trajectory with thrust $T = 93.6$ mN; *c*—trajectory with thrust $T = 95$ mN; *d*—trajectory with thrust $T = 100$ mN; *e*—trajectory with thrust $T = 120$ mN

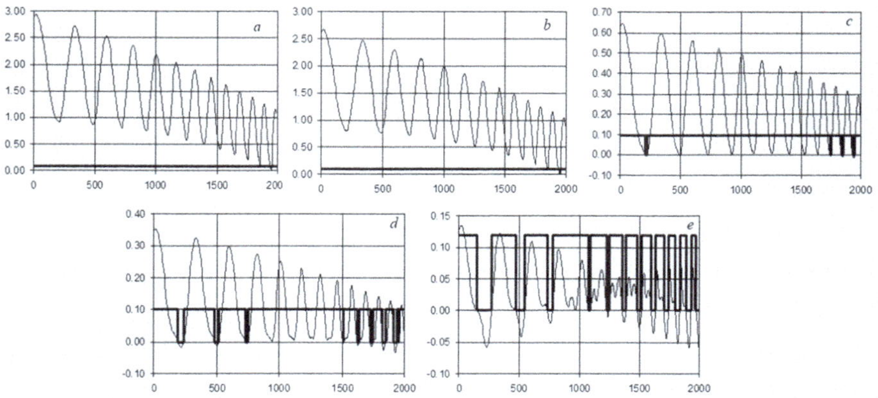

Fig. 9.25 The dependence of thrust (thick line) and switching function (thin line) on the flight time. The abscissa axis is the flight time in days; the ordinate axis is the thrust in Newtons, and the switching function in dimensionless units. *a*—trajectory with minimum thrust $T = 93.547$ mN; *b*—trajectory with thrust $T = 93.6$ mN; *c*—trajectory with thrust $T = 95$ mN; *d*—trajectory with thrust $T = 100$ mN; *e*—trajectory with thrust $T = 120$ mN

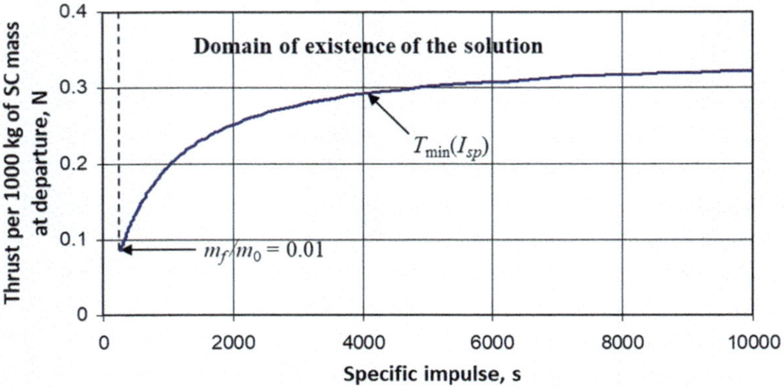

Fig. 9.26 Minimum thrust as a function of the specific impulse for a 350-day Earth-Mars trajectory with a departure date on 01.01.2035

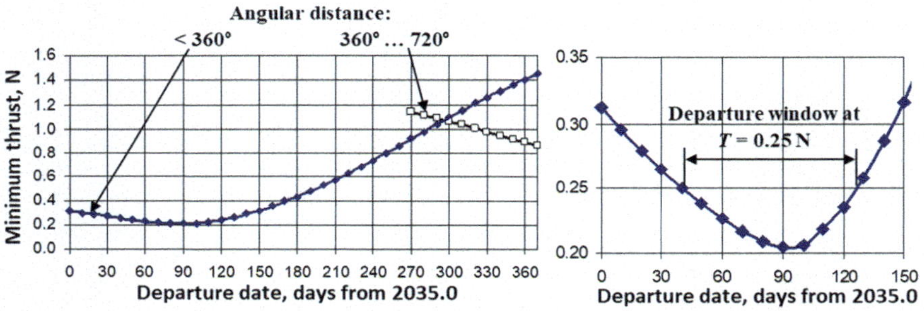

Fig. 9.27 The dependence of the minimum thrust on the departure date for the Earth-Mars trajectory at the specific impulse of 7000 s. On the right plot, the launch window is marked on this dependence with an available thrust of 0.25 N

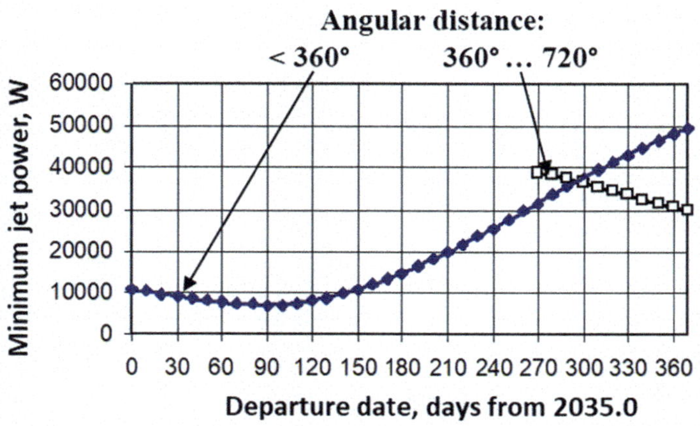

Fig. 9.28 Dependence of the minimum jet power on the departure date for the Earth-Mars trajectory at the specific impulse of 7000 s

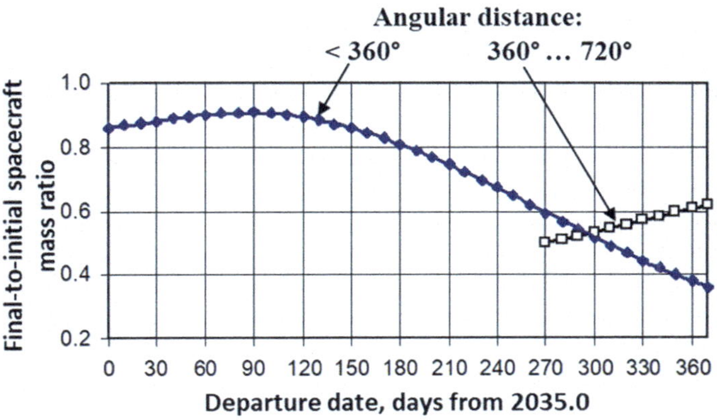

Fig. 9.29 Dependence of the final-to-initial spacecraft mass ratio on the departure date for the Earth-Mars trajectory at a specific impulse of 7000 s

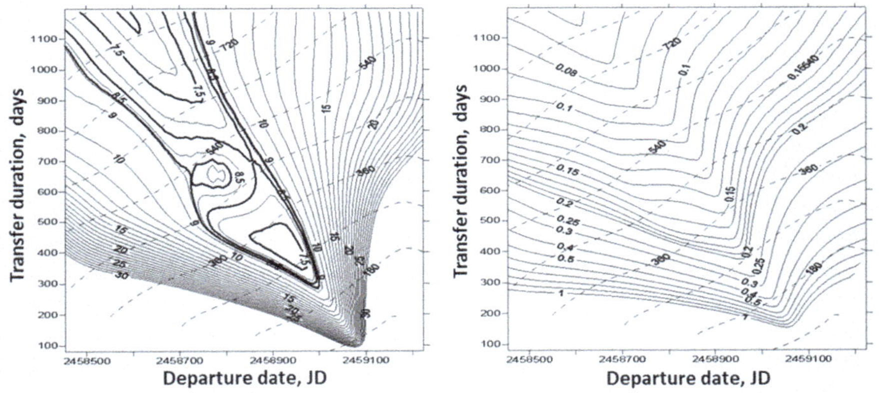

Fig. 9.30 Contour plot of the characteristic velocity [km/s] on trajectories with minimum thrust (left) and of the values of the minimum thrust [N] (right) in the "launch date- flight duration" field (trajectory to Mars, specific impulse is 3100 s, $V_{\infty 0} = 0$)

by t_0 and n_2 nodes by Δt, then $n_1 \cdot n_2$ minimum thrust trajectories should be calculated by solving $n_1 \cdot n_2$ LP-problems (or Lambert problems), $n_1 \cdot n_2$ problems of the continuation of LP- or Keplerian trajectory to the trajectory with the minimum thrust acceleration, and $n_1 \cdot n_2$ problems of the continuation of trajectories with minimum thrust acceleration to the minimum thrust trajectory.

In order to reduce the computational costs, a method has been developed that consists in calculating a minimum thrust trajectory for the minimum flight duration for each t_0 value, and its subsequent continuation until the maximum flight duration is reached. Thus, it is necessary to compute only n_1 LP- or Lambert trajectories, to continue these trajectories into trajectories with minimum thrust acceleration and minimum flight time with subsequent continuation to the trajectories with minimal thrust and minimal flight time. As a result of the last step, for each trajectory with minimum thrust and minimum time, a family of trajectories parameterized by the exhaust velocity c (or specific impulse I_{sp}) is calculated. From this family, a trajectory with a given specific impulse I_{sp} is selected (the parameters of such a trajectory can be calculated using the interpolating polynomials within the integration step when it is continued along the mass flow rate to the minimum thrust trajectory), and then it is continued along the flight time to the maximum value using the continuation method in the form of (7.3.18). As a result of one such continuation, the parameters of all minimum thrust trajectories with a given start date t_0, a given specific impulse I_{sp} are calculated in a given range of flight duration variation. Of course, to construct a field of isolines, it is necessary to solve n_1 such continuation problems over the flight duration, one for each of the considered departure dates.

In a typical case, to construct a field of isolines within one synodic period, about 100 points are required by launch date and 100 points by flight duration. Therefore, the use of the algorithm described here reduces the computation time by about 2 orders of magnitude, from $3n_1n_2$ to $4n_1$ continuation problems. For the typical one- or two-orbit heliocentric flights, the time to calculate one minimum thrust trajectory using a modern multiprocessor personal computer is 5–10 s. In this case, the direct calculation of minimum thrust trajectories at each node of the 100×100 grid by t_0 and Δt takes 14–28 h, and the use of continuation by flight time allows you to reduce the calculation time to 10–20 min.

It should be noted that the flight time continuation method, although it provides a significant acceleration of calculations of the isoline field, has a number of disadvantages in comparison with the direct calculation of the minimum thrust trajectories at each node of a given mesh. First of all, the flight time continuation method allows one to determine solutions only for a predetermined value of the specific impulse of the electric propulsion system, while the direct method gives characteristics for all trajectories in the permissible range of the specific impulse variation. Second, the angular range of the flight calculated at a point with a certain departure date t_0 and flight duration Δt is uniquely defined as the angular range of the minimum time flight with $\Delta t_{\min}$ plus the angular range of displacement of the target planet over the time interval $\Delta t - \Delta t_{\min}$. Thus, to analyze extremals with a different number of entire orbits around the primary when using the method of continuation with respect to flight time, it is necessary to repeat a series of continuation procedures, starting with another extremal of the LP-problem (or for another solution to the Lambert problem). In some cases, this can lead to computational difficulties associated with calculating multi-revolution trajectories for short flight times. When using the direct method, all necessary extremals can be calculated independently at each mesh node.

Figures 9.30 and 9.31 show the isolines of the characteristic velocity on the trajectory to Mars (rendezvous problem) with the minimum thrust (on the left) and the isolines of this minimum thrust (on the right) in the field departure date—flight duration. The spacecraft initial mass is assumed to be 1000 kg, and the specific impulse is 3100 s. The isolines presented in Fig. 9.30 correspond to the departure hyperbolic excess velocity $V_{\infty 0} = 0$, and in Fig. 9.31—to $V_{\infty 0} = 1$ km/s. The dotted lines in

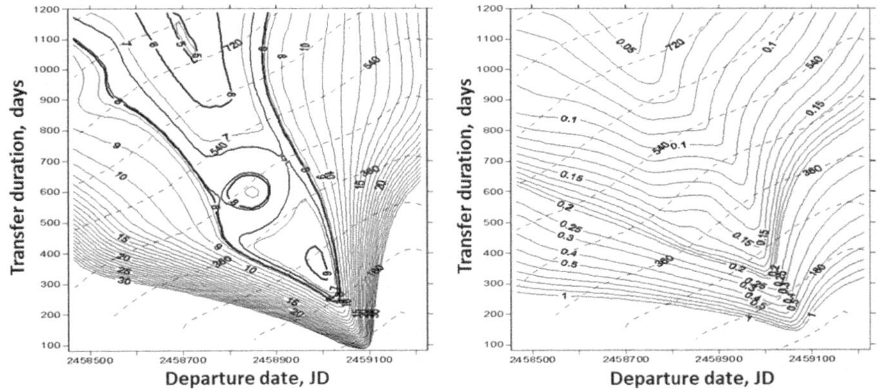

Fig. 9.31 Contour plot of the characteristic velocity [km/s] on trajectories with minimum thrust (left) and the values of the minimum thrust [N] (right) in the "launch date- flight duration" field (trajectory to Mars, specific impulse 3100 s, $V_{\infty 0} = 1$ km/s)

these figures show the isolines of the angular distance of the flight; the values of the angular distance for these lines in the figures are given in degrees.

As can be seen from Figs. 9.30 and 9.31, the characteristic velocity has local minima corresponding to the optimal departure dates and flight durations for different numbers of full orbits around the Sun. The presence of a flat local maximum between these minima, bounded by two saddle points, was unexpected. The minimum thrust required to complete the flight has a minimum only by the departure date. According to these figures, the minimum required thrust decreases monotonically with increasing flight duration.

In the third group of numerical examples, we consider the results of applying the method from Sect. 9.4 to solve the problem of joint optimization of the main trajectory and design parameters of a spacecraft with an electric propulsion system.

In the first example, we consider the problem of optimizing the flight of a spacecraft with an EPS to Mars for the date of departure from Earth 20.4.2035, zero departure hyperbolic excess velocity, flight duration $\Delta t = 350$ days and the initial mass of the spacecraft $m_0 = 1000$ kg. Figure 9.32 shows an example of an optimal trajectory with optimal parameters of the electric propulsion system for the following values of design parameters that determine the level of technology used: $\eta = 0.5$, $\gamma_3 = 40$ kg/kW, and $a_t = 0.1$. As a result of solving the optimization problem, in addition to the optimal control program, the optimal values of the specific impulse $I_{sp} = c/g_0 = 2374$ s, thrust $T = 220.39$ mN and electric power $P = 5131$ W were computed. The propellant consumption on the optimal trajectory is $m_p = 224.89$ kg, and the useful mass of the spacecraft is $m_u = 547.39$ kg.

The upper part of Fig. 9.32 shows the projection of the optimal trajectory onto the ecliptic plane (the thrusting sections are indicated by thick lines, and the coasting section is indicated by a dotted line). Thin lines indicate the orbits of Earth and Mars. Round markers indicate the position of the Earth on the departure date and

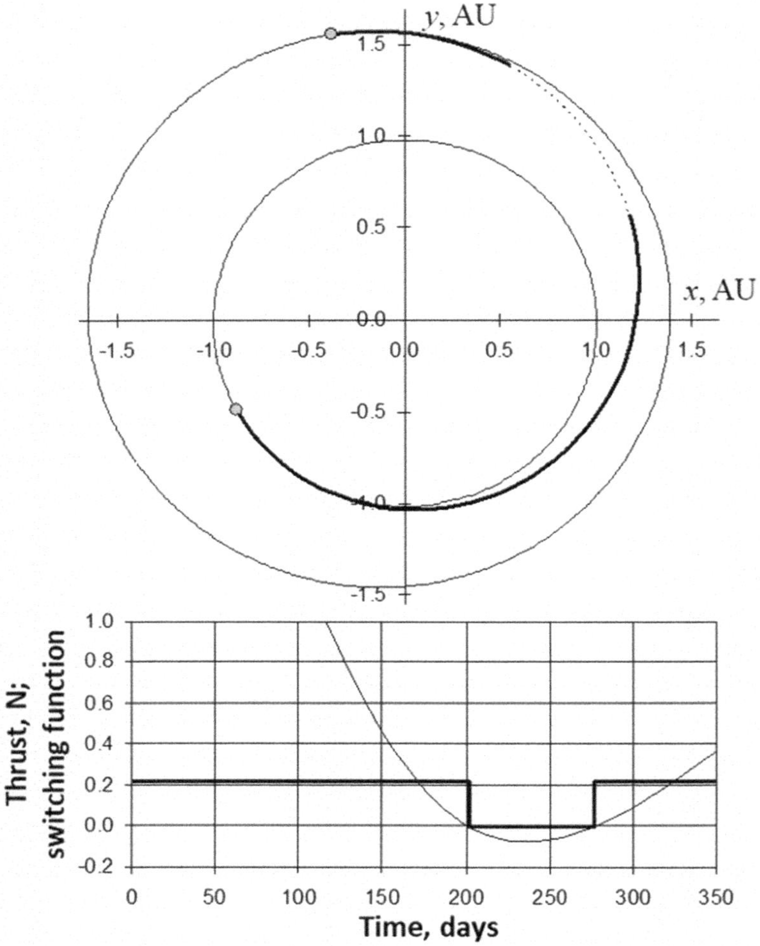

Fig. 9.32 An example of an optimal heliocentric trajectory to Mars

the position of Mars on the arrival date. The lower part of the figure shows the time dependences of the switching function S (thin line) and of the thrust T (thick line).

Table 9.1 shows the results of the joint optimization of the trajectory to Mars and the main design parameters of the electric propulsion system for various values of the design parameters η, γ_3, and a_t characterizing the level of technology used. The optimal specific impulse increases with the decrease of γ_3 and the increase of η and a_t, and the optimal value of electric power of the electric propulsion system increases with the decrease of η and γ_3 and increase of a_t. Obviously, the maximum value of the useful spacecraft mass is reached at the maximum value of EPS efficiency and the minimum specific mass of EPS and PSFS.

Table 9.1 Results of the joint optimization of the flight trajectory to Mars and the main design parameters of the electric propulsion system for various values of design parameters characterizing the level of technology used

No.	η	γ_3	a_t	I_{sp}	T	$m(t_f)$	m_p	m_{EPS}	m_u	P
	–	kg/kW	–	s	mN	kg	kg	kg	kg	W
1	0.7	40	0.1	2869	222.66	809.95	190.05	178.99	611.96	4474
2	0.5	40	0.1	2374	220.39	775.11	224.89	205.23	547.39	5131
3	0.5	80	0.1	1580	214.05	681.77	318.23	265.28	384.67	3316
4	0.5	40	0.2	2494	221.02	784.69	215.31	216.24	525.39	5406

In Fig. 9.33, square markers indicate the results of optimization of the main design parameters of the electric propulsion system, presented on the specific impulse—thrust plane. The marker numbers correspond to the options from Table 9.1. The thick line in this figure represents the dependence of the minimum thrust on the specific impulse, limiting from below the solution existence domain, the dotted line represents the asymptotic value of the minimum thrust as the specific impulse tends to infinity. The minimum thrust curve was obtained using the methodology given in [3, 4] and Sect. 9.3.

Table 9.2 shows the results of the joint optimization of the main design parameters of the electric propulsion system and the spacecraft trajectory for $\eta = 0.7$, $\gamma_3 = 40$ kg/kW and $a_t = 0.1$, flight duration $\Delta t = 350$ days and the initial spacecraft mass $m_0 = 1000$ kg for different values of the departure hyperbolic excess velocity $V_{\infty 0}$. The optimal electric power of the electric propulsion system decreases monotonically with the increase of $V_{\infty 0}$, the optimal thrust has a minimum, and the optimal specific impulse has a maximum in $V_{\infty 0}$. The useful mass of the spacecraft increases

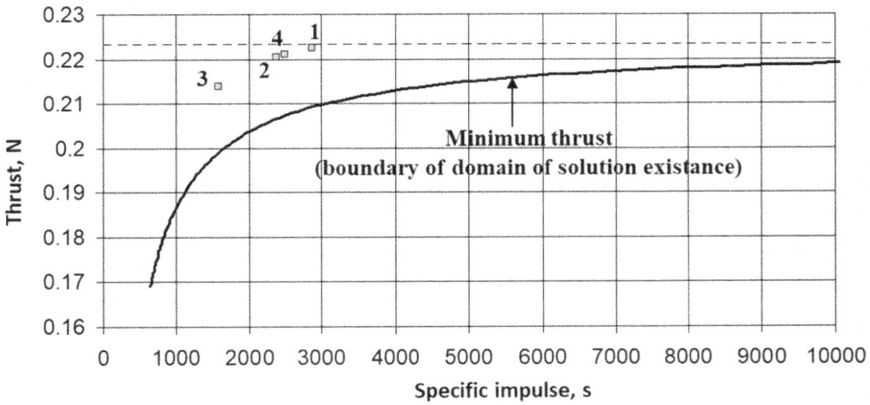

Fig. 9.33 The optimal values of the specific impulse and thrust of the EPS (square markers) for the trajectory to Mars at different values of the design parameters characterizing the level of technologies used

monotonically with increasing $V_{\infty 0}$, since in the case under consideration the initial mass of the spacecraft on the trajectory departing from the Earth does not decrease with the increase of $V_{\infty 0}$.

Figure 9.34 shows contour plot of the useful mass of the spacecraft on the plane specific impulse—thrust of the electric propulsion system for the case corresponding to the first rows of Tables 9.1 and 9.2. The maximum useful mass (about 612 kg) is reached at the point marked with a round marker. This maximum is rather flat—if the specific impulse deviates from the optimum by 300 s (by 10.5%) or the thrust deviates from the optimal value by up to 7 mN (by 3.1%), the useful mass decreases by only 2 kg (0.3%). The useful mass level lines are limited at the bottom by the minimum thrust curve (thick line in the Fig. 9.34). At values of thrust and specific impulse lying on this curve, there are no coasting sections on the trajectory, therefore any decrease in the thrust value below this line leads to the impossibility of performing the flight in a given time [3, 4].

Table 9.2 Dependence of the optimal values of the main design parameters on the value of the hyperbolic excess velocity of departure from the Earth on the trajectory to Mars

$V_{\infty 0}$	T	I_{sp}	$m(t_f)$	m_p	m_{EPS}	m_u	P
m/s	mN	s	kg	kg	kg	kg	W
0	222.66	2869	809.95	190.05	178.99	611.96	4474
1000	160.77	3206	849.56	150.44	144.44	690.08	3611
2000	137.08	3214	869.71	130.29	123.43	733.25	3086
3000	137.43	2997	879.46	120.54	115.40	752.01	2885
3500	140.02	2896	881.94	118.06	113.63	756.49	2841

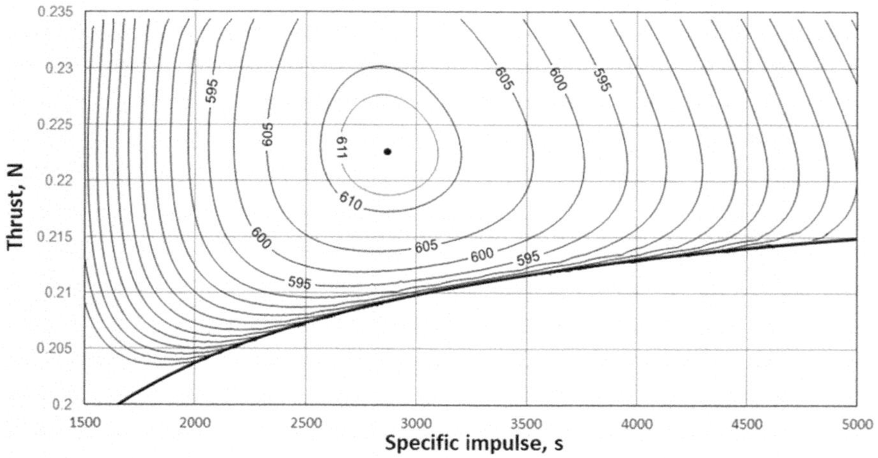

Fig. 9.34 Contour plot of the useful mass of the spacecraft

Table 9.3 Expansion coefficients for the approximating dependences of the initial spacecraft mass on the departure hyperbolic excess velocity

i	"Soyuz-2.1B" + "Fregat"	"Proton-M" + "Breeze-M"
0	$2.033309 \cdot 10^{3}$	$6.474400 \cdot 10^{3}$
1	$-5.623431 \cdot 10^{-2}$	$2.820072 \cdot 10^{-2}$
2	$5.265562 \cdot 10^{-5}$	$-1.869981 \cdot 10^{-4}$
3	$-8.146719 \cdot 10^{-8}$	$5.229323 \cdot 10^{-8}$
4	$3.631484 \cdot 10^{-11}$	$-1.776752 \cdot 10^{-11}$
5	$-7.845370 \cdot 10^{-15}$	$3.061888 \cdot 10^{-15}$
6	$6.651509 \cdot 10^{-19}$	$-1.902993 \cdot 10^{-19}$

To optimize the $V_{\infty 0}$ value, the following polynomial approximation of the dependence $m_0(V_{\infty 0})$ is used:

$$m_0 = \sum_{i=0}^{6} a_i V_{\infty 0}^{i}.$$

Table 9.3 presents the coefficients a_i determined by the least squares method from the data presented in the user manuals for the Soyuz-2.1B launch vehicle (LV) with the Fregat upper stage [8] and the Proton-M LV with the Breeze-M upper stage [9].

The results of the joint optimization of the main spacecraft design parameters, the departure hyperbolic excess velocity and the EPS thrust vector control are presented in Table 9.4. As above, the values of EPS efficiency, EPS specific mass and PSFS specific mass from the first row of the Table 9.1 are used, $t_0 = 20.\text{IV}.2035$, and $\Delta t = 350$ days.

Optimal values of specific impulse and $V_{\infty 0}$ appeared to be close for the cases of using the Soyuz-2.1B LV with the Fregat upper stage and the Proton-M LV with the Breeze-M upper stage. It turned out that in both cases the heliocentric trajectories do not contain coasting sections, i.e. the optimal thrust value turned out to be equal to its minimum value. In this context, the question arises to what extent this situation is typical for the joint optimization of the main spacecraft design parameters and the departure hyperbolic excess velocity.

Let's consider a spacecraft with a single-mode electric propulsion system and with the following parameters: $\gamma_3 = 50$ kg/kW, $a_t = 0.1$, $\eta = 0.7$. Tables 9.5, 9.6 and 9.7 present the optimal trajectory and design parameters as a function of the departure date offset Δt_0 from the nominal date of 20.IV.2035 for a flight duration

Table 9.4 Optimal values of the main design parameters and the departure hyperbolic excess velocity

LV	$V_{\infty 0}$	T	I_{sp}	m_0	m_p	m_{EPS}	m_u	P
	m/s	mN	s	kg	kg	kg	kg	W
"Soyuz-2.1B"	1376.4	287.51	3257	1939.16	272.17	262.40	1377.38	6560
"Proton-M"	1474.2	906.41	3254	6212.56	858.93	826.42	4441.32	20,661

of 250, 300 and 350 days, respectively. The Soyuz-2.1B LV with the Fregat upper stage is being considered for spacecraft insertion onto the departure trajectory.

Figure 9.35 shows the ratio of the electric propulsion operating time τ_{EPS} to the total flight time Δt as a function of the offset of the date of departure from the Earth from the date 20.IV.2035 for $\Delta t = 250$, 300 and 350 days. It can be seen that for sufficiently large Δt (300 and 350 days) the optimization of thrust, exhaust velocity, hyperbolic excess velocity and launch date results in the optimal trajectories without coasting sections, and the range of launch dates in the vicinity of their optimal value, at which the thrust of the electric propulsion system is minimum, expands with increasing flight duration.

Heliocentric flight trajectories to Mars are characterized by a relatively small angular distance. It is interesting to compare the properties of such trajectories with optimal trajectories involving a large number of orbits around the Sun. The maximum number of orbits is required for direct flight trajectories to Mercury, so let us consider

Table 9.5 Dependence of optimal parameters on the offset of the departure date (flight duration 250 days)

Δt_0 (days)	T (N)	c (m/s)	$V_{\infty 0}$ (m/s)	m_0 (kg)	m_p (kg)	P (W)	m_u (kg)
10	0.5845	18,181.0	1740.7	1891.8	492.7	7590.8	970.3
20	0.5370	18,732.0	1734.1	1892.8	450.4	7185.7	1038.0
30	0.4894	19,297.1	1770.1	1887.6	409.5	6745.8	1099.8
40	0.4428	19,822.6	1885.4	1870.4	369.3	6269.5	1150.7
50	0.3995	20,300.8	2115.1	1833.4	332.4	5792.9	1178.1
60	0.3633	20,630.3	2413.8	1780.1	300.6	5353.5	1181.8
70	0.3367	20,739.6	2822.6	1697.8	276.1	4988.3	1144.7
80	0.3219	20,632.1	3312.9	1585.9	261.6	4743.6	1060.9
90	0.3154	20,288.2	3857.1	1451.4	254.4	4570.3	943.0

Table 9.6 Dependence of optimal parameters on the offset of the departure date (flight duration 300 days)

Δt_0 (days)	T (N)	c (m/s)	$V_{\infty 0}$ (m/s)	m_0 (kg)	m_p (kg)	P (W)	m_u (kg)
– 10	0.4044	23,280.0	1403.1	1936.0	401.8	6725.3	1157.7
0	0.3714	23,955.6	1436.2	1932.1	371.1	6354.7	1206.1
10	0.3384	24,668.8	1478.4	1926.9	341.2	5963.4	1253.4
20	0.3075	25,399.5	1571.4	1915.0	313.5	5578.8	1291.2
30	0.2849	25,685.3	1689.7	1899.0	287.5	5227.3	1321.4
40	0.2653	25,899.0	1863.6	1873.7	265.6	4908.7	1336.2
50	0.2495	26,041.5	2125.8	1831.6	248.3	4640.3	1326.5
60	0.2382	26,093.6	2419.7	1779.0	236.6	4439.3	1296.8
70	0.2374	25,372.1	2802.6	1702.1	228.6	4301.7	1235.6
80	0.2442	24,417.1	3228.5	1606.0	227.4	4259.9	1142.9
90	0.2572	23,439.6	3659.8	1500.5	233.3	4306.3	1028.6

Table 9.7 Dependence of optimal parameters on the offset of the departure date (flight duration 350 days)

Δt_0 (days)	T (N)	c (m/s)	$V_{\infty 0}$ (m/s)	m_0 (kg)	m_p (kg)	P (W)	m_u (kg)
− 50	0.3431	27,021.6	1198.8	1958.6	383.9	6621.3	1205.3
− 40	0.3266	27,410.1	1249.6	1953.3	360.3	6394.1	1237.3
− 30	0.3105	27,764.1	1319.1	1945.7	338.2	6157.8	1265.8
− 20	0.2943	28,069.9	1383.7	1938.3	317.0	5900.2	1294.6
− 10	0.2786	28,330.6	1478.5	1926.9	297.4	5637.5	1317.9
0	0.2636	28,534.0	1568.1	1915.5	279.3	5372.3	1339.6
10	0.2489	28,676.9	1651.0	1904.4	262.4	5098.0	1360.8
20	0.2355	28,771.8	1768.6	1887.8	247.5	4840.3	1373.5
30	0.2243	28,811.2	1869.1	1872.9	235.4	4615.8	1383.1
40	0.2150	28,793.2	1991.0	1853.8	225.8	4421.9	1384.4
50	0.2076	28,743.7	2188.2	1820.9	218.4	4261.4	1367.7
60	0.2021	28,653.6	2415.5	1779.8	213.3	4137.1	1338.3
70	0.1980	28,516.8	2737.5	1715.8	210.0	4033.3	1283.2
80	0.2022	27,732.1	3123.8	1630.5	210.4	4004.4	1198.9
90	0.2188	26,300.8	3480.7	1545.0	218.9	4110.9	1098.6

the joint optimization of the thrust vector control program, the main trajectory and design parameters of a spacecraft with a single-mode electric propulsion system using the example of solving the problem of flight from Earth to Mercury. The hyperbolic excess velocity of the spacecraft at the approach to Mercury is set to zero (the rendezvous problem is considered), the departure hyperbolic excess velocity is $V_{\infty 0}$ and the number of full orbits of the spacecraft around the Sun n_{rev} are optimized, the departure date t_0 and the duration of the flight Δt are varied. As in the previous

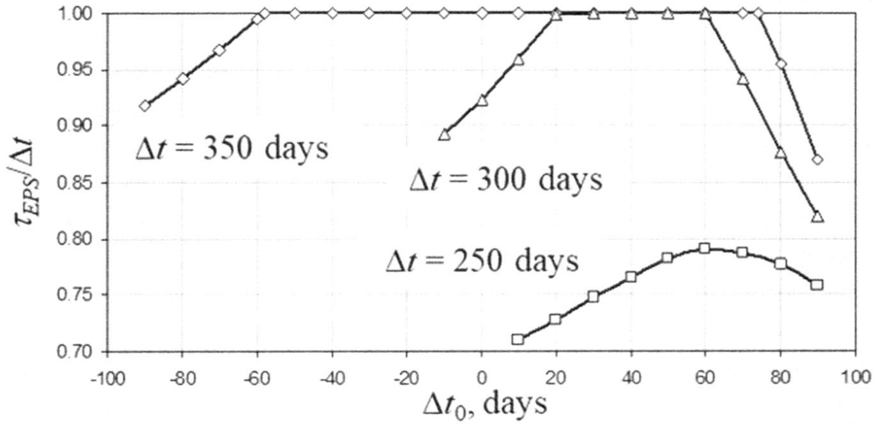

Fig. 9.35 Ratio of EPS operating time to total flight time as a function of departure date offset with an optimal departure hyperbolic excess of velocity

example, we assume $\gamma_3 = 50$ kg/kW, $a_t = 0.1$, $\eta = 0.7$. The Proton-M LV with the Breeze-M upper stage is considered for the spacecraft insertion onto the departure trajectory.

Table 9.8 presents the main parameters of the optimal flight trajectories to Mercury for a flight duration in the range of 1000–3600 days, and Fig. 9.36 shows the projections of the optimal flight trajectories to Mercury onto the ecliptic plane and the time dependence of thrust T and switching function S for the flight durations of 1000, 2400 or 3000 days.

From the data presented in the table it is clear that in the considered range of flight duration, the hypothesis about the optimality of the minimum thrust at the optimal value of the departure hyperbolic excess velocity and the launch date for the flight to Mercury close to the optimal one cannot be confirmed, although the obtained values of the optimal thrust turned out to be close to its minimum value. As the flight duration increases, the total duration of the coasting sections decreases and reaches

Table 9.8 Main parameters of optimal flight trajectories to Mercury

t_0	Δt	n_{rev}	T	I_{sp}	$V_{\infty 0}$	m_0	m_p	P	m_u	$\tau_{EPS}/\Delta t$
–	day	–	mN	s	m/s	kg	kg	W	kg	–
23.IV.2025	1000	6	1021.7	4394.1	1904.6	6044.9	2007.9	31,447.1	2263.9	0.98017
04.V.2025	1200	7	870.4	4895.3	1883.5	6054.0	1839.0	29,845.2	2538.8	0.97833
13.V.2025	1800	11	614.3	6207.2	1621.3	6159.5	1553.8	26,710.5	3114.8	0.99000
10.V.2026	2400	15	481.5	7291.1	1375.3	6245.7	1391.3	24,589.9	3485.8	0.99641
13.V.2025	3000	19	397.1	8244.3	1219.0	6294.0	1270.9	22,931.3	3749.5	0.99829
13.V.2025	3600	23	339.7	9096.2	1078.5	6333.0	1182.8	21,645.7	3949.6	0.99855

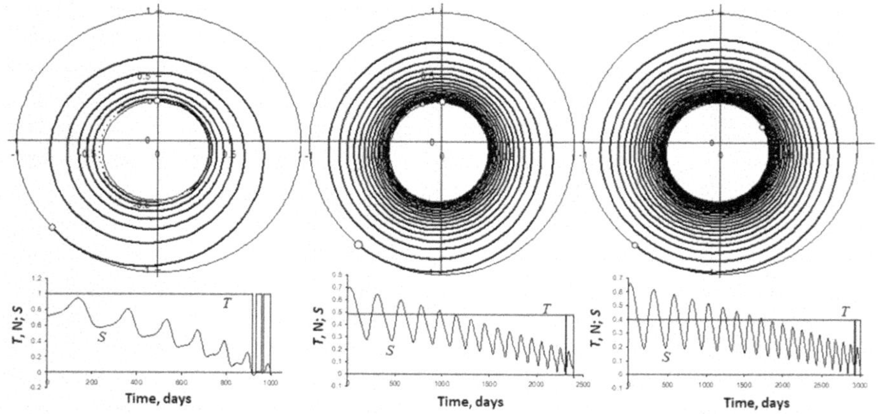

Fig. 9.36 The optimal trajectories to Mercury (top) and the time dependence of thrust T and the switching function S (bottom) for the duration of the flight 1000 days (left), 2400 days (middle), and 3000 days (right)

0.145% of the total flight duration at $\Delta t = 3600$ days. In fact, at $\Delta t \geq 2400$ days, the trajectory has only one small coasting section on the last orbit, the magnitude of which decreases with increasing Δt. It is quite possible that with further growth of Δt the trajectory will not contain any coasting sections and, therefore, the minimum thrust value will be optimal.

In conclusion, we will consider the joint optimization of the control and the main trajectory and design parameters for a flight to Mars when using the Soyuz-2.1B LV with the Fregat upper stage to launch the spacecraft onto the departure trajectory. Let's consider the following values of the main design parameters of the first category: specific mass of the solar panels $\gamma_1 = 10$ kg/kW, specific mass of the solar panel drives and other elements of the solar power supply system and electric propulsion system $\gamma_2 = 15$ kg/kW, $a_t = 0.1$, $\eta = 0.7$.

As a result of solving the boundary value problem with a given flight duration $\Delta t = 400$ days and with the initial approximation for the launch date of January 1, 2031, the following optimal values of the main trajectory and design parameters were obtained: departure hyperbolic excess velocity $V_{\infty 0} = 1144.144$ m/s; departure date—December 17, 2030, 6 h 32 min 22.3 s; $P_{BOL} = 12741.533$ W, $c = 33102.249$ m/s, and $r_{\min} = 1.36252356$ AU. The optimal spacecraft departure mass is 1964.147 kg, the propellant mass $m_p = 261.372$ kg, the PSFS final mass is 26.137 kg, the mass of the solar power supply system and electric propulsion system is 244.092 kg, and the spacecraft useful mass is $m_u = 1432.546$ kg.

The optimal trajectory from Earth to Mars is shown in Fig. 9.37. Figure 9.38 shows the optimal control, including the time dependencies of thrust and switching function (the upper left plot), the time dependencies of the pitch and yaw (the upper right plot), the time dependence of the spacecraft heliocentric offset (the lower left plot), and the time dependencies of the conjugate variables $p_{P\,BOL}$, p_c, and $p_{r\min}$ (the lower right plot). The functions $p_{P\,BOL}(t)$ and $p_c(t)$ have corner points at the moments of thrust switching t^*: $S(t^*) = 0$, and the function $p_{r\min}(t)$ has a corner point at t^{**}: $r(t^{**}) = r_{\min}$.

The relatively high optimum value of $r_{\min}$ was unexpected. In fact, the thrust is reduced in the last quarter of the flight only and as a result is throttled to only 93.2% of the maximum value of 0.290272 N. The maximum power consumption of the electric propulsion system is 6863.39 W, which is a little more than half of the power P_{BOL} generated by the solar panels at the heliocentric distance of 1 AU.

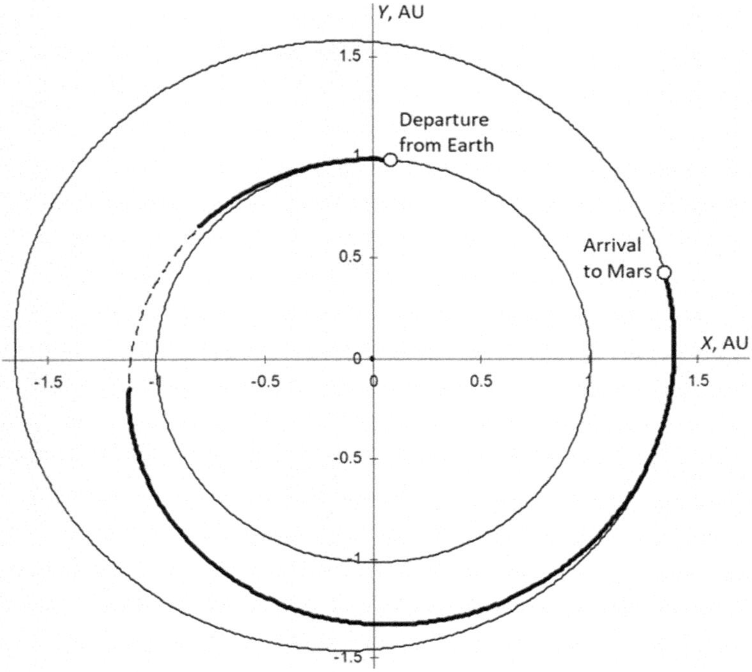

Fig. 9.37 The optimal trajectory from Earth to Mars

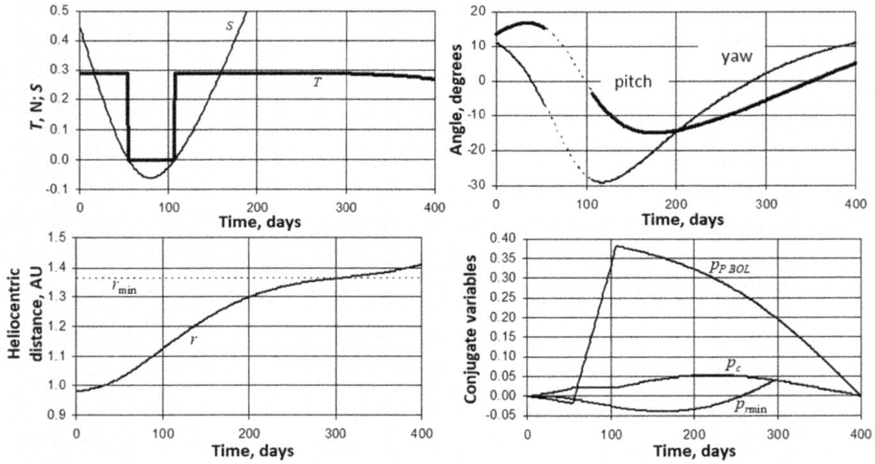

Fig. 9.38 Time dependencies of thrust T, switching function S (top left), pitch and yaw angles (top right), heliocentric spacecraft removal (bottom left), and variables associated with P_{BOL}, c and $r_{\min}$ (bottom right) on the optimal trajectory to Mars

9.7 Conclusion

This section presents a number of fairly effective numerical continuation methods (the homotopy methods) for optimizing low-thrust trajectories based on the maximum principle. Using the developed algorithm for sequential calculation of the LP and LT problems, typical heliocentric low-thrust trajectories can be optimized without choosing an initial guess for the initial values of conjugate variables. The methods have been implemented in the form of numerical algorithms and appropriate software allowing a practical assessment of their computational stability and efficiency. Based on the results of the solution of a number of typical problems of interplanetary transfer, it was revealed that a variant of the method of continuing LP-trajectories into an LT-trajectory with calculation of the sensitivity matrix by the complex step method works stably if the level of thrust acceleration in the LT-problem exceeds the level of thrust acceleration in the LP-problem by no more than 3–5 times (at a greater ratio of thrust acceleration levels in the LT problem and in the LP problem generating it, the ODE system (9.1.15) apparently becomes too stiff as the continuation parameter τ approaches 1, and to solve such problems it is necessary to use special methods of numerical integration). The maximum angular distance, for which a solution was obtained, was about 50 orbits. The developed method for continuing LP-trajectories and LT-trajectory has demonstrated its performance on the "near-critical" trajectories, which are characterized by the presence of very short thrusting or coasting sections that appear or disappear with a small change in the boundary conditions or spacecraft parameters.

Section 9.3 presents a method for calculating the boundary of the domain of existence of trajectories with finite thrust on the specific impulse (or mass flow rate)—thrust plane. The method is based on the sequential solution of the problem of calculating trajectories with the minimum thrust acceleration and with the minimum thrust, each of which is guaranteed to have a solution.

The application of the developed method makes it possible to calculate:

- The minimum thrust as a function of the available specific impulse;
- The minimum allowable specific impulse;
- The lower limit of the final spacecraft mass as a function of the available specific impulse;
- The lower limit of the jet power of the electric propulsion system as a function of the available specific impulse.

In addition, for known values of thrust and specific impulse and for a given range of flight times, the method allows one to estimate the width of the launch window.

A method has been developed for continuing the minimum thrust trajectory with over the flight duration, which makes it possible to effectively construct contour plots of the main characteristics of such trajectories on the launch date—flight time plane.

The use of the developed method for calculating the minimum thrust trajectories allows us to diagnose the existence of solutions to the LT problem, which opens the way to the construction of almost regular methods for the optimization of LT trajectories with the possibility of complete automation of calculations.

This section presents the necessary conditions for the joint optimality of the control of the thrust vector of the electric propulsion system, the departure date, the vector of the departure hyperbolic excess velocity and the main design parameters of the spacecraft with the electric propulsion system for the direct flight between two planets in the central Newtonian gravitational field of the Sun in a fixed time. Variants of a single-mode electric propulsion system with fixed values of thrust and specific impulse and of an electric propulsion system with controlled thrust (power consumption) and constant specific impulse were considered.

In typical cases, the dependence of the useful mass of a spacecraft with electric propulsion system on the magnitude of thrust and specific impulse in the vicinity of the maximum was found to be rather flat: for a flight to Mars lasting about a year, a change of 10% in the specific impulse and of 3% in thrust from the optimal values leads to a decrease of 0.3% in the useful mass.

It was found that in a number of cases, with a sufficiently long flight duration close to the optimal departure date, the optimal departure hyperbolic excess of velocity and specific impulse, the optimal value of thrust is equal to its minimum value. On such trajectories, the electric propulsion system operates constantly (there are no coasting sections in the trajectory). Such optimal trajectories have been found for Earth-Mars flights of 300 days and longer, but not for Earth-Mercury flights of up to 3600 days with an optimal angular distance. All of the found optimal flight trajectories to Mercury contain at least one coasting segment of short duration. This may be due to the relatively large values of the eccentricity and inclination of Mercury's orbit or to the insufficient total flight duration.

The obtained necessary optimality conditions, in addition to other design parameters of the spacecraft, make it possible to compute the optimal installed power of the solar power supply system and the optimal maximum power consumed by the adjustable electric propulsion system. The considered example of the optimization of the flight to Mars of a spacecraft with an adjustable solar electric propulsion system showed that with a small specific mass of the PSS and EPS, the optimal installed power of the solar PSS can be twice the optimal maximum power consumed by the EPS, while the optimal current value of thrust of the adjustable electric propulsion system does not decrease below 90% of the optimal maximum value. In this case, replacing a power-controlled electric propulsion system with a single-mode uncontrolled electric propulsion system will not result in a noticeable decrease in the useful mass of the spacecraft.

References

1. Petukhov, V.G.: Method of continuation for optimization of interplanetary low-thrust trajectories. Cosmic Res. **50**, 249–261 (2012). https://doi.org/10.1134/S0010952512030069
2. Petukhov, V.G.: Homotopic approach to low-thrust trajectory optimization: numerical technique and tools. In: 4th International Conference on Astrodynamics Tools and Techniques, 3–6 May 2010, ESA/ESAC, Madrid, Spain, p. 8. ESA Proceedings WPP-308
3. Petukhov, VG.: Minimum-thrust problem and its application to trajectory optimization with thrust switchings. IAC-13-C1.6.2, Beijing, 2013, p. 9
4. Ivanyukhin, A.V., Petukhov, V.G.: The thrust minimization problem and its applications. Cosm. Res **53**(4), pp. 300–310 (2015). https://doi.org/10.1134/S0010952515040048. EDN UFAWMP
5. Petukhov, V.G.: Joint optimization of the low-thrust trajectory and the main design parameters of electric propulsion system. IAC-15-C1.1.5, Jerusalem, p. 9 (2015)
6. Petukhov, V.G., Ivanyukhin, A.V., Sang Wook, W.: Joint optimization of control and main trajectory and design parameters of an interplanetary spacecraft with an electric propulsion system. Cosm. Res. **57**(3), 188–203 (2019). https://doi.org/10.1134/S0010952519030079. EDN IDNXJZ
7. Petukhov, V.G.: Optimization of interplanetary trajectories for spacecraft with ideally regulated engines using the continuation method. Cosmic Res. **46**, 219–232 (2008). https://doi.org/10.1134/S0010952508030052
8. Proton launch system mission planner's guide. Revision 7. International Launch Services, p. 395 (2009)
9. Soyuz user's manual. Arianespace, Issue 2, Revision 0, March 2012, p. 244
10. Egorov, V.A., Gusev, L.I.: Dynamics of flights between the earth and the moon. Nauka, p. 544 (1980)
11. Mishin, V.P. (ed.): Mechanics of space flight. Mech. Eng. 408 (1989)
12. Bate, R.R., Mueller, D.D., White, J.E.: Fundamentals of Astrodynamics. Dover Publications, New York (1971)

Chapter 10
Optimization of Complex Schemes of Interplanetary Low Thrust Transfers

10.1 Introduction

Let us consider the problem of computing the optimal flight trajectory of a spacecraft with an electric propulsion system from the departure planet to the target planet using one or several gravity assist maneuvers (GAM) at some internal points of the trajectory. In the general case, it is necessary to determine the optimal program for controlling the thrust vector of the electric propulsion system (thrust direction, number and location of the thrusting sections on the trajectory), the optimal conditions for the GAM (including the time of the GAM, the magnitude of the hyperbolic excess velocity, the direction of the vectors of the approach and departure hyperbolic excess velocities) and optimal conditions for the departure from the departure planet and approach to the target planet (magnitude and direction of the hyperbolic excess velocity, departure and arrival times). Maximization of the final spacecraft mass or minimization of propellant is considered as the optimization criterion.

On the heliocentric transfer section, a mathematical model of motion is used that is described in the Sect. 7.6, including the control characteristics of the electric propulsion system (7.5.1)–(7.5.3), cost function (7.5.5), optimal motion Eq. (7.5.10), optimal control (7.5.7)–(7.5.9) and boundary conditions of the form (7.4.12)–(7.4.15). Of course, due to the presence of GAMs, this model must be supplemented with specific boundary conditions.

10.2 Mathematical Model of the Gravity Assist Maneuver

The possibility of using a gravity assist maneuver to form the trajectory of a spacecraft with an electric propulsion system is being considered. It is assumed that the spacecraft motion in the sphere of influence of the planet during a gravity assist maneuver occurs with the electric propulsion system turned off. In the context of

A. S. Filatyev and V. G. Petukhov, *Through Optimization of Aerospace Vehicle Trajectories by the Pontryagin Maximum Principle*, Studies in Systems, Decision and Control 579, https://doi.org/10.1007/978-3-031-80756-5_10

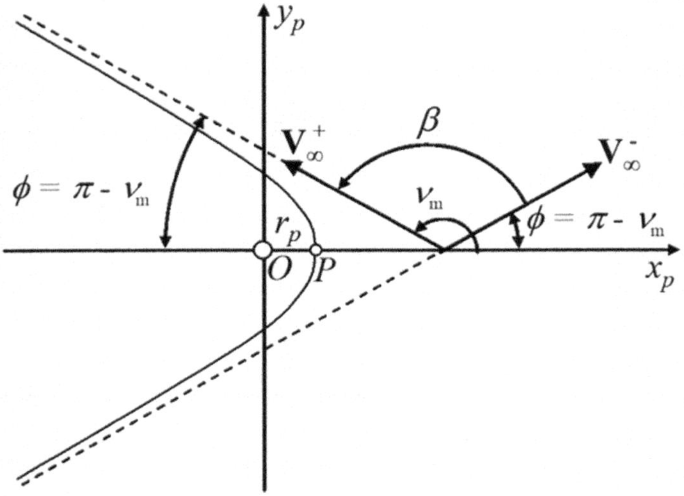

Fig. 10.1 The scheme of rotation of the hyperbolic excess velocity vector in a gravity assist maneuver

the ZSOI method, as a result of a gravity assist maneuver, an instantaneous change in the heliocentric spacecraft velocity vector occurs that is equivalent to a rotation of the hyperbolic excess velocity vector by a certain angle limited by the minimum permissible altitude of the planet flyby. The diagram of the hyperbolic excess velocity vector rotation during a passive gravity maneuver is shown in Fig. 10.1.

Angle between the vectors of approach $\mathbf{V}_\infty^-$ and departure $\mathbf{V}_\infty^+$ hyperbolic excess velocities is given by the following relation:

$$\beta = 2\arccos\left(-\frac{1}{1+\frac{r_p V_\infty^2}{\mu_{pl}}}\right) - \pi,$$

where r_p is the radius of the pericenter of the hyperbolic planetocentric orbit, $V_\infty = |\mathbf{V}_\infty^-| = |\mathbf{V}_\infty^+|$ is the absolute value of the hyperbolic excess velocity, and μ_{pl} is the gravitational parameter of the planet. Since the distance of the planet flyby is limited by a certain minimum permissible value $r_{p\min}$, the angle of rotation of the hyperbolic excess velocity vector cannot be greater than $\beta_{\max}$:

$$\beta_{\max} = 2\arccos\left(-\frac{1}{1+\frac{r_{p\min} V_\infty^2}{\mu_{pl}}}\right) - \pi.$$

We will call a GAM with a flyby at a planetocentric distance greater than $r_{p\min}$ as an unlimited gravitational maneuver, and a gravitational maneuver with a flyby at a distance equal to $r_{p\min}$ as a limited gravitational maneuver.

The plane of the GAM, defined by the vectors of the approach and departure hyperbolic excess velocities, can be rotated to an arbitrary angle around the x_p axis in Fig. 10.1 by choosing the orientation of the aiming range vector in the flight plane with the help of small corrective actions on the heliocentric section of the trajectory before the GAM.

10.3 Continuation Method for Solving the Problem of Interplanetary Transfer of a Low-Thrust Spacecraft Using a Gravity Assist Maneuver

Let us consider the boundary conditions of a heliocentric trajectory between two planets with a GAM at some internal point of the trajectory. We will consider the problem within the framework of the ZSOI method, considering that at the moment of the GAM, the hyperbolic excess of the spacecraft velocity instantly rotates through a certain angle $\beta \leq \beta_{\max}$. As an initial approximation, we will use the solutions of the Lambert's problems on two sections of the heliocentric trajectory separated by the point of the GAM, while we will allow the possibility of a discrepancy between the values of the approach and departure hyperbolic excess velocities during a GAM on such an initial approximation trajectory and will not consider restrictions to the angle of rotation of the hyperbolic excess vector on the obtained solutions of Lambert problems. Of course, as a result of solving such Lambert problems, some initial (for the planet of departure) and final (for the planet of arrival) hyperbolic velocity excesses will be obtained, which, in the general case, do not coincide with the required ones. The goal of the work presented in this section is to construct an algorithm for continuing this initial trajectory, glued together from the Keplerian arcs (solutions of the Lambert problem), into an optimal solution to the problem of interplanetary flight of a spacecraft with limited thrust and GAM.

10.3.1 Boundary Conditions at the Initial and Final Points of the Trajectory

When using the ZSOI method in a problem with a given departure hyperbolic excess velocity, the initial conditions at time t_0, together with the transversality conditions, coincide with the system (7.4.13), supplemented by the initial conditions for the conjugate variables:

$$\mathbf{r}(t_0) = \mathbf{r}_0(t_0),$$
$$\mathbf{v}(t_0) = \mathbf{v}_0(t_0) + \mathbf{V}_{\infty 0}\frac{\mathbf{p}_{v0}}{p_{v0}},$$
$$m(t_0) = m_0,$$

$$\begin{aligned}\mathbf{p}_\mathbf{r}(t_0) &= \mathbf{p}_{\mathbf{r}0},\\ \mathbf{p}_\mathbf{v}(t_0) &= \mathbf{p}_{\mathbf{v}0},\\ p_m(t_0) &= p_{m0},\end{aligned} \tag{10.2.1}$$

where $\mathbf{r}_0$, $\mathbf{v}_0$ are the vectors of coordinates and velocity of the departure planet, m_0 is the initial spacecraft mass at the moment t_0 of leaving the sphere of influence of the planet, and $\mathbf{p}_{\mathbf{v}0}$, $\mathbf{p}_{\mathbf{r}0}$, p_{m0} are the initial values of the conjugate variables that are the unknown parameters of the boundary value problem.

The final conditions at the moment t_f of the spacecraft arrival to the boundary of the sphere of action of the target planet within the framework of the method of point spheres of action for the tracking problem are the conditions for equality of coordinates and velocity components of the spacecraft and the target planet, supplemented by the transversality condition:

$$\begin{aligned}\mathbf{r}(t_f) &= \mathbf{r}_f(t_f),\\ \mathbf{v}(t_f) &= \mathbf{v}_f(t_f),\\ p_m(t_f) &= 0,\end{aligned} \tag{10.2.2}$$

and for the flyby problem they have the form:

$$\begin{aligned}\mathbf{r}(t_f) &= \mathbf{r}_f(t_f),\\ \mathbf{p}_\mathbf{v}(t_f) &= 0,\\ p_m(t_f) &= 0,\end{aligned} \tag{10.2.3}$$

where $\mathbf{r}_f$, $\mathbf{v}_f$ are the coordinate and velocity vectors of the target planet. Thus, the system (10.2.2) or (10.2.3) defines 7 scalar boundary conditions at the end point of the trajectory.

It should be noted that the initial conditions in the form of (10.2.1), which are traditionally used to analyze the trajectories of a low-thrust spacecraft within the framework of the ZSOI method, have a number of disadvantages. The first of them is the presence of the modulus of the Lawden primer-vector $p_{v0} = |\mathbf{p}_{\mathbf{v}0}|$ in the denominator in the initial conditions for the spacecraft velocity. This makes it impossible to use such conditions when continuing the spacecraft trajectory from the Keplerian arc, since $p_v(t) \equiv 0$ on it. Second (which is practically unimportant for the initial boundary conditions, but may turn out to be significant for the conditions at the point of the GAM), the transversality condition for the velocity with a non-zero hyperbolic excess of velocity, written in the second line of (10.2.1) is true only for the values of the hyperbolic excess velocity modulus less than a certain critical value. To overcome this difficulty, the approach described in Sect. 9.2 is used. In particular, we can use (9.2.10) instead of (10.2.1) as the initial conditions.

10.3.2 Boundary Conditions at the Gravity Assist Maneuver Point

In this study, boundary conditions at the GAM point are used in the form obtained in [1–3]. These conditions are immersed in a one-parameter family depending on the continuation parameterτ to implement the capability of smooth continuation of a piecewise Keplerian trajectory into an optimal LT trajectory. At the moment of the GAM, within the framework of the ZSOI method, the hyperbolic excess vector rotates through an angle $\beta \leq \beta_{\max}$, while the absolute value of the hyperbolic excess velocity remains unchanged.

If the required rotation angle of the hyperbolic excess velocity is less than the maximum, the boundary conditions together with the transversality conditions at the GAM time t_1 have the form:

$$\begin{aligned}
&\mathbf{r}(t_{1-}) - \mathbf{r}_1(t_1) = 0,\\
&\mathbf{p}_\mathbf{v}(t_{1-}) - \lambda_1 \mathbf{V}_{\infty 1-} = 0,\\
&\mathbf{p}_\mathbf{v}(t_{1+}) - \lambda_1 \mathbf{V}_{\infty 1+} = 0,\\
&\mathbf{V}_{\infty 1+}^{\mathrm{T}} \mathbf{V}_{\infty 1+} - \left(\mathbf{V}_{\infty 1-}^{\tau}\right)^{\mathrm{T}} \mathbf{V}_{\infty 1-}^{\tau} = 0,
\end{aligned} \tag{10.2.4}$$

where $\mathbf{V}_{\infty 1-} = \mathbf{v}(t_{1-}) - \mathbf{v}_{pl1}(t_1)$ is the approach hyperbolic excess velocity, $\mathbf{V}_{\infty 1+}$ is the departure hyperbolic excess velocity, $\mathbf{V}_{\infty 1-}^{\tau} = (1-\tau)\mathbf{V}_{\infty 1-}^{lamb} + \tau \mathbf{V}_{\infty 1-}$, $\mathbf{V}_{\infty 1-}^{lamb}$ is the approach hyperbolic excess velocity obtained from the solution of the Lambert problem, $\mathbf{r}_1$, $\mathbf{v}_1$ are the coordinate and velocity vectors of the planet, near which the GAM is performed, and λ_1 is the Lagrange multiplier. In total, at the point of the GAM, 10 equalities (10.2.4) must be satisfied, which are included in the boundary conditions of the boundary value problem, and 10 variables are added to the unknown parameters of the boundary value problem:

$$\lambda_1, \mathbf{p}_{\mathbf{r}1+}, \mathbf{p}_{\mathbf{v}1+}, \mathbf{V}_{\infty 1+}. \tag{10.2.5}$$

Here, the "+" sign in the subscript means that the corresponding values are calculated immediately after the GAM is performed, at $t = t_{1+}$.

The initial conditions for integrating the second section of the heliocentric trajectory (after the GAM) will have the form:

$$\begin{aligned}
&\mathbf{r}(t_{1+}) = \mathbf{r}_1(t_1),\\
&\mathbf{v}(t_{1+}) = \mathbf{v}_1(t_1) + \mathbf{V}_{\infty 1+},\\
&m(t_{1+}) = m(t_{1-}),\\
&\mathbf{p}_\mathbf{r}(t_{1+}) = \mathbf{p}_{\mathbf{r}1+},\\
&\mathbf{p}_\mathbf{v}(t_{1+}) = \mathbf{p}_{\mathbf{v}1+},\\
&p_m(t_{1+}) = p_m(t_{1-}).
\end{aligned} \tag{10.2.6}$$

If the minimum flyby altitude limit is reached ($\beta = \beta_{\max}$), the boundary conditions will take the form:

$$\begin{aligned}
&\mathbf{r}(t_{1-}) - \mathbf{r}_1(t_1) = 0,\\
&\mathbf{V}_{\infty 1+}^T \mathbf{V}_{\infty 1-} + \tau \cos 2\phi \mathbf{V}_{\infty 1-}^T \mathbf{V}_{\infty 1-} - (1-\tau)\big(\mathbf{V}_{\infty 1-}^{lamb}\big)^T \mathbf{V}_{\infty 1+}^{lamb} = 0,\\
&p_{v1+}^{||} - p_{v1-}^{||} - 2Ap_{v1-}^{\perp} = 0,\\
&p_{v1+}^{\perp} - p_{v1-}^{\perp} = 0,\\
&p_{v1-}^{b} = 0,\\
&p_{v1+}^{b} = 0,\\
&p_{v1-}^{||} - \lambda_1 V_{\infty 1-} = 0,\\
&\mathbf{V}_{\infty 1+}^T \mathbf{V}_{\infty 1+} - \big(\mathbf{V}_{\infty 1-}^{\tau}\big)^T \mathbf{V}_{\infty 1-}^{\tau} = 0,
\end{aligned} \tag{10.2.7}$$

where $\mathbf{V}_{\infty 1+}^{lamb}$ is the departure hyperbolic excess velocity obtained from the solution of the Lambert problem, $p_{v1\pm}^{||}, p_{v1\pm}^{\perp}, p_{v1\pm}^{b}$ are the projections of the approach (with the index " $-$ ") and departure (with the index " $+$ ") primer-vectors onto the directions along the current vector of the hyperbolic excess velocity, along the direction perpendicular to the vector of the hyperbolic excess velocity in the GAM plane and along the normal to the GAM plane, respectively, $\cos 2\varphi = 2\cos 2\varphi - 1$, $\cos\varphi = \frac{v_p^2}{v_p^2 + V_{\infty 1-}^2}$, $v_p^2 = \frac{\mu_{pl}}{r_{p\,\min}}$, $A = \frac{2}{tg\varphi}\frac{V_{\infty 1-}^2}{v_p^2 + V_{\infty 1-}^2}$, and $\mathrm{tg}\varphi = +\frac{\sqrt{1-\cos^2\varphi}}{\cos\varphi}$.

The primer-vector projections are calculated using the following expressions:

$$\begin{aligned}
p_{v1\pm}^{||} &= \mathbf{p}_{v1\pm}^T \mathbf{e}_{\pm}^{||},\\
p_{v1\pm}^{\perp} &= \mathbf{p}_{v1\pm}^T \mathbf{e}_{\pm}^{\perp},\\
p_{v1\pm}^{b} &= \mathbf{p}_{v1\pm}^T \mathbf{e}^{b},
\end{aligned} \tag{10.2.8}$$

where $\mathbf{e}^b = \frac{\mathbf{V}_{\infty 1-} \times \mathbf{V}_{\infty 1+}}{|\mathbf{V}_{\infty 1-} \times \mathbf{V}_{\infty 1+}|}$, $\mathbf{e}_{\pm}^{||} = \frac{\mathbf{V}_{\infty 1\pm}}{|\mathbf{V}_{\infty 1\pm}|}$, $\mathbf{e}_{\pm}^{\perp} = \frac{\mathbf{e}^b \times \mathbf{V}_{\infty 1\pm}}{|\mathbf{e}^b \times \mathbf{V}_{\infty 1\pm}|}$.

The scheme for changing the primer-vector during a GAM with a limited flyby altitude is presented in Fig. 10.2.

Here, the form of representation of the conditions for the primer-vector at a limited GAM is mainly borrowed from [3], taking into account the amendment from the previous section. In [3] it is shown that on the optimal trajectory: $p_{v1\pm}^{\perp} \geq 0$, otherwise the flyby altitude must be greater than the minimum, i.e. it is necessary to use a GAM model with an unlimited flyby altitude. Equations (10.2.7), like (10.2.4), define 10 boundary conditions at the GAM moment, only for the case of a flyby at a minimum altitude. At this point, the variables (10.2.5) remain to be the unknown parameters of the boundary value problem, and the initial conditions for integrating the next section of the heliocentric trajectory are still determined by the system (10.2.6).

If the GAM time t_1 is not fixed, it becomes an additional unknown parameter of the boundary value problem, and another transversality condition is added to the

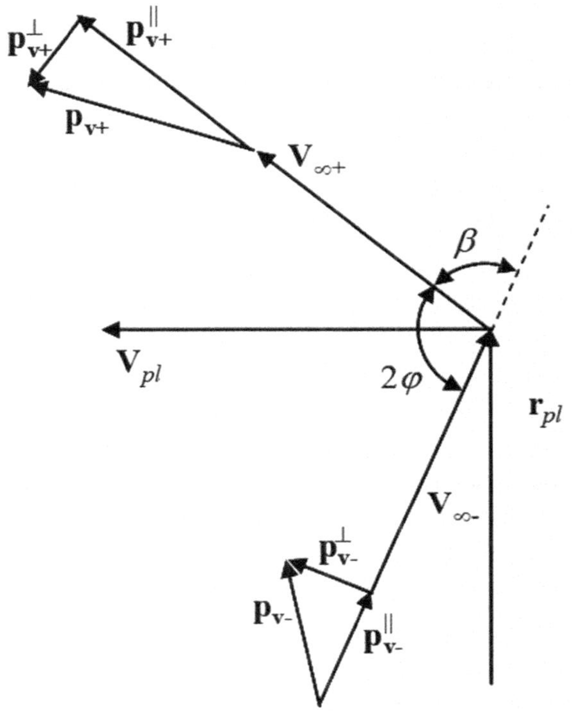

Fig. 10.2 A scheme for changing the primer-vector in a gravity assist maneuver with a constrained flyby altitude

number of boundary conditions:

$$\mathbf{p}_{\mathbf{r}1+}^{T}\mathbf{V}_{\infty 1+} - \mathbf{p}_{\mathbf{r}}(t_{1-})^{T}\mathbf{V}_{\infty 1-} + \left[\mathbf{p}_{\mathbf{v}}(t_{1+}) - \mathbf{p}_{\mathbf{v}}(t_{1-})\right][\Omega_{\mathbf{r}}(t_1) - \dot{\mathbf{v}}_1(t_1)] = 0. \tag{10.2.9}$$

Of course, if the motion of a planet is modeled as the motion of a massless point in the same force field Ω in which the motion of the spacecraft is modeled, the last term in (10.2.9) becomes 0.

10.3.3 Features of the Method of Solving the Boundary Value Problem

Thus, the boundary value problem of the optimization of trajectory with one gravitational maneuver has 18 boundary conditions (9.2.12), (10.2.4) or (10.2.7), (10.2.2) or (10.2.3) and 18 unknown parameters $\mathbf{z} = (\mathbf{V}_{\infty 0}, \lambda_0, \mathbf{p}_{\mathbf{r}0}, p_{m0}, l_1, \mathbf{p}_{\mathbf{r}1+}, \mathbf{p}_{\mathbf{v}1+}, \mathbf{V}_{\infty 1+})^T$ for a fixed time of such a maneuver. With a free time of the gravity maneuver, the number of boundary conditions and unknowns increases by one: (10.2.9) is added to the boundary conditions, and t_1 is added to the unknown parameters.

As an initial approximation, as noted above, the solution of two Lambert problems is used (the flight from the departure planet to the GAM planet and the flight from the GAM planet to the target planet). In the accepted notation, this initial approximation has the form:

$$\mathbf{z}_0 = \mathbf{z}|_{t=0} = \left(\mathbf{V}_{\infty 0}^{lamb}, 0, 0, 0, 0, 0, 0, \mathbf{V}_{\infty 1+}^{lamb}\right)^{\mathrm{T}}, \tag{10.2.10}$$

i.e. as an initial approximation, zero values are used for all parameters, except for the vectors of departure hyperbolic excess velocity from the departure planet and the GAM planet. As an initial approximation for these parameters, the corresponding vectors obtained from solving the Lambert problems for the first and second heliocentric sections, respectively, are used. For the free gravity maneuver time t_l, the initial approximation for t_1 at $\tau = 0$ is chosen to be equal to the value used in solving the Lambert problems. This choice of initial approximation makes it possible to guarantee the fulfillment of all boundary conditions at the beginning of the continuation (at $\tau = 0$). The use of a correctly constructed continuation procedure from this initial approximation allows, for a smooth problem, to maintain the equality of residuals of the boundary value problem to zero (fulfillment of the boundary conditions) over the entire continuation interval from$\tau = 0$ to $\tau = 1$.

Of course, using the above method it is possible to optimize trajectories with any number of GAMs n_{gm}. In this case, the total number of boundary conditions and unknown parameters of the boundary value problem is equal to $8 + 10n_{gm}$ for the fixed times of GAMs or $8 + 11n_{gm}$ for the free times of GAMs. It is also possible to optimize trajectories without GAMs, in this case $n_{gm} = 0$, the number of boundary conditions and unknown parameters of the boundary value problem is 8, and in fact the problem of continuation from the trajectory of coasting motion, which is a solution to the Lambert problem, into the optimal trajectory of a spacecraft with an electric propulsion system is solved.

Since the problem of fuel minimization for a fixed flight time is not smooth due to the relay nature of the thrust function δ, a smooth approximation δ_3 of the thrust function (see Sect. 9.2, (9.1.13)), in the limit converging to $\delta_2 \approx \delta$, is required.

As the equations of optimal motion, a combination (homotopy) of differential equations of motion is used for the LP- and LT-problems in the form (9.1.15). These equations of motion coincide with the equations of motion of a spacecraft for the LP-problem at the continuation parameter $\tau = 0$ and go over into the equations of motion for the LT-problem at $\tau = 1$. This allows us to avoid singularity when using the patched-conic initial approximation at$\tau = 0$, if the vector conjugate to the velocity is 0.

To ensure universality, let's replace the continuation parameterτ in the system (9. 1.15) with the parameter τ_1:

$$\left.\begin{aligned}
\frac{d^2\mathbf{r}}{dt^2} &= \Omega_{\mathbf{r}} + \frac{(1-\tau_1)\cdot\eta P_{j0}+\tau_1\cdot\delta_3 T}{(1-\tau_1)\cdot\left(P_{j0}+m^2 p_m\right)+\tau_1\cdot m p_v}\mathbf{p_v}, \\
\frac{dm}{dt} &= -\frac{(1-\tau_1)\cdot\eta P_{j0} m^2 p_v^2+\tau_1\cdot\delta_3 T}{(1-\tau_1)\cdot 2\left(P_{j0}+m^2 p_m\right)^2+\tau_1\cdot c}, \\
\frac{d^2\mathbf{p_v}}{dt^2} &= \Omega_{\mathbf{rr}}\mathbf{p_v} + \frac{(1-\tau_1)\cdot P_{j0} p_v^2+\tau_1\cdot\delta_3\left(\frac{\partial T}{\partial\eta}\psi+T\frac{1+p_m}{w^2}\frac{\partial c}{\partial\eta}\right)}{(1-\tau_1)\cdot\left(N_{j0}+m^2 p_m\right)+\tau_1}\frac{\partial\eta}{\partial\mathbf{x}}, \\
\frac{dp_m}{dt} &= \frac{(1-\tau_1)\cdot\eta P_{j0} m p_m p_v^2+\tau_1\cdot\delta_3 T p_v}{(1-\tau_1)\cdot\left(P_{j0}+m^2 p_m\right)+\tau_1\cdot m^2}.
\end{aligned}\right\} \quad (10.2.11)$$

If the parameterτ_1 is set to 0, the LP-trajectory is obtained as a result of solving the boundary value problem by the continuation method. If, the relation $\tau_1 = \tau$ is used when calculating the right-hand sides of (10.2.11), then the LT-trajectory is obtained. The second option is considered the main option, although in the general case it makes sense to first obtain a solution to the LP-problem in order to assess the possibility of obtaining a solution with limited thrust using the obtained thrust profile of the LP-problem (i.e. to assess the probability of the existence of a solution with limited thrust). In fact, as noted above, the sufficient conditions for the existence of a solution with limited thrust under given boundary conditions are the following two conditions: (1) the maximum thrust on the LP- solution does not exceed the specified maximum value of the limited thrust and (2) the characteristic velocity on the LP- solution does not exceed the maximum value provided by a limited thrust engine when using a given maximum mass of the propellant. Of course, a solution to a problem with limited thrust may exist even with some violation of the listed conditions, but if they are met, such a solution is guaranteed to exist.

The third option is to set τ_1 equal to 1 in the right-hand sides of (5.2.24). This option corresponds to the continuation between two problems with the limited thrust and can be used either to refine the solution (to reduce the absolute values of the residuals of the boundary conditions), or to determine trajectories with other (sufficiently close) design parameters of the spacecraft and electric propulsion, or with other boundary conditions. This option works stably until the changed parameters lead to a change in the structure (sequence of coasting and thrusting sections) of the optimal trajectory.

10.4 Numerical Examples

Let us give examples of optimal trajectories with several GAMs, calculated using the method described in Sect. 12.3.

As a first example, let us consider the problem of transferring a spacecraft with an EPS to Jupiter. We assume that the EPS thrust is 0.36 N, the specific impulse is 2700 s, the initial spacecraft mass is 1924.2 kg, and the departure hyperbolic excess velocity is 2300 m/s. Let's consider the trajectory from Earth to Jupiter with gravitational maneuvers near Venus and Earth. The date of departure from Earth was assumed to be 05/16/2015, the duration of the flight along the Earth-Venus section was 200 days, the duration of the flight along the Venus-Earth section was 343 days, and that of the Earth-Jupiter section was 956 days (total flight duration was 1499 days or ~ 4.1 years). The minimum altitude for the flyby of Earth and Venus during the GAMs is assumed to be 400 km.

The projection of the optimal trajectory onto the ecliptic plane is presented in Fig. 10.3 (measurement units along the axes are AU, thick lines correspond to active sections of the trajectory). Figure 10.4 shows the time dependences of the switching and thrust functions, and Fig. 10.5 shows the time dependence of the heliocentric distance of the spacecraft. There were three thrusting sections on the optimal trajectory—one on each section of the flight between the planets (the first section at $14.814952 \leq t \leq 96.372149$ days, the second at $275.180814 \leq t \leq 280.031292$ days, and the third at $543.0 \leq t \leq 643.324874$ days). The last active segment ends at the spacecraft heliocentric distance of 1.375522 AU. The hyperbolic excess velocity at the Venus approach was 5402.112 m/s, at the Earth approach—9613.396 m/s, and at the Jupiter approach—5932.850 m/s. The mass of the spacecraft during the Jupiter approach was 1704.843 kg. In both GAMs, the minimum flyby altitude of 400 km proved to be optimal.

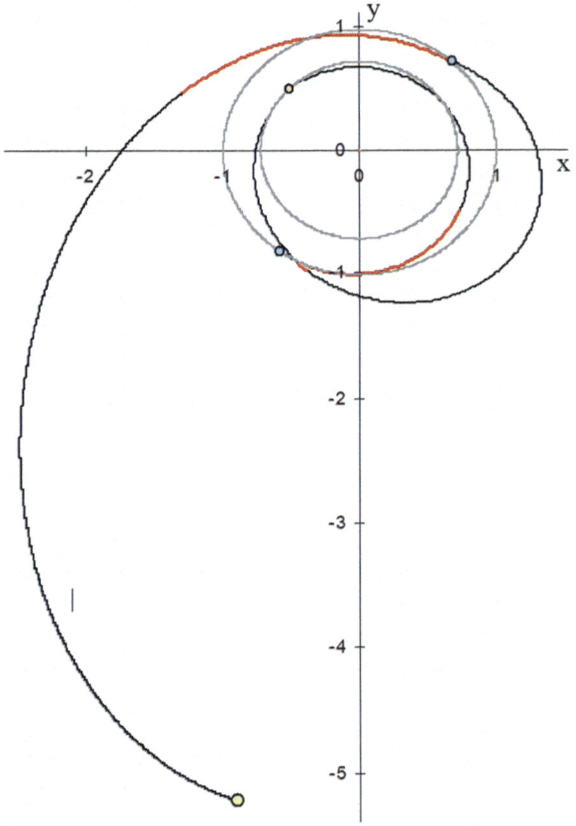

Fig. 10.3 Projection of the optimal Earth-Venus-Earth-Jupiter trajectory onto the ecliptic plane (thick lines are the trajectory sections with the operating EPS)

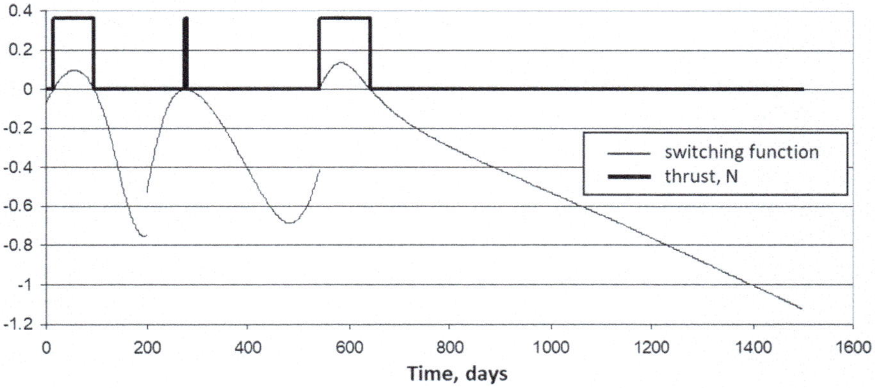

Fig. 10.4 Time dependencies of the switching function and thrust on the optimal trajectory Earth-Venus-Earth-Jupiter

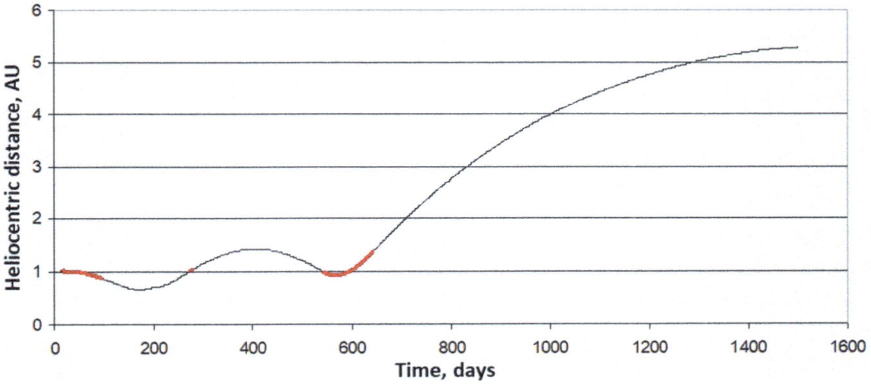

Fig. 10.5 Time dependency of heliocentric distance on the optimal trajectory Earth-Venus-Earth-Jupiter (thick lines correspond to the trajectory sections with the operating EPS)

In particular, the trajectory developed optimization methods have been used in the space mission design and analysis for the heliophysical spacecraft project [4–6]. The task of such a spacecraft is to study the Sun from a short distance and to observe the circumpolar regions of the Sun from out-of-ecliptic positions. The main ideas for designing the trajectory of a heliophysical spacecraft with low thrust and GAMs are as follows:

(1) The injection trajectory is divided into an ecliptic phase and an orbital plane rotation phase. The main task of the ecliptic phase is to reduce the perihelion radius to 30–60 solar radii, and the main task of the orbital plane rotation phase is to increase the orbital inclination. As a result, the required perihelion radius and the required orbital inclination are achieved at different times. The choice

of the phase sequence is defined by the priority of scientific problems and the results of trajectory optimization.

(2) To ensure a final inclination of about 30°, the required hyperbolic excess velocity at the approach of a planet, near which a series of GAMs should be made, should be about half the velocity of the planet's orbital heliocentric motion.
(3) To ensure the required high value of the hyperbolic excess velocity, the trajectories with low thrust and with GAMs near the Earth and Venus are used.
(4) To reduce the perihelion radius and change the spacecraft orbit inclination, it is advisable to use a series of successive GAMs near the same planet, separated by a period of time that is a multiple of the period of revolution of that planet around the Sun. Thus, in the intervals between GAMs, the heliocentric trajectory of the spacecraft should be close to an orbit that is in orbital resonance with the orbit of the planet near which the GAMs are performed.
(5) To reduce the transfer time, it is advisable to use a massive planet with the shortest possible orbital period. Such conditions are best met by Venus and Earth.
(6) To reduce the insertion time, orbital resonances of the lowest possible order should be used (1:1, 2:1, 3:2, 4:3, 5:3).
(7) To reduce the insertion time, low thrust should be used in the trajectory sections between GAMs.
(8) To reduce the mass of the power supply system, it is necessary to minimize the maximum heliocentric distance of the spacecraft in the sections with the electric propulsion system in operation.

Let us consider the heliocentric trajectory of a low-thrust heliophysical spacecraft with the first three GAMs near Venus, Earth and Venus (the EVEV route). Everywhere, unless otherwise specified, the minimum altitude of the flyby over the surfaces of the Earth and Venus during the GAMs is assumed to be 450 km.

General diagram of the EVEV-1 trajectory (its projection onto the ecliptic plane) is presented in Fig. 10.6. The launch windows for such a trajectory are repeated every Venus synodic period, i.e. with a period of 583.92 days (1.6 years).

Let's consider a trajectory option with a moderate value of the minimum heliocentric distance (50–60 solar radii). The departure date of the EVEV-1 trajectory shown in Fig. 10.6 is 05/20/2015, 00:00:00.0 UTC. The initial hyperbolic excess velocity is 2500 m/s. It should be noted that such a hyperbolic excess is close to the minimum allowed in the considered case of using an electric propulsion system with a thrust of 0.275 N and a specific impulse of 4650 s (at a hyperbolic velocity excess of 2400 m/s, there is no solution for a given segment of the flight to Venus). The initial mass of the spacecraft at Earth departure is 1914.3 kg.

The duration of the transfer along the Earth-Venus section is 123 days. This section of the trajectory consists of an initial thrusting and a final coasting sections (the time dependence of thrust is presented in Fig. 10.7). A spacecraft weighing 1872.477 kg approaches Venus with an approaching hyperbolic excess of 4708.845 m/s. The heliocentric orbit of the spacecraft at the time of its first approach to Venus (09/20/

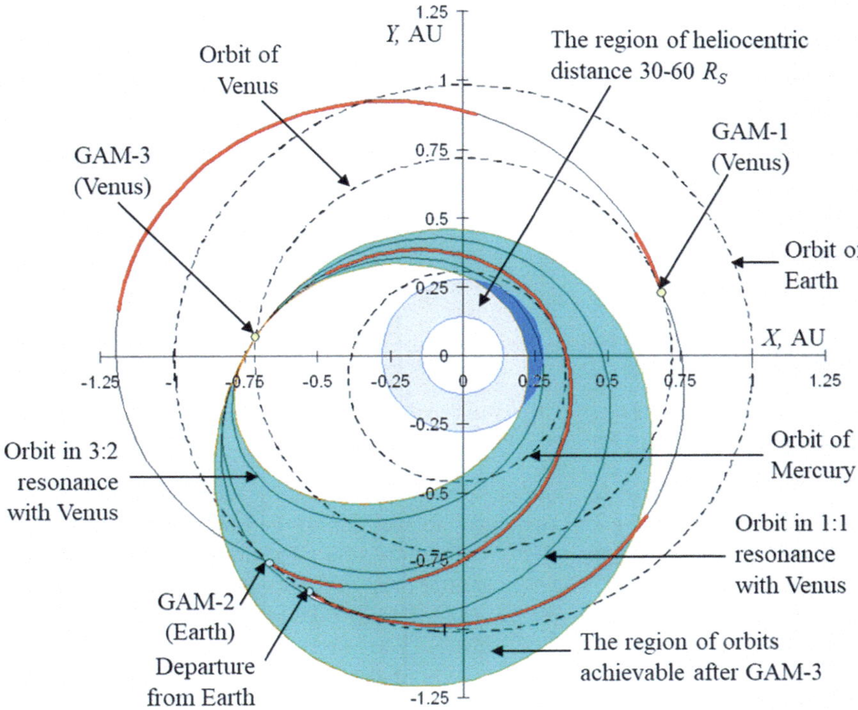

Fig. 10.6 Projection onto the ecliptic plane of the heliocentric trajectory of the heliophysical spacecraft with EPS with the successive gravity assist maneuvers near Venus, Earth and Venus (trajectory EVEV-1). The domain of heliocentric orbits achievable after GAM-3 is highlighted in color

2015 00:00:00.0 UTC) has a perihelion radius of 0.691661 AU, an aphelion radius of 1.007815 AU, and an inclination of 3.5374°.

To perform the first GAM (GAM-1), the spacecraft makes a flyby of Venus at a minimum altitude of 450 km. As a result of GAM-1, the spacecraft will be transferred to a heliocentric orbit with the following parameters: perihelion radius 0.721112 AU, aphelion radius 1.223421 AU and inclination 3.7567°.

The flight from Venus to Earth begins with a short thrusting section and includes a sequence of two thrusting and two coasting sections of the trajectory. In this trajectory, the spacecraft reaches its maximum heliocentric distance of about 1.2 AU. The spacecraft approaches Earth with a mass of 1812.144 kg and an approach hyperbolic excess velocity of 9251.403 m/s. The heliocentric orbit of the spacecraft at the time of Earth approach (05/09/2016 00:00:00.0 UTC) has a perihelion radius of 0.634313 AU, an aphelion radius of 1.203073 AU, and an inclination of 3.6056°.

As a result of the near-Earth GAM (GAM-2, flight altitude 450 km), the perihelion radius is 0.348833 AU, the aphelion radius is 1.022739 AU, and the inclination is 4.5714°.

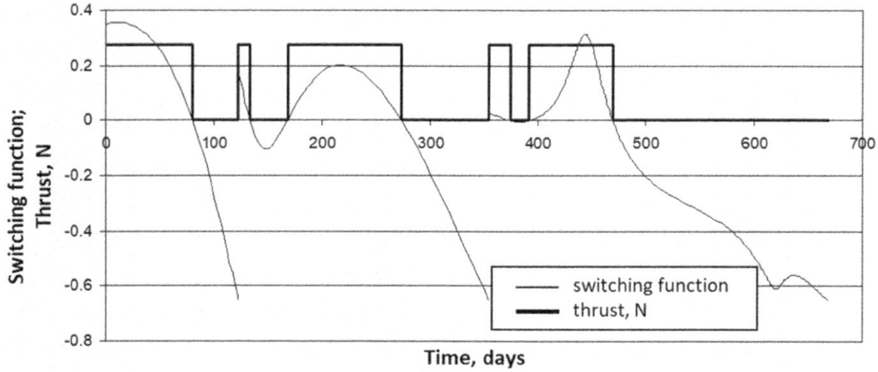

Fig. 10.7 The time dependencies of the thrust and the switching function

The flight from Earth to Venus begins with a thrusting section and includes a sequence of two thrusting and two coasting sections of the trajectory. A spacecraft weighing 1760.733 kg approaches Venus with an approaching hyperbolic excess velocity of 16,527.817 m/s. The heliocentric orbit of the spacecraft at the time of the second approach to Venus (03/19/2017 00:00:00.0 UTC) has a perihelion radius of 0.325508 AU, an aphelion radius of 0.940483 AU and an inclination 4.2665°.

If we assume that the minimum altitude of the Venus flyby in the third gravity assist maneuver (GAM-3) is 450 km, as in the previous ones, we can obtain the region of achievable heliocentric orbits of the spacecraft after GAM-3. In Fig. 10.6, this is the shaded area between two elliptical orbits.

After the GAM-3, the minimum achievable perihelion radius is 0.221587 AU (47.63 solar radii). In this case, the aphelion radius is 0.817183 AU, the inclination is 4.326°, the period of revolution around the Sun is 136.7201 days (0.608435 of the orbital period of Venus).

The maximum achievable perihelion radius after the GAM-3 is 0.455080 AU (97.81 solar radii). In this case, the aphelion radius is 1.231411 AU, the inclination is 4.326°, the period of revolution around the Sun is 282.8326 days (1.258668 of the orbital period of Venus).

It is clear from Fig. 10.6 that the orbits in 1:1 and 3:2 resonances with Venus are in the region achievable after GAM-3. A 1:1 orbital resonance with Venus is achieved at a perihelion radius of 0.410571 AU (88.25 solar radii) and at an inclination of 11.051°, while a 3:2 orbital resonance is achieved at a perihelion radius of 0.264614 AU (56.88 solar radii) and at an inclination of 11.017°.

To perform a further sequence of GAMs near Venus in order to increase the inclination of the spacecraft's orbit, it is necessary to place the spacecraft into a resonant orbit after GAM-3, so that after an integer number of revolutions of Venus and an integer number of revolutions of the spacecraft around the Sun, the next GAM could be performed at the same point in space as the previous one (this is necessary, because, as a result of GAMs, the plane of the spacecraft's orbit rotates about the heliocentric radius vector of Venus at the moment of the GAM). The 1:1 orbital resonance with Venus, with the hyperbolic excess velocity located on the considered trajectory, results in a perihelion radius of the order of 0.4 AU after GAM-3, which is significantly greater than 60 solar radii. Therefore, the insertion of the spacecraft into the 3:2 orbital resonance with Venus as a result of GAM-3 is further considered.

With a 3:2 orbital resonance, the spacecraft approaches Venus every 3 revolutions of the spacecraft and every 2 revolutions of Venus around the Sun, that is, every 449.4 days (1.23 years). The main characteristics of the coasting trajectory of the spacecraft after GAM-3 in a 3:2 orbital resonance with Venus with a sequence of GAMs near Venus to increase the inclination of the spacecraft orbit at a minimum Venus flyby altitude of 450 km are given in Table 10.1. This table uses the following notations: t is the flight time, m is the spacecraft mass, V_∞ is the hyperbolic excess of velocity, α_∞ is the right ascension of the hyperbolic excess velocity in the inertial heliocentric ecliptic coordinate system, δ_∞ is the declination of the hyperbolic excess velocity in the inertial heliocentric ecliptic coordinate system, i is the inclination of the spacecraft's orbit to the ecliptic plane, r_p is the perihelion radius of the spacecraft's orbit, T is the period of the spacecraft's orbit, T_v is the orbital (sidereal) period of Venus, and RS is the radius of the Sun.

From Table 10.1 it is clear that 4 consecutive GAMs near Venus, separated by sections of spacecraft motion in a 3:2 resonance with Venus, allow inclination of the spacecraft orbit to be increased to 28.342° 5.5 years after departure. Leaving the resonant orbit after GAM-6 allows us to increase the inclination by only 0.12°, to 28.462°. The minimum perihelion radius (56.88 solar radii) is reached after GAM-3, 1.83 years after launch.

The important features of the considered trajectory are, first of all, the relatively early termination of the last thrusting section—on the 469th day of the flight, more than one orbit before performing the GAM-3 at Venus. At the end of this section, the electric propulsion system can be separated from the spacecraft if necessary. The second feature is the relatively large minimum heliocentric distance of the spacecraft when flying with an operating EPS (at the level of the heliocentric distance of Mercury), which makes it easier to solve the problem of ensuring thermal conditions during the cruise phases of the flight.

A general view of the trajectory with a series of resonant orbits is shown in Figs. 10.8 and 10.9.

Let us consider the possibility of constructing the EVEV-2 heliocentric trajectories with a smaller perihelion radius (30–50 solar radii) using the same EPS (thrust 0.275 N, specific impulse 4650 s).

The departure date of the trajectory shown in Figs. 10.10 and 10.11 is 05/20/2015, 00:00:00.0 UTC. The initial hyperbolic excess velocity is 2500 m/s, the mass of the spacecraft on the trajectory leaving Earth is 1914.3 kg. The first section of the heliocentric trajectory (the flight from Earth to Venus) practically coincides with the first section of the EVEV-1 trajectory considered earlier.

The transfer from Venus to Earth takes 242 days. The GAM-2 near the Earth is performed on May 19, 2016. The mass of the spacecraft at the Earth approach is 1846.570 kg, and the hyperbolic excess velocity is 8422.340 m/s.

Table. 10.1 Main characteristics of the passive flight trajectory of the spacecraft after GAM-3 on the EVEV-1 trajectory in the 3:2 orbital resonance with Venus with a sequence of GAMs near Venus to increase the inclination of the spacecraft orbit at a minimum Venus flyby altitude of 450 km

		t	m	V_∞	α_∞	δ_∞	i	r_p		T/T_V
		day	kg	km/s	deg	deg	deg	AU	RS	–
1		2	3	4	5	6	7	8	9	10
GAM-3	Before	669.000	1760.733	16.52782	151.35883	– 0.51008	4.266	0.325508	69.96	0.8186
	After		1760.733		138.66182	– 13.0882	11.017	0.264614	56.88	0.6667
GAM-4	Before	1118.416	1760.733	16.52782	138.66182	– 13.0882	11.017	0.264614	56.88	0.6667
	After		1760.733		133.16168	– 30.1360	19.896	0.300795	64.65	0.6667
GAM-5	Before	1567.831	1760.733	16.52782	133.16168	– 30.1360	19.896	0.300795	64.65	0.6667
	After		1760.733		120.68890	– 44.9816	25.798	0.348751	74.96	0.6667
GAM-6	Before	2017.247	1760.733	16.52782	120.68890	– 44.9816	25.798	0.348751	74.96	0.6667
	After		1760.733		97.06901	– 54.3212	28.342	0.382163	82.14	0.6667

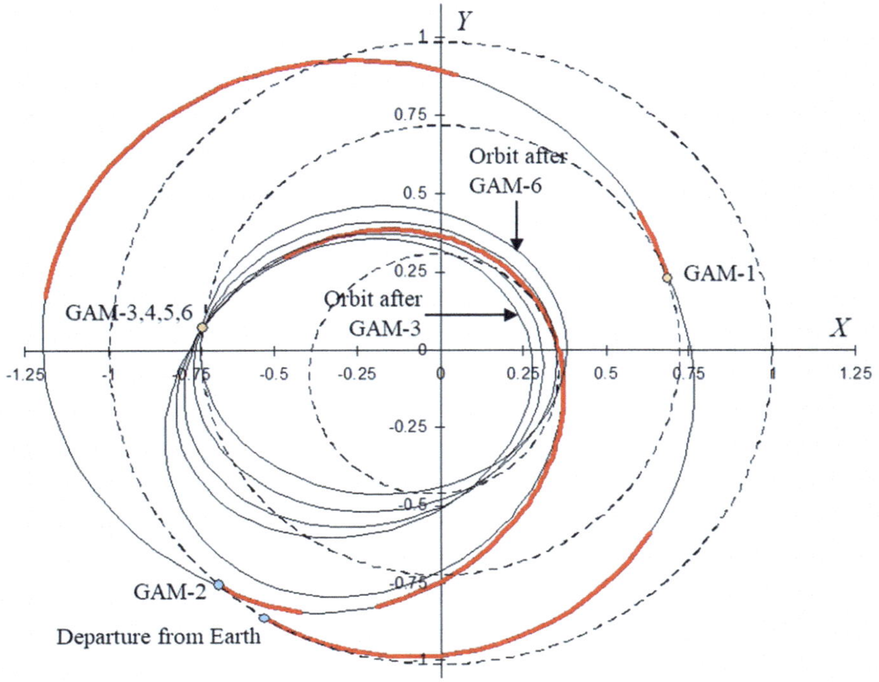

Fig. 10.8 Projection onto the ecliptic plane of a variant of the heliocentric trajectory of a spacecraft with a flight along the EVEVVVV route and with the entry into a 3:2 orbital resonance with Venus after GAM-3

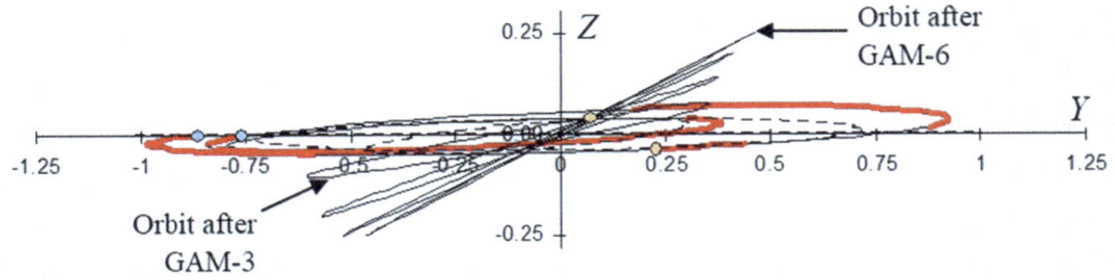

Fig. 10.9 Projection onto the YZ plane of an inertial heliocentric ecliptic coordinate system for a variant of the heliocentric trajectory of a spacecraft with a flight along the EVEVVVV route and with the entry into a 3:2 orbital resonance with Venus after GAM-3

The next leg of the flight from Venus to Earth takes 311 days. The spacecraft approaches Venus on March 26, 2017 with a hyperbolic excess velocity of 16,420.58 m/s. Further motion of the spacecraft is assumed to be coasting. The GAM-3 transfers the spacecraft to an orbit that is in a 5:3 resonance with Venus (that is, the spacecraft makes 5 revolutions around the Sun, while Venus makes 3 revolutions). In this orbit, a perihelion radius of 47.10 solar radii is reached. This orbit is

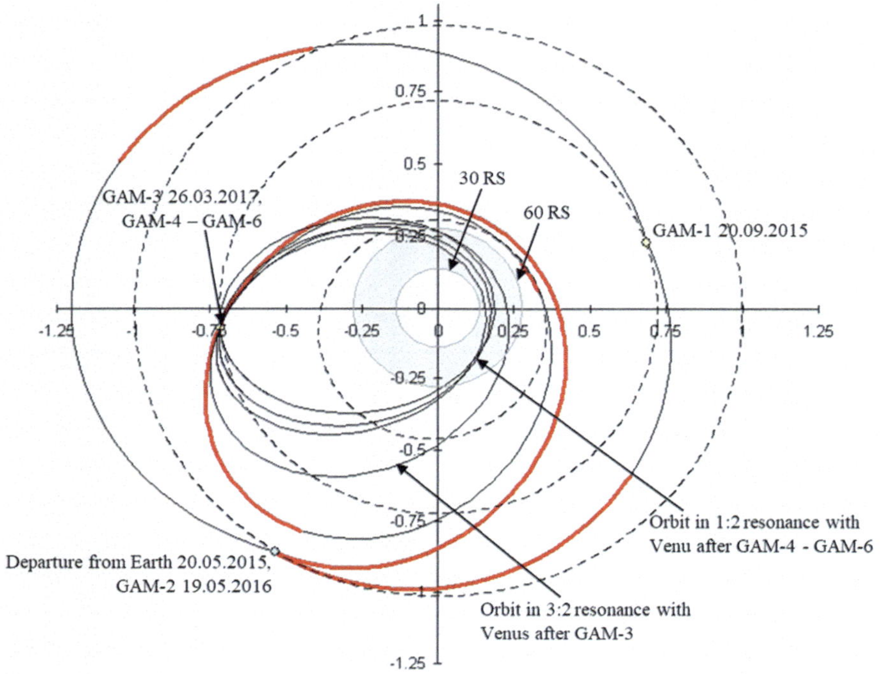

Fig. 10.10 Projection of a variant of the heliocentric trajectory of a spacecraft onto the ecliptic plane with a flight along the EVEVVVV route and using 5:3 and 2:1 orbital resonances with Venus after GAM-3

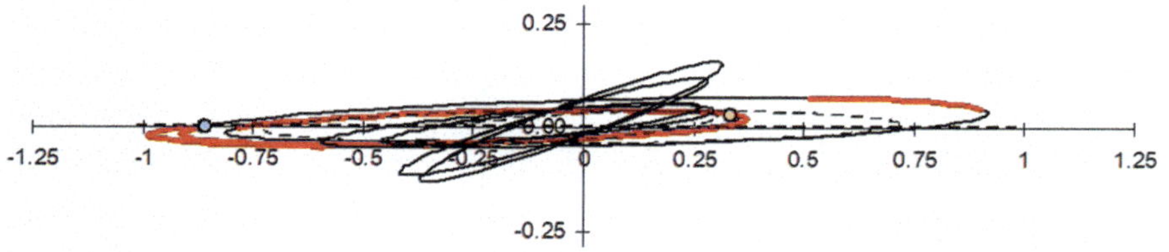

Fig. 10.11 Projection of a variant of the heliocentric trajectory of a spacecraft onto the YZ plane of an inertial heliocentric ecliptic coordinate system with a flight along the EVEVVVV route and using 5:3 and 2:1 orbital resonances with Venus after GAM-3

reached after 676 days (1.85 years) of flight. The 4th GAM near Venus (GAM-4) on the 1350th day (3.7 years after launch) of the flight transfers the spacecraft into a 2:1 orbital resonance with Venus, in which the spacecraft makes 2 revolutions around the Sun during 1 revolution of Venus. After GAM-4, the perihelion radius is 34.21 solar radii. GAM-5 at Venus (1574th day of flight) maintains the 2:1 resonance, and increases the inclination to 19.1° and the perihelion radius to 38.16 solar radii. The final gravity assist maneuver at Venus, GAM-6, is performed on the 1799th day of the flight (4.9 years after launch). As a result of GAM-6, the inclination increases to 23.15°, while the perihelion altitude increases to 41.02 solar radii.

The main characteristics of the coasting trajectory of the spacecraft after GAM-3 in the 5:3 orbital resonance with Venus, and then in the 2:1 orbital resonance with it with a sequence of GAMs near Venus to reduce the perihelion radius and increase the inclination of the spacecraft orbit are given in Table 10.2.

The final EPS operation on this trajectory is performed before GAM-3. The final mass of the spacecraft is 1743.643 kg.

From the results obtained it is clear that a slight variation in the time of the GAMs on the ecliptic phase of the trajectory and the transition to another sequence of resonant orbits after GAM-3 can significantly reduce the minimum perihelion radius.

10.5 Conclusion

This section considers the problem of continuing a heliocentric trajectory with GAMs, obtained as a result of solving a sequence of Lambert problems, into a LT trajectory with GAMs. A mathematical model of a GAM, boundary conditions for GAMs (including transversality conditions), and a description of the continuation method variant used are presented. Numerical examples of heliocentric low-thrust spacecraft trajectories with GAMs are given.

Table. 10.2 Main characteristics of the spacecraft coasting trajectory after GAM-3 on the EVEV-2 trajectory in the 5:3 orbital resonance with Venus and further in the 2:1 orbital resonance with it with a sequence of GAMs near Venus to reduce the perihelion radius and increase the inclination of the spacecraft orbit at the minimum Venus flyby altitude of 450 km

		t	m	V_∞	α_∞	δ_∞	i	r_p		T/T_V
		Day	kg	km/s	deg	deg	deg	AU	RS	–
1		2	3	4	5	6	7	8	9	10
GAM-3	Before	676.000	1743.643	16.42058	160.5652	0.893227	4.0980	0.311978		0.7863
	After		1743.643	16.42058	143.3620	– 4.41	6.9678	0.219111	47.10	0.6000
GAM-4	Before	1350.126	1743.643	16.42058	143.3620	– 4.41	6.9678	0.219111	47.10	0.6000
	After		1743.643	16.42058	125.3978	– 6.35	8.6053	0.159165	34.21	0.5000
GAM-5	Before	1574.834	1743.643	16.42058	125.3978	– 6.35	8.6053	0.159165	34.21	0.5000
	After		1743.643	16.42058	116.3550	– 22.0766	19.1096	0.177542	38.16	0.5000
GAM-6	Before	1799.543	1743.643	16.42058	116.3550	– 22.0766	19.1096	0.177542	38.16	0.5000
	After		1743.643	16.42058	98.23409	– 29.1538	23.1538	0.190845	41.02	0.5000

References

1. Casalino, L., Colasurdo, G.: Indirect methods for the optimization of spacecraft trajectories. In: Modeling and Optimization in Space Engineering. Springer Science + Business Media, New York, pp. 141–158 (2013)
2. Casalino, L., Colasurdo, G., Pastrone, D.: Optimization of ΔV earth-gravity-assist trajectories. J. Guidance Control Dyn. **21**(6), 991–995 (1998)
3. Casalino, L., Colasurdo, G., Pastrone, D.: Optimization procedure for preliminary design of opposition-class mars missions. J. Guidance Control Dyn. **21**(1), 134–140 (1998)
4. Konstantinov, M.S., Loeb, H.W., Petukhov, V.G.: Application of the high-frequency ion engine RIT-22 in the Interhelio-Zond project. Electron. J. Proc MAI 60, 10
5. Loeb, H.W., Petukhov, V.G., Popov, G.A.: Heliocentric trajectories of a spacecraft with ion engines for solar research. Electron. J. Proc. MAI 42, 22 (2011)
6. Konstantinov, M.S., Petukhov, V.G.: One version of a space transport system for research of the sun. In: IAC-11.C4.6.10, 62nd International Astronautical Congress, Cape Town, South Africa, 3–7 Oct 2011

Chapter 11
Optimization of Multi-revolution Flights Between Elliptical Non-complanar Orbits

11.1 Introduction

The problem of optimizing the trajectories of spacecraft with EPS around the planets, in particular around the Earth, differs significantly from the problem of optimizing heliocentric trajectories. This is due to the significantly lower value of Sun's gravitational acceleration on heliocentric trajectories compared to Earth's gravitational acceleration on the near-Earth trajectories at typical spacecraft distances from the center of gravity. The gravitational acceleration from the Sun at a distance of 1 AU is 5.93 mm/s^2, and the gravitational acceleration from the Earth at the altitudes of 200–36,000 km varies from 9.232 to 0.224 m/s^2. As a result, with thrust acceleration of 0.1–1.0 mm/s^2, the characteristic ratio of thrust acceleration to gravitational acceleration is 0.017–0.17 on heliocentric trajectories, and 10^{-5}–$4.5 \cdot 10^{-3}$ on geocentric ones. The difference of 2–4 orders of magnitude in the ratios of thrust acceleration to gravitational acceleration for geocentric and heliocentric trajectories means that typical geocentric trajectories of spacecraft with EPS include a significantly larger number of orbits than typical heliocentric trajectories.

For this reason, to optimize geocentric interorbital trajectories, it is advisable to use the osculating elements as a state vector. To ensure the absence of singularities in the right-hand sides of the differential equations of motion in the vicinity of circular and equatorial orbits, it is advisable to use the equinoctial osculating elements. A mathematical model of the motion of a spacecraft with an EPS in the modified equinoctial elements for the LP and LT problems was presented in Sect. 7. In addition, Sect. 11.2 discusses the formulation of the thrust minimization problem in modified equinoctial elements, and Sect. 11.3 presents a general scheme for solving the LT problem in equinoctial variables, including the sequential solution of the LP, $T_{\min}$ and LT problems. Numerical examples of the use of the presented methods for optimizing multi-orbit trajectories of spacecraft with EPS are given in Sect. 11.4.

A. S. Filatyev and V. G. Petukhov, *Through Optimization of Aerospace Vehicle Trajectories by the Pontryagin Maximum Principle*, Studies in Systems, Decision and Control 579, https://doi.org/10.1007/978-3-031-80756-5_11

11.2 Diagnostics of the Solution Existence: Thrust Minimization Problem

Before we start optimizing the LT trajectory, it is useful to make sure that a solution to this problem exists. The question of the existence of a solution can be answered by considering the problem of thrust minimization [1–3], i.e. the problem of calculating the trajectory of a spacecraft with a minimum thrust and a given specific impulse. For interplanetary trajectories (using Cartesian coordinates and velocity components as the state vector), the thrust minimization problem was considered in Sect. 9.4. In this section, we consider the thrust minimization problem using the equinoctial elements as the state vector and the auxiliary longitude as the independent variable, mainly following the work presented in [3].

Let us consider an unperturbed problem with a fixed angular range and free flight time. In the following, we will designate the problem of minimizing thrust as a $T_{\min}$ problem, and we will refer to trajectories with minimal thrust as $T_{\min}$-trajectories. LT-trajectories with a fixed angular range or flight time exist only if the available thrust is not less than the minimum thrust.

Various representations of the $T_{\min}$-problem cost function are possible. We consider the problem of minimizing the cost function:

$$J = \frac{1}{2}\int\limits_{L_0}^{L_f} T^2 dL = \frac{1}{2}\int\limits_{K_0}^{K_f} T^2 \cdot \frac{dL}{dK} dK = \frac{1}{2}\int\limits_{K_0}^{K_f} T^2 \cdot \left(1 + \frac{dL_K}{dK}\right) dK$$
$$= \frac{1}{2}\int\limits_{K_0}^{K_f} T^2 \cdot \left(1 + \delta\frac{T}{m} k_1 k_2 \xi \sin\psi\right) dK \qquad (11.1.1)$$

for a system of differential Eqs. (7.1.10) with initial conditions (7.1.13) and final conditions (7.1.12). We expand the state vector by the magnitude of thrust T. To do this, we will expand the system (7.1.10) by the formal differential equation

$$dT/dK = 0, \qquad (11.1.2)$$

which corresponds to the constancy of the thrust value during the periods of EPS operation.

Thus, the minimized thrust value is considered as an additional state variable of the considered dynamic system. The minimization of the cost function (11.1.1) is equivalent to the minimization of thrust, since it is constant over the entire trajectory, while the angular flight distance $\Delta L = L_f - L_0$ is fixed.

The Pontryagin function of the $T_{\min}$-problem has the form:

$$H = H_{T\min} + H_t, \qquad (11.1.3)$$

where

$$
\begin{aligned}
H_{T\min} &= -\frac{1}{2}T^2 \cdot \left(1 + \delta\frac{T}{m}k_1k_2 \sin\psi\right) \\
&+ \delta T k_1 \cdot \left[\frac{k_2}{m} \cdot (A_r \sin\vartheta \cos\psi + A_t \cos\vartheta \cos\psi + A_n \sin\psi) - \frac{p_m}{c}\right] \\
&= -\frac{1}{2}T^2 + \delta T k_1 \cdot \left[\frac{k_2}{m} \cdot (A_r \sin\vartheta \cos\psi\right. \\
&\left. + A_t \cos\vartheta \cos\psi + A_{n1} \sin\psi) - \frac{p_m}{c}\right],
\end{aligned}
\tag{11.1.4}
$$

$A_{n1} = \xi \cdot \left(-e_y p_{ex} + e_x p_{ey} + p_{LK} - \frac{T^2}{2}\right) + \frac{s^2}{2}\left(\cos L \cdot p_{ix} + \sin L \cdot p_{iy}\right)$, and the expression for H_t is given in (7.5.14).

It is obvious that only the term $H_{T\min}$ (11.1.4) of the Pontryagin function (11.1.3) explicitly depends on the δ,ϑ, ψ control. By maximizing (11.1.4) by δ,ϑ, ψ we obtain expressions for the optimal control:

$$
\sin\vartheta = A_r/A_{rt}, \cos\vartheta = A_t/A_{rt}, \sin\psi = A_{n1}/A_1, \cos\psi = A_{rt}/A_1, \tag{11.1.5}
$$

$$
\delta = (1 + \mathrm{sign} S_{T\min})/2, \quad S_{T\min} \neq 0; \quad \delta \in [0; 1], \quad S_{T\min} = 0, \tag{11.1.6}
$$

Here $S_{T\min} = \frac{k_2 A_1}{m} - \frac{p_m}{c}$ is the switching function, and $A_1 = \sqrt{A_r^2 + A_t^2 + A_{n1}^2}$. Since k_2, A_1, m, and c are positive along the entire trajectory, and p_m is non-positive (at the end point $p_m = 0$ due to the necessary optimality conditions, and $dp_m/dK = -\partial H/\partial m \geq 0$), then $S_{T\min} > 0$ along the entire trajectory. Consequently, along the entire minimum thrust trajectory:

$$
\delta(K) \equiv 1. \tag{11.1.7}
$$

Substituting the optimal control expressions (11.1.5) and (11.1.7) into (11.1.4), we obtain the following expression for $H_{T\min}$:

$$
H_{T\min} = -\frac{1}{2}T^2 + Tk_1 \cdot \left[\frac{k_2 A_1}{m} - \frac{p_m}{c}\right] \tag{11.1.8}
$$

If we substitute the latter expression in (11.1.3) instead of (11.1.4), we obtain an expression for the optimal Hamiltonian H.

Taking into account (11.1.3)–(11.1.5) and (11.1.7), the optimal motion equations have the form (7.5.18), but must be supplemented by differential equations for thrust (11.1.2) and for the variable p_T conjugated to it:

$$
\frac{dp_T}{dK} = T - k_1 \cdot \left[\frac{k_2 A}{m} - \frac{p_m}{c}\right] + T^2 k_1 k_2 \frac{A_{n1}}{A_1}\frac{\xi}{m}. \tag{11.1.9}
$$

The boundary conditions also are the same as in the problem of minimizing propellant consumption (7.5.19), (7.5.20), but must be supplemented by transversality conditions for p_T at the left and right ends of the trajectory:

$$p_T(0) = 0, p_T(K_f) = 0. \tag{11.1.10}$$

To solve the boundary value problem of the maximum principle (7.5.18), (11.1.2), (11.1.9), (7.5.19), (7.5.20), (11.1.10), it is necessary to determine 8 components of the vector $\mathbf{p}_{\mathbf{x}1}(0)$, the final auxiliary longitude K_f and $p_T(0)$, for which the last equation in (7.5.19), 8 equations in (7.5.20) and the last equation in (11.1.10) are satisfied. Since the final flight time is free, the time-conjugate variable at the final value of the auxiliary longitude $p_t(K_f)$ is 0. In addition, $dp_t/dt = 0$ due to the independence of the Hamiltonian from time. Therefore, $p_t(K) \equiv 0$, hence $p_t(0) = 0$, and the last equation from (7.5.20) can be excluded from consideration along with the need to choose the value $p_t(0)$. Thus, the order of the boundary value problem is reduced to 9. The last equation of (7.5.19) is reduced to a quadratic equation with respect to $p_{LK}(0)$ and can be solved analytically, which allows us to exclude $p_{LK}(0)$ and the last equation of (7.5.19) from the boundary value problem. Thus, the solution of the boundary value problem of the maximum principle is reduced to the calculation of such 6 initial values of the conjugate variables $p_p, p_{ex}, p_{ey}, p_{ix}, p_{iy}, p_m$, the final value of the auxiliary longitude K_f and the thrust T, at which the first three equations of (7.5.20) (two scalar equations and one vector equation with the dimensions 5) and the second equation of (11.1.10) are satisfied.

11.3 General Scheme for Solving the Problem of Optimizing Trajectories with a Constant Exhaust Velocity Thruster

The main problems in the implementation of the numerical method for solving the boundary value problem of the maximum principle (7.5.18), (7.5.19), (7.5.20) are related to ensuring its convergence and stability. The convergence and stability of numerical methods is closely related to the problem of the choice of the initial approximation and with the non-smooth dependence of the residuals of the considered boundary value problem on its unknown parameters (corner points in this dependence appear when new thrust switching points are born or old ones disappear). In a number of publications [3–6], it is proposed to solve similar problems using continuation methods based on the homotopy between the known solution of a given problem and the desired solution. Two-stage methods, in which the optimal LP trajectory is first calculated and then it is continued into the optimal LT trajectory, have proved to be very efficient [3, 7, 8], where at both stages the continuation method is used to solve the corresponding boundary value problem of the maximum principle. With this approach, a zero initial guess can be used for the initial values of the conjugate variables: $\mathbf{p}(0) = 0$, $p_{LK}(0) = p_m(0) = 0$. The initial guess for the final value of the auxiliary longitude can be set equal to the specified angular range of flight: $K_f = \Delta L$. These initial guesses correspond to coasting motion along the initial orbit.

The problem of the non-smooth dependence of the residuals of the boundary value problem on its unknown parameters at the birth and disappearance of the thrust switching points is solved by smoothing the thrust step function $\delta(S)$ [3–8]. In this section an analytical approximation of the following form is used:

$$\delta(S) \approx \frac{1}{2}\left(1 + \frac{S}{\sqrt{S^2 + c\varepsilon^2(\tau)}}\right)$$

Here $c = 2 \cdot \left(\sqrt{2} + 1\right) \approx 4.828427$, $\varepsilon(\tau)$ is the positive regularization term that controls the degree of closeness of the smoothed approximation to the step function: the smaller ε, the closer the approximation is to a step function. The meaning of ε itself: it is a linear function of the continuation parameter τ: at the beginning of the continuation, at $\tau = 0$, the value of ε is high enough, as a result of which the value of $d\delta/dS$ remains quite small in the vicinity of $S = 0$. At the end of the continuation, at $\tau = 1$, the value of ε becomes small enough to provide an acceptable approximation of the thrust step function. For the calculations in this study, we used the dependence $\varepsilon(\tau) = 1 - \tau + \varepsilon_f \tau$, where $\varepsilon_f = 10^{-4} \ldots 10^{-5}$.

The following scheme is used for solving the trajectory optimization problem.

(1) The initial data are specified, including the initial orbit elements $\mathbf{x}_0 = \left(p_0, e_{x0}, e_{y0}, i_{x0}, i_{y0}\right)^{\mathrm{T}}$, the initial true longitude L_0, the final orbital elements $\mathbf{x}_f = \left(p_f, e_{xf}, e_{yf}, i_{xf}, i_{yf}\right)^{\mathrm{T}}$, the angular flight distance ΔL (the final true longitude is calculated by the relationship $L_f = L_0 + \Delta L$), the initial spacecraft mass m_0, the thrust T, and the EPS exhaust velocity c.

(2) The boundary value problem (7.4.27), (7.4.28), (7.4.22) is solved by the continuation method to optimize the LP trajectory. As an initial guess for the unknown initial values of the conjugate variables, zero values are used: $p_p(0) = p_{ex}(0) = p_{ey}(0) = p_{ix}(0) = p_{iy}(0) = p_{LK}(0) = 0$ (which corresponds to the spacecraft coasting motion along the initial orbit), while the initial approximation for the unknown final value of the auxiliary longitude is set equal to the angular range of the flight: $K_f = \Delta L$.

(3) If it is not known whether the specified thrust value is sufficient to complete the transfer, it is necessary to solve the $T_{\min}$-problem. As an initial guess, the decision vector of the boundary value problem obtained for the optimal LP-trajectory is used. The initial approximation for the minimum thrust value $T_{\min}$ is set to $\frac{m_0 c}{\Delta t^{LP}} \cdot \left(1 - e^{-V_{id}^{LP}/c}\right)$, where Δt^{LP}, V_{id}^{LP} are the flight time and the characteristic velocity on the LP-trajectory, respectively. The initial guess for p_m is set to 0 and $p_t(K) \equiv 0$ due to the time independence of the Hamiltonian.

The continuation parameter τ is introduced into the right-hand sides of the differential equations for state and conjugate variables in such a way that when $\tau = 0$, taking into account the initial guess for the conjugate variables, the differential equations coincide with the equations of the LP problem (7.4.27), and when $\tau = 1$—with equations for the $T_{\min}$-problem (with Hamiltonian (11.1.3)) and with additional equations for thrust and its conjugate variable (11.1.2), (11.1.9). Unfortunately, the most obvious approach, which consists in using the Hamiltonian form of writing for these equations with a Hamiltonian of the form $(1-\tau)H_{LP} + \tau H_{T\min}$, leads to an unlimited growth of the initial values of the conjugate variables as τ tends to 1. Therefore, a somewhat more complex parameterization of the equations of motion of the following form is used:

$$\frac{d\mathbf{x}}{dK} = \frac{(1-\tau)k + \tau T_{\min}k_1k_2/m}{1-\tau+\tau B}\frac{\partial \mathbf{B}^T}{\partial \mathbf{p}}\mathbf{B},\quad \frac{dL_K}{dK} = \frac{(1-\tau)k + \tau T_{\min}k_1k_2/m}{1-\tau+\tau B}\frac{\partial \mathbf{B}^T}{\partial p_{LK}}\mathbf{B},$$

$$\frac{dt}{dK} = k_1,\quad \frac{dm}{dK} = -\tau k_1\frac{T_{\min}}{c},\quad \frac{dT_{\min}}{dK} = 0,$$

$$\frac{d\mathbf{p}}{dK} = -\frac{(1-\tau)\left(\frac{1}{2}\frac{\partial k}{\partial \mathbf{x}}A^2 + k\frac{\partial \mathbf{A}^T}{\partial \mathbf{x}}\mathbf{A}\right) + \tau T_{\min}\cdot\left[\frac{\partial k_1}{\partial \mathbf{x}}S_{T\min}A_1 + \frac{k_1}{m}\cdot\left(\frac{\partial k_2}{\partial \mathbf{x}}A_1^2 + k_2\frac{\partial \mathbf{A}_1^T}{\partial \mathbf{x}}\mathbf{A}_1\right)\right]}{1-\tau+\tau B},$$

$$\frac{dp_{LK}}{dK} = -\frac{(1-\tau)\left(\frac{1}{2}\frac{\partial k}{\partial L_K}A^2 + k\frac{\partial \mathbf{A}^T}{\partial L_K}\mathbf{A}\right) + \tau T_{\min}\cdot\left[\frac{\partial k_1}{\partial L_K}S_{T\min}A_1 + \frac{k_1}{m}\cdot\left(\frac{\partial k_2}{\partial L_K}A_1^2 + k_2\frac{\partial \mathbf{A}_1^T}{\partial L_K}\mathbf{A}_1\right)\right]}{1-\tau+\tau B}$$

$$\frac{dp_m}{dK} = \tau T_{\min}k_1k_2\frac{A_1}{m^2},\quad \frac{dp_T}{dK} = \tau\cdot\left[T_{\min} - k_1\cdot\left[\frac{k_2A_1}{m} - \frac{p_m}{c}\right] + T_{\min}^2k_1k_2\frac{A_{n1}}{A_1}\frac{\xi}{m}\right],$$

where $\mathbf{B} = (1-\tau)\mathbf{A} + \tau\mathbf{A}_1$. As a result of solving this problem, the minimum thrust $T_{\min}$ is computed. If the given thrust value T is less than $T_{\min}$, then the flight with the given angular range is impossible. In this case, it is necessary to increase the thrust T or the angular flight distance ΔL. If $T \geq T_{\min}$, then there is a solution and we can proceed to the last stage of solving the problem.

(D) The boundary value problem (7.5.18), (7.5.19), (7.5.20) is solved by the continuation method to calculate the optimal LT trajectory. As an initial guess for the unknown initial values of the conjugate variables $p_p, p_{ex}, p_{ey}, p_{ix}, p_{iy}, p_{LK}$ and the final auxiliary longitude K_f, the values obtained by optimizing the LP-trajectory are used. The initial guess for $p_m(0)$ is set to 0, and the initial guesses for $\mathbf{p}(0)$, $p_{LK}(0)$ and K_f are set to the corresponding values from the solution of the LP-problem. To implement the homotopy between the LP-trajectory and the LT-trajectory, the following system of differential equations is used:

$$\frac{d\mathbf{x}}{dK} = \frac{(1-\tau)k + \tau\delta\frac{T}{m}k_1k_2}{1-\tau+\tau A}\frac{\partial \mathbf{A}^T}{\partial \mathbf{p}}\mathbf{A},$$

$$\frac{dL_K}{dK} = \frac{(1-\tau)k + \tau\delta\frac{T}{m}k_1k_2}{1-\tau+\tau A}\frac{\partial \mathbf{A}^T}{\partial p_{LK}}\mathbf{A},\quad \frac{dm}{dK} = -\tau\delta k_1\frac{T}{c},$$

$$\frac{d\mathbf{p}}{dK} = -\frac{(1-\tau)\left(\frac{1}{2}\frac{\partial k}{\partial \mathbf{x}}A^2 + k\frac{\partial \mathbf{A}^T}{\partial \mathbf{x}}\mathbf{A}\right) + \tau\delta T\cdot\left[\frac{\partial k_1}{\partial \mathbf{x}}SA + \frac{k_1}{m}\cdot\left(\frac{\partial k_2}{\partial \mathbf{x}}A^2 + k_2\frac{\partial \mathbf{A}^T}{\partial \mathbf{x}}\mathbf{A}\right)\right]}{1-\tau+\tau A},$$

$$\frac{dp_{LK}}{dK} = -\frac{(1-\tau)\left(\frac{1}{2}\frac{\partial k}{\partial L_K}A^2 + k\frac{\partial \mathbf{A}^T}{\partial L_K}\mathbf{A}\right) + \tau\delta T\cdot\left[\frac{\partial k_1}{\partial L_K}SA + \frac{k_1}{m}\cdot\left(\frac{\partial k_2}{\partial L_K}A^2 + k_2\frac{\partial \mathbf{A}^T}{\partial L_K}\mathbf{A}\right)\right]}{1-\tau+\tau A},$$

$$\frac{dp_m}{dK} = \tau \delta T k_1 k_2 \frac{A}{m^2}, \ \frac{dt}{dK} = k_1.$$

If $\tau = 0$, this system coincides with the system of differential Eq. (7.4.27), and if $\tau = 1$—with the system (7.5.18).

11.4 Numerical Examples

This section presents numerical examples of optimization of multi-orbit LP-, $T_{\min}$, and LT-trajectories, which were earlier given in papers [3, 9].

Let's consider the application of the scheme proposed in Sect. 11.3 to solve the problem of optimizing the LT-trajectory. As an example, let's consider the problem of optimizing a flight from a high elliptical orbit to a geostationary orbit. For the initial orbit, we will take the following parameters: perigee altitude 500 km, apogee altitude 60,000 km, $i = 7°$, $\omega = 178°$, $\Omega = 0°$. The final orbit has an altitude of 35,793 km and a zero inclination. The initial true longitude L_0 and the angular distance of the flight ΔL will vary. The initial mass of the spacecraft is assumed to be equal to 1000 kg, the specific impulse of the EPS is assumed to be equal to 1520 s, and T will vary. At $L_0 = 300°$, $\Delta L = 720°$ cost function is 7.945323 m^2/s^3, the characteristic velocity is 1850.792 m/s, and the optimal flight time is 2.761054 days.

The calculated values of the characteristic speed and flight time on the LP trajectory allow us to estimate the required thrust value of 7.96023 N. This is the thrust value, the final auxiliary longitude $K_f = 719.958547°$, the initial values of the conjugate variables $\mathbf{p} = (0.0601712907, -0.0569745841, 0.0016263456, -0.0068812764, 0.0004556386)^T$, and $p_{LK} = -0.0001706370$ (dimensionless: the unit of length is equal to the radius of the final orbit, the unit of velocity is the circular velocity in the final orbit), which are taken from the calculated solution LP-problem were used as an initial guess for solving the $T_{\min}$-problem. For the optimal $T_{\min}$-trajectory, it turned out that $T_{\min} = 7.853849$ N, the optimal flight duration is 2.926468 days, $K_f = 719.960624°$, $\mathbf{p} = (0.0360544680, -0.0408377881, 0.0026177997, -0.0026549863, 0.0001313567)^T$, $p_{LK} = 0.0001100354$ and $p_m = -0.0154193714$. The final mass of the spacecraft is 866.778046 kg.

After calculating the minimum thrust, you can proceed to solving the LT problem. Let's choose $T = 15$ N, which is almost twice the minimum thrust. The same initial approximation is used for the decision vector of the LT-problem as for solving the $T_{\min}$-problem. The optimal LT-trajectory corresponds to $\mathbf{p} = (0.1827583099, -0.0866805090, 0.0021479554, -0.0354353837, -0.0116091643)^T$, $p_{LK} = 5.4714834255 \times 10^{-11}$, $p_m = -0.1116559277$, while the optimal flight time is 2.476998 days, and $K_f = 719.967117°$. The final mass of the spacecraft is 902.000747 kg (6 kg more than for the $T_{\min}$-trajectory). Projections on coordinate planes of optimal LP-trajectories (bold lines), $T_{\min}$-trajectories (thin solid lines) and LT-trajectories with a thrust of 15 N (dotted lines for sections with the EPS turned off and thick solid red lines for sections with the EPS turned on) are shown in Fig. 11.1

on the left. On the right, this figure shows the dependencies on time of thrust acceleration, pitch angle, and yaw angle. The optimal flight time turned out to be the maximum for the T_{min}-trajectory and the minimum for the LT-trajectory.

Figure 11.2 shows optimal trajectories with the same boundary conditions and the same specific impulse, but with an angular distance of 100 revolutions. For the LP-trajectory, the optimal flight duration was 114.944643 days. The minimum thrust calculated as a result of solving the T_{min}-problem is 0.19171759 N, the optimal flight time along the trajectory of the minimum thrust is 144.785558 days, and the final mass of the spacecraft is 839.107175 kg. The optimal LT-trajectory was calculated for a thrust of 0.25 N. For such a trajectory, the optimal transfer time is 109.324364 days, and the final mass of the spacecraft is 888.438782 kg. The black lines in Fig. 11.3 indicate the thrusting sections of the trajectory, and the gray lines

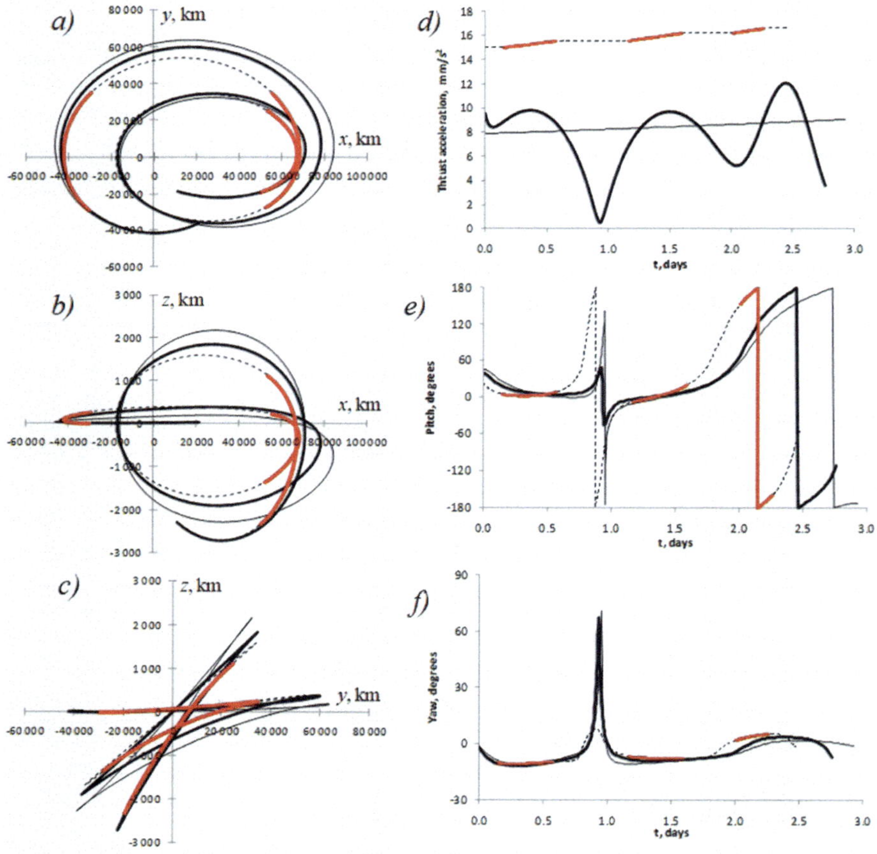

Fig. 11.1 Projections of optimal two-revolutions trajectories on coordinate planes (*a*, *b*, *c*) and the time dependence of optimal control (*d*, *e*, *f*)

indicate the coasting sections. The optimal control on these 100-revolution trajectories is shown in Fig. 11.3. The upper row of graphs shows the optimal control for the LP-problem, the middle row for the $T_{\min}$-problem and the lower row for the LT-problem with a thrust of 0.25 N. The dependences of thrust acceleration on time are presented at left. For the LT-problem, a time dependence of the switching function has been added, indicated by a red line. The middle and right columns show the dependence of pitch angle and yaw angle on time, respectively. Figure 11.4 shows the time dependence of the optimal thrust acceleration and the switching function in the middle section of the optimal LT trajectory on a larger scale. Figure 11.5 shows the dependences of eccentricity, semi-major axis and inclination on time and angular range for optimal LP-, $T_{\min}$- and LT-trajectories.

The dependences of the minimum thrust and optimal transfer duration on the number of orbits n_{rev} and L_0 for the flight between the same initial and final orbits are shown in Tables 11.1 and 11.2 respectively.

Let us consider the dependencies of the main parameters of optimal LT-trajectories with fixed angular range and free flight time on the thrust value. We will study the distinctive features of these dependencies using the example of a flight between the same orbits as before. Let's assume that the initial value of the true longitude is 300°, the initial mass of the spacecraft is 1000 kg, the specific impulse of the propulsion system is 1520 s, the number of orbits is 100, and the thrust T is varied from the minimum value (solution of the $T_{\min}$-problem) to 1 N. The main parameters of the

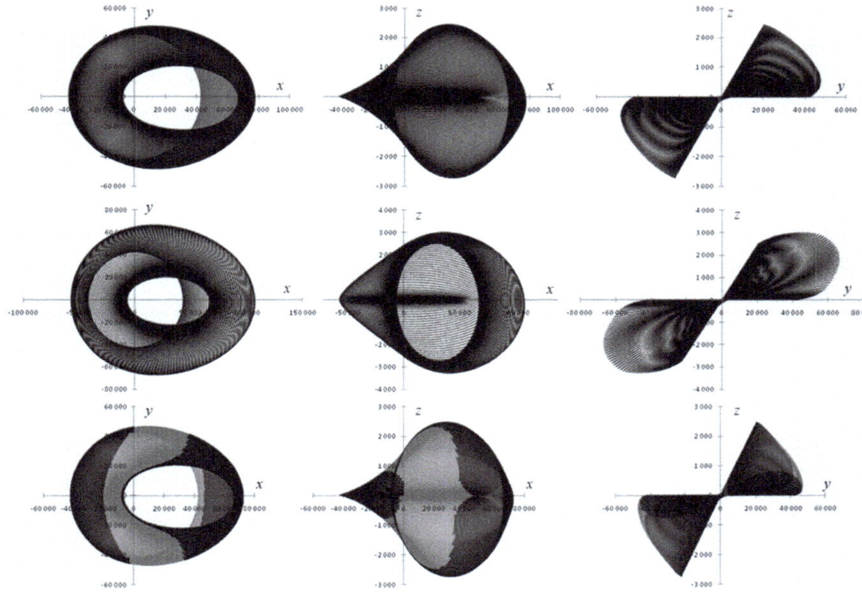

Fig. 11.2 Projections of optimal 100-revolutions trajectories on coordinate planes. The upper row is the LP-trajectory, the middle row is the $T_{\min}$ trajectory, the lower row is the LT-trajectory with the thrust of 0.25 N

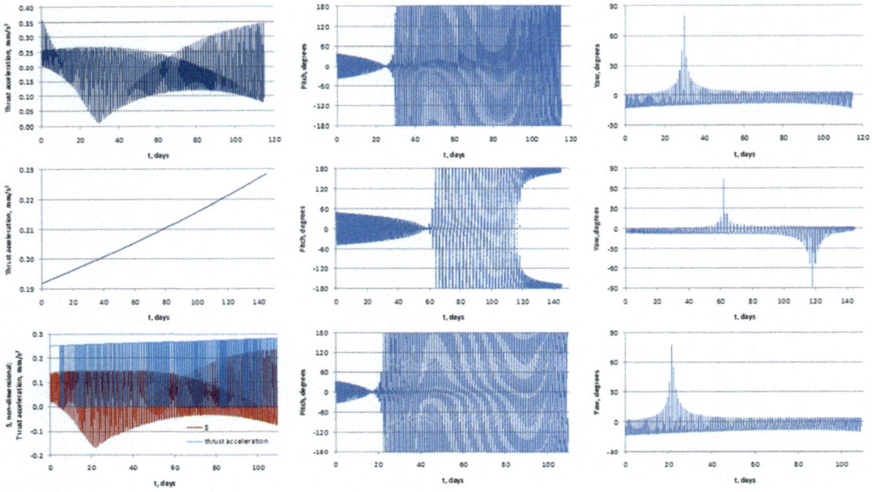

Fig. 11.3 Optimal control on the 100-revolutions LP-, T_{min}- and LT-trajectories

Fig. 11.4 Time dependencies of the optimal thrust acceleration and the switching function in the middle section of the optimal trajectory

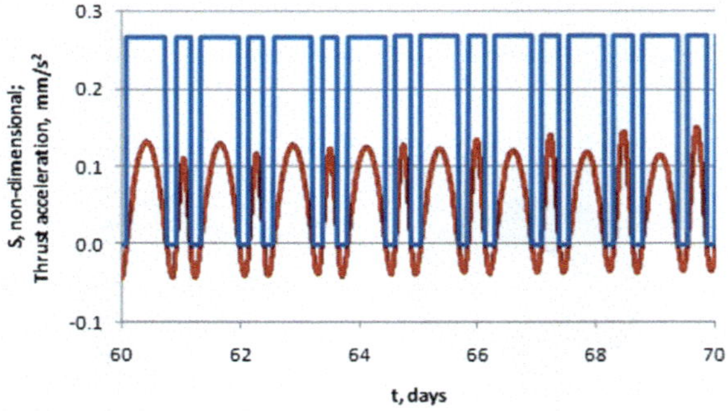

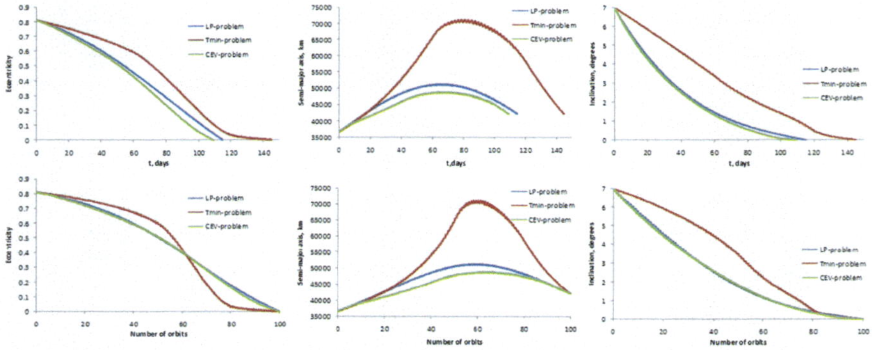

Fig. 11.5 Dependencies of eccentricity, semi-major axis and inclination on time and angular range on the optimal LP-, T_{min} and *LP*-trajectories

Table 11.1 Minimum thrust [N] as a function of number of orbits and initial true longitude of the spacecraft

n_{rev}	L_0, degrees									
	0	10	50	100	150	200	250	300	350	360
2	8.7224	9.2615	10.7203	10.9541	10.3064	9.2950	8.4206	7.8539	8.1978	8.7224
20	0.9612	0.9653	0.9730	0.9709	0.9646	0.9567	0.9499	0.9468	0.9567	0.9612
50	0.3846	0.3853	0.3866	0.3862	0.3852	0.3839	0.3828	0.3823	0.3839	0.3846
100	0.1923	0.1925	0.1928	0.1927	0.1924	0.1921	0.1918	0.1917	0.1921	0.1923
200	0.0961	0.0962	0.0963	0.0962	0.0962	0.0961	0.0960	0.0960	0.0961	0.0961
300	0.0641	0.0641	0.0642	0.0641	0.0641	0.0641	0.0640	0.0640	0.0641	0.0641

optimal LT-trajectories for such initial data as a function of the thrust are given in Table 11.3. The following notation is used in this table: Δt_{opt} is the optimal flight time along the LT-trajectory, m_f is the final mass of the spacecraft at the point with the true longitude L_f, Δt_b is the total thruster operating time on the LT-trajectory, $S_{\min}$ and $S_{\max}$ are the dimensionless minimum and maximum values of the switching function S on the LT-trajectory, respectively. Figure 11.6 shows the projections of optimal LT-trajectories onto the equatorial plane for the following thrust values: 0.191718 N (minimum thrust), 0.195, 0.2, 0.3, 0.4 and 1.0 N. It is clear from Fig. 11.6 that as the thrust increases, the maximum apogee height and the maximum semi-major axis on the optimal LT-trajectory decrease. In the limit, in the example considered, if the thrust is increased, the maximum apogee height becomes equal to the apogee height of the initial orbit, and the maximum semi-major axis becomes equal to the radius of the final orbit.

The dependencies of the optimal flight time, the final mass of the spacecraft and the ratio of the EPS operating time to the total flight time on the thrust in optimal LT-trajectories are presented in Fig. 11.7 based on the data from Table 11.3. It is clear from the Table 11.3 and Fig. 11.7 that the flight time on an LT-trajectory with a fixed angular range has a maximum value at minimum thrust. As the thrust increases, the optimal flight time quickly decreases to a minimum value and then begins to increase again, asymptotically approaching a certain limit value. In the example considered, the maximum flight time at minimum thrust is 144.8 days, the minimum optimal flight time of about 106.2 days is reached at a thrust of about 0.3 N, and the asymptotic value of the optimal flight time with increasing thrust is close to 110 days (left graph in Fig. 11.7). The final mass of the spacecraft increases monotonically with the increase of thrust and approaches a certain asymptotic value (about 910 kg in the considered example, the middle plot in Fig. 11.7). An unexpected discovery turned out to be that on the optimal LT-trajectories with the thrust close to the minimum value, the EPS operates constantly (right graph in Fig. 11.7). In the example considered, the EPS does not shut down on the optimal trajectories with the thrust varying from 0.191718 to 0.1925 N. This property distinguishes the optimal LT-trajectories with a fixed angular range and free flight time from the optimal LT-trajectories with a fixed flight time. On the optimal LT-trajectories with a fixed time, an increase of the thrust

Table 11.2 Flight time [days] along the optimal $T_{\min}$-trajectory as a function of the number of orbits and the initial true longitude of the spacecraft

n_{rev}	L_0, degrees									
	0	10	50	100	150	200	250	300	350	360
2	2.828	2.673	2.255	2.182	2.225	2.352	2.590	2.926	2.999	2.828
20	29.644	29.622	29.390	29.341	29.134	28.955	28.888	28.872	28.744	29.644
50	72.449	72.481	72.576	72.845	72.757	72.490	72.357	72.478	72.418	72.449
100	144.911	144.982	145.035	145.140	145.045	144.828	144.707	144.786	144.833	144.911
200	289.797	289.883	289.916	289.991	289.894	289.686	289.566	289.627	289.694	289.797
300	434.676	434.764	434.795	434.867	434.770	434.562	434.443	434.502	434.571	434.676

Table 11.3 Main parameters of optimal LT-trajectories as a function of thrust

T, N	m_f, kg	Δt_{opt}, day	$S_{\min}$	$S_{\max}$	Δt_b, day	$\Delta t_b/\Delta t_{opt}$
0.191718	839.1072	144.7856	–	–	144.7856	1
0.1919	845.4337	138.9601	0.2729	18.5901	138.9601	1
0.192	846.8409	137.6233	0.1681	14.3748	137.6233	1
0.1925	851.2919	133.2768	0.0054	7.4922	133.2768	1
0.193	854.1752	130.5754	– 0.0426	5.4252	130.3541	0.9983
0.195	861.3472	123.984	– 0.1052	2.7345	122.6717	0.9894
0.2	869.6394	116.1961	– 0.1318	0.9026	112.452	0.9678
0.25	888.4388	109.3105	– 0.1688	0.2324	76.9881	0.7043
0.3	895.6003	106.2226	– 0.175	0.1373	60.0383	0.5652
0.4	901.4687	106.7937	– 0.1793	0.0649	42.4976	0.3979
0.5	903.8702	107.4782	– 0.1815	0.0391	33.1695	0.3086
1	906.8862	108.6846	– 0.187	0.0088	16.0644	0.1478

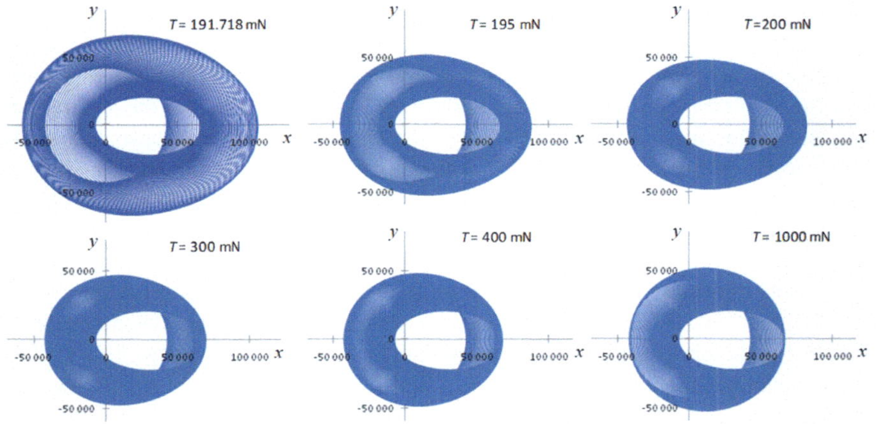

Fig. 11.6 Projections of optimal 100-revolutions trajectories onto the equator plane for various thrust values

with respect to the minimum value leads to the appearance of a coasting trajectory section, since the minimum value of the switching function S is equal to zero at the minimum thrust. The analysis of the considered optimal LT-trajectories with the fixed angular range and free flight time showed that within the thrust variation interval from 0.191718 to 0.1925 N the minimum value of the switching function is positive and decreases monotonically with the thrust increase. The minimum value of the LT-problem switching function also decreases monotonically with the thrust growth beyond this interval, tending to a negative limit as the thrust tends to infinity.

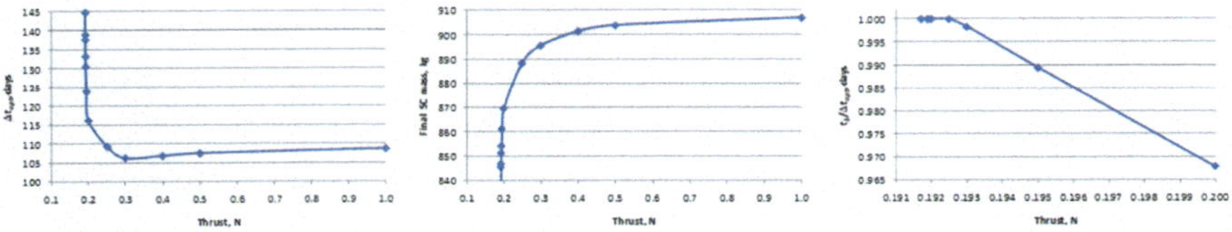

Fig. 11.7 The dependence of the optimal flight time (on the left), of the final spacecraft mass (in the middle) and of the ratio of the thruster operating time to the total flight time (on the right) on thrust at optimal LT-trajectories

The maximum value of the switching function also decreases monotonically, tending to zero as the thrust tends to infinity.

11.5 Conclusion

The study uses the decomposition of the true longitude L into the auxiliary longitude K and its deviation from the true longitude L_K. The advantages of using the auxiliary longitude as an independent variable in the problems of spacecraft trajectory optimization as compared to the true longitude are: (1) a simpler form of the differential equations of motion; (2) a simpler form of the optimal control expressions, and (3) monotonicity of the time dependence of the auxiliary longitude for arbitrary values of the thrust acceleration. The previously obtained numerical results allowed us to formulate the hypothesis that in the problems of optimization of trajectories with a fixed angular distance, there is a single optimal flight time for each type of optimal trajectories, as opposed to the existence of many locally optimal angular ranges for a fixed flight time. This property makes it possible to solve a rather difficult problem of choosing the optimal relation between angular distance and flight time.

The problem of optimizing LP-, $T_{\min}$- and LT-trajectories with a fixed angular distance and free flight time is considered. A rigorous mathematical formulation of the optimal control problem in the extended state space using differential equations of motion with an angular independent variable K and appropriate boundary conditions is presented. For this formulation, the necessary optimality conditions are obtained and analyzed in the form of Pontryagin's maximum principle. The necessary optimality conditions are obtained for the cases of flight with a fixed and free initial true longitude L_0. The first integral of the system of equations of optimal motion in the extended state space has been discovered, which makes it possible to reduce the dimension of the boundary value problem by one, i.e. to the same number of unknowns, as would be in the case of the traditional formulation without extending the state space and using time as an independent variable.

It is convenient to solve the boundary value problem obtained in this way by the method of continuation with respect to a parameter, which reduces its solution to the solution of the Cauchy problem. In this case, the calculation of optimal multi-orbit LP-trajectories with a fixed angular range and optimal flight time is limited only by the capabilities of the modern numerical methods for integrating ordinary differential equations. It should be noted that it is possible to use zero initial values of the conjugate variables as an initial guess, which corresponds to the trajectory of the coasting motion of the spacecraft along the initial orbit.

Thus, we believe that the formulation considered compares favorably with those previously known and will be useful for solving practical problems. In order to demonstrate the efficiency of the method, a number of examples have been considered and, as a result of their solution, the existence of several LP-trajectories with a given flight time and free angular distance, satisfying the necessary optimality conditions, has been verified. The results of computational experiments allow us to make the assumption that, in contrast to the problem with a fixed flight time and free angular distance, the considered problem has a unique extremal in a given family of trajectories that satisfy the necessary optimality conditions. It is shown that for flights in a central Newtonian gravitational field and in the case of time-independent perturbations, the use of the auxiliary longitude as an independent variable allows one to automatically satisfy the necessary conditions for the flight time optimum.

The given numerical examples demonstrate the efficiency, stability and good performance of the numerical method based on the proposed approaches. Numerical examples allow us to state that the characteristic velocity and the product of the cost function and the flight duration tend to asymptotic values as the angular flight range increases.

The high performance of the developed method makes it possible to calculate a large number of trajectories for further analysis of their characteristics with variations in the number of orbits N_{rev} and the initial true longitude L_0. It was found that within the limits of a change in the angular distance by one orbit, there is a local minimum and a local maximum of the flight duration and the characteristic velocity. The cost function decreases monotonically with increasing angular distance of the flight, but has a local minimum and a local maximum in the initial true longitude at a fixed angular distance of the trajectory.

The chapter considers an approach to the optimization of LT-trajectories with fixed angular range and free flight time, based on the sequential solution of LP- and LT-problems. The maximum principle is used to derive the necessary optimality conditions, and the maximum principle boundary value problem is reduced to the Cauchy problem using the continuation method. A method based on solving the LP-problem with the subsequent continuation of the optimal LP-trajectory into the optimal LT-trajectory is presented. A method for diagnosing the existence of an LT-trajectory with a given thrust value using the solution of the $T_{\min}$-problem is presented. The $T_{\min}$-trajectory and minimum thrust are also calculated by continuation from the optimal LP-trajectory. Thus, the proposed method allows us to automate the process of calculating the optimal LT-trajectory with fixed angular distance and optimal flight time. During the calculation process, the existence of a solution is checked and, in case of its absence, it is possible to propose recommendations for changing the initial data (thrust or angular flight distance) to the values for which a solution exists.

The numerical analysis of the optimal $T_{\min}$- and LT-trajectories with fixed angular distance and optimal flight time was performed using the developed method. As a result, the following properties of such trajectories were discovered:

1. The product of the minimum thrust and the angular flight range is almost constant for a sufficiently large angular range. This property makes it easy to estimate the required minimum number of orbits $n_{rev}{*}$ for a given thrust value $T{*}$ after calculating the minimum thrust $T_{\min 1}$ on a $T_{\min}$-trajectory with an arbitrary (sufficiently large) number of orbits n_{rev1}: $n_{rev}{*} = n_{rev1} \cdot T_{\min 1}/T{*}$.
2. In the minimum thrust problem with a fixed angular range and optimal flight time, there is one local minimum and one local maximum of thrust along the initial true longitude in the interval from 0 to 360°, but, depending on the angular range of the flight, there may be one or two local minima and maxima of the optimal flight time in this interval.
3. The optimal flight time in the $T_{\min}$-problem with fixed angular distance is almost proportional to the angular distance of the flight. This means that after solving the $T_{\min}$-problem with any sufficiently large number of orbits n_{rev1}, and calculating the flight time Δt_1 using this trajectory, we can estimate the flight time Δt along the $T_{\min}$-trajectory with any other, sufficiently large number of orbits n_{rev}: $\Delta t = \Delta t_1 n_{rev}/n_{rev1}$.
4. In the LT-problem with fixed angular distance and initial true longitude, the dependence of the optimal flight time has a minimum in terms of thrust. The maximum value of the optimal flight time occurs at minimum thrust. In this case, the required fuel consumption decreases monotonically with increasing thrust.
5. There is a finite thrust interval, limited on the left by the minimum value of thrust and on the right by some value close to it, in which trajectories with minimum fuel consumption, fixed angular range and free flight time do not have coasting sections. In this interval, the minimum value of the switching function is positive and decreases monotonically with increasing thrust. The optimal flight time in this interval also decreases monotonically with increasing thrust.
6. The minimum value of the LT-problem switching function decreases monotonically with increasing thrust, tending to some negative limit value as the thrust tends to infinity. The maximum value of the switching function also decreases monotonically, tending to zero as the thrust tends to infinity.

References

1. Petukhov, V.G.: Minimum-thrust problem and its application to trajectory optimization with thrust switchings. In: IAC-13-C1.6.2, pp. 1–9 (2013)
2. Ivanyukhin, A.V., Petukhov, V.G.: The thrust minimization problem and its applications. Cosm. Res. **53**(4), 300–310 (2015)

3. Petukhov, V., Ivanyukhin, A., Popov, G., et al.: Optimization of finite-thrust trajectories with fixed angular distance. Acta Astronaut. **197**, 354–367 (2022)
4. Bertrand, R., Epenoy, R.: New smoothing techniques for solving bang-bang optimal control problems—numerical results and statistical interpretation. Optimal Control Appl. Meth. **23**(4), 171–197 (2002)
5. Taheri, E., Kolmanovsky, I., Atkins, E.: Enhanced smoothing technique for indirect optimization of minimum-fuel low-thrust trajectories. J. Guid. Control Dyn. **39**(11), 2500–2511 (2016)
6. Petukhov, V.G.: Application of the angular independent variable and its regularizing transformation in the problems of optimizing low-thrust trajectories. Cosm. Res. **57**(5), 351–363 (2019)
7. Haberkorn, T., Martinon, P., Gergaud, J.: Low thrust minimum-fuel orbital transfer: a homotopic approach. J. Guid. Control Dyn. **27**(6), 1046–1060 (2004)
8. Petukhov, V.G.: Method of continuation for optimization of interplanetary low-thrust trajectories. Cosm. Res. **50**(3), 249–260 (2012)
9. Ivanyukhin, A.V., Petukhov, V.G.: Optimization of multi-revolution limited power trajectories using angular independent variable. J. Optim. Theory Appl. **191**, 575–599 (2021). https://doi.org/10.1007/s10957-021-01853-8
10. Petukhov, V.G.: Optimization of multi-orbit transfers between noncoplanar elliptic orbits. Cosm. Res. **42**(3), 250–268 (2004)

Chapter 12
Suboptimal Feedback Control for Multi-orbit Transfer into a Given Circular Orbit and Its Applications

12.1 Introduction

Chapter 11 considered the general case of optimal low-thrust multi-orbit trajectories between non-coplanar elliptical orbits. A particular but practically important problem of inserting a spacecraft from a given initial elliptical orbit to a final circular orbit is analyzed in detail in this section using the numerical methods presented previously. More precisely, we consider a minimum-time low-thrust multi-orbit trajectory between non-coplanar elliptical and circular orbits in a central Newtonian gravitational field. The thrust and specific impulse are assumed to be constant during the EPS operation. The direction of the thrust vector is unconstrained and is chosen from the optimality conditions. It is assumed that in the unperturbed problem the line of apses of the initial elliptical orbit lies in the plane of the final circular orbit. The problem of transfer between such orbits is typical when delivering a spacecraft into high circular orbits, especially into GEO. Of course, the considered problem is reversible, so the results obtained here are also applicable for the transfer from an initial circular orbit to a final elliptical orbit.

The problems considered are characterized by long flight durations and high sensitivity to the errors in the determination of the initial state vector, in the implementation of the control program, and to external perturbances. Therefore, to perform the flight, a periodic correction of the thrust vector control program is required to compensate for the deviation of the trajectory under the influence of various perturbances. Of course, such a correction can be made using a periodically repeated solution to the trajectory optimization problem with a current evaluation of the spacecraft's state vector, and of the thrust and specific impulse of its propulsion system. However, this approach requires large computational costs and is therefore difficult to implement as part of the on-board software of a spacecraft. Performing such calculations by ground means with subsequent transfer of the results to the spacecraft reduces the autonomy of the spacecraft and places additional burden on ground-based mission

A. S. Filatyev and V. G. Petukhov, *Through Optimization of Aerospace Vehicle Trajectories by the Pontryagin Maximum Principle*, Studies in Systems, Decision and Control 579, https://doi.org/10.1007/978-3-031-80756-5_12

control services. Therefore, it is quite relevant to develop a thrust vector control algorithm with feedback that provides a close-to-optimal injection trajectory, tolerance to errors in the determination of the spacecraft state vector and in the execution of the control program, and tolerance to external perturbances of the trajectory.

The main objective of the study presented in this section is the synthesis of a feedback control program, close to the optimum in time and ensuring the spacecraft insertion into a given final circular orbit not only in the central Newtonian gravitational field, but also in the presence of perturbances. At the same time, a number of other problems are considered—a description is given for the methods for quick estimation of the characteristic velocity of flight between non-coplanar elliptical and circular orbits, for optimization of the space mission profile for inserting a spacecraft into circular orbits using a combination of high and low thrust, and for calculation of suboptimal trajectories of spacecraft with EPS to the Moon and libration points.

12.2 Features of Time-Optimal Trajectories Between Non-coplanar Elliptical and Circular Orbits with a Free Line of Apses of the Initial Orbit

Consider the problem of minimizing the transfer time $\Delta t = t_f - t_0$ for a dynamic system described by a system of differential Eq. (7.1.4) with the following boundary conditions:

$$L(t_0) = L_0, \mathbf{x}(t_0) = \mathbf{x}_0, m(t_0) = m_0, \mathbf{x}\left(t_f\right) = \mathbf{x}_f \tag{12.1.1}$$

The functional of the considered optimal control problem can be obviously written as:

$$J = \int_{t_0}^{t_f} dt \tag{12.1.2}$$

We will consider an unperturbed problem, that is, we will assume that $a_{pt} = a_{pr} = a_{pn} = 0$. The Pontryagin function of the considered optimal control problem has the following form:

$$H = -1 + \sqrt{\mu p}\left(\frac{q}{p}\right)^2 p_L + \delta\frac{T}{m}\sqrt{\frac{p}{\mu}}(A_\tau \cos\vartheta \cos\psi$$
$$+A_r \sin\vartheta \cos\psi + A_n \sin\psi) - \delta\frac{T}{c}p_m \tag{12.1.3}$$

Optimal control $\delta(\mathrm{t})$, $\vartheta(\mathrm{t})$, $\psi(\mathrm{t})$ is computed from the Hamiltonian maximum condition (12.1.3):

$$\sin\vartheta = A_r/A_{rt},\ \cos\vartheta = A_t/A_{rt},\ \sin\psi = A_n/A,\ \cos\psi = A_{rt}/A, \tag{12.1.4}$$

$$\delta \equiv 1. \tag{12.1.5}$$

The conditions (12.1.5) follow from the non-positivity of $p_m(t)$ on the interval $[t_0; t_f]$, which follows from the transversality condition $p_m(t_f) = 0$ and the non-negativity of the derivative dp_m/dt.

Substituting the optimal control expressions in (12.1.3), we obtain an expression for the Hamiltonian of the minimum-time problem:

$$H = -1 + \sqrt{\mu p}\left(\frac{q}{p}\right)^2 p_L + \frac{T}{m}\sqrt{\frac{p}{\mu}}A - \frac{T}{c}p_m \tag{12.1.6}$$

By virtue of (12.1.5), the differential equation for the spacecraft mass can be eliminated from the system (7.1.4) using the explicit time dependence of the spacecraft mass:

$$m = m_0 - (T/c)\cdot(t - t_0) \tag{12.1.7}$$

In this case, there is no need to consider p_m, and the Hamiltonian will take the form:

$$H_1 = -1 + \sqrt{\mu p}\left(\frac{q}{p}\right)^2 p_L + \frac{T}{m}\sqrt{\frac{p}{\mu}}A \tag{12.1.8}$$

The dynamic system under consideration has a single fast phase variable L. According to (12.1.7), this system is non-autonomous, but the time-dependent terms on the right-hand sides of the differential equations are small due to the low mass flow rate T/c. At a low value of thrust acceleration T/m, the changes in all orbital elements (except for the true longitude) per one revolution will be small. Therefore, the averaging method can be applied to a dynamic system with the Hamiltonian (12.1.8) in the following form:

$$H_{av} = \frac{1}{\Pi}\int\limits_{t_0}^{t_0+\Pi} H_1(\mathbf{x}, \mathbf{p}, L, p_L, t)dt = \frac{n}{2\pi}\int\limits_{L_0}^{L_0+2\pi} H_1(\mathbf{x}, \mathbf{p}, L, p_L, t)\frac{dt}{dL}dL \tag{12.1.9}$$

where H_{av} is the averaged Hamiltonian, $\Pi = 2\pi\sqrt{\frac{1}{\mu}\left(\frac{p}{1-e_x^2-e_y^2}\right)^3}$ is the osculating orbital period, and $n = 2\pi/\Pi$ is the mean motion. The averaged Hamiltonian (12.1.9) does not depend on the true longitude, thus in the averaged equations of motion $dp_L/dt = -\partial H_{av}/\partial L = 0$. If an optimal interorbital transfer is considered, in which the final value of the true longitude is not specified, then the transversality condition $p_L(t_f) = 0$ must be satisfied at the final point. Together with the condition $dp_L/dt = 0$, this leads to the identity $p_L(t) = 0$. Thus, the averaged functional depends only on the slow state and conjugate variables and on time: $H_{av} = H_{av}(\mathbf{x}, \mathbf{p}, t)$, and the system of differential equations of optimal motion for the averaged system becomes:

$$\frac{d\mathbf{x}}{dt} = \frac{\partial H_{av}}{\partial \mathbf{p}}, \frac{d\mathbf{p}}{dt} = -\frac{\partial H_{av}}{\partial \mathbf{x}} \tag{12.1.10}$$

For convenience of calculations, we will take the opportunity to change the order of performing the operations of calculating partial derivatives and integrating (averaging) the Hamiltonian. As a result, we obtain the following system of differential equations, equivalent to the system (12.1.10):

$$\frac{d\mathbf{x}}{dt} = \frac{n}{2\pi} \int\limits_{L_0}^{L_0+2\pi} \frac{\partial H_2(\mathbf{x}, \mathbf{p}, L, t)}{\partial \mathbf{p}} \frac{dt}{dL} dL, \frac{d\mathbf{p}}{dt} = -\frac{n}{2\pi} \int\limits_{L_0}^{L_0+2\pi} \frac{\partial H_2(\mathbf{x}, \mathbf{p}, L, t)}{\partial \mathbf{x}} \frac{dt}{dL} dL \tag{12.1.11}$$

where $H_2 = -1 + \frac{T}{m}\sqrt{\frac{p}{\mu}}A$.

Thus, the problem of minimizing the interorbital transfer time for an averaged low-thrust trajectory is reduced to solving a boundary value problem for a system of differential equations (12.1.11) with the following boundary conditions:

$$\mathbf{x}(t_0) = \mathbf{x}_0, \tag{12.1.12}$$

$$\mathbf{x}(t_f) = \mathbf{x}_f, \frac{n}{2\pi} \int\limits_{L_0}^{L_0+2\pi} H_2[\mathbf{x}(t_f), \mathbf{p}(t_f), L, t_f] \frac{dt}{dL} dL = 0 \tag{12.1.13}$$

where the latter equation is the transversality condition. The unknown parameters of the boundary value problem (12.1.11), (12.1.12), (12.1.13) are: the vector of initial values of the conjugate variables $\mathbf{p}$ (5 scalar parameters) and the transfer termination time t_f. These quantities are chosen so that 6 scalar Eq. (12.1.13) are satisfied. This boundary value problem is solved by the method of continuation with respect to a parameter of the form (7.3.17); the derivatives in (12.1.11) are calculated analytically in a closed form, while the integration is performed by the trapezoidal method with the integration interval divided into 100–200 equal intervals.

Let us consider the problem of minimizing the time of transfer between non-coplanar elliptical orbit and circular orbits in the central Newtonian gravitational field. If the line of apses of the initial elliptical orbit is free, it follows from the transversality conditions that it must coincide with the line of intersection of the planes of the initial and final orbits [1].

In this section we will use dimensionless variables, having chosen the radius r_k of the final orbit and the reciprocal of the mean motion $\left(r_k^3/\mu\right)^{1/2}$ along the final orbit, as the units of length and time, respectively. As follows from the problem formulation, there are only three essential parameters of the initial orbit that completely determine the boundary value problem of the maximum principle up to the rotation of the coordinate system: the dimensionless perigee radius r_p, the dimensionless apogee radius r_a and the inclination i of the initial orbit. In fact, since the line of apses coincides with the line of intersection of the planes of the initial and final orbits, it is always possible to define a coordinate system, in which the longitude of the ascending node and the argument of the periapsis of the initial orbit are equal to zero. To do this, it is necessary that the X-axis of this coordinate system is directed towards the pericenter of the initial orbit, and the Y-axis is in the plane of the final orbit.

Due to the symmetry of the problem, the optimal control is defined by three essential parameters: these are the initial values of p_p, p_{ex}, p_{ix}. The initial values of p_{ey} and p_{iy} are equal to 0. In this case, the optimal control on the orbit is symmetric with respect to the line of apses. Under such conditions, the convergence of the continuation method when solving the boundary value problem (12.1.11), (12.1.12), (12.1.13) can usually be achieved by using an initial guess of the simplest form, e.g. $p_p(t_0) = 1, p_{ex}(t_0) = 0, p_{ey}(t_0) = 0, p_{ix}(t_0) = 0, p_{iy}(t_0) = 0$, and $\Delta t = t_f - t_0 = 1$.

The application of the averaging method leads to the asymptotic nature of the found optimal solution. This means that the averaged optimal solution is applicable to various values of thrust, specific impulse and spacecraft mass as long as the assumption of a small relative change of the orbital elements per orbit is met. In this case, for close values of the "slow" orbital elements and a small change in these elements during one revolution, the dependencies of the optimal angles of the thrust vector orientation on the true anomaly will be close for the spacecraft with different values of the current thrust-to-mass ratio and specific impulse.

Finally, the optimal solution found within the framework of the formulation under consideration is scalable with respect to the planetocentric distance and the gravitational parameter of the central celestial body. In fact, let us consider the transfer from an elliptical orbit with periapsis radius r_{p0}, apocentric radius r_{a0} and inclination i_0 to a final circular orbit with radius r_k and inclination i_k in the central Newtonian gravitational field of a celestial body with the gravitational parameter μ. As a result of solving the boundary value problem of optimal control, we obtain the dimensionless values of the conjugate variables at the initial moment of time $p^*_{p0}, p^*_{ex0}, p^*_{ix0}$ and the dimensionless flight time Δt^*. The resulting optimal trajectory can be scaled for a transfer in the Newtonian gravitational field of a celestial body with a gravitational parameter μ_1 from an elliptical orbit with periapsis radius r_{p01}, apocentric radius r_{a01} and inclination i_{01} to a final circular orbit with radius r_{k1} and inclination i_{k1}, if $r_{p01} = r_{p0}r_{k1}/r_k, r_{a01} = r_{a0}r_{k1}/r_k, i_{k1} = i_{01} + (i_k - i_0)$; as this takes place, the transfer time is: $\Delta t_1 = \left(r^3_{k1}/m_1\right)^{1/2} \Delta t^*$.

12.3 Synthesis of Suboptimal Feedback Control

In order to develop an algorithm for the quick evaluation of the main design parameters of the spacecraft and its trajectory [2, 3], as well as for the synthesis of the feedback control [4, 5], a numerical analysis of the minimum-time averaged multi-orbit flights between the non-coplanar elliptical orbit and GEO was performed for a wide range of variation of the initial elliptical orbit parameters. For this purpose, a three-dimensional grid of initial values of the perigee radius, apogee radius and inclination was considered. The value $r_0 = 6571$ km is used as the initial value for the perigee and apogee radii, the j-th grid node has the value $r_j = \exp\left[\ln r_0 + \frac{j}{20}\ln\frac{r_{GEO}}{r_0}\right]$, where j varies from 0 to 39, and $r_{GEO} = 42{,}164$ km. Thus, a grid of 40×40 nodes is defined with a maximum perigee and apogee radii of ~ 246,539.565 km. The initial inclination range considered is from 0 to 90° in 5° increments. The total number of nodes on the considered grid is thus $40 \times 40 \times 19$, which, taking into account the limitation $r_p \leq r_a$, results in the definition of $40 \times (40 + 1) \times 19/2 = 15{,}580$ different initial orbits on the considered grid.

For each initial orbit, the parameters of which are determined by the nodes of the introduced 3D grid, the problem of computing the minimum-time trajectory was solved and the values of the required characteristic velocity, the minimum and maximum geocentric distance of the spacecraft during the flight and the initial values of the conjugate variables were obtained. The values of the characteristic velocity on the grid of initial orbital parameters were used to quickly estimate the final spacecraft mass and the flight time [2, 3] for a space mission design purpose, and the initial values of the conjugate variables—for the synthesis of suboptimal feedback control (SOFC).

The initial values of the conjugate variables p_p, p_{ex}, p_{ix}, calculated on a three-dimensional grid of initial orbit parameters, can be used in the linear interpolation procedure to estimate the initial values of these variables for an arbitrary initial elliptical orbit with a line of apses coinciding with the line of intersection of the planes of the initial and final orbits. The initial values of p_p, p_{ex}, p_{ix} obtained in this way, supplemented by zero initial values of the remaining conjugate variables, can be used in the integration of non-averaged equations of motion. Numerical modeling has shown that this approach gives good results for unperturbed trajectories in a Newtonian gravitational field—the osculating elements of the unaveraged trajectory perform small short-period oscillations in the vicinity of the averaged solution.

The situation becomes more complicated when trying to integrate the trajectories with such a control, which are perturbed by external accelerations, errors in the determination of the initial conditions and in the execution of the control program. In this case, the errors accumulate over time, eventually leading to unacceptably large discrepancies at the right end of the trajectory. As an example, consider the problem of a low-thrust spacecraft transfer from an initial elliptical orbit with a perigee altitude of 500 km, an apogee altitude of 30,000 km and an inclination of 62.8° to GEO (the altitudes are indicated relative to the average radius of the Earth of 6371 km). The initial mass of the spacecraft is assumed to be 1000 kg, the engine thrust to be 0.2 N, and the specific impulse to be 1500 s. Figure 12.1 shows the time dependences of the semimajor axis, apogee and perigee radii: (1) on the averaged trajectory (dashed line); (2) on an unperturbed, non-averaged trajectory (solid thin black line); (3) on a non-averaged trajectory, taking into account the perturbances from the geopotential up to the 10th degree and 10th order and the lunar-solar gravity, and (4) on a non-averaged trajectory, on which, in addition to the listed perturbances, the EPS does not operate during eclipses. It can be seen that the perturbances under consideration lead to a deviation in the final radii of perigee and apogee by 4–5 thousand km. The deviation in the inclination of the final orbit reaches 13–14°.

Another option for using the calculated values of the conjugate variables is to compute the current values of the conjugate variables from the current values of the perigee and apogee radii and inclination. In this case, the current values of p_p, p_{ex}, p_{ix} are calculated by linear interpolation on a three-dimensional grid as a function of the current values of r_p, r_a and i. As before, the values of p_L, p_{ey}, p_{iy} are assumed to be zero. The calculated values of the conjugate variables are substituted into the expressions (12.1.4) to calculate the direction of the thrust vector. This approach ensures control stability with respect to perturbances in the semimajor axis, eccentricity and inclination, but remains unstable with respect to perturbances in the perigee argument and in the right ascension of the ascending node. As a result, on perturbed trajectories, due to the rotation of the line of apses and of the line of nodes, the errors accumulate in all orbital elements, although to a lesser extent than in the previous case.

In order to ensure the control stability with respect to arbitrary perturbances, the following transformation of the current conjugate variables was used:

$$
\begin{aligned}
p_h^* &= p_h(r_a, r_p, i),\\
p_{ex}^* &= p_{ex}(r_a, r_p, i)\cos(\omega + \Omega),\\
p_{ey}^* &= p_{ex}(r_a, r_p, i)\sin(\omega + \Omega),\\
p_{ix}^* &= p_{ix}(r_a, r_p, i)\cos(\Omega),\\
p_{iy}^* &= p_{ix}(r_a, r_p, i)\sin(\Omega),
\end{aligned}
\tag{12.2.1}
$$

where p_p, p_{ex}, p_{ix} are still calculated by 3D linear interpolation on a 3D grid as a function of the current values of r_p, r_a and i. The control (12.2.1) does not explicitly depend on time, clearly depends on all 6 current orbital elements, and has the form of a feedback control. It is easy to show that the use of the transformed values of the current conjugate vector in expressions (12.1.4) leads to the fact that local extrema of the yaw angle are achieved at the nodal points, and the pitch angle is 0° or 180° at perigee and apogee. As a result, even if the lines of the nodes and apses diverge under the influence of perturbances, the control (12.2.1) ensures the convergence of the solution to the final orbit. In the absence of perturbances, the control (12.2.1) will coincide with the optimal one (up to the errors of averaging and interpolation), while in the presence of such perturbances it will differ from the optimal one, the more, the

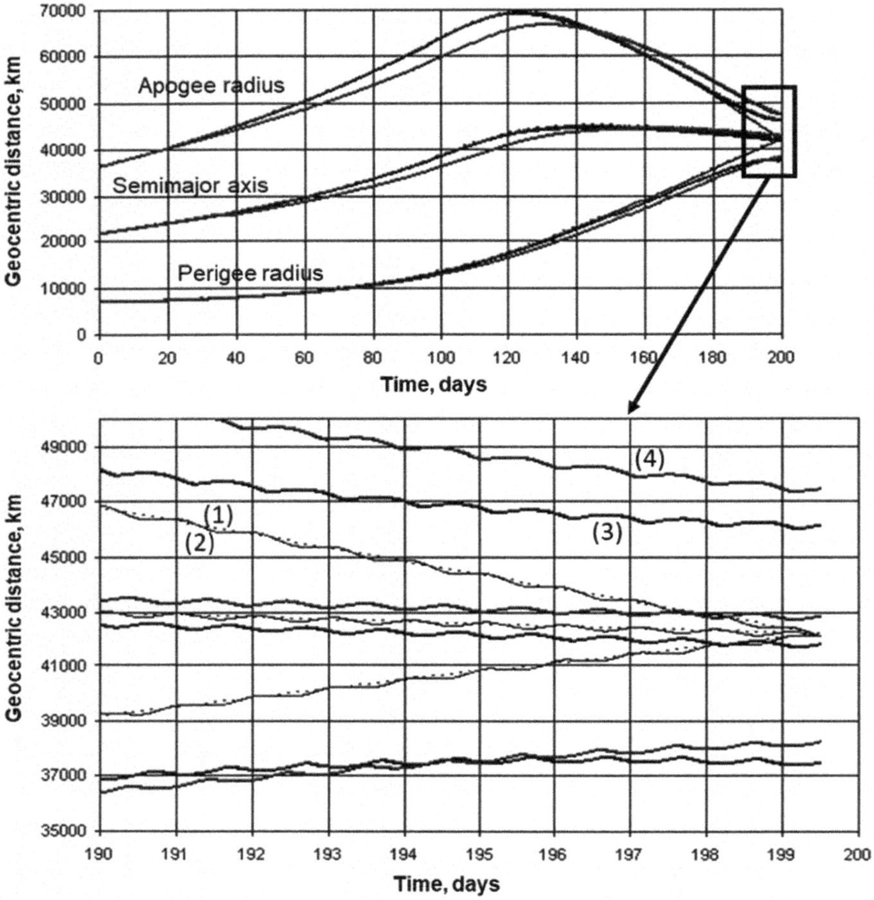

Fig. 12.1 Time dependencies of the semimajor axis, perigee radius, and apogee radius: (1) the dotted line is the averaged optimal solution for the central gravity field; (2) the thin solid line is the integration of the non–averaged equations of optimal motion in the central gravity field with the initial values of the conjugate variables from the averaged problem; (3)—the same as (2), but taking into account the geopotential of the Earth 10×10 and the gravity of the Moon and the Sun; (4)—the same as (3), but taking into account the shutdown of the thrusters during eclipses

stronger the perturbances. We call the control (12.2.1) suboptimal feedback control (SOFC).

Figure 12.2 shows the time dependence of the perigee and apogee radii and of the semimajor axis in the final section of the trajectory in the previously considered transfer to GEO for an unperturbed averaged trajectory and two trajectories using SOFC (12.2.1). In one case, the perturbances include perturbances from the geopotential (matrix 10×10) and the gravity of the Moon and the Sun, and in the

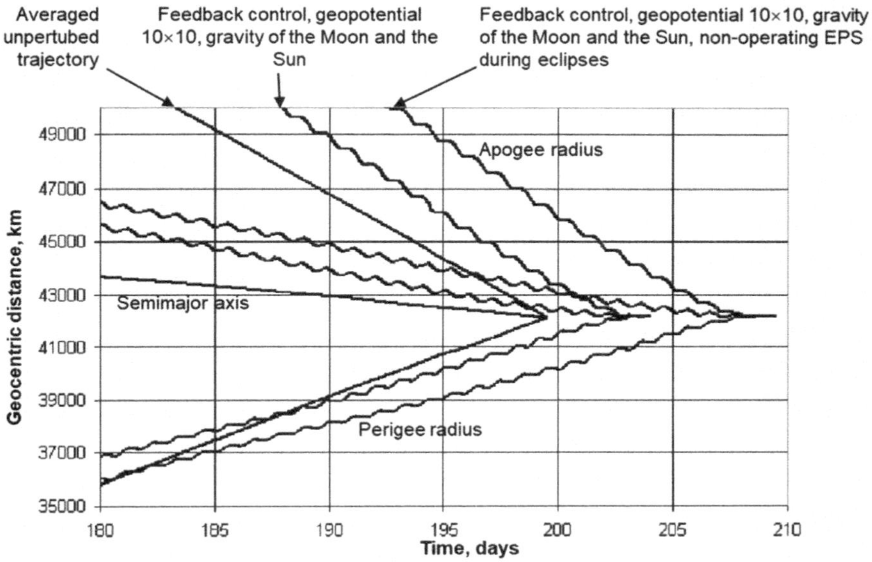

Fig. 12.2 The effect of perturbations on trajectories with suboptimal feedback control

other case—the EPS shutdown in the shaded sections of the orbit is added to these perturbances.

It is clear from Fig. 12.2 that the use of the feedback control ensures that the solution converges to the final orbit and is quite close to the optimal solution.

The control (12.2.1) appeared to be stable not only with respect to perturbing accelerations, but also with respect to large deviations of the initial conditions. Figure 12.3 shows the dependence of the radii of perigee, apogee and semi-major axis for a transfer from a high elliptical orbit of the Molniya type to GEO using the optimal and suboptimal feedback control. An initial orbit with the following parameters was considered: perigee altitude 1000 km, apogee altitude 40,000 km, inclination 63.4°, and perigee argument 250°. Despite the fact that the SOFC was obtained based on the solution to the problem of transfer with zero initial value of the perigee argument, it ensures, thanks to the transformation (12.2.1), the convergence to the final orbit even for large deviations of the apsidal line from the nodal line.

The SOFC algorithm is constructed on the basis of an array of averaged solutions to the minimum-time problem. Due to the use of the averaging method, this algorithm can produce a significant error on the last orbit (or orbits) of the trajectory, if the assumptions of the averaging method are no longer satisfied. If the thrust value is such that the required changes in the orbital elements on the last orbit significantly exceed the maximum permissible values of the guidance errors in these elements, there is a high probability that the SOFC will not be able to provide the specified guidance accuracy. The magnitude of the guidance errors depends on the parameters of the initial and final orbits, the level of thrust acceleration, the composition of external

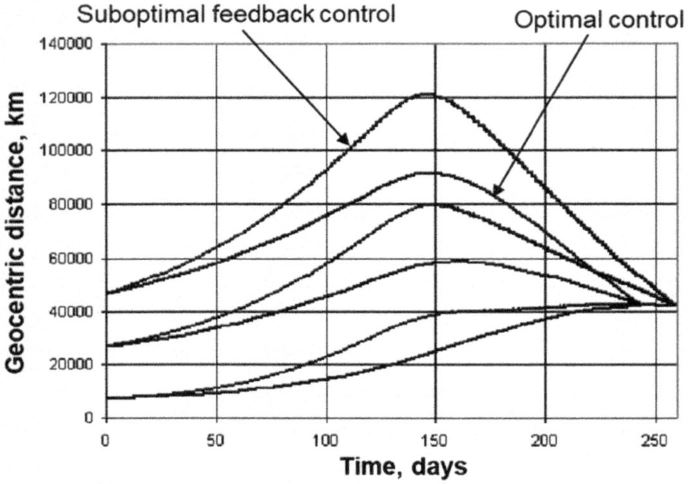

Fig. 12.3 Transfer from a Molniya-type orbit to the GEO

perturbing factors, and the errors in the implementation of the required thrust vector. For a flight to GEO, in some cases, the guidance error with respect to the semimajor axis can exceed 100 km, to the orbital period—360 s, to the eccentricity—0.005, and to the inclination—0.1°.

In order to improve the accuracy of spacecraft guidance, the SOFC algorithm was modified for the final part of the trajectory, as a result of which the SOFC+ algorithm was developed.

The difference between the SOFC+ and SOFC algorithms lies in the redefinition of the control tables in the final control section, the use of characteristic velocity tables to determine the degree of proximity of the spacecraft's osculating orbit to the final orbit, and the modification of the algorithm for calculating the thrust vector orientation parameters in the final section of the flight.

To reduce the interpolation errors of the conjugate variables, additional tables are introduced for the control in the vicinity of the final orbit. The values of the conjugate variables are calculated at the nodes of a three-dimensional grid, defined by 7 values of the perigee radius, 7 values of the apogee radius, and two values of the inclination. In terms of the perigee and apogee radii, these values are: 41,864, 41,964, 42,064, 42,164, 42,264, 42,364 and 42,464 km (in case of a flight to GEO), and for the inclination they are: 0 and 0.2°. The total number of different paths defined on the new 3D mesh is $2 \times 7 \times (7 + 1)/2 = 56$.

The transition to new control tables when using SOFC+ occurs if the osculating perigee and apogee of the spacecraft differ from the GEO radius by no more than 300 km, and the osculating inclination becomes less than 0.2°.

The value of the characteristic velocity that a spacecraft located in the current orbit must provide to achieve the final circular orbit, as well as the values of the conjugate variables, can be calculated by interpolating the characteristic velocity $\Delta V_{\text{ch}}(r_p, r_a,$

i) using the three-dimensional tables. To do this, of course, the corresponding tables must be completed when solving an array of optimization problems on a three-dimensional grid of r_p, r_a, i values, similarly to completing the arrays for the values of conjugate variables.

The value of the remaining characteristic velocity obtained by linear interpolation on a three-dimensional grid of perigee and apogee radii and inclination $\Delta V_{\text{ch}}(r_p, r_a, i)$ is used in the SOFC+ algorithm as the norm of the distance between the osculating spacecraft orbit and the final circular orbit.

The main difference between the SOFC+ and SOFC algorithms, which makes it possible to achieve the required guidance accuracy, lies in the modification of the algorithm for calculating the thrust vector orientation parameters in the final transfer section. Let us present an algorithm for calculating the thrust vector direction within the framework of the SOFC+ algorithm for the case of transfer to GEO.

If $\Delta V_{\text{ch}}\big(r_p, r_a, i\big) < \Delta V^*_{\text{ch}}$, the values from (12.1.4) are redefined as:

$$A_\tau = -\text{sgn}(a - a_k)|A_\tau|, A_r = -\text{sgn}(v_r)|A_r|, A_n = -\text{sgn}(v_z)|A_n|,$$

where a is the osculating semimajor axis of the spacecraft orbit, a_k is the semimajor axis of the final orbit (GEO), v_r is the radial velocity of the spacecraft, v_z is the projection of the spacecraft velocity onto the normal to the true Earth equator on the date.

If $|a - a_k| < \Delta a^*$, A_τ is set equal to 0 (control along the semimajor axis is blocked), and if, in addition, the spacecraft latitude argument is within $[90° - \Delta u^*; 90° + \Delta u^*] \cup [270° - \Delta u^*; 270° + \Delta u^*]$, A_n is also set to 0, blocking the inclination control in inefficient sections of the orbit (otherwise, the thrust can cause the spacecraft to "hover" in the region of low inclinations). In the numerical examples below: $\Delta V^*_{\text{ch}} = 10\,\text{m/s}$, $\Delta a^* = 500\,\text{m}$, $\Delta u^* = 36°$.

In the case of transfer to a circular orbit other than GEO, the modification of the algorithm remains the same, but the reference plane for determining the current orbital elements should be the plane of the final orbit, and the values of a, r_p, r_a, i and ΔV_{ch} should be scaled according to the rules given at the end of Sect. 12.2.

To assess the SOFC stability with respect to thrust vector implementation errors, it is necessary to perform statistical modeling. The thrust vector implementation errors include many components, in particular:

- deviation of thrust and specific impulse due to dynamic processes in the EPS;
- deviation of thrust and specific impulse associated with the end of thruster life;
- thrust vector deviation associated with the on-board orbit determination errors;
- thrust vector deviation associated with spacecraft attitude errors;
- deviation of the thrust vector associated with the control system logic.

Even such an incomplete list of possible reasons for the deviation thrust vector from the nominal value shows the extreme complexity of modeling the real variation of the thrust vector over time.

The uncertainty of statistical characteristics of the variation in the magnitude and deviation of the direction of the thrust vector over time leads to the need to use

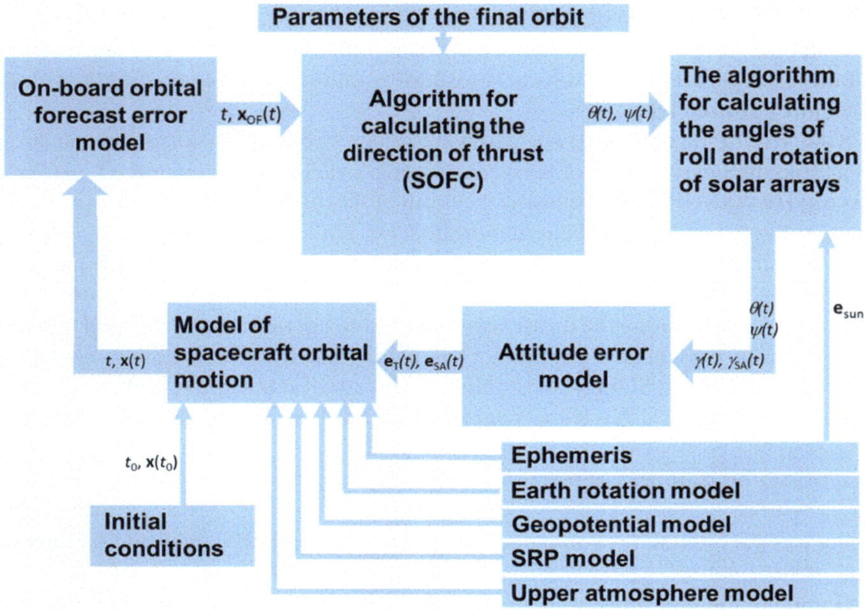

Fig. 12.4 The scheme of modeling the trajectory of spacecraft with EPS

simplified models to assess the stability of the SOFC with respect to errors in the implementation of the thrust vector.

To assess the insertion errors, a statistical modeling of the trajectory of the final insertion of a spacecraft with EPS into the GEO was performed. The modeling diagram is shown in Fig. 12.4.

The statistical model of errors in the onboard trajectory prediction is given by the standard deviation of errors in the determination of coordinates σ_R and the standard deviation of errors in the determination of components of the spacecraft orbital velocity σ_V in the orbital coordinate system. The entire launch trajectory is divided into correlation intervals of a given duration $\Delta t_{corr.tr}$ each. At each correlation interval, N_{tr} nodes are uniformly specified, in each of which the error in determining the components of the state vector is represented by a random variable with a normal distribution and a given standard deviation. At intermediate points within the correlation interval, the errors are calculated by linear interpolation between the nearest nodes.

Statistical models of errors for thrust, specific impulse, and thrust vector orientation angles are specified by N_T nodes at the correlation intervals of $\Delta t_{corr.T}$ each. The RMS of EPS thrust (σ_T), EPS specific impulse (σ_I), and thrust vector orientation angles (i.e. pitch and yaw—σ_{att}) are given.

In the general case, the mathematical expectations of the errors of all considered quantities can be specified at the boundaries of the correlation intervals, while the mathematical expectations of the errors at the nodes are calculated by linear interpolation.

Table 12.1 presents the results of statistical modeling of the trajectories of final spacecraft insertion into GEO for $\sigma_R = 5$ m, $\sigma_V = 0.3$ m/s, $\Delta t_{corr.tr} = 24$ h, $N_{tr} = 25$, $N_T = 101$, $\Delta t_{corr.T} = 1000$ h, $\sigma_T = 10$ mN, $\sigma_I = 25$ s, and $\sigma_{att} = 0.5°$.

The following initial conditions were assumed for the moment of ignition of the electric propulsion system (06/01/2016 00:00:00 UT):

- perigee altitude 20,600 km;
- apogee altitude 58,000 km;
- argument of perigee 180°;
- inclination 4.4°;
- right ascension of the ascending node 270°;
- true anomaly 200°.

The initial mass of the spacecraft is assumed to be 4429 kg, the thrust of the electric propulsion system is 0.548 N, and the specific impulse of the electric propulsion system is 1790 s.

Statistical modeling was performed on a sample of 1000 implementations of the final insertion trajectory with the same initial conditions for each option considered. Geopotential harmonics of the up to the 10th degree and 10th order, the gravity of the Moon and the Sun, perturbances from solar radiation pressure, precession and nutation of the earth equator were taken into account; the EPS operated continuously throughout the entire transfer. Integration was carried out until the eccentricity reached a value of 0.0003.

The significant standard deviation in the perigee argument and in the right ascension of the ascending node is obviously explained by the small eccentricity and inclination of the final orbit. A significant standard deviation in the final geocentric longitude is explained by errors in the implementation of the thrust vector and

Table 12.1 Estimation of average values and standard deviation of the main parameters at the end point of the trajectory (suboptimal control program)

Parameter	Average value	RMS
Transfer duration, days	78.88696	0.12264
Semimajor axis (error), km	42,162.770 (1.400)	14.163
Eccentricity	0.0003[a]	0[a]
Inclination, deg.	0.046	0.014
Argument of perigee, deg.	249.212	43.837
Right ascension of the ascending node, deg.	282.472	10.463
Geocentric longitude, deg.	9.206[b]	5.097
Spacecraft mass in GEO, kg	4105.200	0.363

[a]End of simulation when eccentricity reaches the value 0.0003
[b]Depends on the geocentric longitude of the point of the electric propulsion system firing

can be minimized by planning the final transfer with the minimum EPS thrust and subsequent introduction of the phasing coasting sections (if necessary).

12.4 Universal Tables of Characteristic Velocity in the Design of Combined Profiles for SC Insertion into a Circular Final Orbit

The main advantage of using EPS over traditional chemical propulsion systems for the inter-orbital transfers is the ability to significantly increase in the spacecraft mass in the final orbit due to the high specific impulse of the electric propulsion system. The main disadvantage of using electric propulsion is the long flight time due to the low thrust-to-mass ratio of a spacecraft with EPS. The use of a combined profile for delivering a spacecraft into geostationary orbit is a compromise between the requirements of providing the maximum spacecraft mass in the final orbit and performing an inter-orbital transfer in the shortest possible time.

The combined profile for delivering a spacecraft into a target orbit includes the stages of insertion into a low Earth (parking) orbit using a launch vehicle, formation of an elliptical intermediate orbit (IO) using the upper stage, and the transfer of the spacecraft from the IO to the target orbit (i.e., GEO) using the EPS. A typical profile for delivering a spacecraft to GEO using EPS is shown in Fig. 12.5.

In a typical flight profile, a flight between the parking and intermediate orbits is performed using a two-impulse scheme: the first velocity increment at the parking orbit node forms a transfer orbit with an apogee altitude equal to the apogee altitude of

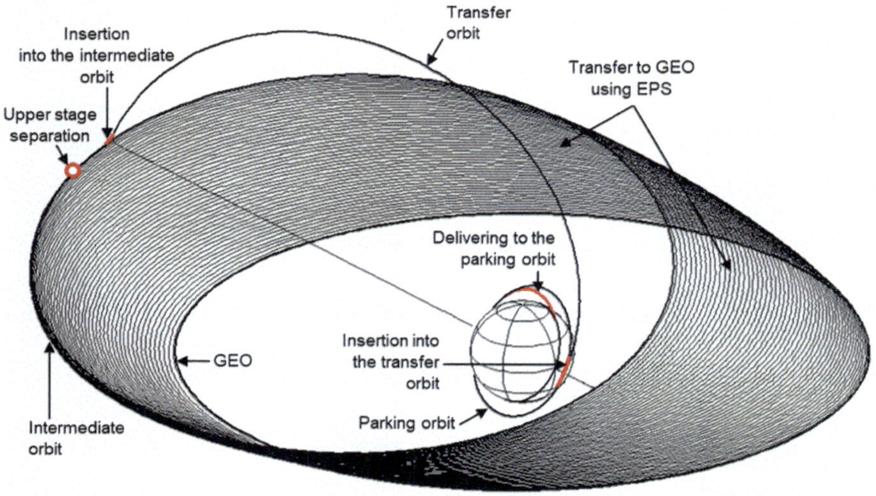

Fig. 12.5 A typical flight profile for spacecraft insertion into GEO using EPS

the IO, and the second velocity increment—at the apogee of the transfer orbit—raises the perigee altitude. The change in inclination is carried out during both impulses in order to minimize the characteristic velocity loss. Most of the inclination change occurs in the second impulse.

The optimization of the combined flight profile includes calculation of the optimal trajectory from IO to the target orbit and the optimization of the orbital parameters of the IO. In this section, GEO is considered as the target orbit; however, due to the properties of symmetry and similarity properties of multi-orbit low-thrust trajectories (see Sect. 12.2), the results are easily applicable to the optimization of transfers to other circular target orbits. The goal of optimization can be either to minimize the duration of transfer of a spacecraft with a given mass into GEO, or to maximize the mass of spacecraft on GEO for a given transfer duration.

Within the framework of the proposed approach (Refs. to [2, 3]), the inclination, perigee altitude and apogee altitude of the IO are considered as external parameters to be optimized. For each set of these parameters, the mass of the spacecraft delivered by the upper stage to the IO is calculated, and then the trajectory of the spacecraft with the EPS from the IO to the GEO is optimized in order to compute the trajectory that delivers the spacecraft to GEO in the minimum time. It can be shown that the use of minimum-time control during the flight phase of a spacecraft with an electric propulsion system, provided that optimal IO is used and given the characteristics of the launch vehicles, ensures the delivery of a spacecraft with a greater mass to the GEO in a given time than with minimum-fuel control [6]. This is due to the fact that the derivatives of the final mass of the spacecraft on the GEO in terms of IO parameters in the vicinity of the optimum are superior to the derivatives of the final mass of the spacecraft on the GEO in terms of the duration of the transfer from IO to GEO in the vicinity of the minimum-time solution. In this regard, the time-optimal control is the best for a combined flight profile.

To optimize the combined flight profile, we used pre-calculated values of the characteristic velocity for optimal trajectories from inclined elliptical orbits to GEO on a three-dimensional grid of perigee radius, apogee radius and IO inclination values presented in Sect. 12.3—$V_{ch}^{EPS}(i, r_p, r_a)$. This is possible, since at a sufficiently low thrust acceleration of the spacecraft (less than 2–4 mm/s^2) the considered optimal multi-orbit trajectories have the asymptotic property—on such trajectories the characteristic velocity is practically independent of the level of thrust acceleration, therefore the problem of calculating the spacecraft mass (parametric problem) can be solved after solving the problem of calculating the trajectory (dynamic problem). Taking into account the possibility of scaling the values of the characteristic velocity for various final circular orbits and for various values of the gravitational parameter (see Sect. 12.2), as well as the noted asymptotic properties, the three-dimensional tables of the characteristic velocity can be called universal.

As a result, the problem of assessing the characteristics of the optimal combined flight profile for the given IO parameters is reduced to the following:

(1) Calculation (in the patched-conic approximation, taking into account the losses) of the characteristic velocity of transfer from the parking orbit to the IO

using the upper stage in order to calculate the mass of the spacecraft on the IO $m_{SC}^{IO}(i, r_p, r_a)$. During the calculation process, the change in inclination in impulses is optimized, and the velocity losses are either specified as a constant relative value or calculated using approximation dependencies;

(2) Interpolation of the EPS characteristic velocity on a three-dimensional grid of perigee radius, apogee radius and inclination of the IO—$V_{ch}^{EPS}(i, r_p, r_a)$ (either a three-dimensional linear interpolation or a more accurate three-dimensional interpolation with B-splines is used [7]);

(3) Calculation of the final mass of the spacecraft on GEO:

$$m_k(i, h_p, h_a) = m_{\mathrm{SC}}^{\mathrm{IO}}(i, r_p, r_a) \times \exp\left[-V_{\mathrm{ch}}^{\mathrm{EPS}}(i, r_p, r_a)/I_{\mathrm{sp}}^{\mathrm{EPS}} g_0\right]$$

and of the transfer time:

$$\Delta t(i, r_p, r_a) = \left[m_{\mathrm{SC}}^{\mathrm{IO}}(i, r_p, r_a) - m_\kappa(i, r_p, r_a)\right]/\left(T^{\mathrm{EPS}}/I_{\mathrm{sp}}^{\mathrm{EPS}} g_0\right),$$

where ${I_{\mathrm{sp}}}^{\mathrm{EPS}}$ is the EPS specific impulse, T^{EPS} is the EPS thrust, and $g_0 = 9.80665$ m/s^2.

The optimal low-thrust trajectories calculated in advance on the IO parameter grid allow, thanks to the asymptotic property of the optimal solutions, to avoid solving the complex boundary value problem of the maximum principle when optimizing a specific combined flight profile.

Optimization of a combined profile reduces to the problem of calculating such IO parameters i, r_p, r_a, which would give a maximum $m_k(i, r_p, r_a)$ for a given $\Delta t(i, r_p, r_a)$, and both functions, $m_k(i, r_p, r_a)$ and $\Delta t(i, r_p, r_a)$, are calculated using simple high-speed algorithms. This method for optimizing the combined flight profile was proposed in [2, 3], has been used many times and has shown its reliability and efficiency.

In practice, in order to determine optimal IO parameters i, r_p, r_a, the functions $m_k(i, r_p, r_a)$ and $\Delta t(i, r_p, r_a)$ were calculated using the given algorithm on a three-dimensional grid of i, r_p, r_a values with a sufficiently small step specified by the user. For each given flight time Δt^*, the optimal set of IO parameters i, r_p, r_a was accepted for which the following condition is satisfied:

$$\{i, r_p, r_a\} = \underset{\Delta t(i, r_p, r_a) \le \Delta t^*}{\operatorname{argmax}} \left[m_k(i, r_p, r_a)\right].$$

The results obtained with the described method have methodological errors associated with ignoring perturbances, inaccuracy in determining the loss of the characteristic velocity during the upper stage operation, errors in the three-dimensional interpolation, and ignoring the peculiarities of the spacecraft operation sequence at different stages of the flight. Therefore, it is necessary to clarify the main trajectory parameters using more adequate mathematical models at both stages of the flight—with high and low thrust.

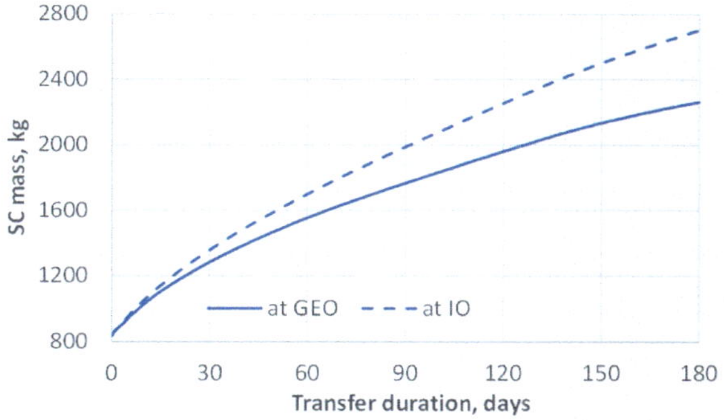

Fig. 12.6 The spacecraft initial mass on the intermediate orbit (immediately after the separation from the upper stage) and its mass on GEO as a function of the flight duration

As an example, let us consider the problem of analyzing a combined launch profile for the following initial data: a launch vehicle inserts a spacecraft and an upper stage with a total mass of 8200 kg into a circular orbit with an altitude of 200 km and an inclination of 51.8°. The final mass of the upper stage is 1000 kg, the maximum mass of the upper stage fuel is 6400 kg, the thrust of the upper stage propulsion system is 4000 kgf, the specific impulse of the upper stage propulsion system is 332 s, the thrust of the spacecraft EPS is 0.5 N, and the specific impulse of the spacecraft EPS is 1800 s. With such initial data, the upper stage can deliver a spacecraft weighing 837 kg to GEO. When using a combined launch scheme, a spacecraft of with a mass of 1768 kg can be delivered to GEO with a flight duration of 90 days, and with a flight duration of 180 days, its mass can be as high as 2262 kg. The dependence of the initial mass of the spacecraft on the IO (immediately after separation from the upper stage) and the mass of the spacecraft on the GEO on the flight duration is shown in Fig. 12.6. The dependence of the optimal IO parameters (perigee altitude, apogee altitude and inclination) on the flight duration is presented in Fig. 12.7.

12.5 Application of Suboptimal Feedback Control to Calculate Low-Thrust Trajectories to the Moon and Libration Points of the Earth-Moon and Earth-Sun Systems

The optimization of low-thrust spacecraft trajectories in the Earth-Moon system has been considered in many studies, for example, in [8–14]. However, the problem of end-to-end optimization of low-thrust spacecraft trajectories between Earth and Moon orbits remains a challenging problem to date for which no regular methods have been developed. In particular, the difficulty of the problem lies in the complex nature of the external perturbing forces that determine (along with the EPS thrust) the trajectory of the spacecraft. One of the peculiarities of the considered problem is its significant multi-extremal nature, so that even if a trajectory satisfying the

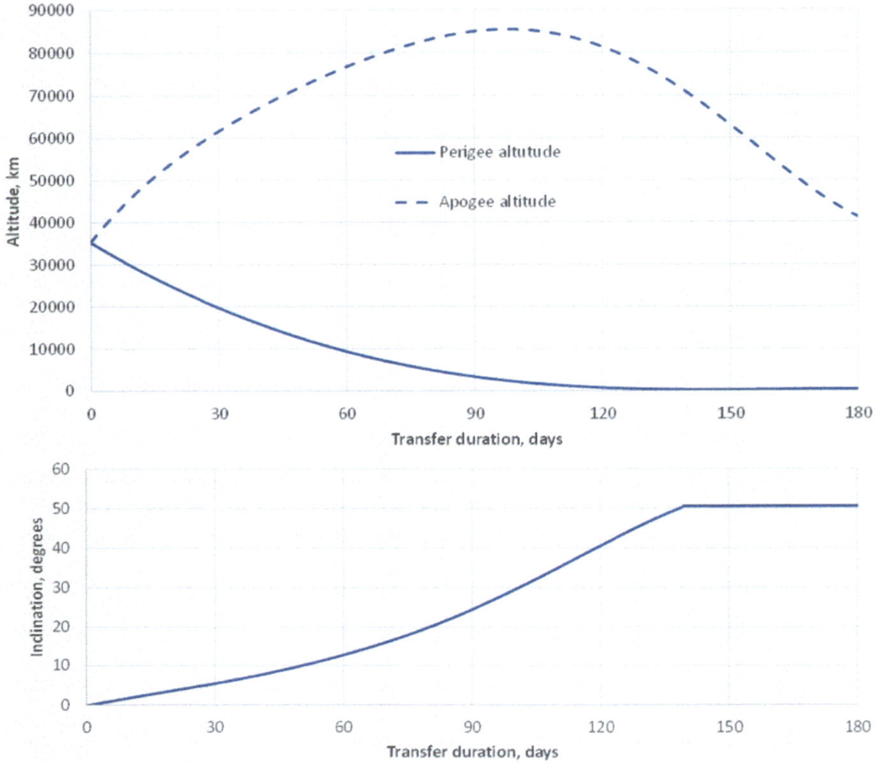

Fig. 12.7 The optimal parameters of the intermediate orbit (perigee altitude, apogee altitude and inclination) as a function of the flight duration

optimality conditions is calculated (for example, by indirect methods based on the maximum principle or by direct optimization of the cost function using the nonlinear programming methods on any class of controls), only the local optimality of the found control can be guaranteed and there is no guarantee that there is no better solution with significant variation in the control.

Since it is practically impossible to calculate globally optimal trajectories on a regular basis, it seems advisable to use simplified methods to calculate trajectories that are close to the locally optimal in the sense of satisfying the necessary optimality conditions of the maximum principle. One way to calculate such suboptimal trajectories is to use the SOFC algorithm and its modifications.

The possibility and feasibility of using SOFC for calculating spatial transfers between circular Earth and Moon orbits follows from the following considerations.

First of all, the low-thrust flight trajectories from the Earth orbit to the Moon orbit and back pass through the vicinity of the libration point L1. Therefore, sections of geocentric and selenocentric motion can be joined at point L1 with minor losses in cost function (flight time or EPS propellant mass) compared to the optimal trajectory.

More precisely, the possibility of joining the geocentric and selenocentric sections of the trajectory at the libration point L1 is based on the following reasoning:

1. The typical duration of life of a spacecraft moving along a hyperbolic trajectory in the Moon's activity sphere is on the order of a day.
2. The typical change in the spacecraft velocity due to the EPS operation with thrust acceleration of ~ 0.1 mm/s^2 is ~ 10 m/s.
3. The width of the throat in the vicinity of the libration point with an excess relative velocity of the spacecraft on the Hill's sphere of 10 m/s, through which a spacecraft with an EPS can enter the region of selenocentric satellite motion, is ~ 60,000 km (see Fig. 12.8).

The Hill's sphere is a closed surface characterized by a critical value of the Jacobi integral (analogous to the energy integral in the restricted three-body problem). This surface separates the region of satellite motion around the Moon from the region of geocentric motion. The Hill's sphere contains L1. In Fig. 12.8, the Hill's sphere section by the plane of the Moon's orbit is shown by a thick line that limits the region of satellite motions.

Therefore, to implement the capture of a spacecraft into the region of satellite motion using an EPS, the velocity of the spacecraft relative to the libration point

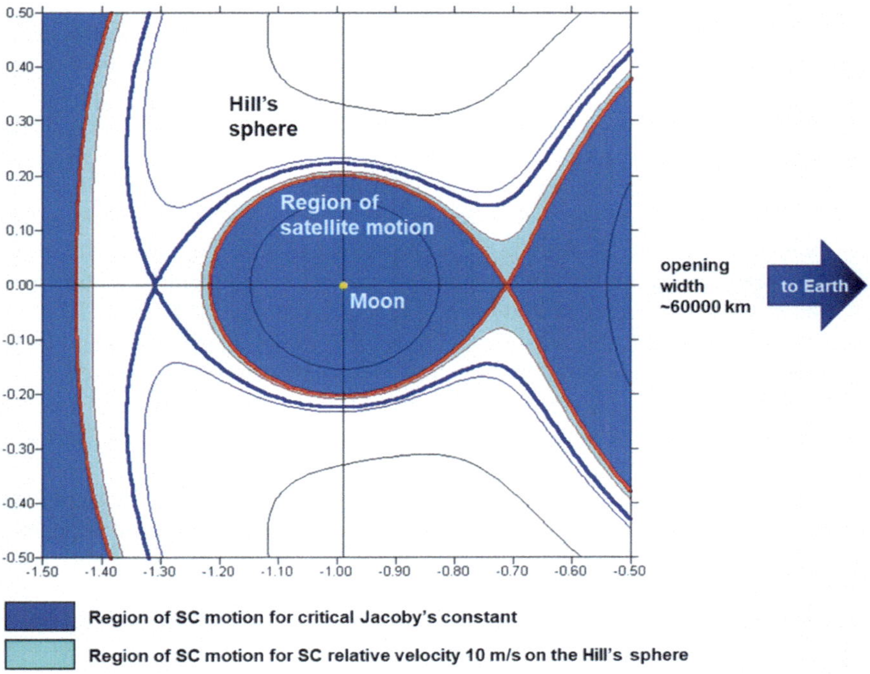

Fig. 12.8 The Hill sphere and the domains of possible spacecraft motion in the restricted three-body problem

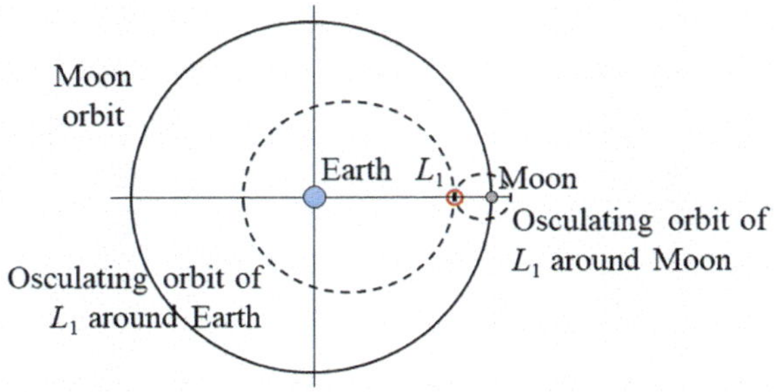

Fig. 12.9 A diagram of the osculating geocentric and selenocentric orbits of the libration point L1

when crossing the Hill's sphere should not exceed ~ 10 m/s, and the distance from the libration point should not exceed ~ 30,000 km (less than 8% of the distance between the Earth and the Moon). Therefore, we can expect that the low-thrust trajectories passing through the point L1 with zero relative velocity will be close to the optimal trajectories.

Another factor that makes it convenient to use the libration point L1 to join the geocentric and selenocentric sections of the trajectory, is the ellipticity of the osculating orbit of L1 when considering both geocentric and selenocentric motion. The diagram of the osculating geocentric and selenocentric orbits of L1 is shown in Fig. 12.9.

Tables 12.2 and 12.3 present the limits of change in the values of the osculating elements of the geocentric and selenocentric orbits of the libration point L1 in the geocentric and selenocentric earth-equatorial inertial coordinate systems J2000, respectively, in the interval of 2008…2028. Figure. 12.10 shows a plot of the evolution of the osculating perigee radius, and Fig. 12.11—a plot of the evolution of the osculating inclination of the geocentric orbit of the libration point L1 in this time interval. The evolution of the osculating apogee radius and of the eccentricity and true anomaly of the libration point L1 is of the same nature as the evolution of the osculating apogee radius. The periapsis argument varies from 0 to 360° with a period equal to the orbital period of the Moon. The orbital elements of L1 orbit are subject to short-period oscillations with a period of about one month and long-period oscillations with a period of more than six months; moreover, the inclination and longitude of the ascending node are subject to long-period oscillations with a period of 18.6 years, which is equal to the precession period of the line of the nodes of the Moon's orbit.

It is clear from Tables 12.2, 12.3 and Figs. 12.10, 12.11 that the point L1 always remains on the osculating ellipse, all elements of which (except for the pericenter argument) do not have a pronounced secular drift, but are subject only to the short- and long-period oscillations. The line of apses, generally speaking, has a secular drift (rotates), but, as was shown earlier, the divergence of this line has virtually no effect on the convergence of the SOFC algorithm.

Thus, to calculate the trajectory of a spacecraft with an EPS from a given initial circular Earth orbit to a circular Moon orbit, it is necessary to calculate the geocentric

Table 12.2 Extreme values of the osculating elements of the geocentric orbit of the libration point L1 in the geocentric earth-equatorial inertial coordinate system J2000 in the interval of 2008–2028

Orbit parameter	min	max
Pericenter radius, km	139,958.67	151,605.69
Apocenter radius, km	302,756.80	345,305.84
Eccentricity	0.332666	0.420817
Inclination, deg.	18.1351	28.7223
Right ascension of the ascending node, deg.	−13.3854	13.3571
True anomaly, deg.	173.2454	186.7365

Table 12.3 Extreme values of the osculating elements of the selenocentric orbit of the libration point L1 in the selenocentric earth-equatorial inertial coordinate system J2000 in the interval of 2008–2028

Orbit parameter	min	max
Periapsis radius, km	9299.85	9674.29
Apocenter radius, km	53,817.53	61,381.12
Eccentricity	0.695259	0.735513
Inclination, deg	18.1351	28.7223
Right ascension of the ascending node, deg	−13.3854	13.3571
True anomaly, deg	178.3732	181.6206

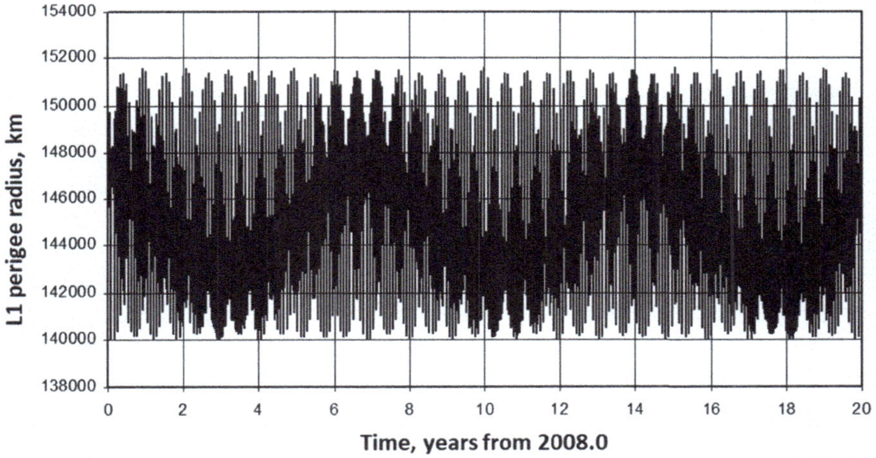

Fig. 12.10 Time dependence of the perigee radius of the libration point L1

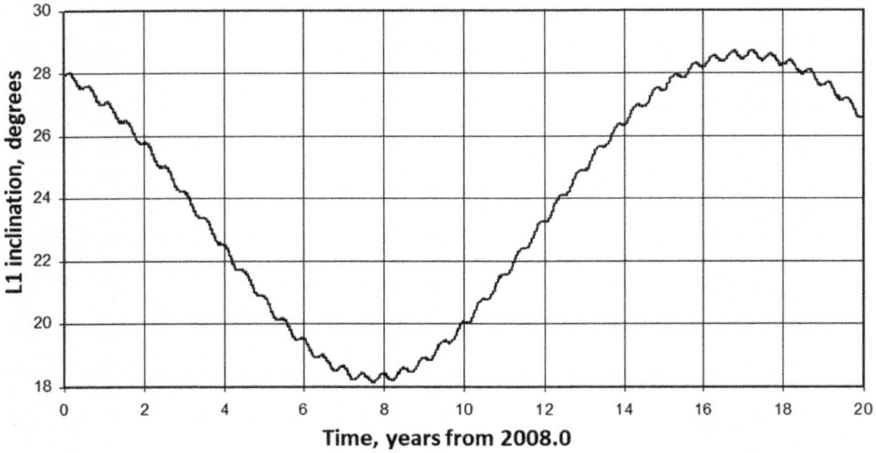

Fig. 12.11 Time dependence of inclination of the libration point L1 geocentric orbit to the J2000 earth equator

trajectory from the Earth orbit to L1 and the selenocentric trajectory from L1 to a given final Moon orbit. If there is an algorithm (e.g. SOFC) to compute the thrust direction vector **e** as a function of the current state variables of the trajectory $\mathbf{x}(t)$: $\mathbf{e}[\mathbf{x}(t)]$, which ensures the convergence of the trajectory to a given final circular orbit, the listed problems can be solved as follows:

(1) The time t_{L1} at which the spacecraft passes point L1 is fixed and the initial guess for the spacecraft mass m_{L1} at this point is set;
(2) Differential equations of the geocentric spacecraft motion with the control $-\mathbf{e}[\mathbf{x}(t)]$ are integrated in the reverse time direction from L1 until a small neighborhood of the initial circular geocentric orbit is reached. The result of this integration is the mass m_{01} of the spacecraft on the initial Earth orbit;
(3) Points 1–2 above define the dependence $m_{01}(m_{L1})$. The root of the equation $\mathrm{m}_{01}(m_{L1}) - m_0 = 0$, where m_0 is the given initial mass of the spacecraft on the initial circular Earth orbit, is calculated by some standard method (modified Newton's method or others). As a result, in addition to m_{L1}, the trajectory from the initial Earth orbit to the libration point L1 is computed.
(4) The differential equations of the selenocentric spacecraft motion with control $\mathbf{e}[\mathbf{x}(t)]$ are integrated in the forward time direction from L1 until a small neighborhood of the final Moon orbit is reached. As a result of this integration, the spacecraft mass m_k on the final Moon orbit and the trajectory from L1 to the final Moon orbit are computed.

As a result, the solution to the problem of the transfer trajectory calculation is reduced to the problem of synthesizing a stable feedback control of the form $\mathbf{e}[\mathbf{x}(t)]$, for which it is proposed to use the SOFC.

The presented methodology for calculating low-thrust flight trajectories from Earth orbit to Moon orbit assumes that both the initial and final orbits are circular. This requirement is related to the SOFC structure, which ensures convergence only to circular orbits, and is the most serious limitation of the methodology under consideration. Therefore, if the initial Earth orbit is not circular, it is necessary to introduce a second, circular, intermediate Earth orbit. The geocentric low-thrust flight segment in this case will consist of a trajectory segment from the initial elliptical orbit to the intermediate circular orbit, and then from the intermediate circular orbit to L1.

Another reason for the need to introduce a geocentric circular intermediate orbit with a sufficiently large radius is the limitation of the range of changes in the geocentric radius vector of the spacecraft during the flight resulting from the grid of parameters of the initial orbits implemented in the SOFC interpolation tables.

The methodological necessity of introducing a circular intermediate orbit when using suboptimal control to calculate the transfer from an elliptical initial earth orbit to a circular Moon orbit naturally leads to a further deviation of the flight trajectory from the optimal one. Fortunately, it follows from the results of numerical modeling that with typical parameters of the initial Earth orbit, the osculating eccentricity during the optimal flight has a minimum in the geocentric section of the trajectory, reaching values less than 0.1. This means that during the optimal flight, the osculating orbit becomes nearly circular at some point in time, so that the introduction of a circular intermediate orbit into consideration can be justified provided that its altitude and inclination are optimized. On the other hand, the requirement to minimize the radiation load during the insertion process leads to the necessity that the spacecraft leaves the region of the radiation belts as quickly as possible, and this problem is well satisfied with a minimum-time transfer from the initial Earth orbit to a sufficiently high circular Earth orbit with a minimum change in inclination.

As an example, let us consider the problem of preliminary space mission design and trajectory analysis for a reusable lunar tug (RLT) with a nuclear electric propulsion system. For RLT, the low Earth orbit should be near-circular, so the presented technique can be efficiently used to solve the considered problem.

In this example, a circular geocentric orbit with an altitude of 20,000 km and an inclination equal to the inclination of the initial Earth orbit (IEO)—the RLT orbit—was chosen as the intermediate orbit (IO). The altitude of the IEO is assumed to be 800 km and the inclination is 51.6°. It is assumed that the transfer from IEO to IO is performed with a constant circumferential thrust of the EPS.

Thus, the transfer from IEO to the circular final lunar orbit (FLO) is divided into three sections:

1. The flight section from IEO to IO with an altitude of 20,000 km and an inclination equal to the inclination of IEO;
2. The flight section from IO to L1;
3. The flight section from L1 to the FLO with a given altitude and inclination.

To calculate the trajectory, an initial guess value for the spacecraft mass at L1 is chosen, and the differential equations of spacecraft motion are integrated in Sect. 12.3 in reverse time using SOFC to determine the direction of the thrust vector, and then in Sect. 12.2, also in reverse time, but with a circumferential direction of the thrust. As a result, the flight time and mass of the spacecraft on IEO are calculated. By repeating these calculations many times, the mass of the spacecraft at L1 is chosen to ensure the given initial mass of the spacecraft on the IEO. The differential equations of the spacecraft motion are then integrated over the selenocentric section of the spacecraft transfer from L1 to the FLO in direct time using SOFC to compute the direction of the thrust vector.

The return flight from FLO to IEO is calculated in reverse order. First, an initial guess value for the spacecraft mass at L1 is chosen and the differential equations of spacecraft motion on the selenocentric section of the transfer from FLO to L1 are integrated in reverse time using the SOFC to compute the direction of the thrust vector. As a result of repeated calculations of this trajectory, a value of the spacecraft mass at L1 is calculated that ensures the given spacecraft mass at the FLO. Then, in direct time, the differential equations of spacecraft motion are sequentially integrated over the geocentric sections of the L1-IO transfer (using SOFC) and the IO-IEO transfer (with circumferential thrust direction).

It should be noted that at a large radius of the osculating orbit apocenter, the orbital velocity of the spacecraft is low, and therefore, at a sufficiently high thrust acceleration, the averaged equations used in the SOFC synthesis may lead to a large deviation of the control from the optimal one. In addition, the use of the SOFC in the region of the apocenters of highly elongated orbits at high thrust-to-weight ratio can result in the orbital velocity of the spacecraft being almost completely reduced at some time moment, which can lead to an unacceptable reduction in the periapsis radius. To compensate for these drawbacks, a modification of the SOFC algorithm was developed and applied that provides for the EPS shutdown when the current orbital velocity of the spacecraft is less than $k\ (T/m)\ \Pi$, where T is the EPS thrust, m is the spacecraft mass, Π is the osculating orbital period, and k is the control coefficient. Under this condition, the use of the EPS shutdown leads to the appearance of coasting sections in the region of the apocenters of the orbits with the greatest eccentricity, and the duration of such coasting sections can be regulated by varying the control coefficient. To ensure the transfer from the point L1 in the desired direction (to the Moon or to the Earth), a certain (short enough) time interval Δt_0 counted from the moment of L1 flyby is introduced, during which the EPS shutdown is prohibited.

Tables 12.4 and 12.5 show the duration and characteristic velocity of the transfer between IO and FLO as a function of the specific impulse and thrust of the electric propulsion system for the following L1 flyby dates: 06/01/2015 (minimum inclination of the Moon's orbit to the Earth equatorial plane) and 01/01/2025 (maximum inclination of the Moon's orbit to the Earth equatorial plane). The FLO altitude is assumed to be 100 km, the inclination—90°; and the mass of the spacecraft at the moment of L1 flyby—100 tons.

Table 12.4 Duration and characteristic velocity of the transfer between IO and FLO depending on the specific impulse and thrust of the EPS for the L1 flyby on 01.06.2015

I_{sp}, s	T, N	"Minimum time"		"Minimum V_{ch}"			
		Δt, days	V_{ch}, m/s	T, days	V_{ch}, m/s	k	Δt_0, days
10,000	20	198.36896	3369.268	230.32827	3318.475	10.0000	0
	30	132.52283	3376.201	155.40876	3321.246	6.0000	0
	50	80.80001	3429.875	99.89080	3321.491	4.0000	0
	100	44.41918	3764.624	57.26835	3326.630	2.1500	0
	200	27.63174	4662.165	32.61316	3310.406	0.4000	0
8000	20	199.23707	3369.414	231.16633	3318.739	10.0000	0
	30	133.08332	3375.835	155.93100	3320.756	6.0000	0
	50	81.14405	3429.366	100.25891	3322.419	4.0000	0
	100	44.64272	3765.310	57.44867	3326.741	2.1500	0
	200	27.75306	4654.863	32.69727	3307.411	0.3950	0
6000	20	200.70861	3369.876	232.56486	3319.058	10.0000	0
	30	134.03295	3375.427	156.85953	3320.966	6.0000	0
	50	81.73625	3429.114	100.76655	3321.524	4.0000	0
	100	44.95574	3761.350	57.66637	3325.844	2.1500	0
	200	27.98970	4648.098	32.66789	3316.052	0.3793	0
4000	20	203.62014	3369.595	235.30263	3318.300	10.0000	0
	30	135.97188	3374.950	158.74807	3321.612	6.0000	0
	50	82.92127	3427.956	101.89110	3321.563	4.0000	0
	100	45.49985	3746.444	58.22147	3324.598	2.1500	0
	200	28.36497	4618.575	33.01330	3332.297	0.3498	0
2000	20	212.71708	3369.082	244.03327	3318.495	10.0000	0
	30	142.07078	3374.743	164.61826	3323.482	6.0000	0
	50	86.63909	3425.474	105.36176	3321.390	4.0000	0
	100	47.77606	3746.178	59.94872	3322.725	2.1500	0
	200	30.28044	4638.135	33.94008	3330.999	0.3705	0

From the above tables it can be seen that as the thrust-to-mass ratio increases, the characteristic velocity on the "minimum-time" trajectories (without shutting down the electric propulsion system) increases and reaches values 1.5 times higher than the characteristic velocity required on the "minimum-time" trajectories with low thrust-to-mass ratio. This fact is explained by the presence of long thrusting sections with a large trajectory angle on the "minimum-time" trajectories with a high thrust-to-mass ratio, which leads to an increase in the characteristic velocity losses (see Fig. 12.12).

Table 12.5 Duration and characteristic velocity of the flight between IO and FLO depending on the specific impulse and thrust of the EPS for the L1 flyby on 01.01.2025

I_{sp}, s	T, N	"Minimum time"		"Minimum V_{ch}"			
		Δt, days	V_{ch}, m/s	T, days	V_{ch}, m/s	k	Δt_0, days
10,000	20	190.61078	3239.649	211.46754	2983.188	10	0
	30	129.24840	3294.166	144.95323	2995.802	6	0
	50	80.83431	3431.305	94.48493	3013.734	4	0
	100	46.35143	3925.151	56.30122	3002.474	2.15	0
	200	28.72430	4842.029	30.12317	3037.218	1.6	0.14
8000	20	191.40515	3239.661	212.09741	2984.389	10	0
	30	129.80068	3394.290	145.03743	2998.442	6	0
	50	81.17310	3430.567	94.99708	3012.144	4	0
	100	46.60226	3926.516	56.11723	3006.478	2.15	0
	200	28.93632	4847.314	30.18234	3037.266	1.6	0.14
6000	20	192.74968	3239.857	213.23716	2983.790	10	0
	30	130.70725	3293.984	146.00917	2993.525	6	0
	50	81.75990	3430.077	95.03283	3013.955	4	0
	100	46.99390	3926.311	56.50389	3002.660	2.15	0
	200	29.27150	4852.414	30.28201	3037.391	1.6	0.14
4000	20	195.43126	3239.524	215.53893	2981.752	10	0
	30	132.54974	3293.476	147.66597	2995.553	6	0
	50	82.98211	3430.364	96.00996	3012.848	4	0
	100	47.74559	3922.407	57.26360	2997.597	2.15	0
	200	29.88133	4850.809	30.48491	3037.755	1.6	0.14
2000	20	203.83210	3239.354	222.62020	2981.362	10	0
	30	138.33857	3293.127	152.38917	2995.015	6	0
	50	86.69095	3427.355	98.77208	3012.334	4	0
	100	50.20917	3919.083	58.47859	3001.701	2.15	0
	200	31.68380	4828.637	31.08827	3042.780	1.61	0.17

At the same time, when using coasting sections in the area of osculating apocenters, it is possible to maintain the required characteristic velocity losses at approximately the same level for an arbitrary thrust-to-mass ratio due to a slight increase in the transfer time at large values of thrust acceleration compared to a transfer without shutting down the electric propulsion system. Tables 7.7.3 and 7.7.4 show that on trajectories with coasting sections the maximum deviation of the characteristic velocity from the average value does not exceed 0.4% for 2015 and 1.2% for 2025, while the flight duration increases by 7...35% compared to a minimum.

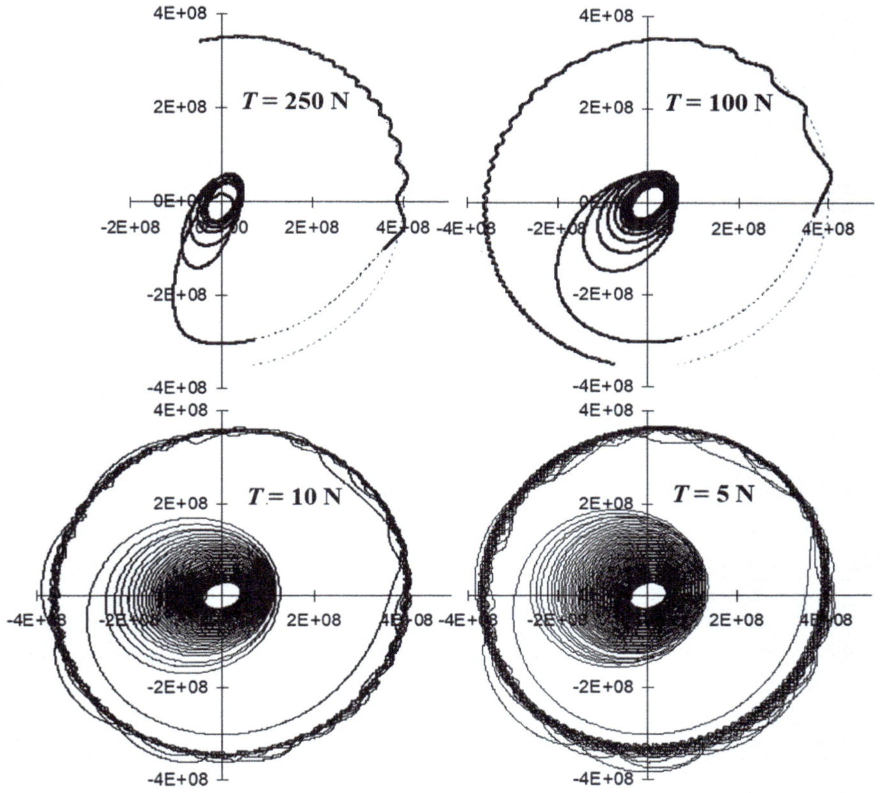

Fig. 12.12 Projections of low-thrust trajectories between the earth and the lunar orbits on the plane of the earth equator in J2000 at different thrust T values

The analysis of the data given in Tables 12.4 and 12.5 allows us to construct the following approximating dependence of the ration of the flight time Δt to the EPS operating time $\Delta t_{\rm m}$ for the case of a transfer from an IEO with an altitude of 800 km and an inclination of 51.6° on to an FLO with an altitude of 100 km and an inclination of 90°:

$$\Delta t/\Delta t_{\rm m} = -0.0531a_0^2 + 0.2693a_0 + 1.0275$$

where a_0 is the initial thrust acceleration on the IEO, while the required characteristic velocity is estimated to be in the range of 8624…8671 m/s for trajectories in 2015 and in the range of 8220…8349 m/s for trajectories in 2025. For the purposes of space mission design, it is proposed to use the maximum value of the characteristic velocity loss (8671 m/s), which is 0.5% greater than the minimum in the same epoch, and 5% greater than the minimum during the precession period of the Moon's orbit.

As with the calculation of the trajectory from the Earth orbit to the libration point L1 of the Earth-Moon system, the SOFC can also be used to calculate the trajectory to the libration points L1 and L2 of the Earth-Sun system.

In fact, the geocentric velocity of L1 and L2 (~ 300 m/s) is less than the circular geocentric velocity at such a distance (> 500 m/s), so the osculating geocentric orbits of L1 and L2 are elliptical, and the apogee radii of these orbits are approximately equal to the geocentric distance of L1 and L2, respectively, and the perigee radii are less than the radius of the lunar orbit (~ 300 thousand km). To calculate the transfer to L1 and L2 using SOFC, the methodology described in Sect. 7.8 is used.

The following flight profile from the parking orbit to L2 is considered:

(1) The upper stage inserts the spacecraft into some elliptical IO, and then it separates from the spacecraft. The perigee and apogee altitudes of the IO are optimized; the inclination of the IO is equal to the inclination of the reference orbit (51.6°).
(2) The spacecraft moves to a geosynchronous orbit (GSO, circular orbit with an inclination of 51.6° and an orbital period of 1 day) using an EPS. The thrust vector control in this section is optimized using the methods from Sect. 6.
(3) The spacecraft moves from GSO to L2 using an EPS. In this section, the same methodology is used to calculate the control, as was used to calculate the control during the flight to the libration point L1 of the Earth-Moon system. A modification of the algorithm is used that allows the introduction of coasting sections on the trajectory. The values of Δt_0 and k were chosen from the condition of minimizing the flight time for a given mass of the spacecraft at L2.

Table 12.6 and Fig. 12.13 present the results of calculating trajectories to L2 using SOFC for different spacecraft masses delivered to L2. The date of arrival at L2 is fixed in all considered trajectories (January 9, 2020 00:00:00 UT). The following initial data were used for the calculations: the parking orbit is circular with an altitude of 200 km and an inclination of 51.6°, the mass of the orbital unit (upper stage + spacecraft) in the parking orbit is 8080 kg, the thrust of the upper stage propulsion system is 2000 kgf, the specific impulse is 335 s, the final mass of the upper stage is 777.3 kg. The EPS thrust is assumed to be 0.56 N, the specific impulse is 1650 s.

Figures 12.14, 12.15 and 12.6 show some results for the trajectory option No. 2 using SOFC from Table 12.6 (delivery of a spacecraft with a mass of 2500 kg to L2 in ~ 427 days). Figure 12.14 shows projections of the trajectory from GSO to L2 onto the coordinate planes of the J2000 Earth equatorial coordinate system, and Figure 12.15 shows the view of such a trajectory in the synodic coordinate system. The solid line shows the thrusting sections of the trajectory, and the dotted line shows the coasting ones. Figure. 12.16 shows the time dependence of the spacecraft's distance from the Moon. It is interesting to note that on day 185.4227 of the flight from the GSO to L2, the spacecraft flies quite close to the Moon (at a distance of 39,739 km). It is obvious that it is possible to use a gravity assist maneuver near the Moon to improve the flight characteristics.

Table 12.6 Main parameters of the trajectories of a spacecraft with an EPS to L2 using SOFC as a function of the spacecraft mass at L2

Option No.	1	2	3	4
Spacecraft mass at L2, kg	2400	2500	2600	2700
Transfer time, days	399.515	426.657	438.046	447.609
incl. IO—GSO	148.293	165.167	180.346	197.904
GSO—L2	251.223	261.490	257.700	249.704
Mass of the spacecraft on the IO after the upper stage separation, kg	3203.914	3376.074	3529.115	3704.111
Spacecraft mass on GSO, kg	2760.494	2882.196	2989.85	3112.341
EPS propellant mass, kg	803.914	876.074	929.115	1004.111
EPS operating time, days	268.851	292.984	310.722	335.803
IO perigee altitude, km	5463	4999	4711	4530
IO apogee altitude, km	9426	8329	7336	6182
Δt_0, days	6.5	6.8	7.04	7.3
k	7	7.6	7	6

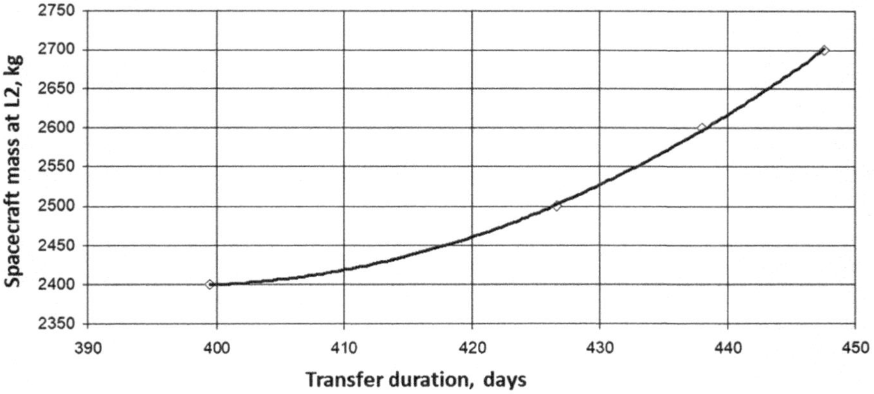

Fig. 12.13 The spacecraft mass at L2 as a function of the transfer duration when using SOFC

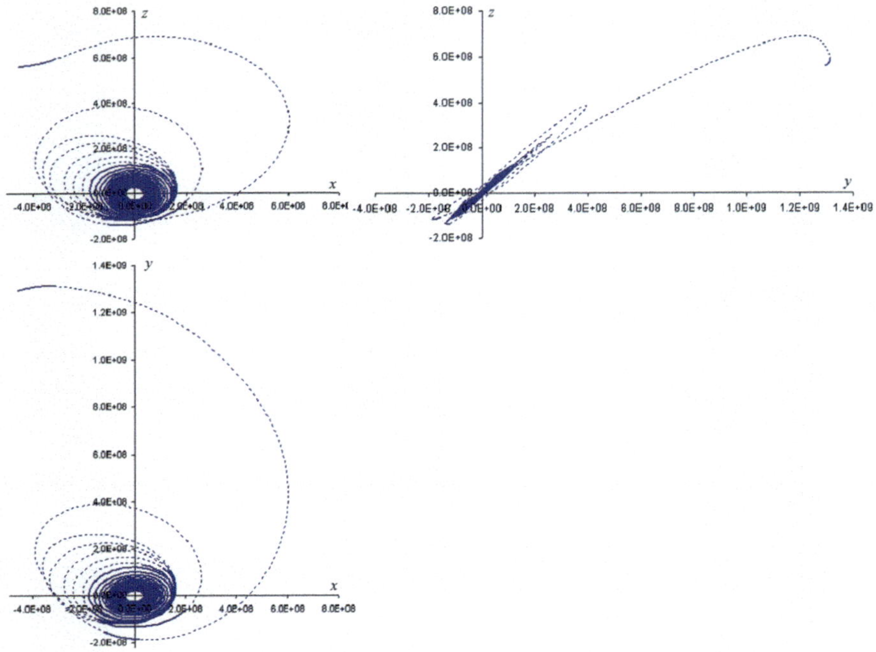

Fig. 12.14 The trajectory of transfer from the circular geosynchronous orbit to L2 using SOFC (projections onto J2000 coordinate planes). Solid lines correspond to the sections with the EPS in operation, dotted lines correspond to the coasting trajectory sections

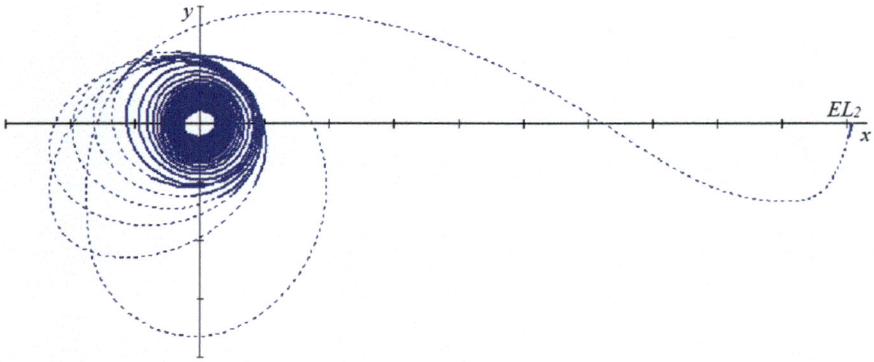

Fig. 12.15 The trajectory of transfer from the circular geosynchronous orbit to L2 using SOFC in the synodic coordinate system

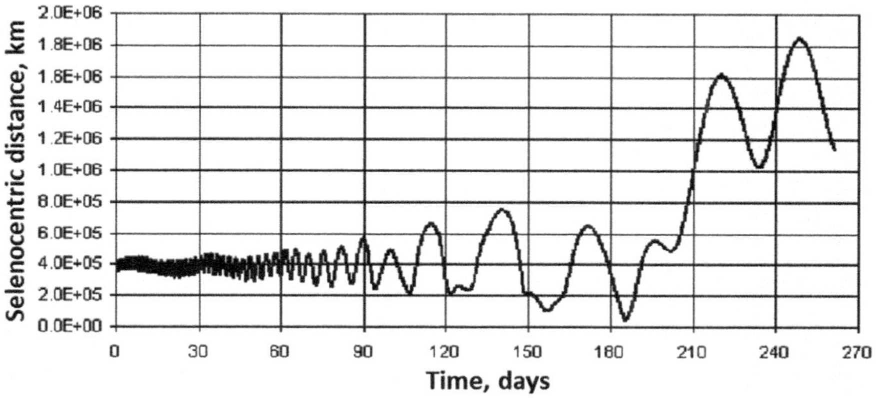

Fig. 12.16 Time dependence of the spacecraft distance from the Moon

12.6 Conclusion

This section considers the problem of minimizing the low-thrust flight time using the averaged equations of motion. A method for computing minimum-time trajectories between arbitrary elliptical orbits is presented. The asymptotic properties and symmetry properties of multi-orbit transfers between non-coplanar elliptical and circular orbits, where the line of apses of the initial elliptical orbit lies in the plane of the final circular orbit, have allowed the development of the suboptimal feedback control that makes it possible to implement transfers between arbitrary elliptical and circular orbits. The calculated dependences of the asymptotic characteristic velocity on the parameters of the initial elliptical orbit made it possible to develop a method for the rapid assessment of the main parameters of the mission profile for delivering spacecraft into high circular orbits with a combination of high and low thrust. It was possible to apply the suboptimal feedback control to calculate the trajectories of a spacecraft with an EPS to the Moon and the libration points L1, L2 of the Earth-Sun system.

References

1. Petukhov, V.G.: Optimization of multi-orbit flights between non-coplanar elliptical orbits. Space Res. **42**(3), 260–279 (2004)
2. Konstantinov, M.S., Petukhov, V.G.: Easy engineering technique of optimal electric propulsion trajectory estimation. IAC-06-C4.4.06 (2006)
3. Petukhov, V.G., Konstantinov, M.S.: Spacecraft insertion into high working orbits using light-class launcher and electric propulsion. In: The 17th International Symposium on Space Flight Dynamics, Moscow, vol. 2. Proceedings (2003)

4. Petukhov, V.G., Svotina, V.V.: Robust suboptimal feedback control for low-thrust transfers between noncoplanar elliptical and circular orbits. In: The 3rd European Conference for Aero-Space Sciences (EUCASS-2009), Versailles, France (2009)
5. Petukhov, V.G.: Quasi-optimal feedback control for low-thrust multi-orbit transfer between non-coplanar elliptical and circular orbits. Space Res. **49**(2), 128–137 (2011)
6. Sinitsin, A.A.: Study of the effectiveness of using a sustainer electric rocket propulsion system for launching a spacecraft into geostationary orbit. Cosmonautics Rocket Sci. 4(57), 95–108 (2009)
7. de Boor, C.: A Practical Guide to Splines, p. 372. Springer, New York (1978)
8. Zakharov, Y.A.: Design of inter-orbital spacecraft. Selection of trajectories and design parameters. M.: Mechanical Engineering (1984)
9. Kuvshinova E.Yu., Sinitsyn A.A.: Efficiency of using interorbital tugs based on nuclear electric rocket propulsion systems in earth-moon-earth transport operations. Cosmonautics Rocket Sci. 3(60) (2010)
10. Betts, J.T., Erb S.O.: Optimal low thrust trajectories to moon. http://www.boeing.com/phantom/socs/extra/smrtppr.pdf
11. Cano, J.L., Schoenmaekers, J., Jehn, R., et al.: SMART-1 mission analysis: collection of notes on the moon mission. S1-ESC-RP-5004 (1999)
12. Kluever, C.A., Pierson, B.L.: Optimal low-thrust three-dimensional earth-moon trajectories. J. Guidance Control Dyn. **18**(4) (1995)
13. Kluever, C.A., Pierson, B.L.: Optimal low-thrust earth-moon transfers with a switching function. J. Astronaut. Sci. **42**(3), 269–284 (1994)
14. Pierson, B.L., Kluever, C.A.: Three stage approach to optimal low-thrust earth moon trajectories. J. Guidance Control Dyn. **17**(6), 1275–1282 (1994)

The manufacturer's authorised representative in the EU is Springer Nature Customer Service Centre GmbH, Europaplatz 3, 69115 Heidelberg, Germany. If you have any concerns regarding our products, please contact ProductSafety@springernature.com

Printed and bound by CPI Group (UK) Ltd, Croydon, CR0 4YY
01/10/2026
02225962-0016